INTRODUCTION TO SOLAR TERRESTRIAL RELATIONS

ASTROPHYSICS AND SPACE SCIENCE LIBRARY

A SERIES OF BOOKS ON THE RECENT DEVELOPMENTS
OF SPACE SCIENCE AND OF GENERAL GEOPHYSICS AND ASTROPHYSICS
PUBLISHED IN CONNECTION WITH THE JOURNAL
SPACE SCIENCE REVIEWS

INTRODUCTION TO SOLAR TERRESTRIAL RELATIONS

PROCEEDINGS OF THE SUMMER SCHOOL IN SPACE PHYSICS HELD IN ALPBACH, AUSTRIA, JULY 15–AUGUST 10, 1963 AND ORGANIZED BY THE EUROPEAN PREPARATORY COMMISSION FOR SPACE RESEARCH (COPERS)

Edited by

J. ORTNER
European Space Research Organisation, Paris

and

H. MASELAND
Sterrewacht 'Sonnenborgh', Utrecht

D. REIDEL PUBLISHING COMPANY / DORDRECHT-HOLLAND

GORDON AND BREACH / SCIENCE PUBLISHERS
NEW YORK

1965

Printed in The Netherlands by D. Reidel, Dordrecht

FOREWORD

The textbook presented in the following is composed of the proceedings of the Summer School in Space Physics held during the summer months of 1963. This Summer School was organized by the Preparatory Commission (COPERS) of the European Space Research Organisation (ESRO). It was the first time that such a summer course was held in Europe on a subject of space physics. Thanks to an invitation from the Austrian Government these lectures were given in the College House of Alpbach, Tyrol.

Eight outstanding European scientists each presented five two-hour lectures on topics covering the region between the Sun and the Earth. The courses contained the physics of the Sun, the Interplanetary Medium and Trapped Radiation, the Ionosphere and High Latitude Phenomena. Furthermore, a course on space instrumentation was given. Sixty students were selected to attend the courses. They represented the eleven Member States of COPERS/ESRO in the following way: Austria (3), Belgium (6), Denmark (4), France (10), Germany (9), Italy (4), The Netherlands (3), Spain (2), Sweden (6), Switzerland (4), and the United Kingdom (9).

This volume provides an introduction to solar terrestrial relationships which form a part of the great field of space physics. It is intended as a textbook for students and a reference volume for general education for post-graduate students and space engineers who are investigating the interrelations of the various subjects in this field.

This book consists of two parts. The first part contains the lectures on the various topics of solar terrestrial relations. The order in which the lectures are reproduced was chosen in such a way that it starts with the physics of the Sun, and later covers the phenomena occurring in the interplanetary medium and finally ends with the effects produced in the ionosphere.* This part concludes with a few lectures in space instrumentation.

The second part of this textbook contains the seminars given during the summer course which were given by experts in the various topics who visited the Summer School only for a short period, and also by some of the participants in the course. This part of the book should thus be regarded as papers on more specialized topics in the fields covered by the lectures in the first part of the book.** The seminars deal not only with theoretical aspects of this field; many of them describe experiments carried out by satellites, rockets and balloons, the instrumentation used and recent results obtained.

* The lectures given by Prof. N. Herlofson on Trapped Radiation are not published in this volume. They are covered by the review article of C. G. Fälthammar 'Some aspects of particles and fields in the magnetosphere' in the proceedings of the COSPAR Panel Symposium on *High Latitude Particles and their effects on the Polar Ionosphere* (Logos Press, London 1965).

** A few seminars given at the summer course have been published elsewhere, namely: A. W. Lines: 'Some technical possibilities for astronomical satellites' *Space Sci. Rev.* **1** (1963) 781–785; J. A. Ratcliffe: 'The ionospheric F-region and space research; a critical review' *Space Research* **4** (ed. P. Müller), North Holland Publishing Co., Amsterdam 1964, pp. 3–7.

As already indicated above, this textbook covers the first part of the field of space physics. The second part would include stellar astronomy, the physics of the planets and the moon and galactic cosmic radiation which might be the subject of future summer courses organized by ESRO.

The editors would like to thank all who helped to make this Summer School a success and who assisted in preparing this textbook. These thanks are especially due to all the students participating who wrote up the first draft of the lecture notes.

Paris, September 1964

J. ORTNER AND H. MASELAND

TABLE OF CONTENTS

LECTURES

PART III / IONOSPHERE AND HIGH LATITUDE PHENOMENA

PART IV / SPACE INSTRUMENTATION

PART V / RETROSPECT

SEMINARS

PART VI / SOLAR PHYSICS

PART VII / INTERPLANETARY MEDIUM

PART VIII / SOLAR COSMIC RAYS

PART IX / IONOSPHERE AND PLANETARY ATMOSPHERE

OPENING LECTURE ON THE PLANS OF THE EUROPEAN SPACE RESEARCH ORGANIZATION

R. LÜST

COPERS and Max-Planck-Institut für Physik und Astrophysik, München

Herr Hofrat, Professor Cap, Bürgermeister, ladies and gentlemen, on behalf of COPERS I would like to welcome everybody here to the Summer School. Professor Auger, the Secretary General of COPERS sends you his best regards and he is very sorry that he cannot be here to-day. We would like to thank the Austrian Government and we would like to express particular thanks to Hofrat Bachmann that we can have the first Summer School of COPERS in these wonderful surroundings. I think we may take this invitation as an indication of the interest of the Austrian Government in space research and in the work of COPERS in particular. Therefore, we hope that any difficulties will be overcome and Austria will finally join ESRO. Also, I would like to thank very much the town of Alpbach and, in particular, the Bürgermeister that we can have this Summer School here in this wonderful place. I am quite sure that everybody will like this place with its quiet atmosphere for our work. We should not forget that here in Alpbach we are already somewhat nearer to the object of our Summer School, that is 900 or 1000 m nearer to space.

Since COPERS and ESRO should be scientific organizations, we are especially glad and thankful that Professor Cap is here not only as a representative of the University of Innsbrück, but also at the same time as the Austrian representative for COPERS. Professor Cap is really the driving force here in Austria for space research and I also believe that Austrian scientists can be most thankful to him for what he has done so far in all the work for COPERS. I should indicate that the whole work of COPERS and particularly that of ESRO can only be successful if research is done at the universities. I will try to explain later that the main idea of ESRO is that ESRO will be a service organization to research institutes in the Member States. Therefore, I am especially glad that Professor Cap is here from the University of Innsbrück. Last not least I would like to express a most hearty welcome to all the lecturers who are coming here and we are really extremely thankful that they could come and that they are able to spend this time here for giving the lectures. A number of them have already participated very actively in the work of COPERS and I would like to mention particularly Professor de Jager and Dr. Hultqvist who have been with COPERS from the start as Chairmen of working groups preparing different work for ESRO.

Of course, I am also very glad that all the participants could come and participate in this Summer School and I am not quite sure if the term "students" is really appropriate, since I think most of you are already grown-up scientists. I should indicate that it was not very easy for us to make a selection from all those who applied. We had about 100 applicants and, as Dr. Ortner told you, we could only accept 60. But I hope that you will take this also as a certain responsibility to get as much as possible out of the Summer School, since it might not be very easy to get the right combination between the lectures and the wonderful surroundings here.

Since probably quite a number of you are not very familiar with the idea of COPERS and ESRO, it might be useful if I try to tell you a little in this introduction about the future plans of ESRO. I will give this introduction in our traditional language which is not really an official language of COPERS; the official languages are English and French, but other languages can be used. But normally we, on the scientific side, never have any difficulties in coming together and speaking in a language understandable to each other, which is the so-called B.E.: Broken English, the new scientific language.

The great success achieved by the European Organization for nuclear research (CERN) – the first collective scientific undertaking in Europe and perhaps the first real international research undertaking – made it possible to believe that achievements of the same order could be obtained in fields other than high energy physics. Therefore, it was of no surprise that some of the scientists who had helped to create CERN also helped quite a bit in making the first planning for ESRO. Professor Cap has already mentioned the names of those who played a very important role there. But we will also see that there are really some differences in the planned organization of ESRO in comparison to CERN. Space research, in particular, is an undertaking where collaboration is needed. The extent of the resources needed at the various levels – at the scientific levels, in manpower, on the technical and industrial level and also on the financial level, are so large that a single European country could not really carry out a very balanced programme in space research alone.

The first question is what really is space research. It might be a surprise to some people here if I say that space science does not really exist, that is to say space research is not a new branch of science. Space research is only a new kind of technique which enables quite a number of different branches of science to carry out research and obtain information by new means on the objects of interest. Space research, in the strict sense, has already been carried out a long, long time ago and, for instance, astronomy and astrophysics are the classical branches of science for investigating space, that is the objects and the region above the earth's atmosphere. For astrophysics and for astronomy the advantages for observations above the atmosphere are really tremendous, since from the ground one can only observe the objects through two very narrow windows, namely, in the visible region and in the radio frequency region. Furthermore, as far as astrophysics and astronomy are concerned, and you will excuse me for taking this example since my main interest is in astrophysics, one is no longer hampered by the bad seeing conditions of the atmosphere and the astronomers can really improve, for instance, the resolution of their instruments quite a bit, since in most cases the resolution of an optical instrument on the ground is not really determined by the diameter of the optical system, but more by atmospheric conditions.

For other branches of science, like physics, geophysics, meteorology or geodesy, other advantages are offered by space research. This means the use of certain vehicles to carry out observations above the ground, that is carrying out research with the help of balloons, sounding rockets, satellites and space probes. The basic idea for the planned European Space Research Organization is to offer the scientific groups in Europe the possibilities of using these kinds of space vehicles and, in particular, sounding rockets, satellites and space probes. The initiative for the plans of the European Space Research Organization came from scientists. A number of unofficial discussions and meetings took place and these meetings finally led, by the end of 1960, to the so-called Meyrin Agreement by which the Preparatory Commission for the European Space Organization (COPERS) was established. The main task of this Preparatory Commission was to prepare the programme for the future Organisation and to draft a Convention. Twelve European States participated in the preparation of the European Space Research Organization and these twelve countries are: Austria, Belgium, Denmark, France, Germany, Italy, Netherlands, Norway, Spain, Sweden, Switzerland and the United Kingdom. The President of the Preparatory Commission is Sir Harrie Massey of the United Kingdom. The two vice-presidents are Professor H. C. van de Hulst (Netherlands) and Professor L. Broglio (Italy), and the Secretary General of the Preparatory Commission is Professor P. Auger (France). The plans for this organization have been worked out by a Scientific and Technical Working Group under the chairmanship of Professor L. Hulthén (Sweden), and by an Administrative, Financial and Legal Working Group under the chairmanship of Dr. Hocker (Germany). In addition, there were quite a number of different sub-groups and ad hoc working groups participating in the preparation of the work. The work of these different groups then finally led to the Convention of ESRO and this Convention was signed last June by ten European States. A Financial Protocol has also been set up and a number of different Resolutions. And I think there is now a good hope that the Convention and the Financial Protocol will be ratified before the end of the year, and that means that ESRO will enter into force around the end of this year.* Switzerland has been the first country to have already ratified the Convention.

What is the purpose of this planned organization? Generally speaking, the purpose of the planned organization shall be to provide and to promote collaboration among the European States in space research and technology. And, in doing this, the Organization might operate facilities to design and to construct payloads for sounding rockets, for satellites and for space probes. Furthermore, the Organization shall provide the launching vehicles for these payloads, it shall provide for the launchings and the reception, collection, reduction and analysis of the data. This is, in very general terms, what the Organization shall do. But, I should repeat, the basic idea for ESRO is that the scientific work shall not be done inside the Organization, but it shall be done by the research institutes, at the universities and other research institutes in the Member States. Of course, there will also be scientific work inside of ESRO, particularly in the field of applied research. There will also be one scientific laboratory,

* The convention entered into force in March 1964.

but the important point is that the preparation and the building of the scientific instruments which will be flown by the sounding rockets, the satellites and the space probes will be done, and has to be done, by the scientific groups in the Member States. This is the main difference, for instance, to the set-up in CERN where the major part of the scientific work is done within CERN. But, of cource, the putting together and the construction of the payload as a whole and everything which is connected with the payload, such as the telemetry, the stabilization, the power supply, will be provided and will be done by ESRO.

For carrying out this programme, ESRO must and will have a certain number of establishments. The main establishment will be the European Centre of Technology, called ESTEC, and this centre will be located in Delft. The main task of this centre will be to construct the payloads as a whole as I indicated; then to carry out advanced research on certain items which are necessary for carrying out successful research programmes, as indicated, on telemetry, stabilization and power supply and, this is a very important point, setting up test facilities for the different payloads, test facilities on a large scale which could not be set up in the Member States. Connected with ESTEC there will be a number of Tracking Stations which will be distributed over the world for getting the data back from the satellites and space probes. But these tracking stations – one is speaking about four or five tracking stations owned by ESRO – should be incorporated into already existing networks. Furthermore, locally connected to ESTEC, to the Space Technology Centre, there will be a scientific laboratory this is called ESLAB. The purpose of this laboratory is to provide a place for scientists to carry out work which could not be done at home in the Member States. This laboratory will be operated on a fellowship basis so that the scientists from the Member States can come for one or two years and can carry out their programme connected with satellites or space probes and then go back to their home country. In this way, we hope to make sure that ESRO is not a drain on the scientists of the Member States but really bring the scientists back to the Member States.

But, of course, one also has to have a set-up for processing the large amount of data which are coming from the satellites and space probes and the sounding rockets, and this shall be done in the European Space Data Centre called ESDAC. This Data Centre will be located in Darmstadt.

Then there will be a scientific laboratory located in Italy, and the purpose of this laboratory will be to study certain phenomena, which are important for the understanding of certain space experiments. This means certain experiments on atomic parameters, on plasma physics of low density should be carried out in this laboratory.

Then there will be a launching range for sounding rockets, and this launching range will be established in the northern part of Sweden, in Lapland, in Kiruna. The reason for establishing and locating the ESRANGE, the range of ESRO at high latitude, is that we think that Europe could make a very good contribution for studying the auroral phenomena.

And, finally, as in every organisation, there will be a headquarters and the headquarters will be located in Paris. It is organized in such a way that one has a Technical and a Scientific Directorate and an Administrative Directorate and finally the Secretariat of the Council. But I will not go into any detail about the internal part of the future organization.

But, in complement to the Organization itself, there will be an organization for the representation of the Member States and the scientists. The Organization will not run by itself but the Organization will receive instructions from the Member States. The Member States are represented by a Council and all important decisions for ESRO are made by the Council, in particular, for instance, the decision about the budget. But, in doing so, there will be two committees, one the Scientific Committee and one the Finance Committee for helping the Council in carrying out its work.

On the scientific side, that which is of particular interest is how certain experiments will be selected and how one will finally bring the certain payloads together. In doing so the so-called Launching Programmes Sub-Committee (LPSC) has been set up. This is a committee to the Scientific Committee and the LPSC should help the Scientific Committee in making certain recommendations in the selection of certain experiments. But experiments will only be accepted and selected after they have been thoroughly discussed by the scientific experts in different ad hoc working groups. Each scientific proposal for certain experiments for sounding rockets for satellites and space probes will first be discussed by the scientists on its scientific merits in different ad hoc working groups before it can be accepted by ESRO. The main point in accepting an experiment will be its scientific merits. Of course, then, in the later stage the technical and financial implications also have to be taken into account.

A last few words about the over-all programme of ESRO as it is established at the moment. In the Convention it is said that ESRO will finally launch every year in the order of about 65 sounding rockets and will launch about six small satellites or space probes per year; but this will only start after the fourth year. Finally, it will launch, in addition one large satellite each year, but this will only start after the sixth year, since preparation for a large satellite takes about six years.

The first large satellite will be an astronomical satellite and discussions are already on their way for selecting the best scientific object for this large astronomical satellite. The second large project is a second large satellite – it might be a lunar satellite, but just recently the possibility of a lunar station or maybe a large planetary or interplanetary station has been discussed. These plans are under discussion in scientific working groups and we hope that at some stage certain recommendations about this large programme can be made. In the meantime, we have already received quite a number of proposals for scientific experiments, for sounding rockets and for small satellites; we received 78 for experiments for sounding rockets and about 71 proposals for scientific experiments incorporated into small satellites and space probes. This indicates that there is a very large scientific interest for creating such a scientific organization. We hope to commence operations at the end of the year and, if this is the case, we hope that the first launching of sounding rockets would already take place next April or May.

Let me conclude this short outline about the programme of the European Space Research Organization with some remarks about why it seems to be important to carry out space research in Europe, since one still finds from time to time the opinion that we should really leave space research to the two big countries, namely to America and Russia. The reasons why such large efforts are being made for space research in the U.S.A. are certainly very complex and not quite easy to analyse. But it is certainly not right to believe that there are only political or military motives involved in the U.S.A. for carrying out and having such a large programme. The high scientific value of space research is quite well recognized in the U.S.A., and one cannot underestimate how strong the drive in the U.S.A. really is to carry out a scientific programme. Unfortunately, and this has already been pointed out by Professor Cap, space research is in the same position as some other fields of science, like nuclear physics, high energy physics or plasma physics where one has to spend large sums of money to carry out this research effectively. It is not worthwhile to spend only a small amount of money – this will have no effect at all. And, of course, it is sometimes not easy to convince governments to spend large sums of money just for a fundamental research. But, in this connection I may tell the story of Faraday who was once asked by an official of his government what he was doing and what kind of profit one could make with his research, and Faraday's answer was that certainly he would not expect any practical application and profit from his basic fundamental research but he would be quite certain that the next generation would collect tax from his research.

In the case of space research we might have a somewhat better position since we can already prove that it is not only a fundamental part of research, but has a direct and very large impact on technology and industry. Already now certain achievements in technology have been obtained only because they were first requested by space research. I do not think we should be deterred by the fear that space vehicles might carry other things than satellites or just space probes and one should not reject space research just for this reason, since in every country now nuclear reactors are being developed and built and it is also quite certain that one can do other things with nuclear reactors than just produce peaceful kilowatt hours. From experience in other fields we know that it is not possible to leave research only to the two big countries, the U.S.A. and Russia, and use only their results. Experience has shown that after a very short while direct use of these results cannot really be made, due to lack of the right understanding. Furthermore, there is the second effect that probably the best people would have left.

Finally, the branches of science which are participating in space research, like physics, astronomy, astrophysics, geophysics and meteorology, have been studied, and have been carried out already for a very, very long time in Europe. And, of course, quite a number of problems in these fields can still be solved by traditional methods, but it is also quite clear that certain problems can be attacked only by extraterrestrial methods, only by extraterrestrial observations. Therefore, if we did not participate here in Europe in space research, we would after a very short while, only carry out in certain fields second-hand research. It is the aim of this first European Summer School in Space Research to help build up space research in the different European countries and to show the way in space research. I hope that this Summer School will be a successful one to prove that we are on the right road.

LECTURES

PART I

SOLAR PHYSICS

CHAPTER 1

THE SOLAR PHOTOSPHERE

C. DE JAGER
Sterrewacht 'Sonnenborgh', Utrecht, The Netherlands

1. Introduction; Solar Parameters

The sun has a radius $R=7\times10^5$ km, and is situated about $r_E=1.5\times10^8$ km from the earth. Since the mass of the sun is $M=2\times10^{33}$ g the average density can be determined as 1.4 g cm^{-3}. However, the density changes from 70 g cm^{-3} at the solar centre, where the temperature is 14×10^6 °K, to 10^{-7} g cm^{-3} in the photosphere.

Measurements of the *solar constant* S, the total radiation falling in one minute on a square centimeter perpendicular to the direction to the sun, outside the earth's atmosphere, show that the value is close to $S=2$ cal cm^{-2} min^{-1}; this makes the total emission from the sun $L=4\pi r_E^2\times60\times10^7 S=3.6\times10^{33}\pm2\%$ ergs sec^{-1}. Although the sun does not emit radiation as a black body, we *define* the *effective temperature*, being a measure for the radiation flux, by means of Stefan's law: $L=4\pi R^2\sigma T_e^4$, where σ is Stefan's constant. This gives a value $T_e=5800$ °K.

This great amount of energy is produced in the sun's core by the fusion reaction

$$4\text{H}\rightarrow\text{He}$$

the loss of hydrogen being only about 2% of the sun's mass in 10^9 years.

The two main solar problems are:

(a) *Solar structure.* Given the initial solar parameters at the time of its formation (mass, chemical abundances) and its age, it is necessary to predict the actual parameters ($M=M_{\text{initial}}$, R, L). This first problem will perhaps be solved in the coming ten years.

(b) *Solar variability.* Given the actual structure of the sun, the problem is to explain and predict the transient features occurring at and around its surface. This second problem will require the careful attention of a few generations of solar physicists before its final solution.

2. The Quiet Photosphere; Radiative Equilibrium

Three regions of the sun are to be discussed in these lectures. The lowest is the photosphere which produces the greatest part of the solar emission. Above this is a thin layer called the chromosphere, and above that there is the very hot corona which reaches out to beyond the earth's orbit.

The emission of radiation is low in the corona and the chromosphere, but is very strong in the photosphere. However, the density and opacity also increase with depth, with the result that little of the radiation emitted is produced at a depth greater than 400 km below the visible surface.

As a result of the main part of the photosphere being in a state of radiative equilibrium – that is, energy is transferred mainly by radiation – the structure can be represented by means of three parameters, which are the solar constant, the acceleration of gravity at the surface and the chemical composition. Corresponding to the temper-

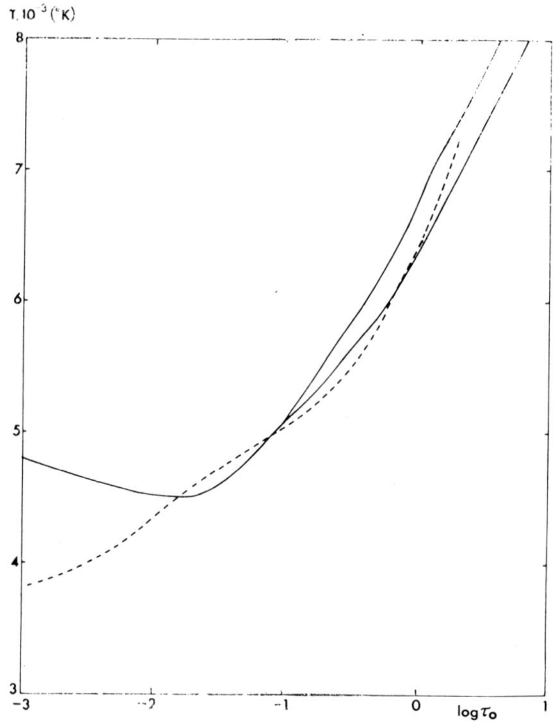

Fig. 1. Relations between temperature and optical depth in the solar atmosphere. *Dashed Line:* $T(\tau_0)$ relation computed for a model photosphere in radiative equilibrium (Böhm, 1953). *Solid line:* The empirical $T(\tau_0)$ relation, derived from various kinds of observations (this model is the Utrecht 1964 reference photosphere). Abscissa: τ_0 = monochromatic optical depth for λ = 5000 Å. Ordinate: temperature.

atures of the emitting layers, there is a distribution $F(\lambda)$ of flux with wavelength. Integration over the whole spectrum gives the total flux observed, $F = \int F(\lambda)\,d\lambda$.

It has already been pointed out that the effective solar temperature can be established, but we want to know the actual temperature $T(z)$ as a function of depth z. To do this it is necessary to assume that local thermodynamic equilibrium holds. That is, we assume that for each atmospheric layer, Kirchhoff's law relating absorption and emission is valid. At this stage it is useful to introduce the optical depth τ which is a dimensionless parameter related to the absorption coefficient $\kappa(\lambda, T, \varrho)$ cm^{-1} by

$$d\tau = \kappa\, dz\,.$$

It is possible to calculate the absorption coefficient as a function of the wavelength λ, the temperature T, and pressure P from the chemical abundances of the elements in the sun; so it is possible to compute for any assumed temperature distribution, at each depth (z), the integrated flux

$$F(z) = \int F(\lambda, z) \, d\lambda ,$$

and to examine whether this flux is constant with depth, and equal to F, as should be

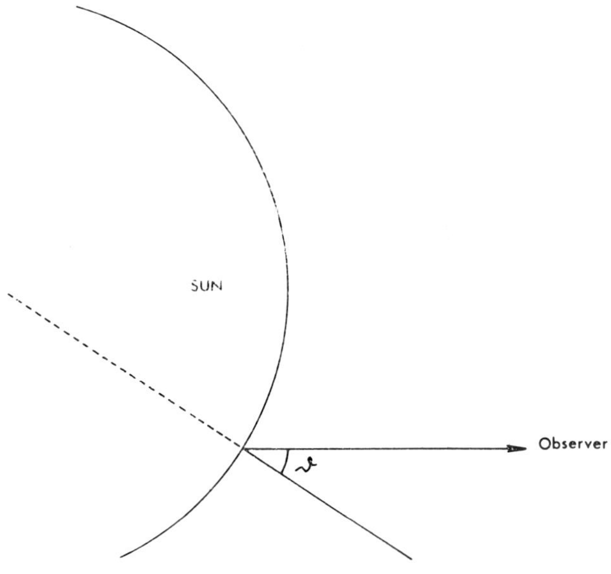

Fig. 2. Definition of angle ϑ.

the case in radiative equilibrium. If this is not so, the procedure can be repeated with different temperature distributions until a satisfactory model is obtained, as has been done for instance by Böhm (dashed line in Figure 1).

When the sun is examined the radiation emitted by the photosphere appears to depend on the angle ϑ between the direction to the observer and the local normal to the emitting surface (Figure 2). The observed intensity decreases toward the limb (one calls this the *solar limb darkening*; see Figure 3); this reflects the outward decrease of the photospheric temperature. To be able to compare theory with observation, it is therefore necessary to calculate the function $I(\lambda; \cos \vartheta)$.

When $S(\tau_\lambda)$ is the source function for radiation of wavelength λ at an optical depth τ_λ the intensity of radiation of this wavelength is

$$I_\lambda = \int_0^\infty S_\lambda(\tau_\lambda) \exp(-\tau_\lambda \sec \vartheta) \sec \vartheta \, d\tau_\lambda ,$$

if we assume that emission occurs from a plane parallel photosphere. It is usual to approximate the depth dependence of the source function, by the expression

$$S = a + b\tau + c\tau^2 .$$

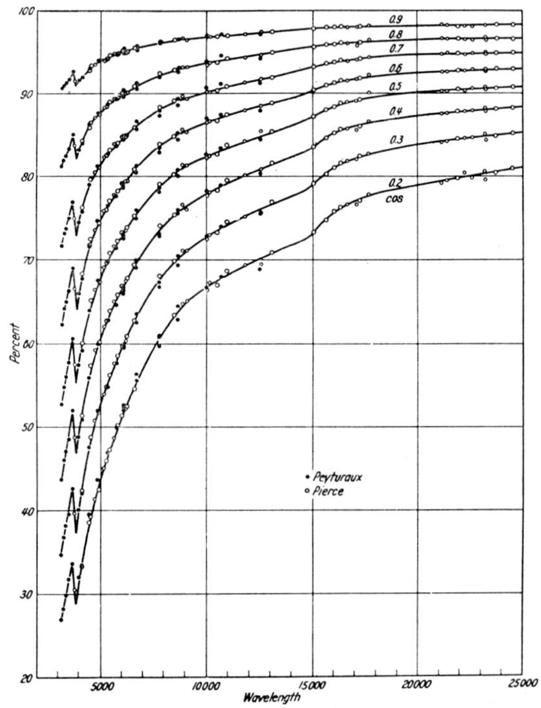

Fig. 3. Limb darkening as a function of the wavelength for selected points of the solar disk, defined by cos ϑ (from GOLDBERG, L. and PIERCE, A. K.: 1959, *Handbuch der Physik* **52,** 7).

By direct integration of the expression (1), it is found that

$$I = a + b \cos \vartheta + 2c \cos^2 \vartheta .$$

Comparison of I_λ with measurements makes it possible to find a suitable source function, thus allowing the determination of the depth dependence of the temperature.

It is even possible to apply this treatment to the center-limb variation of the intensities measured in *spectral lines*. This allows the extension of the empirical $T(\tau)$ curve to smaller values of τ than would have been possible with continuum observations only: The light emitted in the cores of Fraunhofer lines, with their great selective absorption coefficients is emitted by and yields information on the highest photospheric layers, higher than those from where the continuous radiation emerges.

The empirical $T(\tau)$ relation thus found from a combination of continuum and line observations (solid line in Figure 1) should be compared with the theoretical curve (dashed in Figure 1). Apparently they disagree partly.

3. Failure of the Radiative Equilibrium Theory; the Mechanical Flux

The theoretical temperature distribution does not account for certain features of the observed distribution (comparison of the dashed and solid lines in Figure 1).

(1) The observed temperature is found to reach a minimum at τ_{5000}Å ~ 0.02 where the temperature is $T = 4500$ °K. No such minimum occurs in the theory based on the hypothesis of radiative equilibrium.

(2) Granulation is observed in the sun's disc. This indicates convective motions, and, hence, convective energy transport – an observation that is obviously incompatible with the assumption of radiative equilibrium.

To explain these facts, it is necessary to realize that radiation is not the only possible means of energy transport in the photosphere, but that at great depths convection becomes important. At depths greater than about 200 km below the solar limb surface down to about 10^5 km, nearly the whole solar energy flux is transported by convective motions, and radiative transport plays only a negligible part in this region. In the following sections we shall show that this convection zone is the source of a *mechanical energy flux* F_m which has the character of a field of *pressure waves*. The velocity amplitude v_t of these waves increases when propagating outward. When v_t approaches the local velocity of sound the pressure waves transform into *shock waves*, which dissipate energy and heat the medium. The result is that at these small depths the sun is hotter than it would be if radiation and convection were the sole means of energy transport.

To describe granulation one usually represents the relevant photospheric layers by a two-element model. Also empirically, from Fraunhofer line and continuum studies and from studies of the brightness fluctuations in the solar granulation, one finds evidence for the existence of two kinds of photospheric elements. At an optical depth $\tau \simeq 1$ for $\lambda = 5000$ Å the temperatures of the two kinds of elements differ by about 300°. The differences decrease outward and seem to be zero at $\tau_{5000} \simeq 0.1$.

However, the dissipation of mechanical energy, mentioned above, increases outward and leads to the heating of the chromosphere, which begins to be of importance at $\tau_{5000} < 0.05$.

Roughly we may say: The structure of the *convection* zone is defined by the *convective energy flux*; the structure of the main parts of the *photosphere* is defined by the *radiative flux*; the structure of the *chromosphere* is defined by the *mechanical flux*; that of the *corona* by the *mechanical* and *conductive fluxes*.

4. Granules and Convection Observations

The observation of the granules is very difficult since their diameters are of the order of 500 km i.e. about ½″ as seen from the Earth's distance. A resolution of ½″ can be obtained with telescopes of more than 20 cm objective diameter (resolution $\simeq$ wavelength/objective diameter), assuming there are no disturbances in the Earth's atmosphere. Normally such disturbances will wash away the picture completely.

Only observations from high mountain tops and under extremely calm atmospheric conditions or from balloon-borne apparatus in the stratosphere have given satisfactory results.

Rösch, observing from Pic du Midi (2900 m) with a 25 cm and later a 60 cm telescope obtained very good pictures of the granulation during a short interval ($\simeq$ 10 min) just after sunrise, before the radiation from the sun started heating the

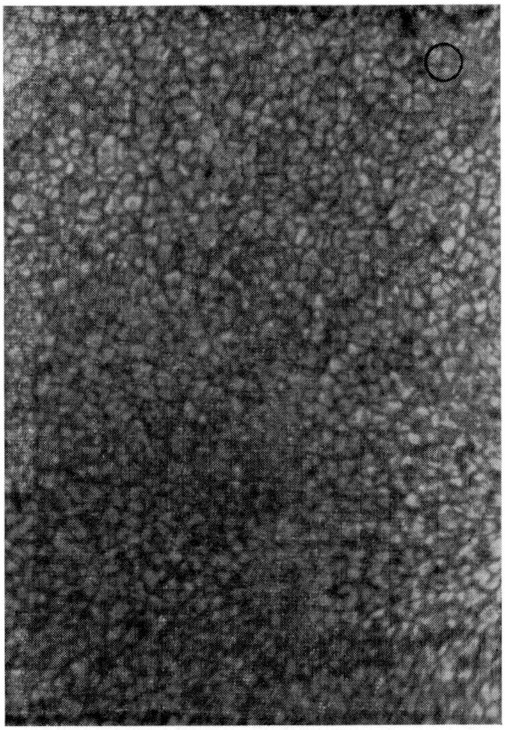

Fig. 4. Granulation photographed by M. Schwarzschild *et al.*, with a balloon-borne instrument at 25 km altitude (Project Stratoscope). The circle has a diameter of 5″.

ground and causing convection in the Earth's atmosphere. Comparable and even slightly better pictures from a single day were obtained by Schwarzschild *et al.* from an unmanned balloon at 25 km altitude (project Stratoscope, Figure 4).

On films taken by Rösch the granules often showed a complicated behaviour, splitting into several small pieces. The lifetime of the granules is of the order 8 min.

The granules can be observed indirectly by measuring asymmetries in the Fraunhofer lines. The granulation is connected with convective motions and shows therefore Doppler effects in the emitted and absorbed light. Some Fraunhofer absorption lines, mainly those arising from high excitation levels, show a slight asymmetry when observed with a high resolution spectrograph. This asymmetry does not occur in the deepest parts of the lines, but rather in the less deep parts, which originate in the deeper parts of the photosphere. One can imagine this asymmetry as arising from the super-position of two absorption lines coming from hot gas masses moving upward and cool, downward moving gases. The deepest part of the lines, which correspond to layers high in the photosphere (since with large absorption we cannot see far down) has no asymmetry, showing that convective motions occur only in the deep layers. The sign of the line-asymmetry depends on the relative strength of the absorption lines in the cool and the hot elements respectively (Figure 5).

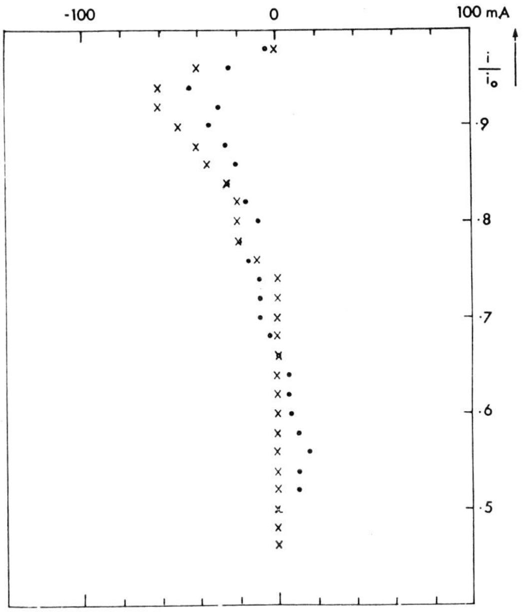

Fig. 5. Asymmetry observed by L. Neven and C. de Jager in an infrared carbon line; the spectra were obtained with the Jungfraujoch spectrograph of the Liege Observatory. The dots show the relative positions of the middle of the lines connecting two points in the line profile with the same intensities. Crosses show the positions of these points, after having corrected the line profile for the influence of the apparatus profile.

Since it is inevitable to observe several granules at the same time with the spectrograph, the velocities found are only mean values, but these run to about 2 km/sec for an optical depth of about $\tau_{5000} \approx 0.5$.

5. Consequences of Convection

(a) The most important consequence of the convection and the turbulence thereby following is the *generation of sound waves*. These sound waves carry a small amount of energy to the upper part of the photosphere, the chromosphere and the corona where it is dissipated, and causes a large increase in the temperature of these outer layers. That fraction of the total energy flux which goes into these sound waves is less than 1%; this is a result of the very low efficiency of a turbulent motion in generating pressure (or sound) waves; a turbulent region acts as a quadrupole source only. The layer where the sound waves are generated and whence they escape out of the turbulent region is only about 100 km thick and lies about a hundred km below the top of the convection zone.

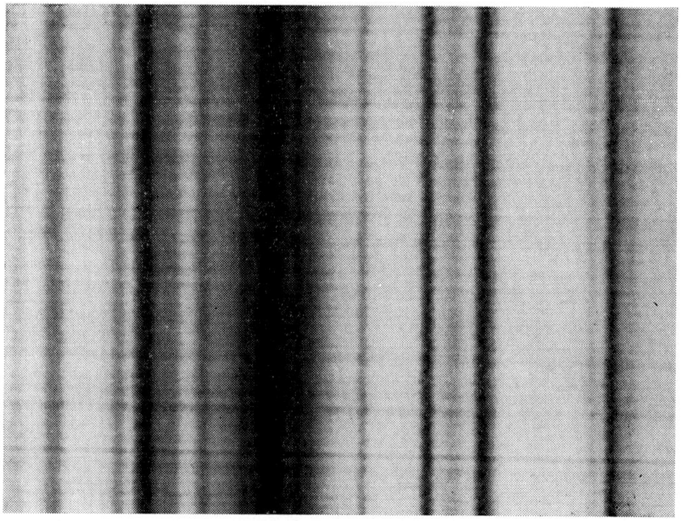

Fig. 6. Wiggling of Fraunhofer lines. The photo shows $H\gamma$ and a surrounding part of the spectrum ($\lambda = 4340$ Å). Verticale scale: 1 mm = 3″.5; wavelength scale: 1 Å = 29.6 mm. Spectrum obtained at the Potsdam Astrophysical Observatory; courtesy Prof. H. Jäger, Potsdam.

To see what happens to a sound wave as it progresses up from the source region, we consider the expression for the energy flux F_{sound} of these waves:

$$F_{\text{sound}} = \tfrac{1}{2}\varrho v_t^2 v_s .$$

Here ϱ is the density, v_t is the velocity amplitude of the wave i.e. amplitude × frequency and v_s is the sound velocity. In the upper part of the photosphere and the chromosphere the temperature and therefore $v_s = \sqrt{\gamma RT/\mu}$ stays fairly constant, whereas ϱ decreases rapidly. In order to preserve F_{sound} (assuming no dissipation) we therefore see that $v_t \sim \varrho^{-1/2}$ i.e. v_t increases. This will, in due course, when v_t becomes comparable to v_s, lead to the formation of a shock wave and the sound wave will rapidly dissipate its energy and heat the solar atmosphere.

(b) Another consequence of the convection is that it causes the upper layers of the photosphere to oscillate. The period of these oscillations is $\simeq 5$ minutes in the outer layers of the photosphere and is a slowly decreasing function of height in the chromosphere. These oscillations do not travel far out in the solar atmosphere but seem to be confined to a layer some few thousands kilometres thick in the upper photosphere and chromosphere.

The experimental evidence seems to indicate that the oscillations in the upper photosphere involve mass motions; spectral lines from this region shows a wiggling which corresponds to velocities of the order 0.5 to 1 km/sec (Figure 6), whereas farther out the existence of oscillations with slightly smaller periods and with a definite phase-relationship to the mass-oscillations is not indicated by Doppler-wiggling of the lines, but by oscillations in the residual brightness in the line centers.

The theory of oscillations in a gas at these low frequencies shows that there exists

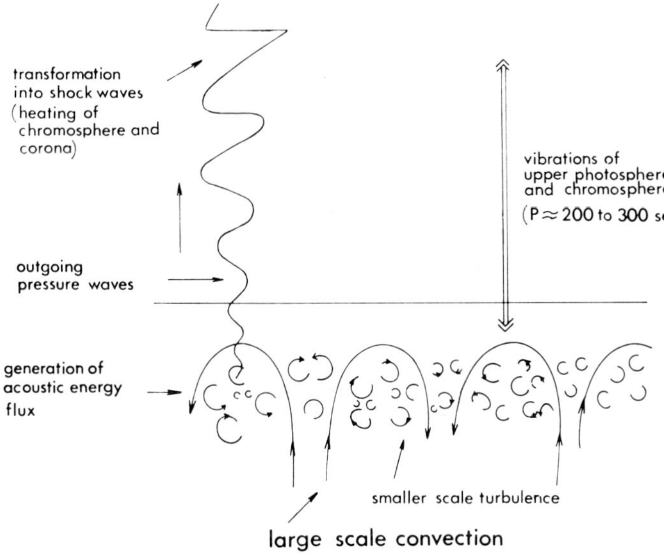

Fig. 7. Schematic drawing of the photosphere.

a critical frequency ω_c, such that all oscillations at higher frequencies will be damped out. The critical frequency is given by

$$\omega_c = \tfrac{1}{2}\sqrt{(\gamma g/H)},$$

where γ is the ratio of specific heats, g is the gravitational acceleration and H is the scale height.

In the upper photosphere these quantities have the following values: $g = 10^{4.44}$ cm/sec^2, $H = 1.2 \times 10^7$ cm and $1 \leqslant \gamma \leqslant 5/3$. We thus find

$$\omega_c = 0.024$$

$$T_r = \frac{2\pi}{\omega_c} = \begin{cases} 260 \text{ sec for } \gamma = 1\,, \\ 200 \text{ sec for } \gamma = 5/3\,. \end{cases}$$

The observations fit in well with these periods since the shortest period observed is $\simeq 200$ sec.

Current ideas on the structure of the solar photosphere are summarised in Figure 7.

Fig. 8. The University of Colorado pointing-control on the top part of an Aerobee-Hi rocket instrumented by the Naval Research Laboratory. Stabilisation against roll of the rocket is obtained by rotation of the whole top part just above the black section. The diameter of the rocket is 38 cm. (Courtesy R. Tousey, Naval Research Laboratory).

6. Temperature Inversion in the Upper Photosphere

Part of the dissipation of shock-wave energy, and its transformation into heat occurs already in the upper parts of the photosphere. Towards higher levels this dissipation becomes increasingly efficient for the heating of the solar gas. As a consequence a temperature inversion takes place in the uppermost part of the photosphere, where

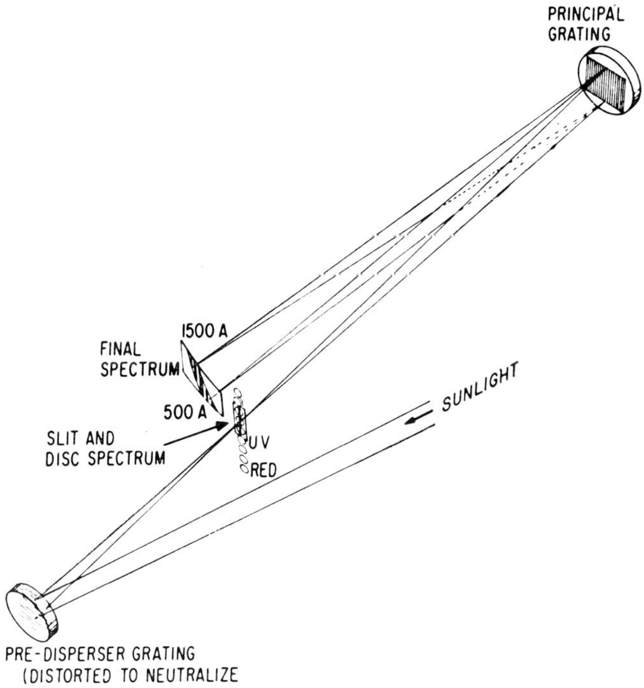

Fig. 9. The double-dispersion spectrograph of the Naval Research Laboratory, first flown on April 19, 1960 (Tousey, 1963).

the kinetic temperature begins to increase outwards. It has been known for a long time that the temperature in the solar corona is much higher than the surface temperature, but the depth in the solar atmosphere where the inversion actually takes place has not been known until the last few years.

By means of the pointing control instrument developed by the University of Colorado (see Figure 8) and placed in Aerobee-Hi rockets, Tousey and co-workers have been able to obtain solar spectra in the wavelength region 45–3500 Å. A drawing of one of the recently used spectrographs is shown in Figure 9. The spectral lines for 1000 Å$\lesssim\lambda\lesssim$2000 Å are predominantly formed around and above the height of the temperature inversion.

In the spectra the inversion reveals itself in two ways. First the *radiation temperature* which can be deduced from the measured absolute intensities in the continuous spectrum shows a minimum around $\lambda = 1400$ Å (Figure 10). Secondly the line spectrum changes from absorption lines to emission lines around $\lambda = 1800$ Å.

The radiation temperature is simply the temperature which one has to substitute in Planck's law to obtain, at the given wavelength, an intensity that is equal to the radiation intensity observed. So the minimum in the observed *radiation temperature* (=spectral intensity parameter) reflects the minimum in the *kinetic temperature* (=local physical parameter of the gas).

The reason for the *minimum in the radiation temperature* is related to the fact that the continuous absorption coefficient of the photospheric gases increases nearly

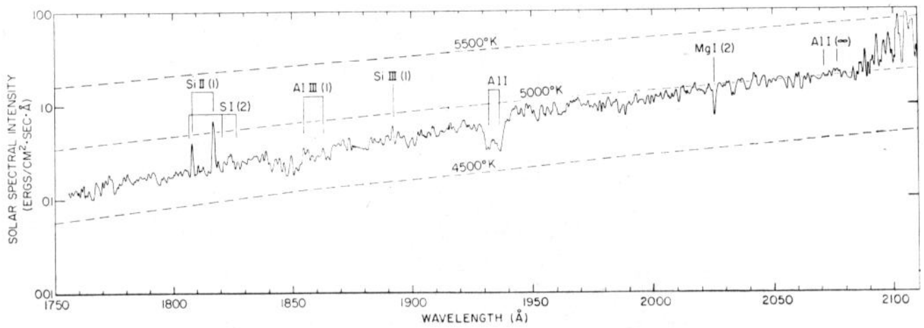

Fig. 10. The intensity distribution in the solar spectrum between 1300 and 1550 Å. The dip at $\lambda = 1510$ Å is instrumental (Tousey, 1963).

monotonically with decreasing wavelength. Hence the radiation from shorter and shorter wavelengths originates from higher and higher regions of the solar outer layers. Furthermore the observed emission intensity at any wavelength τ_λ is roughly equal to the radiation intensity at a monochromatic optical depth $\tau_\lambda \simeq 1$, because

$$I_\lambda = \int_0^\infty S_\lambda e^{-\tau_\lambda} \mathrm{d}\tau_\lambda \simeq S_\lambda(\tau_\lambda = 1) \int_0^\infty e^{-\tau_\lambda} \mathrm{d}\tau_\lambda = S_\lambda(\tau_\lambda = 1),$$

where I_λ is the observed intensity, S_λ the source function and τ_λ the optical depth. Hence the observed intensity as a function of the wavelength reflects the variation of the source function with depth in the outer solar layers.

From the observed radiation temperature as a function of λ it is possible to calculate back and find the real (kinetic) temperature and the optical depth of the temperature-minimum and one finds $T_{\mathrm{kin}} = 4500°$ for $\tau_0 \simeq 0.02$ (monochromatic optical depth for 5000 Å).

The other important observation, mentioned above is, that of the change in the character of the spectral lines, from absorption lines for wavelengths of longer than 1800 Å into emission lines for shorter wavelengths. There is a range of about 300 Å where absorption and emission lines occur simultaneously.

The explanation of this phenomenon is clear if one realises

(a) that the absorption coefficient in a Fraunhofer line is larger than in the adjacent spectral regions, so that the light emitted in a line originates from regions higher up in the solar atmosphere than the light emitted in the adjacent region;

(b) that the continuous absorption coefficient increases greatly with decreasing wavelength. Consequently, the lines emitted at long wavelengths originate in a gas that is cooler than the gas whence the continuous emission comes: these lines are absorption lines. However, lines emitted at short wavelengths are emitted by a gas with a temperature higher than the gas where the adjacent continuous spectrum is emitted: *these* lines appear as emission lines. The wavelength of the change-over corresponds roughly to the same optical depth for the temperature minimum as the value mentioned above.

References

Goldberg, L. and Pierce, A. K.: 1959, *Handbuch der Physik* **52,** 1.
Minnaert, M.: 1952, *The Sun* (ed. by G. Kuiper), The University of Chicago Press, p. 88.
Minnaert, M.: 1960, *Radioastronomia solare*, Varenna Summer School, Corso XII, (ed. by G. Righini), Zanichelli, Bologna, p. 1.
Minnaert, M.: 1965, *The Solar Spectrum* (ed. by C. de Jager), D. Reidel Publishing Co., Dordrecht, p. 3.
Pecker, J.-C.: 1965, *The Solar Spectrum* (ed. by C. de Jager), D. Reidel Publishing Co., Dordrecht, p. 29.
Unsöld, A.: 1955, *Physik der Sternatmosphären*, 2. Aufl., Springer-Verlag, Berlin.
Tousey, R.: 1963, *Space Sci. Rev.* **2,** 3.

CHAPTER 2

THE QUIET CHROMOSPHERE AND CORONA

C. DE JAGER
Sterrewacht 'Sonnenborgh', Utrecht, The Netherlands

1. Chromosphere

A. GENERALITIES

The chromosphere is a very thin layer of plasma, 5000 to 10000 km thick, directly covering the photosphere (Figure 1). It may be studied especially during total solar eclipses and, in fact, was discoverd during one. When the moon has covered the entire photosphere the chromosphere remains for a few seconds as a luminous coloured arch. This colour is the origin of the name "chromosphere".

During these few seconds we can obtain the spectrum (flash spectrum) of the chromosphere, which is constituted by bright emission lines. While the moon is gradually covering the chromosphere we can measure the radiation emitted by a segment of a flash arch and follow how this radiation decreases with time; by differentiating we can determine the contribution of each surface element of the chromosphere. In this way it is possible to study the detailed structure of the chromosphere as a function of height.

At the chromosphere temperatures of 4000 to 40000 °K occur. The chromosphere is a transition layer between a cool and dense, practically neutral, gas (the photosphere) with a temperature at the surface of about 4500 °K, and a hot tenuous plasma (the corona) in which the temperature is 10^6 °K.

Two features may characterise the chromosphere and roughly define its physical state: (1) it is transparent for continuous radiation; (2) it is opaque in the cores of the strongest Fraunhofer lines like Hα and the H and K lines of Ca^+.

From the first feature it can be shown that, at the basis of the chromosphere, the gas density is of the order of 10^{-7} to 10^{-8} g cm^{-3}. In the region of transition between the sun's photosphere and the overlaying chromosphere, the sun's atmosphere changes from opaque to transparent. Because of the rather fast decrease in density with height (the scale-height of the transition layer photosphere-chromosphere is of the order of 150 km) this changing takes place in a rather narrow interval; the "thickness" of the limb is not greater than 150 km. For a temperature of 4500 °K (the estimated photospheric limb temperature) one finds from the well-known tables of E. Vitense at $\lambda = 5000$ Å: $\log \kappa/P_e = -1 g^{-1}$ where κ is the continuous absorption coefficient and P_e the electron pressure. Since at the sun's limb $P_e \simeq 0.1$ to 1, the requirement that the chromosphere be transparent for continuous radiation results in the demand that the total amount of matter in the low chromosphere in a column with an area of one cm^2 and passing from the observer along the limb towards

Ortner & Maseland (eds.), Solar Terrestrial Relations.

infinity in any case should contain no more than 100 to 10 grams of matter. The effective length of this column should be about equal to the *tangential chromospheric scale height*. In the low chromosphere the *vertical* scale height H is about 150 km. The corresponding scale height h in *tangential* direction (Figure 2) is $h=\sqrt{(2R_\odot H)}$, as can easily be verified ($R_\odot$ = solar radius). Hence for $H=150$ km, $h=1.5\times10^4$ km. So, the gas density at the basis of the chromosphere is of the order 10^{-7} to 10^{-8} g cm^{-3}.

Consider the second feature. As the strongest Fraunhofer lines are visible up to distances of about 10000 km from the limb, up to that distance the chromospheric

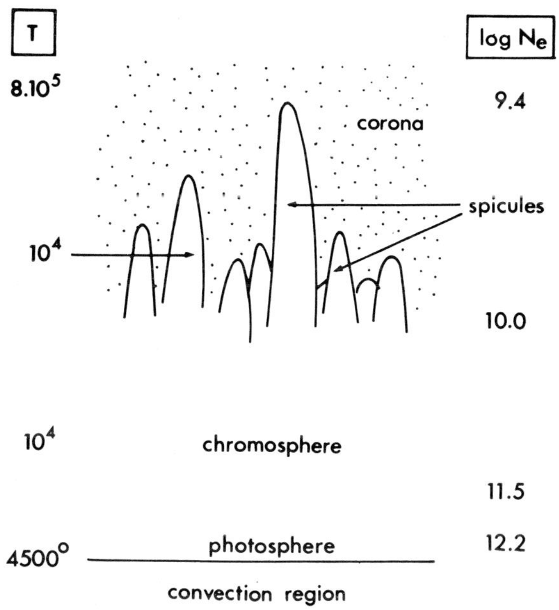

Fig. 1. Schematic drawing of the chromosphere.

ionization temperature must be such that neutral and singly ionized elements occur in sufficient quantities; excitation and ionization temperatures must be of the order of 10^4 °K (not much higher). These excitation and ionization temperatures in gases like the chromosphere can be very different from the kinetic temperatures defined by the velocity distribution of the particles. This implies that deviations from local thermodynamical equilibrium (LTE) arise in the chromosphere, and Boltzmann's and Saha's laws can no longer be used for finding the population of the atomic levels.

B. CHROMOSPHERIC OBSERVATIONS

1. Flash spectrum (Figure 3)

Most of the information about the chromosphere comes from the flash spectrum obtained during eclipses. Because of the short duration of totality this information is

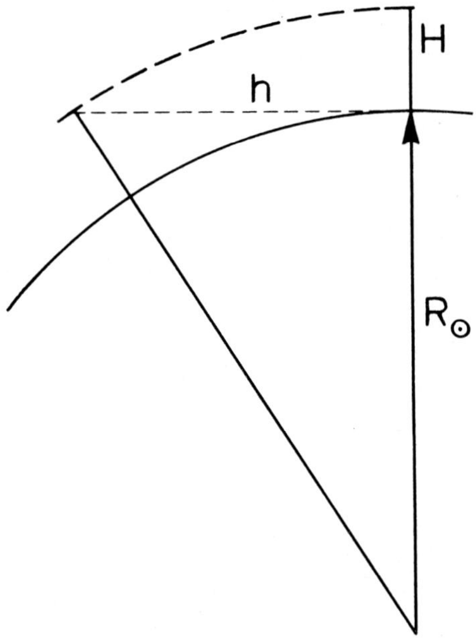

Fig. 2. Illustrating the relation between radial and tangential scale heights H and h.

still very scanty. Table I below gives the duration of visibility of different lines appearing in the flash spectrum.

The observed chromospheric crescent during an eclipse is very thin and therefore the flash spectrum can be obtained with a slitless spectrograph (camera equipped

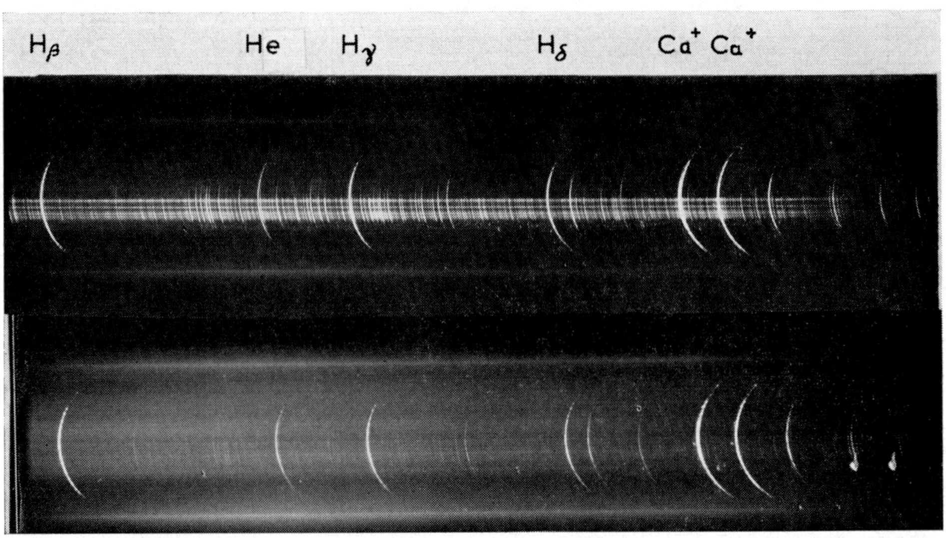

Fig. 3. Flash spectrum taken at the eclipse of February 25, 1952 (Khartoum) by the Utrecht expedition. By courtesy of J. Houtgast.

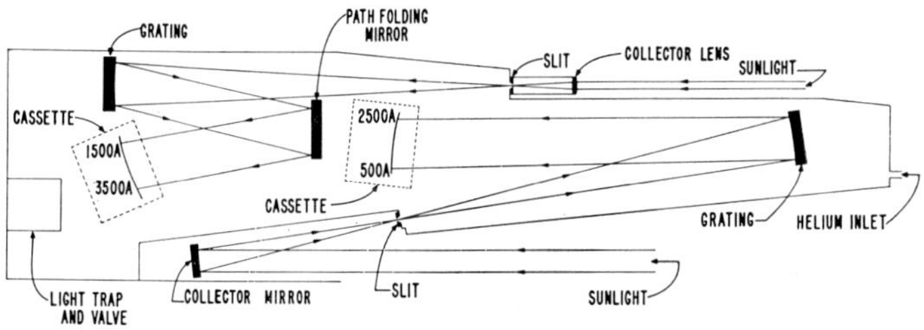

Fig. 4. Two grating spectrographs in one single housing. The instrument was flown by the U.S. Naval Research Laboratory, for the first time on February 21, 1955 (Tousey, 1963).

with an objective prism). The flash spectrum (essential on all programs of eclipse observations) gives information only on the part of the chromosphere seen beyond the sun's limb.

2. *Rocket spectra*

Another way to obtain spectra of the chromosphere is by spectrographs flown in rockets. These devices show the chromospheric spectrum *before the disk*.

These instruments have to fulfil two demands: (1) the pointing-control system should keep the spectroscopic instrument pointed at the sun; (2) the scattered light should be eliminated. For the study of the extreme ultraviolet one has to use reflection type diffraction grating; lenses cannot be used.

An example of such a spectrograph is the one launched by the Naval Research Laboratory on February 21, 1955, with two gratings in the same housing to study two ranges of wavelength (1500–3500 Å and 3000–5000 Å) (Figure 4). With this type of *normal incidence spectrograph* the stray light produced by the intense long wavelength portion of solar spectrum sets a detection limit on the weak emission (Figure 5). This difficulty was solved by Detwiler, Purcell and Tousey (Tousey, 1963) by their use of *two concave diffraction gratings* whose dispersions were at right angles to one another. The first grating served as a predisperser for the second; at the focal plane

TABLE I

DURATION OF VISIBILITY OF SEVERAL LINES IN THE FLASH SPECTRUM

Lines	*Average duration of visibility (sec)*
Weak neutral metals	2
Weak metal ions	2
Strong metal ions	7
Neutral He	10
D3 line of He	15
Hα (Balmer-α)	25
H, K (Ca^+)	30

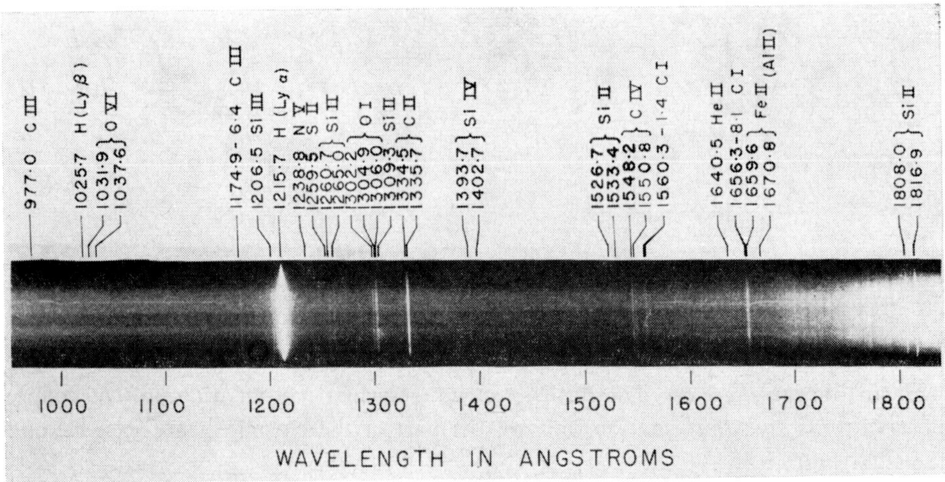

Fig. 5. Solar spectrum recorded at an altitude of 115 km above New Mexico on February 21, 1955, by the U.S. Naval Research Laboratory.

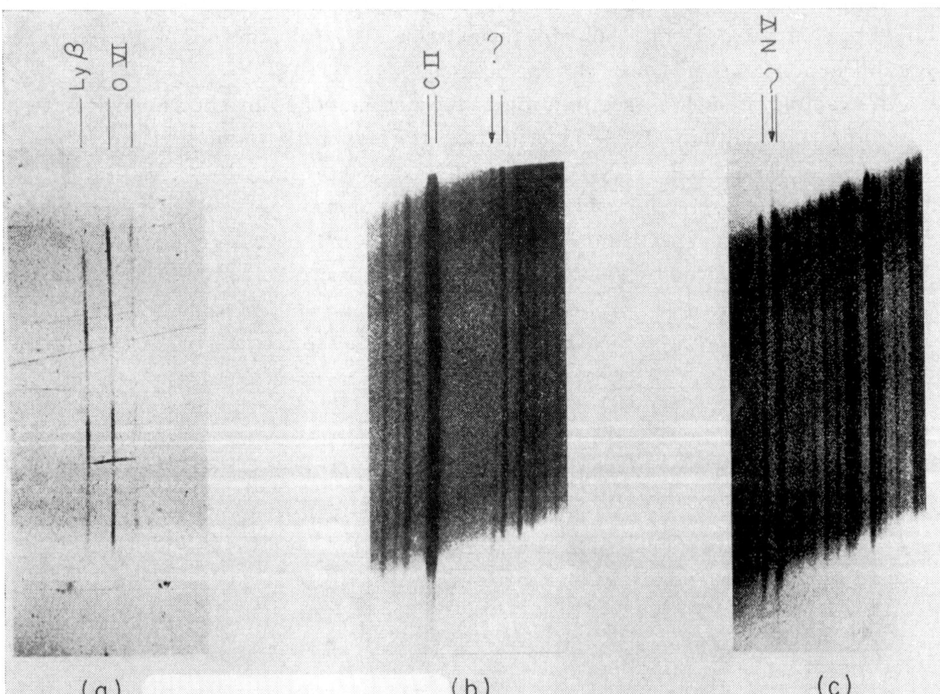

Fig. 6. Details of spectra taken by the U.S. Naval Research Laboratory on April 19, 1961 showing different emission lines (the picture is a negative). The high excitation O VI lines show limb brightening whereas Lyman-β does not. By courtesy of U.S. Naval Research Laboratory, Washington.

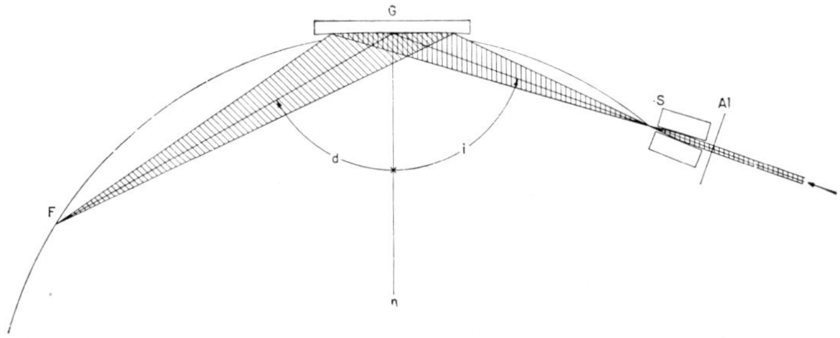

Fig. 7. Grazing incidence concave grating spectrograph used in EUV spectrography (Austen, Purcell and Tousey, 1962).

of the first a diaphragm eliminates all unwanted parts of the spectrum. This double-dispersion instrument was launched first on April 19, 1961 (see Chapter I, Figure 9). The visible and near-ultraviolet radiations fall beyond the end of the slit and do not enter the spectrograph. The final spectrum is inclined at an angle to the ruling of the last grating, but this is not a serious problem. Two versions of the instrument were flown in 1961, one to cover the wavelength region 2000–1200 Å, the other to cover 1250–800 Å. The resolutions were 0.2 and 0.1 Å respectively. It was possible to follow the continuum down to 950 Å, and from the Lyman limit to approximately 800 Å, due to nearly complete elimination of the stray light. Fraunhofer lines were distinguished together with emission lines in the region of 2000–1600 Å. The width of the stronger emission lines like Lyman-α and Lyman-β were determined.

The variation in intensity across the sun can be studied with the spectrograph of Tousey and others. When the first grating is distorted to neutralize astigmatism in the second, a fine spectrum line is obtained which bears a point-to-point conjugate relation with a line across the sun. The double dispersion instrument therefore eliminates stray light and permits the study of intensity variations between the centre and limb of the sun.

Parts of the spectra obtained with the double-dispersion grating spectrograph

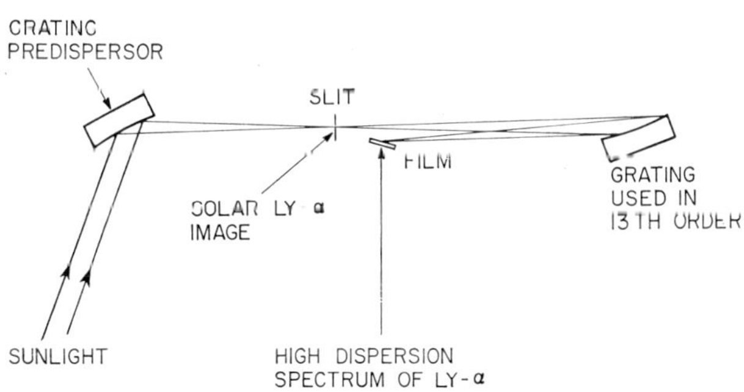

Fig. 8. The stigmatic high-resolution Lyman-α line spectrograph of the U.S. Naval Research Laboratory. The gratings are of 50 cm radius and have 1200 lines/mm (Purcell and Tousey, 1961).

are shown in Figure 6. They show the fine resolution and the different centre-limb behavior of Lyman-β and the O VI lines (the latter show limb-brightening).

Another type of spectrograph is that of the grazing *incidence spectrograph*, based

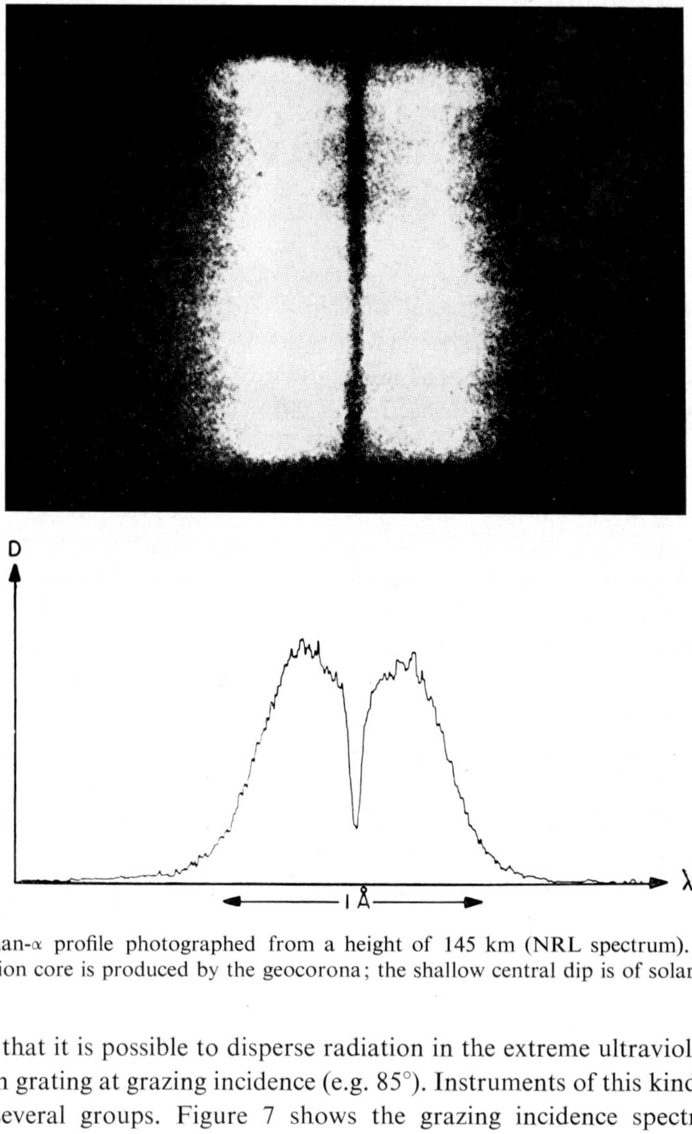

Fig. 9. Lyman-α profile photographed from a height of 145 km (NRL spectrum). The narrow absorption core is produced by the geocorona; the shallow central dip is of solar origin.

on the fact that it is possible to disperse radiation in the extreme ultraviolet by using a diffraction grating at grazing incidence (e.g. 85°). Instruments of this kind have been flown by several groups. Figure 7 shows the grazing incidence spectrograph of Austin, Purcell and Tousey.

Although the grazing incidence spectrograph is more efficient at short wavelengths than the normal incidence one, stray light of long wavelengths is more troublesome. Tousey and others reduce this by employing an aluminum sheet as a filter, approximately 1000 Å thick. Such a sheet only transmits light in the wavelength region 170–830 Å. Some gain in the light gathering power was produced by using a slit provided with reflecting jaws, which were small mirrors inclined at $\frac{1}{2}°$ to each other.

In order to obtain spectra with increased resolving power and of moderate size, it is necessary to go to high orders of interference. A *high resolving-power spectrograph* of normal incidence applying the thirteenth order was used by Purcell and Tousey (see Figure 8) to obtain the profile of the strongest emission line in the ultraviolet, i.e. the Lyman-α line of hydrogen. High resolution spectra of this line (Purcell and Tousey show a central absorption core and two asymmetrical wings, the typical profile of Lyman-α being that of Figure 9.

As has been shown by Purcell and Tousey the sharp central absorption core is produced by atomic hydrogen between the spectrograph and the sun, maybe by an extended "*geocorona*".

C. INTERPRETATION OF LINE INTENSITIES

The flash spectrum enables one to measure in the emission lines the integrated intensity E of a slab of the chromosphere, one cm thick, for chromospheric heights greater than the projected height of the moon's limb.

From this flash spectrum intensity E, it is not difficult to obtain the surface brightness F (per cm^2) since they are related by

$$E(h) = \int_h^\infty F(x)\,\mathrm{d}x \text{ (see Figure 10)}.$$

To find the emission ε (per cm^3) from the surface brightness F is more difficult

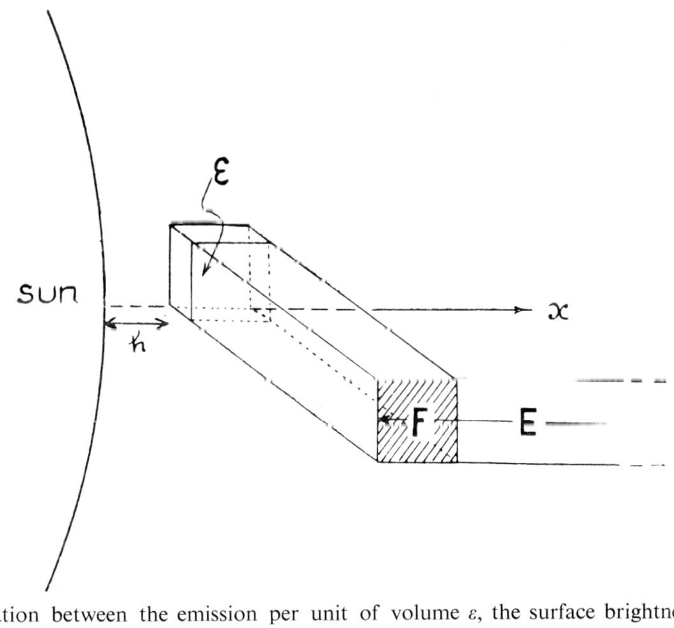

Fig. 10. Relation between the emission per unit of volume ε, the surface brightness F, and the flash spectrum brightness E.

for the following reasons: (1) although we can determine F we do not know the number of emitting atoms in the line of sight because of the self-absorption; (2) nor do we know the variation of ε with the height h. The usual procedure to overcome these difficulties is to assume: (a) the emission ε is given by a sum of exponentials:

$$\varepsilon = \sum_n \varepsilon_n \exp(-\alpha_n h);$$

(b) no self-absorption is considered. Then

$$F(h) = \sum_n \sqrt{\frac{2\pi R_\odot}{\alpha_n}}\, \varepsilon_n \exp(-\alpha_n h)$$

and

$$E(h_0) = \int_{h_0}^{\infty} F(x)\,\mathrm{d}x = \sum_n \sqrt{\frac{2\pi R_\odot}{\alpha_n^3}}\, \varepsilon_n \exp(-\alpha_n h_0).$$

In the particular case that ε can be represented by one exponential only, then E and F also contain one term. In this case E and ε have the same scale-height.

When ε is known, it is possible to obtain the number of atoms per cm^3, N, or physical parameters like T_e, the kinetic temperature of the electron gas. It should be realized that for many atoms the condition of local thermodynamical equilibrium (LTE) is far from being fulfilled, and Boltzmann's and Saha's laws cannot be applied. Then, for each volume element, the population of the atomic levels has to be calculated by steps: (a) a population of the levels is assumed as a first approximation; (b) one calculates the radiation field due to the photospheric, chromospheric and coronal emission reaching the volume element to be studied; (c) one equates for each atomic level the number of transitions to that level and the number of transitions from it, taking into account collisions and radiative transitions, and finds more precise values for the populations of the atomic levels; (d) the calculation is repeated for the chromospheric radiation field, using these improved values and taking radiation transfer into account. This program has not yet been fully worked out but some approximate solutions have been obtained, for hydrogen and helium.

Summarizing, the line of research is:

$$E \to F \to \varepsilon \to N$$

in which (a) in the transition $F \to \varepsilon$ the influence of self-absorption should be eliminated; and (b) in the transition $\varepsilon \to N$ the deviations from LTE should be taken into account.

D. CHROMOSPHERE AT THE LIMB: TWO-ELEMENT MODEL OF CHROMOSPHERE

When the earth's atmosphere is stable and quiet, we say "when the seeing is good", photographs of the limb obtained during eclipses show that the high chromosphere is inhomogeneous, having a spicular structure. This spiky appearance, compared by

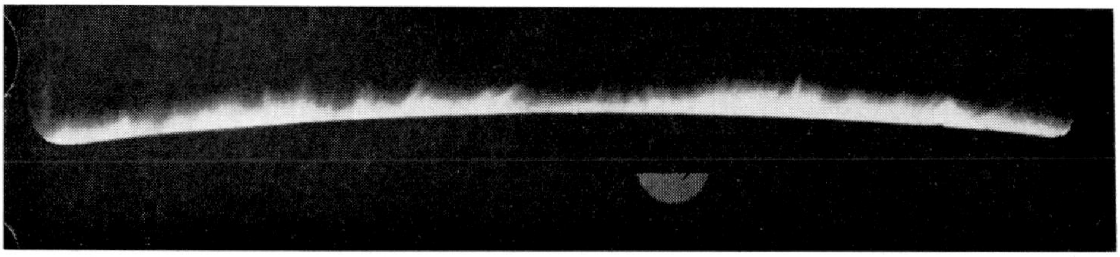

Fig. 11. Solar spicules photographed through an Hα filter at the limb by R. B. Dunn (Sacramento Peak Observatory).

earlier observers with a burning prairie, may also be studied outside of eclipses with monochromatic Lyot filters. With this optical instrument a feebly luminous layer, 5000 km thick, is observed, striated more or less perpendicular to the limb and numerous "spicules" growing above the general level and then disappearing in the course of some minutes (Figure 11).

The principal features of these spicules are:

1. They are mainly observed between 5000 to 10000 km;
2. They have life-times of a few minutes;
3. Their diameters are of 500–1000 km;
4. Within the spicules ascending and descending movements are observed whose velocity is about 20 km/sec;
5. The area occupied by the spicules is: at 2000 km, 10%; at 7000 km, 1%; at 15000 km, 0.01%.

Below the height of 5000 km either the spicules do not exist, or their number is so great that they can no longer be seen individually. Spectroheliographic studies on the disk show, as we shall see, the spicules from above and suggest that they cluster together in larger structures. The smallest structures which may be identified with spicules may occur down to about 1000 km above the photosphere.

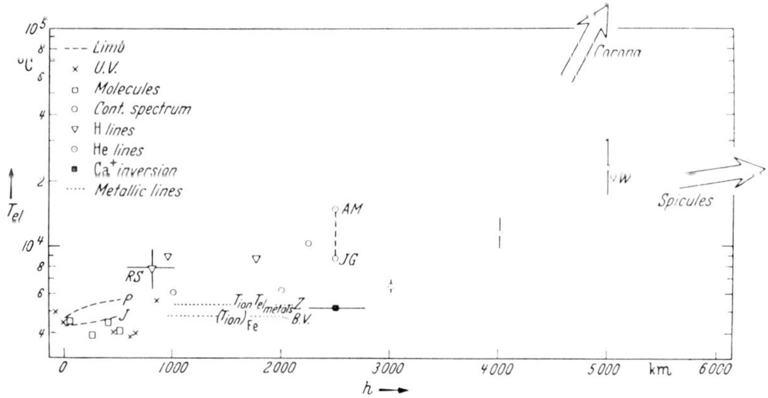

Fig. 12. Temperature determinations in the chromosphere.

The striking feature of the chromosphere is that in medium, and even in a part of the low regions inhomogeneities exist, an observation that also results from spectral observations.

The conclusion seems inevitable that the chromosphere, at least above 1000 km, is inhomogeneous and consists of dense elements in a more teneous medium. In the high levels (>7000 km) the spicules are cold elements in a hot medium (the corona). Near 2000 to 4000 km the situation is very complicated: the chromosphere merges into the corona near that height, but the height of transition to coronal temperatures is still unknown (Figure 12).

E. CHROMOSPHERE BEFORE THE DISK. INSTRUMENTS

It is possible to obtain pictures of the chromosphere at different levels by means of spectroheliograms. These are sharp monochromatic images of the sun, made, generally, in the light of the Fraunhofer lines. Since at these wavelengths the absorption coefficient is so great that the chromosphere is opaque for radiation of that wavelength, a

TABLE II

THE MEAN CHROMOSPHERIC HEIGHTS OF EMISSION (h^*) OF FRAUNHOFER LINES

Line	$\Delta\lambda$	h^* (10^3 km) *from model computations*	h^* (10^3 km) *from observed emission lines*	*Element*	λ	h^* (10^3 km) *from observed emission lines*
Hα	0.0	4.0	5.0			
	0.2	4.0		Mg	3838	1.2
	0.4	3.5			5184	0.3
	0.6	3.0				
	0.8	1.0		Fe	4202	0.1
					4384	0.4
Hβ	0.0	3.0				
	0.2	2.5		Ca	4227	0.8
	0.4	1.0				
				Ca^+	8542	1.4
Hγ	0.0	2.0			8498	
	0.2	1.0				
				Na	5890	0.3
Hδ	0.0	1.5				
	0.2	0.8		Sr^+	4078	0.9
						from model computations
K(Ca^+)	0.0	4.0	3.0	He	10832	2.0
	0.2	3.2			5876	2.0
	0.3(K2)	2.0	1.6			
	0.6(K1)	0.6				
	1.0	0.2				
	1.5	0.1				

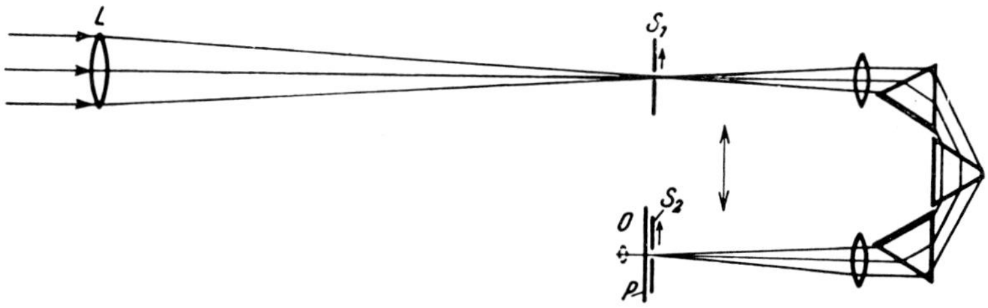

Fig. 13. Principle of a spectroheliograph (from K. O. KIEPENHEUER, *Die Sonne*, 1957).

set of spectroheliograms made in various lines or in various parts of the same line will yield pictures of the sun in various atmospheric levels, thus giving a three-dimensional image of the chromosphere. For its interpretation we should know the heights of the chromospheric layers that contribute mainly to the structures observed in spectroheliograms. These heights are of the same order as the height h^* in the chromosphere where the radial optical depth at the wavelength considered is unity. The heights either calculated for a given chromospheric model, or derived from the observed height where the tangential optical depth is of the order of unity (height of chromospheric emission lines above the limb), are shown in Table II.

Instruments to obtain monochromatic solar image spectroheliograms are:

(a) *Spectroheliographs*. Figure 13 shows a spectroheliograph. If used as a spectro-

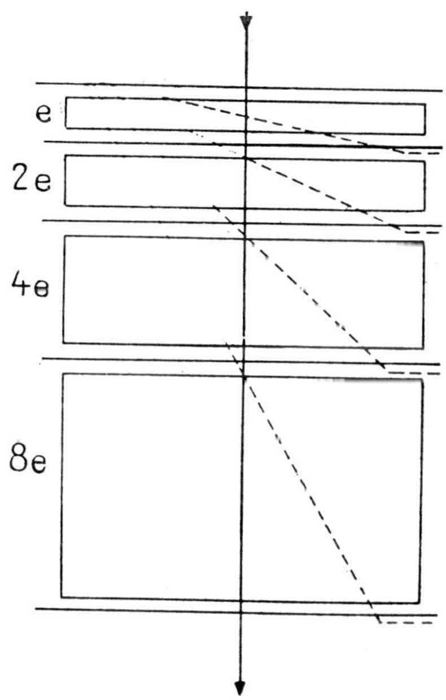

Fig. 14. Principle of the Lyot filter.

graph it is capable of producing a monochromatic line image of a small strip of the sun's disk. An image of the sun is produced in the focal plane of the first lens (L). The sun's spectrum is dispersed and focussed in the focal plane of the third lens, where the unwanted part of the spectrum is removed by a second slit. By scanning the sun's image across the first slit by means of motion of L_1, and by synchronized

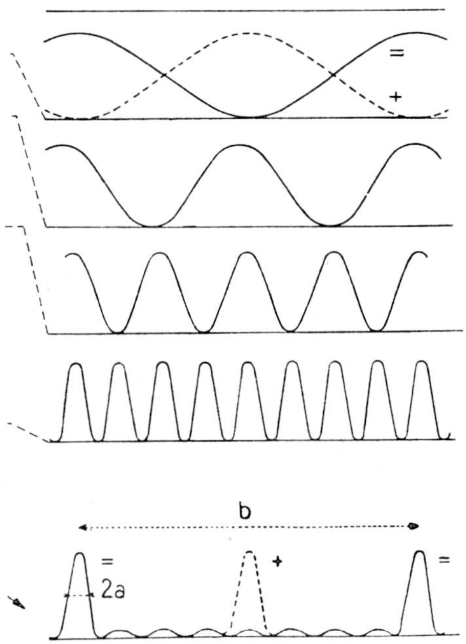

Fig. 15. Graphs of transmission plotted against wavelength for three layers.

movement of the photographic plate, a monochromatic solar image is obtained. The disadvantage of this system is the slow scanning rate which runs into minutes.

(b) *Lyot Filter*. Monochromatic pictures of the sun have been obtained with telescopes fitted with monochromatic polarizing or birefringent (Lyot) filters. The Lyot filter (Figure 14) consists of a series of parallel polaroid plates; between each there is a block of quartz. The thickness of successive quartz blocks bears a two to one relationship. The optical axes of the polaroids are parallel and lie in their plane. The quartz blocks also have optical axes which are parallel to their faces but inclined at 45° to those of the polaroids.

Graphs of transmission plotted against wavelength are shown for three layers in Figure 15. All residual peaks, except the one required, say λ_0, are removed by a normal filter.

The resultant filter may be designed to have a transmission peak of width of about 0.75 Å unit (Lyot filter; Levallois), or 0.5 Å (Halle, German Fed. Rep.) or 0.12 Å (Giovanelli, Sydney).

(c) *Lyman-α camera.* This camera, as used by Tousey and others consists of two concave gratings (Figure 16). The first forms an image of the sun dispersed horizontally. Since the Lyman-α line is more intense than any nearby lines, it is possible to exclude images of other wavelengths with a diaphragm. The second grating disperses the spectrum in the horizontal plane and produces a magnified image of the sun in Lyman-α. The image definition is improved by mechanically distorting the

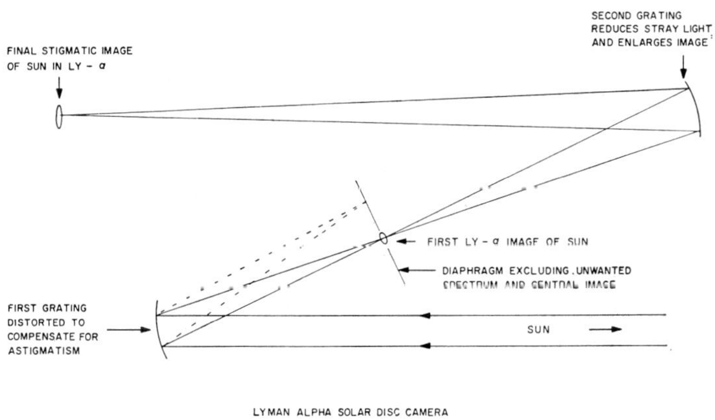

Fig. 16. The Lyman-α spectroheliograph of Purcell, Packer and Tousey (N.R.L.).

first grating so as to neutralize astigmatism produced by the second. Images of the sun in the calcium *K* lines bear some resemblance to those in Lyman-α (Figure 17).

F. RESULTS: CHROMOSPHERIC STRUCTURES

Table II shows that metal-line spectrograms refer to the low chromosphere (i.e. below 1500 km) and that only spectroheliograms obtained in the central parts of the Balmer lines, in the H and K lines and the infrared helium line, refer to higher levels. For Hβ, Hα and Ca^+K the whole central part originates from the high chromosphere.

From the above the following results are reached:

In metal line spectroheliograms referring to the low chromosphere, the images show bright *mottles* on a darker field. The diameters of the mottles are 2000 to 6000 km. A *network* with a characteristic length (mesh-width) of 25000 to 50000 km is also clearly visible. This is, for instance, the picture obtained in K_1 spectroheliograms. Also, in K_2 spectroheliograms (which are taken closer to the line centre) the mottling and the network are very pronounced, with diameters of the coarse mottling being of the order of 8000 km.

In spectroheliograms referring to the medium chromosphere, made in K_3 (centre of K line), *coarse mottles* are visible with diameters between 4000 and 8000 km and the *coarse network*, composed of mottles, with a characteristic length (meshwidth) of 25000 to 50000 km. Moreover, when spectrograms are obtained with the highest resolving power and in the best atmospheric conditions, the mottles are found to

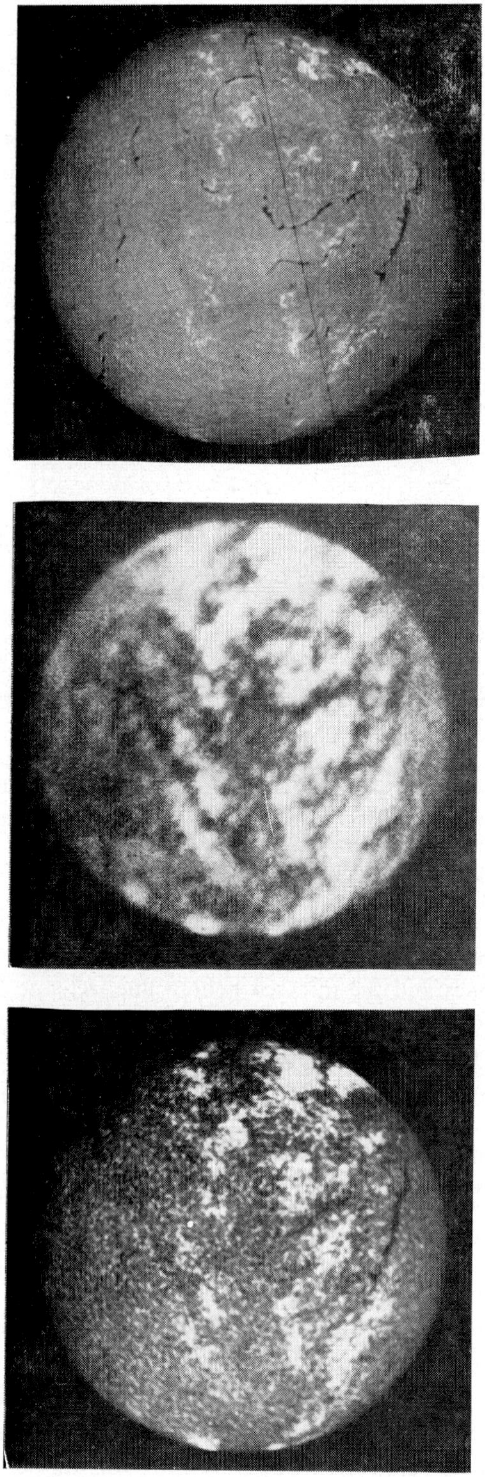

Fig. 17. From right to left: pictures of the sun obtained the same day, in Balmer-α, Lyman-α, and in the K-line of Ca II.

consist of many small elements with diameters of the order of 700 to 1600 km: it is called *fine mottling*. These fine mottles have recently been found to be identical with the spicules seen from above. (Bruzek, 1957; Beckers, 1963, Figure 18).

Summarizing these results on the spectroheliograms three main structures occur, which are described in Table III.

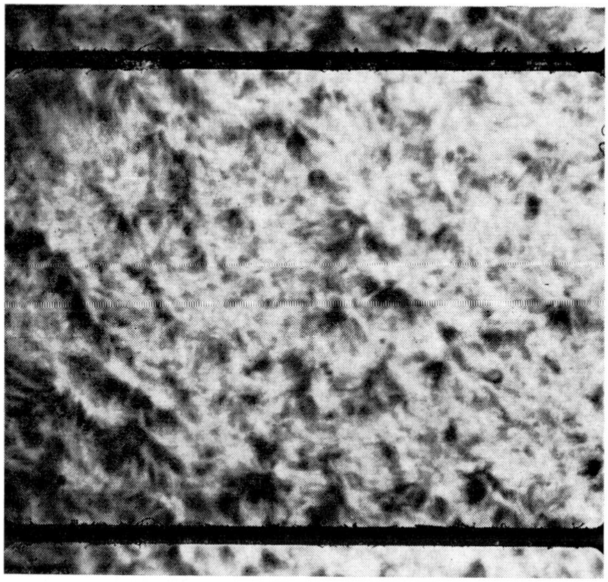

Fig. 18. Disk filtergram obtained August 4, 1963 by J. Beckers at Sacramento Peak, with a 0.5 Å Halle filter tuned to the red part of Hα. Scale 1″.9 per mm; the solar diameter is 1000 mm, the limb is about 1 cm outside the lower left corner. The coarse mottles appear to consist of "bushes" of spicules which are situated more or less radially to the sun's surface. By courtesy of Sacramento Peak Observatory, A.F.C.R.L.

Although spectroheliograms showing these structures can be obtained in various lines, K_3 of Ca II shows them the best. A given place in the sun may appear bright in one line and dark in another; their relative intensity may differ from line to line and differ, also, between the wing and centre of a line profile. The Ca II and Hα spectroheliograms emitted by the same layers are an example of this difference. the mottles are bright in K_3, whereas they are dark in $H\alpha_{1,2}$. This can be explained by assuming that Ca II and $H\alpha_{1,2}$ structures are formed mainly in the region where the ionization and excitation temperatures are near 6000 °K, and that the structures are due to the local temperature variations. For $T > 6000$ °K the values of dN/dT have inverse signs for Ca II and H.

The fine mottling, not being observed in spectroheliograms made in lines emitted below 1500 km, have to originate above this level. This observation confirms the result of the preceding section, that at this height a splitting of the chromosphere into two kinds of elements (cold and hot) occur. If the given average life-time of the

TABLE III

CHROMOSPHERIC STRUCTURES

Structure	*Diameter or characteristic length*	*Life time*	*Doppler velocities*
fine mottling	1000 km	5 min	20 km/sec (?)
coarse mottling	5000–8000 km	20 hours	
network	50 000 km	days (?)	0.1 to 1.0 km/sec

fine mottles is correct, one is tempted to identify them with the spicules as seen from above. This identification leads to the tentative conclusion that the spicules occur only above the 1500 km level.

The coarse mottles and the network are related to upward and downward motions in the chromosphere. At the 4000 km level (inner wings of Hα) the coarse mottles descend with an average velocity of 1 km/sec with respect to the mean level, and the region between the mottles ascend with about half that velocity. This velocity increases with height.

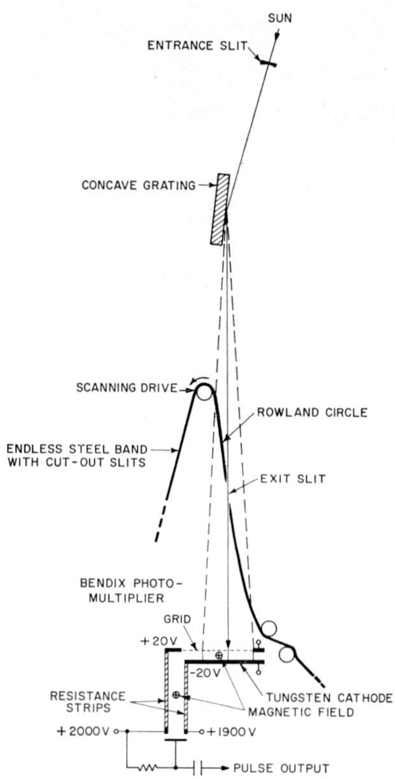

Fig. 19. Hinteregger's scanning grazing incidence monochromator.

The differences in brightness between the coarse mottles and their surroundings are perhaps due to a difference of about 200 °K in their excitation temperature, the mottles being hotter. An alternative possibility is that the microturbulences in the mottles are about 5 km/sec greater than in the surroundings.

Leighton has shown that the network is closely correlated with the magnetic field pattern.

In each "cell" of the network the matter moves predominantly horizontally from the centre to the edge with velocities of 0.3–0.5 km/sec. At the outer boundary the matter descends with velocities of 1.5–2.0 km/sec. Leighton calls these structures *supergranulation*.

2. Transition to the Corona

Hinteregger, and his associates have overcome the problem of stray light in grazing incidence spectrographs by measuring the spectral intensity with a special type of open photomultiplier (Figure 19). The tungsten cathode has a small quantum efficiency above 1300 Å, thus the detector has only a small response to the long wavelength stray light. Later a LiF cathode was used (see page 51).

The spectrum is scanned by a slit cut in a metal band which is constrained to move along the Rowland circle. Four slits of different widths were cut in the band so that the resolving power of the instrument could be changed. With this instrument the chromospheric and coronal lines have been measured down to 44 Å (Figure 19).

These spectra show the many high-excitation lines in the solar spectrum between about 100 and 1000 Å; these lines originate from the transition layer between the chromosphere and the corona; they are even partly of coronal origin. Later, Tousey and co-workers succeeded in obtaining photographic spectra in the same region, see Figure 20.

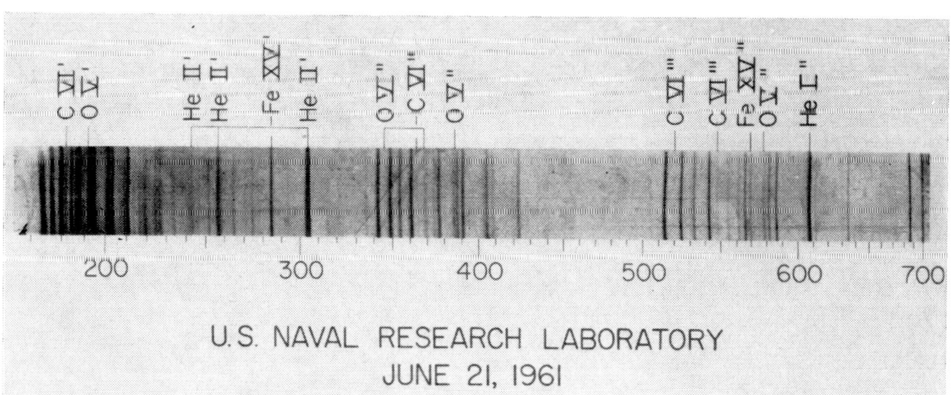

Fig. 20. Photographic spectrum obtained by the N.R.L. between 170 and 700 Å. Note that several orders overlap (superscripts ′, ″, and ‴).

The agreement between the recording curves obtained from Tousey's spectrum and that obtained by Hinteregger is excellent.

The mechanism usually adopted for explaining the heating of the chromosphere has already been explained in Chapter I, and is due to dissipation of shock wave energy. In the low chromosphere, around a height of 1000 km the acoustic waves become shock waves and the dissipation of energy becomes appreciable. This may explain why the chromospheric mottling becomes visible only above 1000 km. On the other hand, the energy radiated by the gas (mainly by Lyman free-bound hydrogen emission) is $5.5 \times 10^{-22} T^{-1/2} N_e^2$ erg cm^{-3} sec^{-1} and decreases with height because of the decreasing electron density. At heights where log $N_e \simeq 9.5$ the dissipation of mechanical energy (i.e. $\frac{1}{2}\varrho v^3 H^{-1}$ where H is the scale height) balances the radiation; at still greater heights it exceeds the rate of emission of the gas and the temperature assumes the coronal values of 10^6 °K. At this temperature the dissipation of mechanical energy is compensated by conductive losses.

At the basis of the transition zone between the chromosphere and corona the electron density is log $N_e \simeq 9.5$.

3. The Corona

A. INTRODUCTION. HEATING OF THE CORONA

The corona, like the chromosphere, was discovered during total solar eclipses as a pale aureole around the moon's disk. Until 1930 it could only be studied during eclipses because it is too faint compared with the brightness of the sky and the stray light in a telescope. The surface brightness of the inner corona is about 10^{-6} of that of the sun's disk and decreases by a factor 10^{-3} over a distance of one diameter.

During eclipses the corona is visible to distances many times the solar radius and it merges gradually into the zodiacal light.

The kinetic temperature of the coronal gas is high, between 5×10^5 and 2×10^6 °K Disturbed regions may attain temperatures up to several millions of degrees.

TABLE IV

CHROMOSPHERIC ENERGY BALANCE

h (*km*)	*Photospheric radiation flux*	*Mechanical flux*	*Emission of radiation*	*Heat conduction*
100	10.8	8.7	8.7	
1000	10.8	7.1		
4000	10.8	5.9 ↓	6.3	
Transition layer	10.8		5.4	4.6 ↑
Corona	10.8 ↓		4.5	—

To get insight into the heating of the corona the chromospheric energy balance must be considered. It is possible to measure the turbulent velocities in the chromosphere, and with the known densities, to calculate the mechanical energy flux. This is expressed in the third column of the Table IV which also shows the photospheric radiation flux, the observed emission of radiation and the computed heat conduction. The figures in the table are the logarithms of the various fluxes in erg cm^{-2} sec^{-1}.

As we have seen in the preceding section, at heights greater than that at which $\log N_e \simeq 9.5 (N_e \simeq 3 \times 10^9$ $cm^{-3})$ the temperature increases because the emission of radiation becomes relatively less important. The radiation of the corona is of the order of 5×10^3 erg cm^{-2} sec^{-1}, whereas the inflow of mechanical energy is about 10^5 erg cm^{-2} sec^{-1}. Hence coronal energy is mainly lost by conduction. Bearing in mind the energy balance the temperature can be determined. Detailed computations give $T = 7 \times 10^5$ °K, when no account is taken of the magnetic field. Considering the presence of magnetic fields the coronal temperature will presumably be higher.

B. OBSERVATIONS

The methods for observing the corona are:

1. Eclipse observations

The only way in which the corona can be observed at great distances of the solar limb is from photographs taken during total solar eclipses.

The form of the corona changes with the 11-year cycle of solar activity. At the minimum of the solar activity the corona shows a nearly uniform atmosphere in the equatorial region, and two systems of polar rays are well defined. Between minimum and maximum the rays are more clearly visible and may appear as long extensions to the corona. During the maximum the density and brightness are greater; the shape is nearly circular and there are frequent disturbances, such as coronal rays, arches, etc. See Figure 21.

Three components can be distinguished:

(a) The *L-corona* shows an emission line spectrum produced by a highly ionized gas (Ca XII–Ca XV; Fe X–Fe XVI; Ni XII–Ni XVI and A X). Such degrees of ionization can be explained only by assuming ionization temperatures between 5×10^5 and several times 10^6 °K.

(b) The *K-corona*, the light is partly polarized; the continuous spectrum consists of photospheric light scattered by free electrons, close to the sun, and with high kinetic temperatures. Because of the great velocities of these fast moving electrons ($v \simeq 6000$ km/sec) the photospheric Fraunhofer lines are washed out in this scattering.

(c) The *F-corona* in which the light is less polarized is due to scattered photospheric light. Just as the photosphere does, it presents a continuous spectrum with Fraunhofer lines; their profiles are similar to those of the photosphere, which shows that these scattering particles move but slowly. These are the zodiacal dust particles between the sun and earth, most of them at great distances from the sun.

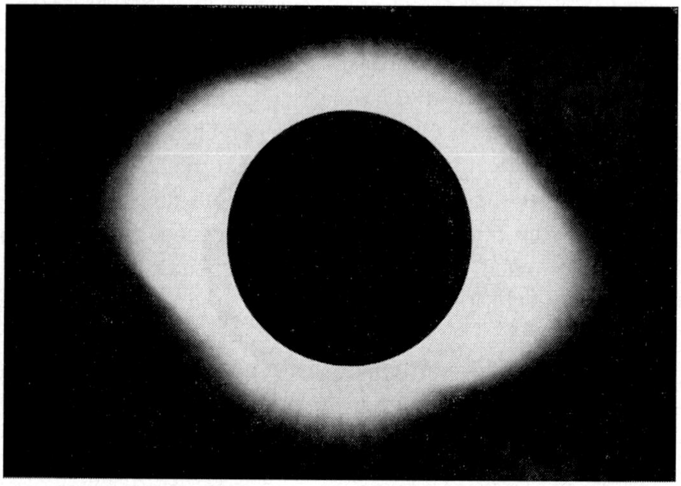

Fig. 21a. Solar corona photographed by A. Wallenquist (Uppsala) at the eclipse of June 30, 1954 (Gotland, Sweden). A typical example of a corona at sunspot minimum.

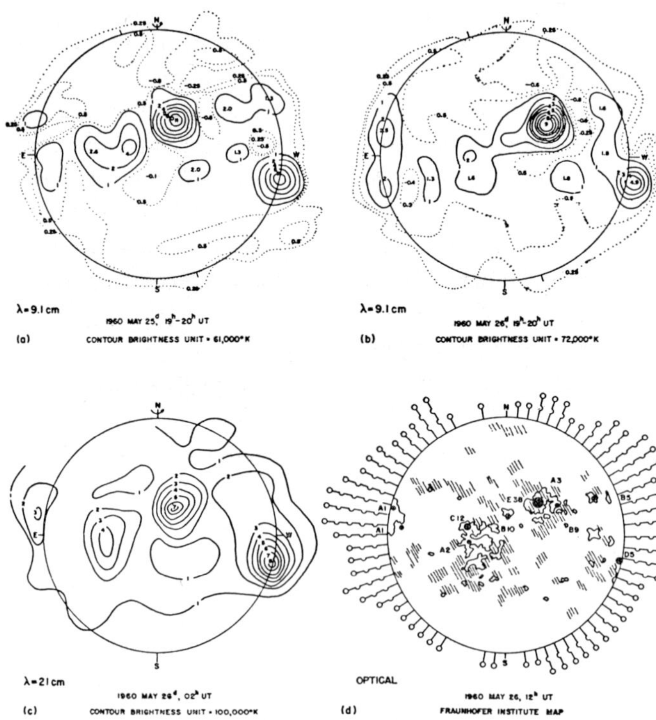

Fig. 21b. Comparison of radio-observations on 9.1 and 20 cm taken by Swarup *et al.*, with the Fraunhofer Institute optical map of the sun, for May 25–26, 1960.

The K- and F-corona form the *white corona*. Their light is nearly the same colour as that of the sun.

2. *Coronagraph and K-coronameter*

In Lyot's coronagraph instrumental scattered light is carefully and ingeniously elimi-

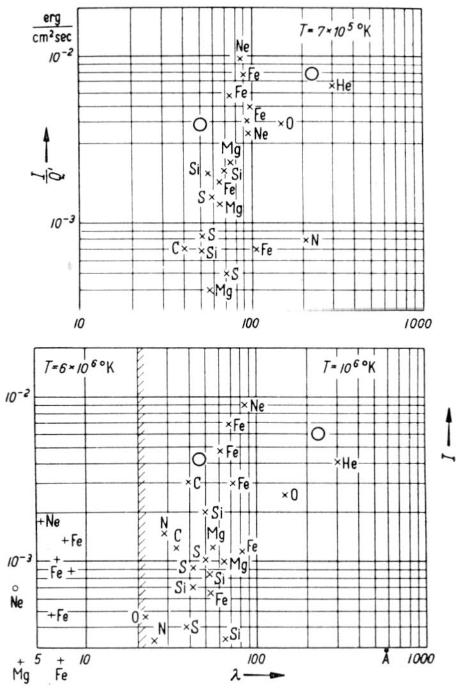

a

Fig. 22. Computed spectra of the corona for $Te_{\rm kin} = 7 \times 10^5$, 10^6 and 6×10^6 °K, after Elwert. Lines are shown by crosses, series limits by circles.

nated. When it is placed on a high mountain so that atmospheric scattered light is reduced and when we observe monochromatically one of the strong coronal lines, the red line: 6374.5 (Fe X, ionization potential 233 eV), the green line: 5303 (Fe XIV, 255 eV), the yellow line: 5694.5 (Ca XV, 814 eV), the instrument will enable us to observe the L-corona up to limb distances of $0.4R_\odot$.

The *K-coronameter*, developed by Dollfus, Newkirk and Wlérick measures the non-atmospheric polarization around the sun (which is due to the electrons in the K-corona), and thus measures the number of electrons in the line of sight beyond the sun's limb. Occasionally coronal rays can be detected with this instrument up to one or 1.5 radii from the sun's limb.

3. *X-ray pictures*

The solar corona is the most important extraterrestrial source of X-rays (see Section

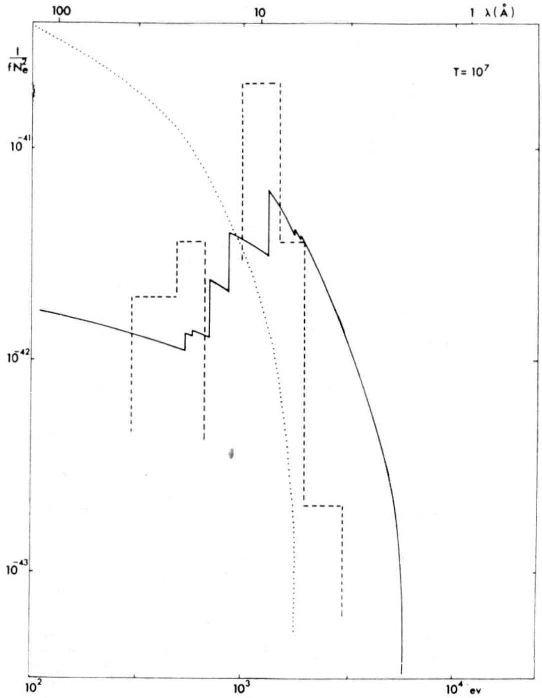

Fig. 23a. Kawabate's computed coronal spectrum for a temperature of 10^7 °K. Dotted lines: free-free emissions; solid lines: free-bound-emissions; dashed lines: line emissions.

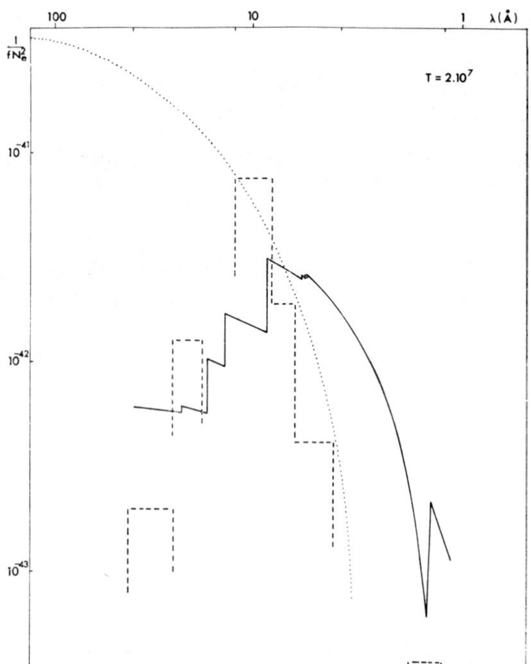

Fig. 23b. Kawabate's computed coronal spectrum for a temperature of 2.10^7 °K. Dotted lines: free-free emissions; solid lines: free-bound emissions; dashed lines: line emissions.

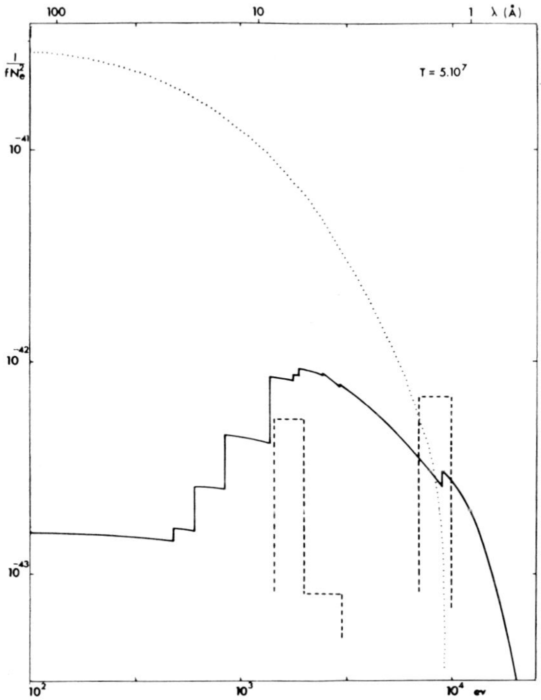

Fig. 23c. Kawabate's computed coronal spectrum for a temperature of 5.10^7 °K. Dotted lines: free-free emissions; solid lines: free-bound emissions; dashed lines: line emissions.

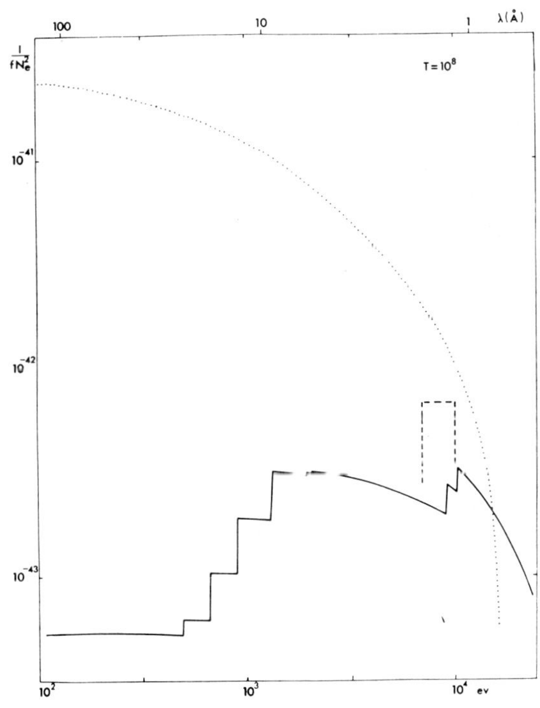

Fig. 23d. Kawabate's computed coronal spectrum for a temperature of 10^8 °K. Dotted lines: free-free emissions; solid lines: free-bound emissions; dashed lines: line emissions.

3D). In the wavelength region below about 200 Å other parts of the sun do not emit any radiation so that observations in this spectral region give information on the corona and are not disturbed by radiations from other sources.

A few X-ray heliograms have up to now been obtained by means of a pinhole camera, the angular resolving power being of the order of some minutes of arc. Improvement of the image-forming device would be a very important means of increasing our information on the fine structure of the corona.

4. Radio heliograms

All solar radio emission at wavelengths above one meter comes from the corona. So the solar radiospectrum and radio-heliograms are sources of information on the corona. A difficulty is the low angular resolving power: at a wavelength of one meter an aperture of 3000 m is needed to obtain a resolving power of one minute of arc only! In Sydney (Australia) daily radioheliograms are obtained at 20 cm, and in Stanford (Berkeley) at 9.7 cm. These do not refer to the corona but to the transition region chromosphere-corona (20 cm) and to the upper chromosphere.

C. THE X-RAY SPECTRUM OF THE QUIET CORONA

There are two, or in fact, three components to the radiation of the corona: the continuous radiation consisting of a *free-free* component (mainly free-free transitions of electrons in the field of protons) and a *free-bound* component (recombinations of electrons with the heavy ions of the coronal plasma), and the *line emissions*. Some of the emission lines are due to a de-excitation of an upper level populated by recombination; the greater part, however, is the result of direct excitation by electron impacts.

Since all the emissions are proportional to N_eN_i, and because in this highly ionized plasma N_i is proportional to N_e, the intensity of the coronal emissions is proportional to N_e^2.

As in the corona N_e decreases outward with a scale height of about 5×10^4 km, the X-ray emission intensity should decrease outward with a scale height of about 2.5×10^5 km; i.e. the intensity would decrease to 0.01 of the limb value over a distance of little more than 10^5 km.

The detailed computations, made with great care by Elwert still involve a number of uncertainties: most ionic levels have not yet been measured and their excitation and ionization energies have to be computed, assuming the ion to be hydrogen-like, or by extrapolation of iso-electronic sequences. Further uncertain factors are the probability of recombination to excited states, and the probability of direct excitation by electron impact. Furthermore the fine structure of the lines could not be taken into account. Notwithstanding these uncertainties Elwert's computations tallied remarkably well with later observations.

Elwert made his predictions for temperatures of 7.10^5, 10^6 and 6.10^6 °K. Later, Kawabate extended them to values of 10^7, 2.10^7, 5.10^7 and 10^8 °K. One feature of the results is that the maximum wavelength of emission shifts to shorter wavelengths for increasing temperatures; another feature is that, in any case for low temperatures, line emissions are predicted to be much more important than the continuous emission. For $T \simeq 10^6$ °K their ratio is of the order of ten. Figure 22 shows Elwert's results, Figure 23 those of Kawabate.

From these Elwert predicted an *integrated X-ray flux* of 0.03 Q erg cm^{-2} sec^{-1} during sunspot minimum; 0.4 Q erg cm^{-2} sec^{-1} during sunspot maximum (Q is an uncertainty factor of the order 2 to 4).

These values are roughly in agreement with observations: during sunspot minimum Friedman observed 0.13 erg cm^{-2} sec^{-1}; during maximum he found values ranging from 0.4 to 1 erg cm^{-2} sec^{-1}.

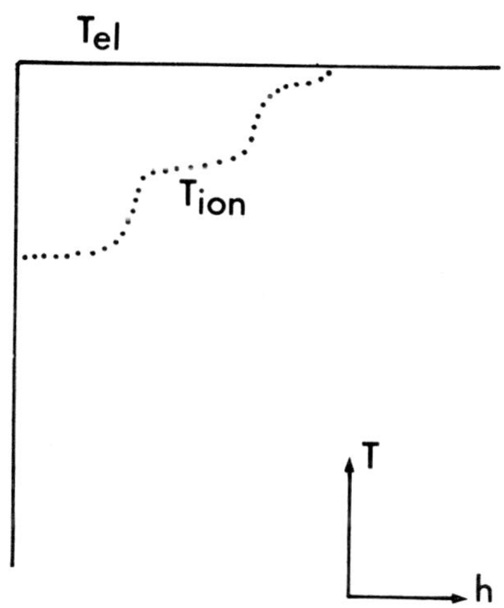

Fig. 24. Neupert's hypothesis to explain the various coronal ionization temperatures.

In 1963 spectral line observations were made with a fairly high spectral resolving power. HINTEREGGER and co-workers flew the scanning grazing-incidence monochromator shown in Figure 19. Scanning was done between 315 and 55 Å with a scanning speed of 2 Å/sec; the detector had a LiF photocathode, which is blind to $\lambda > 1100$ Å, and this greatly reduces the influence of straylight (Lyman-α!). Furthermore, the LiF cathode has a greater quantum efficiency for shorter wavelengths than ordinary tungsten. The instrumental profile had a width of 0.35 Å at 55 Å; lines could be detected down to 65 Å. The intensity drop at 170 Å, observed earlier, appears to be real.

On May 10 1963 Austin, Purcell and Tousey flew the NLR grazing-incidence spectrograph with a new diffraction grazing with increased speed. Lines could be detected down to 44 Å.

In the summer of 1963 Friedman, Blake, Chubb and co-workers flew a crystal spectrometer in a rocket and were able to detect many lines down to a wavelength of 13 Å; a preliminary investigation of the spectrum did not show traces of a continuous spectrum.

The short wavelength tail of the X-ray spectrum ($\lambda \lesssim 20$ Å) appears to be highly variable which is due to the fact that even fairly small activity centers produce an appreciable intensity in this region. Rough spectral scanning is fairly easy in this region, since proportional counters may be used, as has been done on various occasions (Figure 24).

D. PROBLEM OF CORONAL TEMPERATURES

From the various observational results obtained near the minimum of solar activity, the temperature of minimum corona can be calculated. The following two groups of results have been found:

1. Electron temperatures

The non-appearance of Fraunhofer lines in the K-corona indicates that the electron temperature is in any case greater than 10^5 °K.

The ionization of Fe and the comparison of the intensities of the red and green Fe lines has during many years been thought to point to an electron temperature of 700000 °K. Very recently (1964) Burgess has reconsidered the problem of the ionization in the corona, thereby taking into account the influence of di-electronic recombinations. The influence of this mechanism proves to be great; a certain degree of ionization occurs now at higher temperatures than earlier. We may now assume that coronal ionization and excitation indicates temperatures of about 1.5×10^6 °K.

The measured total intensities in the X-ray spectrum of the sun are thought to correspond to an electron temperature of 750000 °K; but also these values may turn out to be higher in view of Burgess's new results.

The radioradiation of the quiet sun at metre wavelengths indicates an electron temperature of 700000 °K, in the lower part of the corona.

2. Kinetic temperature

(a) From the density determinations of the K-corona a kinetic temperature of approximately 1.5×10^6 °K is obtained.

(b) Intensity gradients in the monochromatic corona yield kinetic temperatures of 1.8×10^6 °K to 2×10^6 °K.

(c) The widths of the emission lines indicate similar kinetic temperatures.

Hence, with the *old* values of the ionization cross sections there seemed to be some disagreement between the electron and kinetic temperatures.

Two explanations that were suggested to solve this discrepancy are:

(1) Small-scale non-systematic motion in the corona with velocities of the order of 50 km/sec. These motions will greatly affect the line widths of the ion lines: at $T = 10^6$ °K and Fe-ion has a rms thermal velocity of about 20 km/sec. The thermal electron motions ($\bar{v} \simeq 6000$ km/sec), which determine the excitation and ionization conditions and the radio spectrum, will not be influenced by these "turbulent" motions.

(2) There is a solar wind: the coronal matter streams outward from the sun. Near the earth these velocities are some hundreds of km/sec, but in the low corona they are presumably small and of the order of some km/sec only. Suppose that solar matter streaming out from the chromosphere into the corona passes the level when it suddenly acquires a kinetic temperature of about 2.10^6 °K. Since there is a time lag in the ionization the ionization temperature will be less than 2.10^6 °K in the low

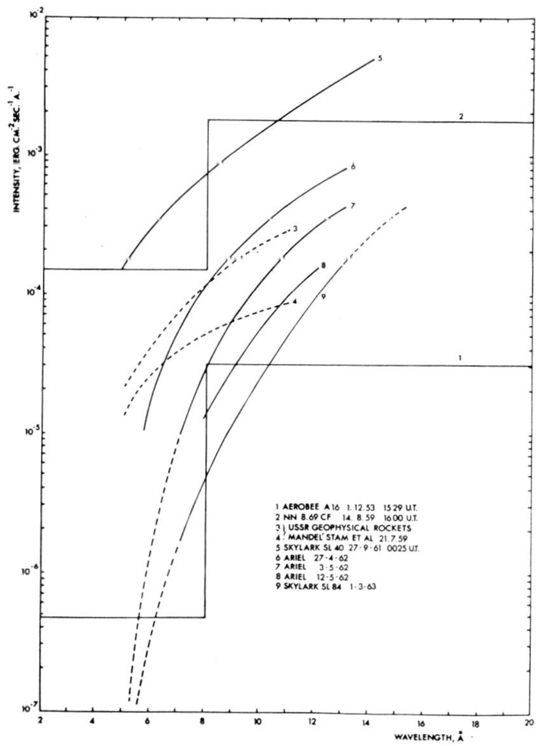

Fig. 25. A comparison of X-ray measurements of the quiet sun (see *Space Research* **4**). By courtesy of K. A. Pounds, Leicester.

corona, and only in the high parts there will be equality between the kinetic and ionization temperatures (Figure 25). This hypothesis, due to Neupert, has not yet been confirmed by the observations.

However, in view of the recent results on the ionization cross sections it now seems natural to assume that the quiet corona has a temperature of about 1500000 °K. The lower temperature found from the meter-wave radio-radiation of the quiet corona may be a consequence of the fact that the corona at meterwaves is a "grey" body: due to the curvature of the ray-paths the optical thickness of the corona does not exceed unity for meter waves.

References

ATHAY, R. G.: 1965, *The Solar Spectrum* (ed. by C. DE JAGER), D. Reidel Publishing Co., Dordrecht, p. 151.

BILLINGS, D. E.: 1965, *The Solar Spectrum* (ed. by C. DE JAGER), D. Reidel Publishing Co., Dordrecht, p. 280.

HINTEREGGER, H. E.: 1965, *The Solar Spectrum* (ed. by C. DE JAGER), D. Reidel Publishing Co., Dordrecht, p. 179.

VAN DE HULST, H. C.: 1952, *The Sun* (ed. by G. KUIPER), p. 207.

DE JAGER, C.: 1959, *Handbuch der Physik* (ed. by M. FLÜGGE), **52,** p. 80.

JEFFERIES, J. G.: 1965, *The Solar Spectrum* (ed. by C. DE JAGER), D. Reidel Publishing Company, Dordrecht, p. 131.

SEATON, M. J.: 1965, *The Solar Spectrum* (ed. by C. DE JAGER), D. Reidel Publishing Co., Dordrecht, p. 273.

TOUSEY, R.: 1963, *Space Sci. Rev.* **2,** 3.

UNSÖLD, A.: 1955, *Physik der Sternatmosphären*, Springer-Verlag, Berlin.

CHAPTER 3

THE SOLAR CENTER OF ACTIVITY

C. DE JAGER

Sterrewacht 'Sonnenborgh', Utrecht, The Netherlands

1. Definition

A center of activity (CA) is the totality of all phenomena accompanying the birth and development of sunspots. It is also concerned with what remains after the sunspots have disappeared.

The complexity of the many optical phenomena which together form a center of solar activity arises from the fact that none of them can apparently be considered as the primary one. However, the arrival of the magnetic disturbance (large magnetic field) at the sun's surface is probably the essential and primary event.

2. General Review of the Different Phenomena Occurring in a CA

Table I lists the transient phenomena associated with solar activity.

TABLE I

SOLAR PHENOMENA AND THEIR DURATION

Phenomena		*Life-time*
Spots (4)*		days to weeks
Faculae (or plages) (5)	photospheric	weeks to months
	chromospheric	
Coronal center of activity (6)		weeks to months
Coronal sporadic condensation (7)		hours to days
Prominences	spot prominences (8)	hours to days
	quiescent prominences (10)	weeks to months
Flares (9)		ten minutes to some hours
Coronal rays	R rays above CA's	weeks
	streamers (connected with prominences)	months

* The numbers between parentheses refer to the location of the features in Figure 1.

3. Schematic Evolution of a CA

The following life-history of a CA represents a typical evolution, though the actual succession of events may often differ considerably from that shown in Figure 2: each activity center is different and has its own character and development.

First phase: pre-spot phase. Development of a bipolar magnetic (BM) field region

Ortner & Maseland (eds.), Solar Terrestrial Relations.

followed (some hours to two days later) by small bright faculae which originate rather suddenly in the BM regions. We also note during this phase the formation of a coronal activity region.

Second phase: spot-phase. The faculae and the magnetic field region increase in size and spots appear (some hours to one day later) and develop as indicated on page 295. Other phenomena occur, such as spot prominences, flares, surges, sporadic condensation, etc. The CA becomes more active: enhancement of radio wave radi-

Fig. 1. Schematic drawing of a center of activity.

ation, emission of X-rays and sometimes cosmic ray particles. This phase may last between one and several weeks.

Third phase: BM post-spot phase. The spots, flares and associated phenomena have disappeared. The facular field decreases in brightness and becomes more and more patchy. The long quiescent filaments are formed during this phase. Duration: one to several months.

Last phase. After five or ten solar rotations, the optical signs of activity have almost disappeared, but it is not impossible that there remain unipolar magnetic (UM) regions which may be visible during a long time (months). The long quiescent filaments are still visible and they seem to be connected with the coronal streamers, which develop above quiescent prominences.

The development of the *coronal part* of the CA is described next (Figure 3): The first indication of a developing CA is a gradual increase in density in the lower part of the corona, above the facular region. In the phase of greatest activity, when the spots are largest and when flares occur, the coronal CA is marked by loop-prominences, spot filaments and other active prominences. In this phase one also observes the sporadic coronal condensations. After this phase when the spots have disappeared, only the enhanced density and the quiescent prominences may persist for some months.

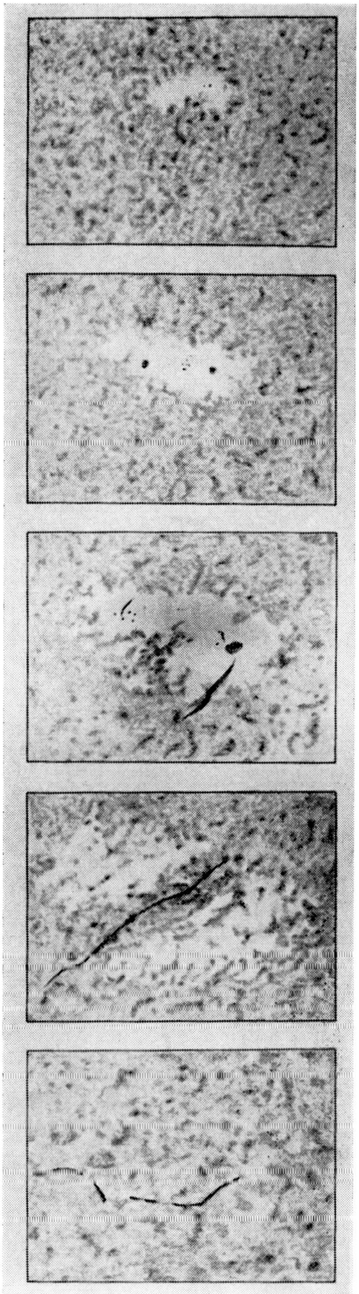

Fig. 2. Development of a typical center of activity, seen on the disk with a spectroheliograph.

Fig. 3a. Schematic development of the coronal part of a center of activity; pre-spot phase: faculae and a coronal condensation.

Fig. 3b. Schematic development of the coronal part of a center of activity. Phase of growing spots, spot prominences, the first flares.

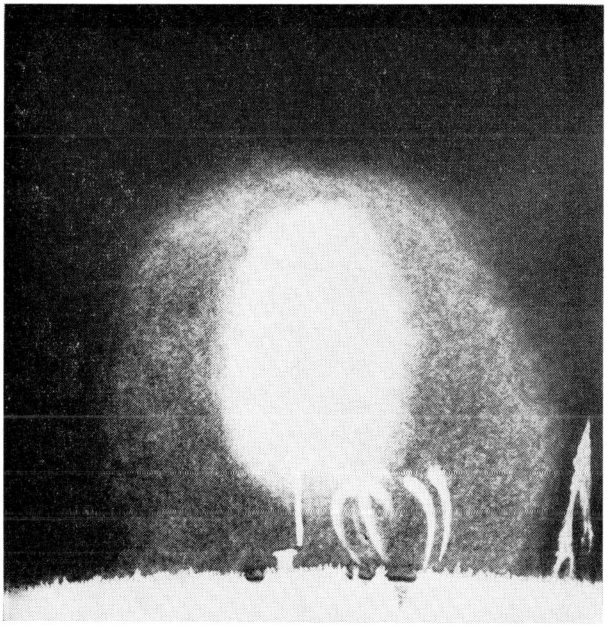

Fig. 3c. Schematic development of the coronal part of a center of activity. Phase of greatest activity. Flares, surges, loop prominences; sporadic coronal condensations and enhanced densification in the center of the coronal condensation.

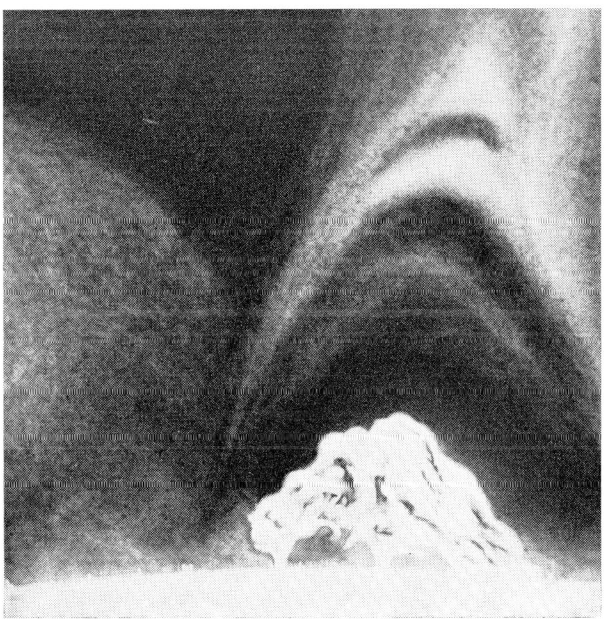

Fig. 3d. Schematic development of the coronal part of a center of activity. BM, non-spot phase.

Fig. 3e. Schematic development of the coronal part of a center of activity. Latest phase; the CA has disappeared; the quiescent prominence dissolves.

The *peculiar emissions* associated with a CA in the absence of solar flares may be summarized as follows:

(a) Enhanced UV radiation, in particular Lyman-α;

(b) Enhanced X-ray radiation, particularly in the region between 5 and 60 Å;

(c) Enhanced particle emission of the solar-wind type (R-rays).

In this Chapter we discuss in some more detail the various aspects of a CA, with the exception of the flares, which are dealt with in Chapters IV, V and VI.

4. Description of Some Transient Solar Phenomena

A. SUNSPOTS

1. Description

A typical spot consists of a dark umbra appearing in the photosphere and surrounded by a brighter penumbra (Figure 4). The intensity of the umbra measured in light of visual wavelengths varies between 5 to 10% of the intensity of the undisturbed photosphere. The intensity of the penumbra is about 75% of the quiet disk intensity. The most important feature of the sunspots is their strong magnetic field. The magnetic fields of spots increase roughly with increasing area and are of the order of 500 to 4500 gauss.

The darkness of the spots is due to the magnetic inhibition of convective motion in the sub-photospheric layers where the spots occur. In the region below the undisturbed photosphere more than 99% of the solar energy flux is transported by convective motion (upward and downward). The inhibition of such motions in a strong magnetic field results in a reduced energy flow and thus in a darkening of the spot region. Observations show that the umbral surfaces of large spots are about 1000 km below the photospheric surface (the Wilson effect).

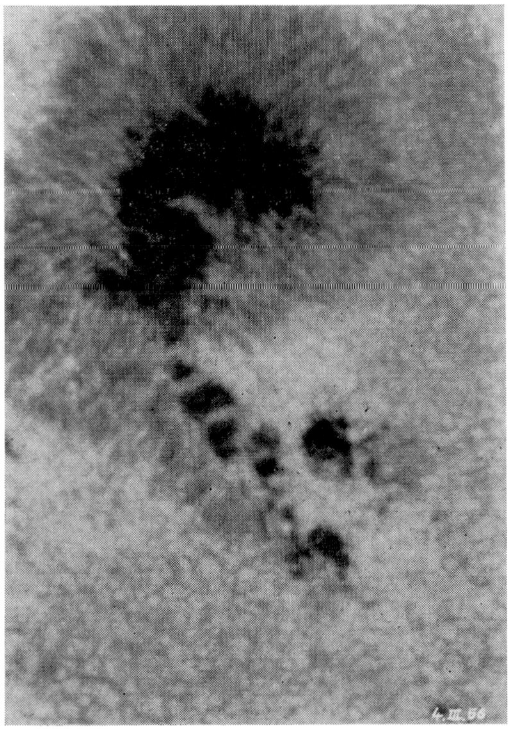

Fig. 4. Sunspot photographed on March 4, 1956 at the Pic du Midi Observatory by J. Rösch.

Laminar motions along the field lines of the spot are not inhibited. Such motions have been observed (the so-called *Evershed effect*) as sideward motions along the divergent field lines with velocities between 2 to 8 km/sec in the higher photosphere, and downward motions with velocities up to 35 km/sec in the lower chromosphere. This Evershed effect is observed by means of the Doppler effect on the spectral lines. It manifests itself directly in the *penumbral striae* (see Figure 4). The Evershed motion is essentially confined to the striae and does not occur outside these structures. In fact the striae must be considered as regions where convective motions occur; gas ascends under an angle of about 10° to 25° with the horizontal plane and thus moves from the lower lying umbrae towards the photosphere, and *vice versa*.

2. Polarity law

About 90% of the spots occur in pairs or groups of spots with a bipolar magnetic field. Sometimes the bipolar structure appears *in* a spot. The famous *polarity law* has three aspects:

(a) The two members of a sunspot group have opposite polarity.

(b) At the same hemisphere all preceding (westward) spots have the same polarity. All spots in northern hemisphere have polarities opposite to those of the symmetric ones in the southern hemisphere.

(c) The distribution of polarity among the two members of a group changes sign with each new solar cycle ($\simeq 11$ years).

Thus a real solar cycle is a cycle of 22 years.

The distribution of polarities in the last few cycles is shown in Table II.

TABLE II

DISTRIBUTION OF POLARITIES OF SPOTS (1933–1964)

	Hemisphere			
Cycle	*Northern*		*Southern*	
	p*	f	p	f
1933–1945	N	S	S	N
1945–1954	S	N	N	S
1954–1964	N	S	S	N

* p = preceding spot; f = following spot.

3. *General development of spots (Waldmeiers's designation)*

The general development of spots is described as follows:

A. Single spot or group; pores without penumbra, no bipolar structure B. Bipolar group	1st day
C. Bipolar group with penumbra D. Larger bipolar group with penumbra	2nd day
E. Large bipolar group, many small spots F. Maximum type	5th–10th day
G. Large bipolar group without smaller companions H. Unipolar spot with penumbra I., J. Spot decreases in size and disappears	10th–15th day

In *most* cases only stage A is reached; in other cases they develop following A–B–A or A–B–C–B–A, . . ., etc. Only some spots follow the whole sequence A–J.

B. FACULAE

1. *Photospheric faculae*

Photospheric faculae appear as a bright structure on the disk around sunspots and

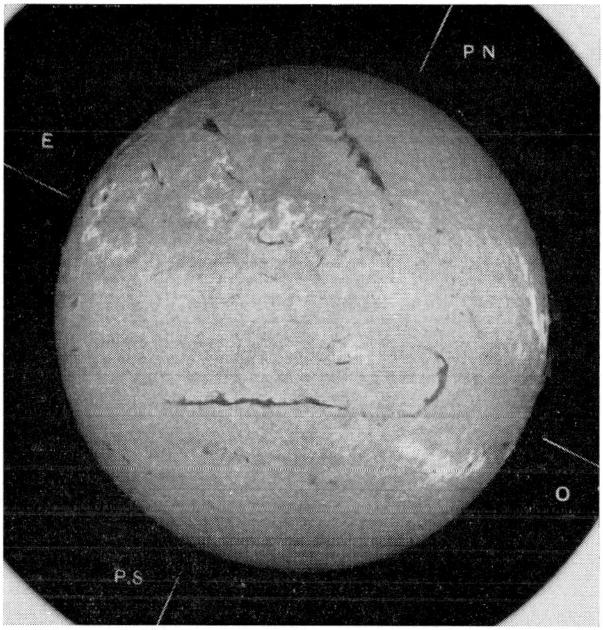

Fig. 5a. Spectroheliogram of the sun of March 25, 1949 in Hα, showing at least four well-developed facular regions and a very long filament (by courtesy of R. Michard, Meudon Observatory).

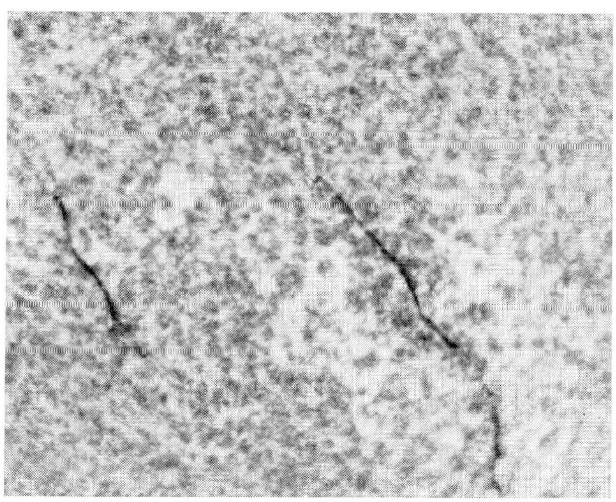

Fig. 5b. Part of a Ca^+ spectroheliogram of May 25, 1922, obtained at the Meudon Observatory. The picture shows a weak facular field split into two by a thin filament.

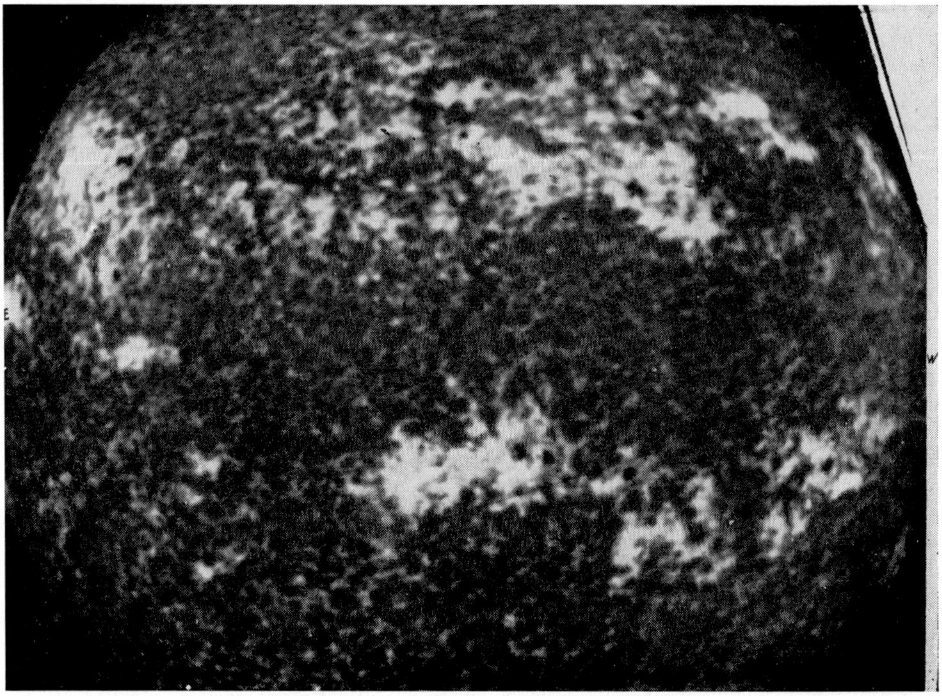

Fig. 5c. Part of a K3 spectroheliogram of May 3, 1958, obtained at the Schauinsland station of the Freiburg Observatory (by courtesy of K. O. Kiepenheuer).

are only visible near the sun's limb. The contrast between photospheric faculae and their surroundings is zero near the disk's center and increases toward the limb. The upper parts of the faculae are hotter than equally high photospheric layers and the lower parts are colder. From the brightness excess, as a function of the distance to the disk center, we derive the temperature excess (faculae *minus* photosphere) as a function of the optical depth. If we plot $T = T_{\mathrm{fac}} - T_{\mathrm{phot}}$ against the optical depth τ we obtain the results as given in Table III.

TABLE III

TEMPERATURE AS A FUNCTION OF OPTICAL DEPTH

$\tau_{4100\text{Å}}$	$\Delta T(fac - phot)$
0.01	+ 80 °K
0.1	+ 250
0.3	+ 250
0.6	+ 60
1.0	− 40
3.0	− 260

The invisibility of the faculae in the disk center is apparently due to the fact that the higher temperature in the upper layers is compensated by a lower temperature in the deep photospheric regions. Another important feature of the faculae is their magnetic field; the faculae, just like the bright mottles of the quiet photosphere, are the seat of magnetic fields.

Different regions of the faculae often have different magnetic polarities corre-

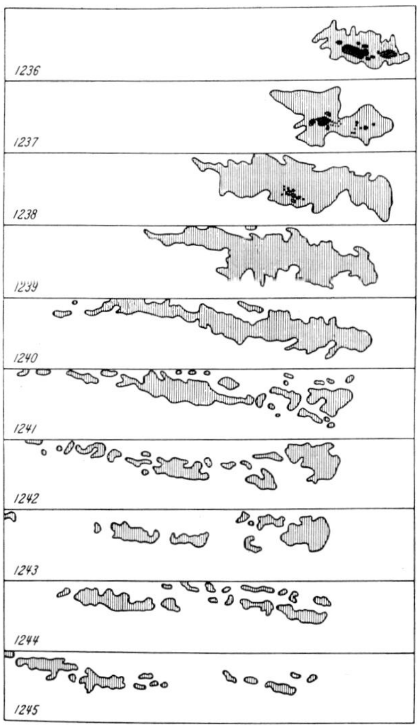

Fig. 6. Life history during ten consecutive solar rotations, of a big spot-group and its facular field. Drawings by M. Waldmeier during solar rotations 1236–1245; February to October, 1946.

sponding with that of the embedded sunspots; the magnetic field in the center of the faculae is about 50 gauss and it is 10 gauss at the borders.

2. *Chromospheric faculae*

Chromospheric faculae (Figure 5), also called "plages" are the chromospheric extension of the photospheric faculae; there is no fundamental difference between these two features. They can be observed in spectroheliograms or in filtergrams of the strong Fraunhofer lines (H, K and Hα) which are formed above the photosphere.

3. *Development*

The development of the faculae is closely related to that of the sunspots. Sometimes they appear together with the sunspots, mostly some hours to some days earlier.

The greatest facular brightness is reached after some days. Later the facular fields become larger and tend to extend in latitude direction, but the brightness diminishes (Figure 6). They separate into different parts and can still be visible a long time after the spots have disappeared. Their life-time varies from weeks to many months.

4. Detailed structure

The faculae are composed of elements with the same physical parameters as the

Fig. 7. Coronal activity region photographed in the green coronal line on February 11, 1959, at the Sacramento Peak Observatory, U.S. Air Force Cambridge Research Laboratories. To the right faint loop structures are visible.

elements of the chromospheric mottling and the network occurring in the quiet chromosphere. The difference is, however, that in faculae the number density of the elements is greater. Profiles and intensity ratios between the weak and strong components in the He "doublet" ($\lambda = 10830$ Å; the doublet is in fact a triplet with two lines nearly coinciding; the lines originate from the transition $2^3S \rightarrow 2^3P$; the lower excitation potential is as high as 19.73 eV; the line originates in the chromosphere) are the same for discrete elements of faculae and the mottling elements of the quiet chromosphere.

This shows that the elements in faculae and in the quiet chromosphere which absorb He light have the same physical parameters (T, N_e), but differ in number density in faculae and outside. In faculae about 1/3 of the matter is in high excitation

Fig. 8a. A very active loop prominence photographed June 28, 1957 at 1337 UT by K. Hermann-Otavsky, Praha, through a Solč filter centered at Hα.

state and can absorb the He lines. In the quiet chromosphere one estimates that about only 1/10 of the matter is in high excitation state. Moreover, about 1/8 of the facular helium absorbing matter must have downward motions with velocities of the order of about 10 km/sec (O. Namba, 1963).

Fig. 8b. Part of the loop prominence shown in Figure 8a, photographed at the Sacramento Peak Observatory (U.S. Air Force Cambridge Research Laboratories).

C. THE CORONAL CENTER OF ACTIVITY (OR PERMANENT CORONAL CONDENSATION)

1. *Optical phenomena*

Situated above the photospheric and chromospheric faculae, the optical phenomena are characterized by an increased intensity of the green coronal line (Fe XIV, 5303 Å), they have a greater particle density (5 to 10 times) and a higher electron temperature (some 10^6 °K) than the quiet corona. A permanent condensation is best developed when the associated spots group has reached the class F. Its life-time will then appear to be of the order of a few months.

It shows (in monochromatic light) stable rays and (or) arches and (or) bushes (see Figure 7). Above a CA, one may see streamers, often several concentric semi-spherical or parabolic elements with sharp boundaries, usually decreasing in brightness from inner to outer filaments and expanding in the course of hours. There is also rapid brightness and sudden development of sharply outlined dark areas, but apparently no material motion. In one case an arched streamer has been observed which expanded very rapidly ($\simeq 500$ km/sec) to become stretched within five minutes; this occurred simultaneously with the appearance of a flare.

There exist also *sporadic coronal condensations* which can be observed as well in the high excitation lines 5303 Å, 5694 Å and in the low excitation line Hα. They occur within the permanent condensation at the top of their development. They are often connected with the so-called loop or funnel or knot-prominences. They seem to be cool prominences matter, surrounded by very hot (some 10^6 °K) coronal matter (Figure 8).

2. *Radio waves emitted by the coronal condensations*

The absorption coefficient for the radiowaves in the corona is given by

$$\kappa = \zeta N_e^2 f^{-2} T_e^{-3/2} n^{-1} \, (\mathrm{cm}^{-1}),$$

where $\zeta = 1.3 \times 10^{-2} \ln (367 \, T N_e^{-\frac{1}{3}})$, f is the frequency, and n the refractive index.

If T increases outward from the quiet chromosphere into the corona, then N_e diminishes; it appears that ζ is nearly constant with height.

For the quiet sun, for $\lambda = 1$ cm, $\zeta = 0.08$, and for $\lambda = 600$ cm, $\zeta = 0.2$. In active regions ζ may deviate slightly from these values.

From the relation $d\tau_\lambda = k_\lambda \varrho \, dh$, we can compute the height **h** of the layer where the radio waves mainly come from (it corresponds with the height where $\tau \simeq 1$). The results are the following: If $\lambda = 1$ cm, then $\mathbf{h} \simeq 5000$ km above the photosphere (chromosphere). If $\lambda = 1$ m, then $\mathbf{h} \simeq 15000$ km above the photosphere (low corona). If $\lambda = 10$ m, then $\mathbf{h} \simeq 10^6$ km above the photosphere (corona). From these results we see that the radio waves are a good means for studying the different outer solar layers, from the chromosphere to the high corona.

Let us now compare the radio emission of a coronal condensation with the radio emission of the normal corona (Figure 9). At centimeter and decimeter waves, there is an enhancement of the solar radio emission in the CA. This corresponds to electron densities in a CA, on the average equal to about 2 to 5 times the value of N_e in the quiet corona and to temperatures slightly larger ($\simeq 50\%$).

3. X-ray emission

X-ray emissions are also enhanced in the coronal condensation. The fluxes of X-rays from CC are on an average about eighty times greater than the fluxes of X-rays

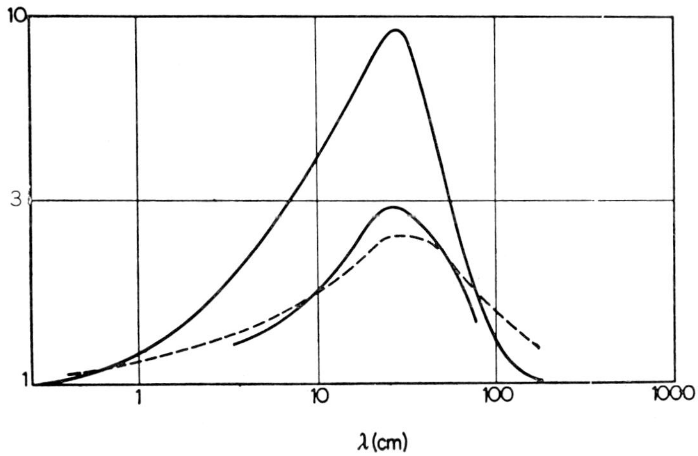

Fig. 9. The solid lines give the intensities of the radio-spectra of a strong, and of an average coronal activity-center, in units of that of the same area of the quiet Sun's disk. The dotted line gives the spectrum computed for an activity center with the same temperatures of the quiet sun, but with twice its density.

emitted by the quiet corona. The X-rays are emitted by collisions between ions and electrons, so that their intensity is proportional to

$$N_{\rm ion} \times N_e .$$

In the plasma of the corona, when the temperatures are very high, we have

$$N_{\rm ion} = {\rm const}\, N_e .$$

Hence we must have

$$I_{\rm X\text{-}rays} = {\rm const}\, N_e^2 .$$

If the electron density is five times greater in a CC, the X-ray emission must be twenty-five times greater in a CC than in the quiet corona. From this point of view, the X-rays are an important tool for the study of coronal densities of electrons. It would be of great interest to know the precise distribution of X-ray radiation over the sun and to compare this with the fine structure of the CA so as to find the relation between X-ray emission and the CA.

Friedman and co-workers have taken some pictures of the sun in the X-ray region, by means of a pinhole camera installed in the nose cone of a rocket. (See Figure 10.) Since the camera was only pointing biaxially, it turned over a certain angle during the exposure, thus smearing out the coronal details over parts of concentric circles. Nevertheless, very useful details were derived from this and later observations. It was found that

$$(I_{\text{X-rays}})_{\text{CC}} \simeq 60 \times (I_{\text{X-ray}})_{\text{QC}}$$

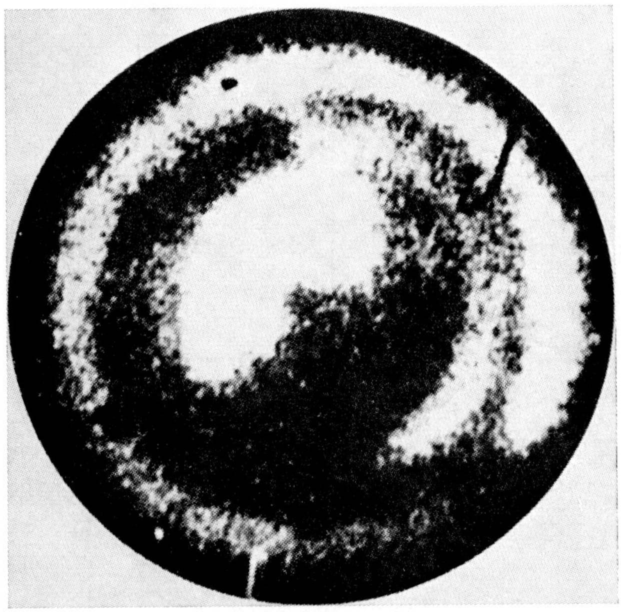

Fig. 10. X-ray pinhole camera photograph of the sun, obtained April 19, 1960 by the U.S. Naval Research Laboratory (by courtesy of H. Friedman).

and since $T_{\text{CC}}/T_{\text{QC}} \simeq 1\frac{1}{2}$, we have

$$(N_e)_{\text{CC}} \simeq 6 \times (N_e)_{\text{QC}} .$$

This is in reasonable agreement with the results found by radio wave observations.

Comparing heliograms obtained with X-rays and with radiowaves ($\lambda \simeq 10$ cm) one finds that these two pictures are very similar (except for a greater limb brightness in radiowaves). This confirms that these two radiations come from the same region of the corona. The best agreement between X-ray and radio heliograms would be expected for $\lambda \simeq 20$ to 30 cm.

To improve the result the following techniques should be applied in future.

(a) Pointing three-axially instead of pointing biaxially.

(b) Shortening of the exposure by the use of Fresnel-Soret zone plates instead

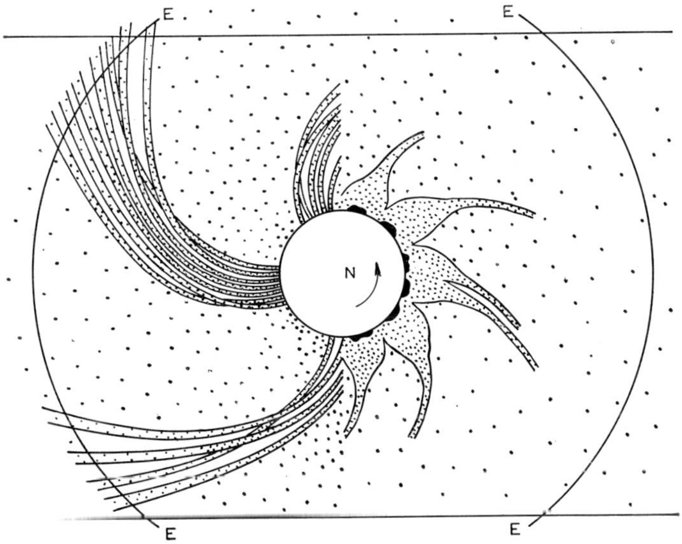

Fig. 11. Schematic drawing of R-rays (left) and classical streamers (right) after Mustel'.

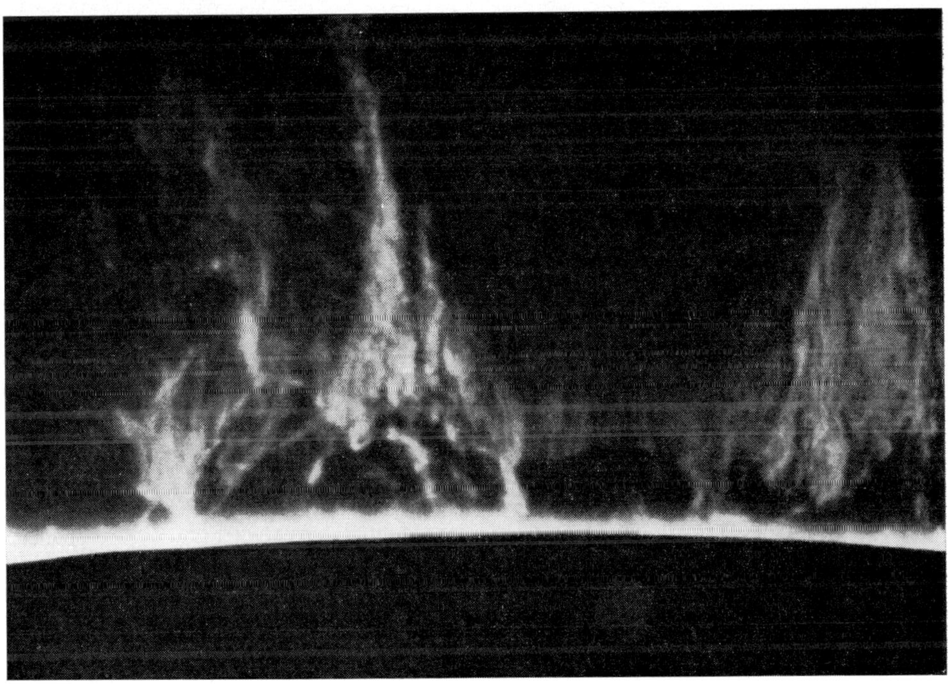

Fig. 12. So-called hedgerow prominence, one of the types of quiescent prominences (courtesy by Sacramento Peak Observatory, U.S. Air Force Cambridge Research Laboratories).

of the very simple pinhole camera, or by an imaging technique using grazing reflection at one or more parabolic mirror elements, or with other image-forming devices. To be really useful the observation technique should have a resolving power better than 10″ (7000 km).

D. CORONAL RAYS

The *R-rays* are the outer coronal extensions of the real activity centers, whereas the *classical streamers* are the structures connected with the long-lived stationary filaments,

Fig. 13. Activated prominence, photographed at the Sacramento Peak Observatory, U.S. Air Force Cambridge Research Laboratories.

and, hence, with regions between activity regions or with the very old (4 to 6 months) activity regions themselves (See Figure 11). The R-rays may consist of many thinner plasma threads, which do not expand sideward, presumably being kept together by the magnetic fields they carry with them. With their fine structure they may extend as far as the earth and cause the *recurrent geomagnetic storms*. The classical streamers are of a more amorphous nature; they may only extend to about $\frac{1}{2}$AU.

E. PROMINENCES

Outside the sun's limb they appear in the light of strong spectral lines, especially Hα; then they are bright on the dark ground. When seen in projection on the disk, they look dark, forming long structures, called filaments.

(a) *Spot-prominences*. These are close to sunspots. They have a short life-time and vary rapidly in form and extent. They are also characterized by a rather low brightness. Their main physical parameters are: $T \simeq 2 \times 10^4$ °K; $N_e \simeq 10^{10}$ cm^{-3}.

(b) *Flare-associated spot prominences*. They occur when the CA is at the top of its development, mostly connected with solar flares. Examples are loop prominences, knot prominences, funnel prominences, etc. (Figure 8).

(c) *Quiescent prominences*. They appear away from sunspots or in the "dark lane" between two magnetic oppositely polarized parts of CA, have grown to a great length, and are stable. The life-time is long, and may extend to many months (Figure 12). Some characteristic numbers are the following: Their thickness is from 3000 to 10000 km; their length reaches up to 10^6 km; $T_e \simeq 5 \times 10^3$ to 1.5×10^4 °K; $N_e \simeq 10^{10}$ to 10^{11} cm^{-3}.

Many stable prominences are associated with the coronal streamers, showing a dark dome around the prominence; the green coronal line starts an increased emission at the border of these regions. The form of these prominences is at least partly defined by the magnetic fields supporting their equilibrium configuration.

(d) *Activated prominences* (Figure 13). Initially quiescent, they are activated by the influence of neighbouring flares. The velocity of progression of the activating events is either that of light (UV or X-ray radiation) or about 1000 to 2000 km/sec (activation by particle streams). There is often a sudden disappearance of the prominences, but very often they may return in the same configuration some hours to days later. Often the disappearance is only apparent and due to the acquisition of a radial velocity by the prominence, so that it disappears from the passband of the filter or spectroheliograph.

5. Parameters for Measuring Solar Activity

A. THE WOLF NUMBER

When classifying observations of sunspots it is customary to use the "Wolf sunspot number" R, where R is a measure of the spot activity of the sun:

$$R = K(10g + f).$$

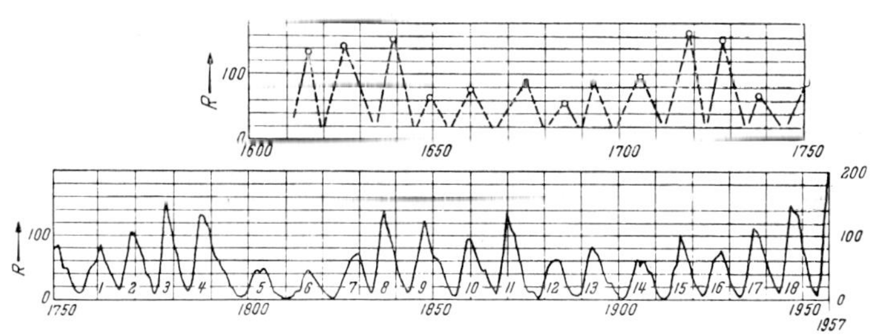

Fig. 14. Sunspot numbers R since 1610. The R-number fluctuates with an average period of 11.0 years. A longer period of slightly more than 80 years is also visible in the graph.

K is a normalization constant and depends on the observing faculties of the observer and type of observation instruments used; f is the number of spots, g the number of spot-groups. (If there is only one spot on the sun and if $K=1$, then $R=11$).

The Wolf number provides no real measure of solar activity (as, for example, sunspot-area or the total magnetic flux would), but it derives its importance from the easiness of the observing technique and from the fact that the Wolf numbers R have

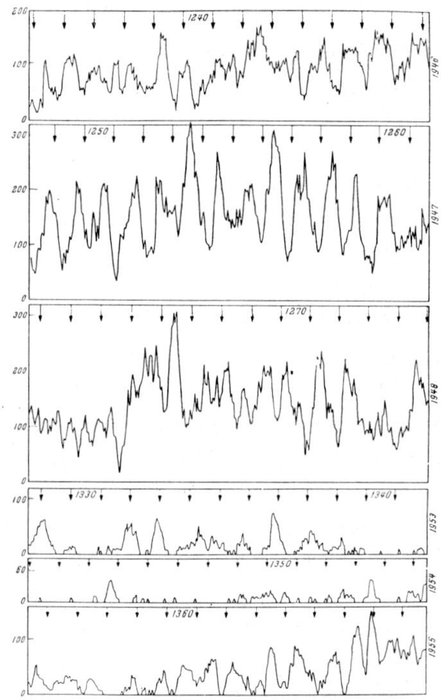

Fig. 15. Daily sunspot numbers for the maximum years 1946 to 1948 and the minimum years 1953 to 1955, taken from Waldmeier's annual reports of solar activity (*Astron. Mitt. Eidgenössische Sternwarte, Zürich*). Each tenth solar rotation is indicated along the upper side of the diagram. The ordinates give the sunspot numbers. The 27-days period is very obvious.

been recorded continuously for more than 100 years. From older observations the monthly mean R numbers can even be traced back to 1750 and cruder estimates are available down to 1610, when Fabricius, Galileo and Scheiner were the first to see the sunspots through a telescope (see Figure 14).

Study of these R records shows average periods during which we get repetitive phenomena of 27 days (see Figure 15), 11 years and 80 years. All these are average periods and actual periods may vary from these by quite a lot. The first of these three periods is obviously the sun's rotation period and is due to the distribution of spots over the disk which is often uneven. The two longer periods are intrinsic to the sun and as yet unexplained.

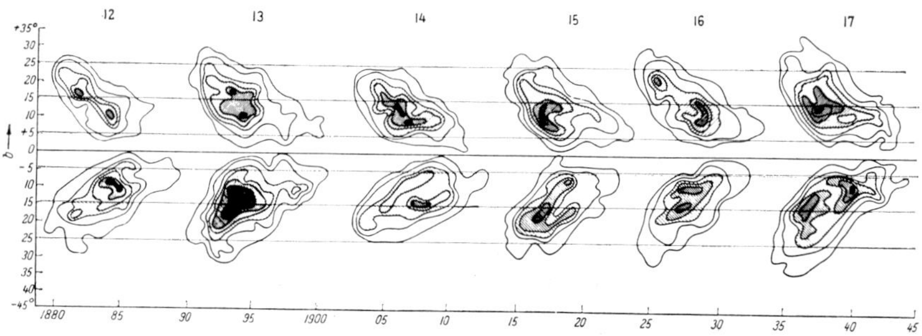

Fig. 16. Butterfly diagrams of sunspots for the years 1880 to 1945. The lines connect latitudes of equal annual spot numbers.

B. DISTRIBUTION OF SPOTS WITH LATITUDE

Almost all spots occur in the region from $+45°$ latitude to $-45°$ latitude. The distribution can be seen from Maunder's "butterfly diagram" (Figure 16). It gives the latitude distribution of all the spots observed, without regard to spot size, during the sunspot cycles.

The butterfly diagram clearly shows that a new cycle starts already before the end of the preceding one. This means that the real duration of the solar cycles is greater than 11.0 years, which is the average time interval between two successive maxima or minima.

The average duration of the true solar cycles appears to be 13.2 years.

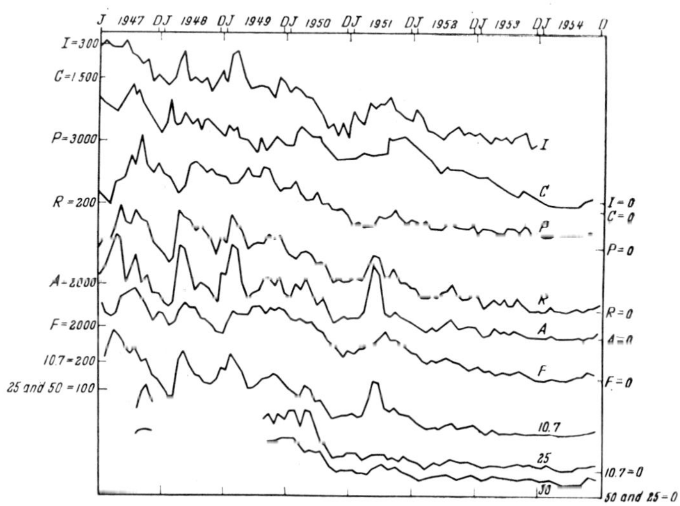

Fig. 17. Solar variability between 1947 and 1954. The symbols have the following meaning: 50, 25 and 10.7: radio emission at 50, 25 and 10.7 cm; F – facular area (unit: 10^{-6} of the visible hemisphere); A – sunspot area (same unit); P – prominence area (same unit); R – Zurich sunspot numbers; C – integrated intensity along the limit of the 5303 Å coronal line (arbitrary units); $I = (f_0E)^4 (\cos z)^{-1}$ – proportional to the X-ray flux ionizing the ionospheric E-layer. All curves are monthly means, except C.

C. OTHER PARAMETERS FOR MEASURING SOLAR ACTIVITY

Figure 17 shows different aspects of solar variability as observed between 1947 and 1954. It illustrates that many solar phenomena, such as the radio-emission at highly different wavelengths, the facular area (F), the total sunspot area (R), the prominence area (P), the integrated coronal intensity (C) and the X-ray flux (I), are closely correlated with the sunspot number (R). Statistically one finds a time lag between the spot number and various other phenomena; the other phenomena all come later: radio emission at 10 cm, 1 to 2 days; faculae: (H and Ca^+) 5 to 20 days; 5303 Å coronal line: 25 days.

These time lags reflect the lag in the evolution of the faculae and the corona with respect to the spots.

6. Solar Rotation

The sun shows a differential rotation, (Figure 18) the equatorial regions rotate faster than the high-latitude regions. The rotation velocities can best be determined by means of position determinations of individual features. Measurements of radial velocities, by means of spectral lines, are decidedly *too imprecise* to determine this effect with sufficient reliability, because of the smallness of the solar rotational velocity

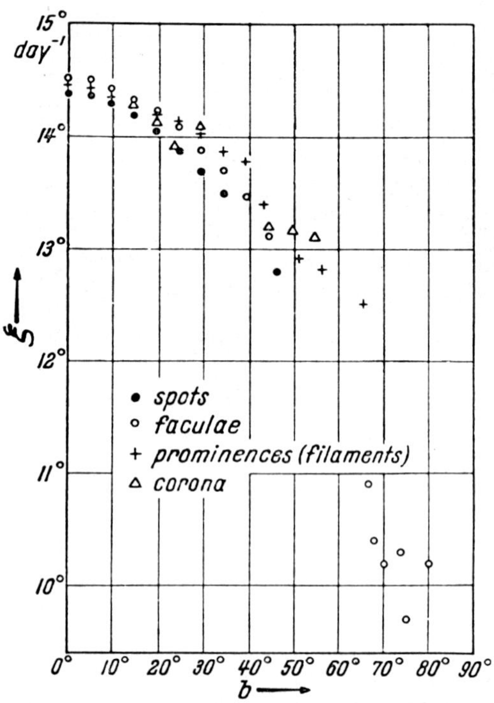

Fig. 18. Rotational velocity of the sun in relation to the heliographic latitude *b*.

(at the equation $v=2$ km/sec, which corresponds to a maximum line displacement at 6000 Å of $\Delta\lambda=0.04$ Å).

7. The Solar Magnetism

The sun has a faint bipolar magnetic field which is restricted to a region within about 30° of the poles. There the field is of the order of one gauss. The solar north pole had a magnetic north pole till about 1957. In the course of that year the north polar field became a magnetic south pole. For about one year (1957–1958) the sun had then two magnetic south poles; one at each heliographic pole. In 1958 the field at the south pole changed into a magnetic north pole so that the field is now like that of the earth. It has been suggested that such a change occurs during *each* sunspot maximum, but, of course, this has still to be observed.

The total poloidal magnetic flux is of the order of 10^{22} Maxwells, comparable to that of the spots.

To this "general" solar magnetic field one should add the transient fields of the spots and the faculae. Roughly the magnetic flux of these fields is equal to that of the general solar magnetic field.

The polarity laws of the spots have already been mentioned (Section 4B). During each solar cycle all the preceding spots at the same hemisphere have the same polarity, opposite to that of their following spots and to that of the preceding spots at the other hemisphere. The fields of the preceding spots change between two consecutive cycles, so that the shortest period of solar variability actually lasts for 22 years, instead of for 11 years.

References

De Jager, C.: 1959, *Handbuch der Physik* (ed. by M. Flügge) **52,** 80.
De Jager, C.: 1960, *Vistas in Astronomy* **4** (ed. by A. Beer), Pergamon.
Kiepenheuer, K. O.: 1952, *The Sun* (ed. by G. P. Kuiper), The University of Chicago Press, p. 322.
Kiepenheuer, K. O.: 1960, *Radioastronomia Solare*, Varenna Summer School, Corso XII, (ed. by G. Righini), Zanichelli, Bologna, p. 39.
Kiepenheuer, K. O.: 1965, *The Solar Spectrum* (ed. by C. de Jager) D. Reidel Publishing Co., Dordrecht, p. 210.
Severny, A.: 1965, *The Solar Spectrum* (ed. by C. de Jager), D. Reidel Publishing Co., Dordrecht, p. 221.

CHAPTER 4

OPTICAL SOLAR FLARES

C. DE JAGER
Sterrewacht 'Sonnenborgh', Utrecht, The Netherlands

1. Description

We have to distinguish between the *optical flare* and what we now tend to call the *radio* or *X-ray flare*. Initially flares have been described and defined after their optical appearance. We now realize that in doing this, we may have excluded the most important part. In this chapter we restrict ourselves to the optical flares.

An optical flare is a sudden and short-lived monochromatic brightening of a small area of the sun's disk (Figure 1). Its life-time may be of the order of 15 minutes to some hours. The flares are best observed in Hα through a monochromatic filter or a spectroheliograph. A plot of Hα intensity against time gives a profile as shown in Figure 2. Some flares show a *flash-phase* which is often reached after some minutes, when the flare almost exclusively reaches a maximum brightness. Not all flares show such a flash phase; most of them have a fairly smooth intensity curve.

In complex flares a consecutive brightening of successive parts of the flare is often observed, so that it seems as if the flare excitation is 'creeping' over the solar surface.

Flares are categorized in the so-called *importance* or *class* by their *area*, as is shown in Table I.

In addition were introduced designations 1^+ (area of 1, greater Hα brightness) and – similarly – 2^+. N.B. Flare classes 2^- and 3^- *do not exist*! After January 1966 this classification will be changed into a new one, consisting of *five* flare types: S (subflare, the old type 1^-), 1, 2, 3, and 4 (=the old 3^+). The classification is strictly according

TABLE I

CHARACTERISTICS OF FLARES

Importance or class	*Area (units of 10^{-6} of the visible hemisphere)*	*Duration (in minutes)*	H *line with* (Å)	H *maximum central intensity (in units of adjacent continuous spectrum)*	*Relative frequency (1^- flares excluded)*
1^-	100		1.5		
1	100– 250	20	30	1.5	0.72
2	250– 600	30	45	2.0	0.25
3	600–1200	60	8	2.5	0.03
3^+	1200	80	15	3	

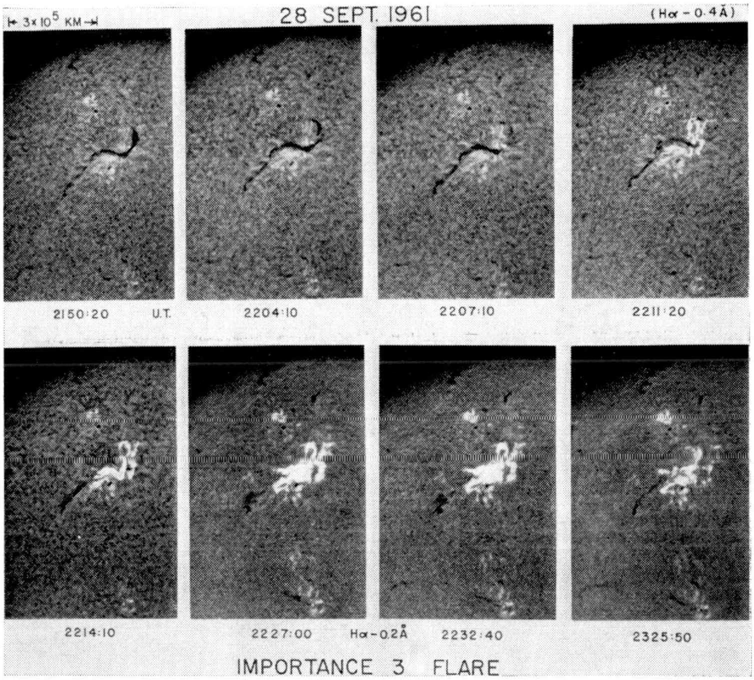

Fig. 1. Flare of September 28, 1961, photographed by G. Moreton at the Lockheed Solar Observatory, Los Angeles. Note the activation of the prominence after 22 h 12 m UT.

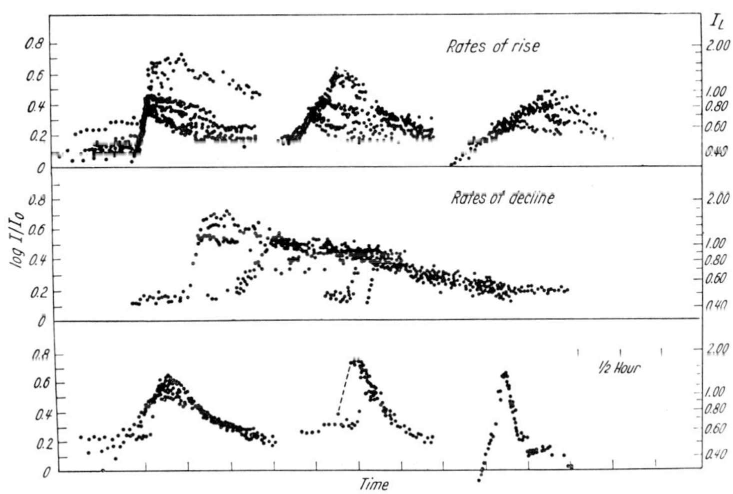

Fig. 2. Hα light curves of solar flares (Hedeman, Dodson, and Mc Math).

to the area. The Hα intensity is indicated by the added letter *f* (faint), *n* (normal) or *b* (bright).

2. Occurrence; Relation to Faculae and Spots

Flares occur nearly always in faculae and close to spots – see Figures 2 and 3. Most

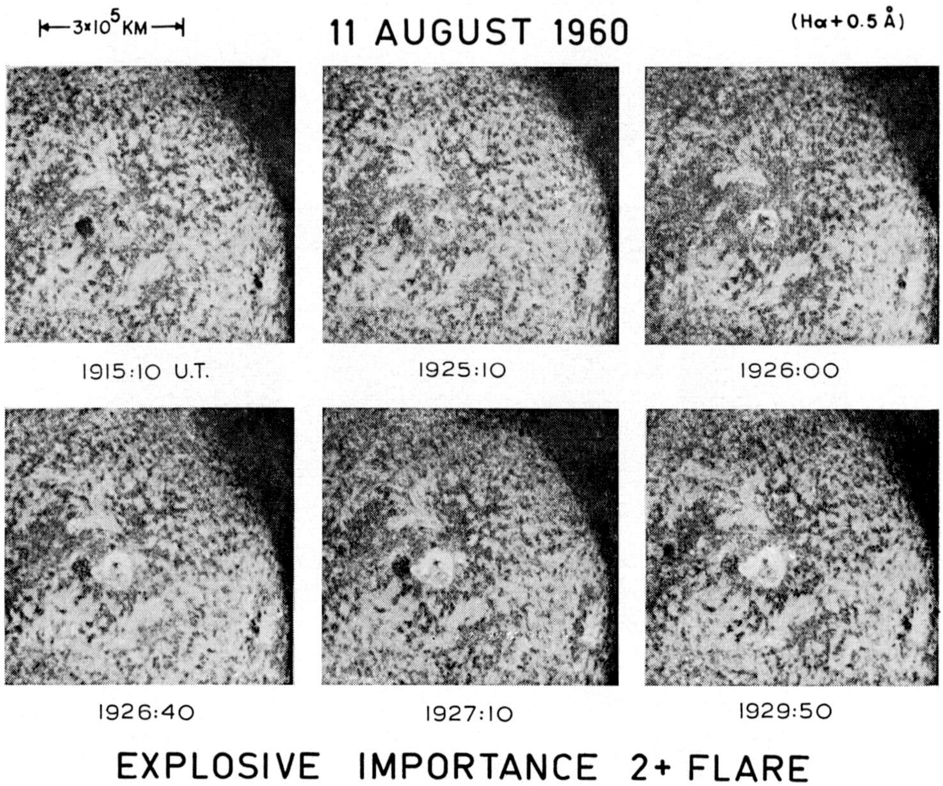

Fig. 3. Flare of August 11, 1960. The Hα filter was centered at a wavelength 0.5 Å longer than Hα. At this wavelength the facular area is a dark, smooth region around the sunspot group. This flare produced a strong cm burst, a cluster of Type-III radiobursts, and, simultaneously with the Type-III bursts, an X-ray burst. Photograph obtained at the Lockheed Solar Observatory by G. Moreton.

flares have a fine structure which follows the initially existing pattern of the facular field. Often a flare is a brightening of previously existing facular structures.

Although the flares occur close to the spots they are seldom found on top of the spot umbrae. They are mostly seen in complex spot groups, close to or on the penumbrae of spots. Those observed on the spots penumbrae are the most interesting for the solar-terrestrial relations: they are often accompanied by Type-IV radiobursts and by the emission of plasma clouds containing sub-relativistic particles (causing the Polar Cap Absorption events), and sometimes by the emission of cosmic ray particle bursts.

There is a correlation between the mean number E of flares per day, and the sunspot number R. We had:

$$E = 0.047\, R \text{ during the 17th solar cycle (1934–1944)}$$
$$E = 0.034\, R \text{ during the 18th solar cycle (1944–1954)}.$$

Hence, if $R = 150$, we may expect about nine flares a day, two of which are of importance 2.

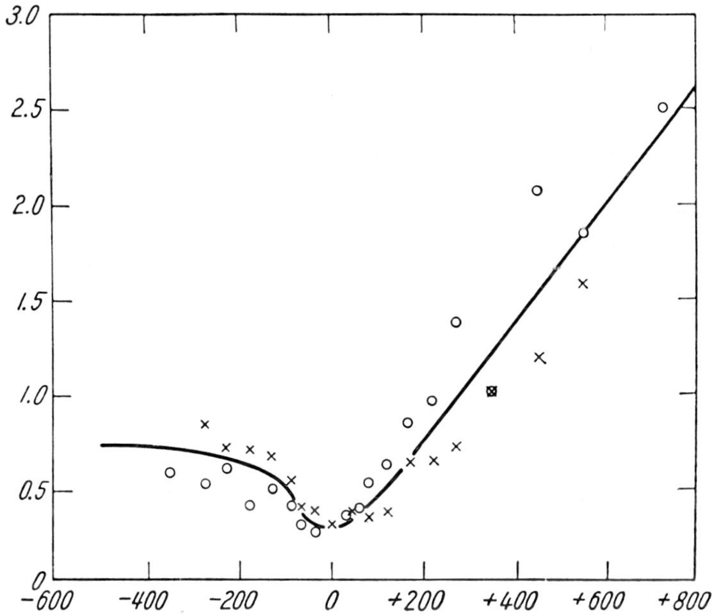

Fig. 4. Probability of the occurrence of solar flares as a function of the rate of growth or decrease of the sunspot area, after Giovanelli. Abscissa: rate of growth in 10^{-6} hemisphere per day; ordinate: E.

There is a further correlation between E and the derivative of the total area of the groups with respect to time:

$$E - A \cdot f\,(\mathrm{d}E/\mathrm{d}t).$$

This relation is shown in Figure 4.

3. Spectra of Flares

Many flare spectra were obtained in Hα. The line profiles show a central emission peak which is faintly developed and merely a filling up of the line in the case of importance 1 flares, and which is intense with broad wings for importance 3 flares (Figures 5 and 6).

A peculiar class is formed by the moustaches. These show *no* central brightening, but only broad emission in the two wings of Hα. The explanation for these latter phenomena lies in the height of origin of the flares. We know that the normal chromosphere up to a height of about 5000 km is opaque for the central parts of the Hα line profile. The fact that no emission is visible in the central part of Hα signifies that moustache emissions take place in the low chromosphere, certainly below a height of 5000 km, but above the photosphere (otherwise the line wings would be invisible too). See further for quantitative data on the average level of emission of chromospheric lines Chapter II, Table II.

The "normal" flares, on the other hand, which *do* show central emission *must*

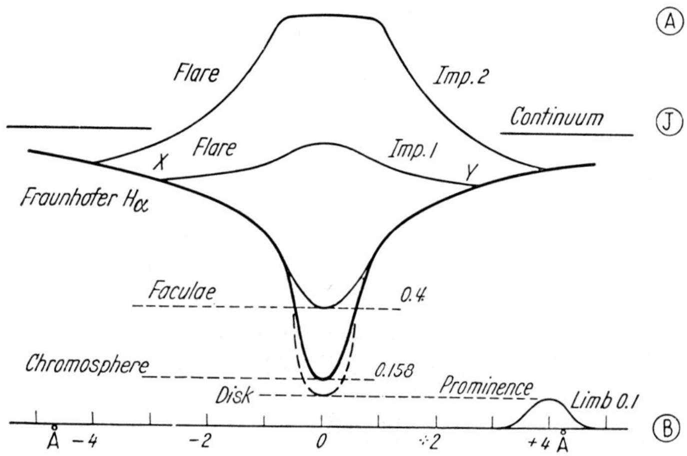

Fig. 5. Average Hα spectra of solar flares, after Ellison.

occur above the chromosphere, at least above 4000 or 5000 km. Hence these phenomena take place either in the transition region chromosphere-corona, or they are even completely coronal.

From a study of flare spectra of Hα and of other lines (especially of the high members of the Balmer series) we obtain information about the electron density in flares. An average value is log $N_e = 13.5$ where N_e is the number of electrons per cm^3. The flare temperatures are of the order of one or two times 10^4 °K.

4. Flare Theories

We compare the densities in flares with that of the surrounding media – the upper chromosphere or the low corona, where the densities are 10^3 to 10^5 times smaller than in the flare. The essential feature of a flare is apparently its great and suddenly acquired density. *A flare is a sudden collapse of the solar plasma.* That we may indeed be dealing with such a collapse is supported by certain observations of Severny, indicating a high degree of internal turbulence in flares. Also an observation by Waldmeier pleads for such a condensation. In a part of the corona above a spotgroup, where two days earlier a large flare occurred followed by a large, slowly moving sporadic coronal condensation, the coronal lines were exceedingly faint, and indicate a low density in that part of the corona.

Very important is an observation made by Suemoto and co-workers, and confirmed by Švestka, who draw attention to the fact that an Hα flare cannot be homogeneous, but must consist of a great number of small knots, or threads, having

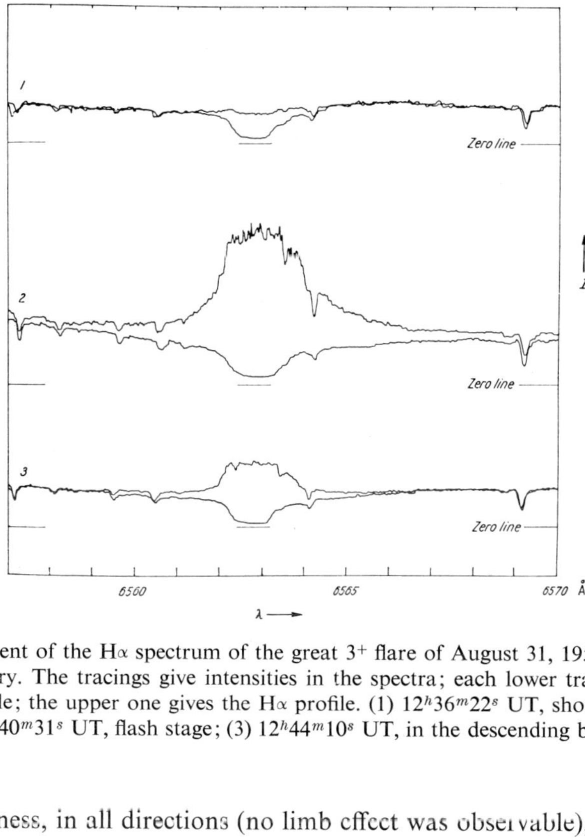

Fig. 6. Development of the Hα spectrum of the great 3+ flare of August 31, 1956, observed at the Utrecht Observatory. The tracings give intensities in the spectra; each lower tracing gives the undisturbed Hα profile; the upper one gives the Hα profile. (1) $12^h36^m22^s$ UT, shortly after the start; (2) $12^h40^m31^s$ UT, flash stage; (3) $12^h44^m10^s$ UT, in the descending branch.

together a thickness, in all directions (no limb effect was observable), of the order of 10 km. This observation will certainly be of great value for the explanation of the flare phenomenon.

To be able to explain the observations of solar flares it is useful to realize that a flare occurs mostly in magnetically complex spotgroups, and that statistically the frequency of occurrence of flares is related to changes in the area of the group. The variation of the area might involve a change of the total magnetic field, but the present observations are not yet conclusive on this last point. It is often thought that a flare might be the result of a compression of coronal (or high-chromosphere) plasma between two approaching magnetic fields carried in turn by two spotgroups (Figure 7).

An average flare emits about 10^{28} ergs per second and lives about 1000 seconds;

the total emission of electromagnetic radiation is, hence, of the order of 10^{31} ergs. A large flare emits more than 10^{32} ergs. The ejected matter, being of the order of 10^{19} gram for a large flare, represents an energy of about 10^{34} ergs.

This energy cannot arise from the cooling of the hot coronal gas which condensed into the flare. The maximum heat, freed per cm^3 of flare matter, cannot be more than NkT with $N=10^{13}$ and $T=10^6$ °K, hence 10^3 erg/cm^3; a flare should therefore have been contracted from an initial volume of 10^{28} cm^3 = $(20000 \text{ km})^3$. Since the densification factor is 10^4 to 10^5, the resulting flare volume should be 10^{23} to 10^{24} cm^3, at variance with the 10^{18} cm^3 resulting from the considerations of Suemoto *et al.*,

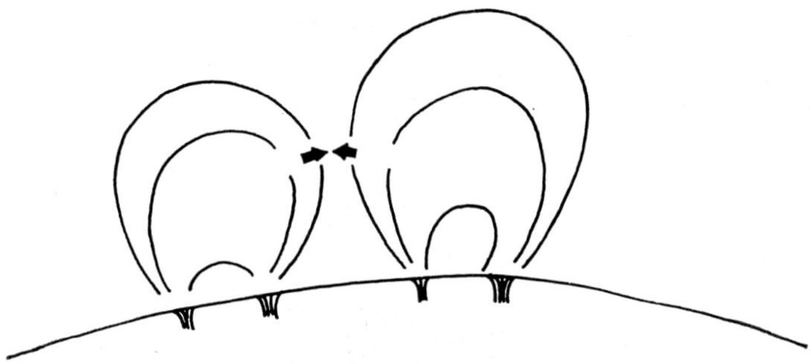

Fig. 7. Schematic picture showing the origin of a flare due to a compression between two approaching magnetic fields, following a plasma instability.

and Švestka. In fact, with an Hα emitting volume of 10^{18} cm^3 and $\log N = 13.5$, the energy release is 300 erg per proton-electron pair!

Neither can the emitted radiation flux of a flare be due to the kinetic energy of the two approaching masses of gas sketched in Figure 7, as a short computation would show.

The often quoted hypothesis that the large energy freed in flares might be the result of a release of magnetic energy is also difficult to justify. Assuming that the initial coronal magnetic field was H gauss, one would have a magnetic energy per cm^3 of coronal matter of $H^2/8\pi$ ergs. In order to explain an emission of 10^{32} ergs by some (unknown) process of transmutation of magnetic into optical and kinetic energy, a coronal volume of $8\pi \times 10^{32}/H^2$ cm^3 would initially be involved. By increasing the density by a factor 10^5 one would get a flare volume of $8\pi \times 10^{27}/H^2$. The actual flare volume is 10^{18} cm^3. H should then be of the order of 10^5 gauss, which is much too high.

The usual way of solving this problem is to suppose that the emitted energy is the magnetic energy of a far larger region than assumed here. Making the region 30 times larger in all directions would reduce the necessary field to 1000 gauss. The region where the field would be annihilated would in that case be 15000 km in diameter signifying that a field or 1000 gauss would be annihilated over the whole central part of a coronal activity region. This is large; it looks improbable, but does not seem wholly impossible.

Another way of accounting for the enormous release of energy in a flare might be the mechanism of nuclear transmutations. An indication that such a process may play a part in the sun may be found from observations of nuclear emulsions and metallic parts of the recovered Discoverer XVII. The satellite Discoverer XVII was launched from the base Vandenbergh on November 12, 1960 at 20h42m UT, and recovered some days later, on November 14, 1960 at 22h39m UT. During its period of orbiting, a large solar flare of importance 3^+ was observed on the sun; it was the source of a great cosmic ray storm. Nuclear emulsion blocks recovered from the satellite revealed a flux of particles with $Z \geq 4$, about a hundred times greater than that observed in a similar satellite (Discoverer XVIII) which flew one month later during a period of quiet sun. Radiochemical analysis of the metallic parts of the discoverer capsule indicates an external flux of tritium particles, presumably originating from the flare.

The idea that the release of energy in connection with solar flares might be owing to nuclear processes has several times been put forward by Severny.

Other arguments for the assumption that nuclear reactions might be of importance during a part of a flare's life history may be found in the high temperatures (10^7 to 10^8 °K) often occurring in connection with – but not inside – solar flares.

Radio- and X-ray observations have indicated the existence, in the vicinity of the flare, of regions with temperatures of 10^7 to (perhaps) 10^8 °K. We call this region the radio flare or X-ray flare. Recent Ariel satellite observations (Pounds and Willmore) confirm the earlier finding (Chubb and co-workers) that the short wavelength's tail of the solar X-ray spectrum increases considerably in intensity during the occurrence of flares, even during rather unimportant ones. This hardening of the X-ray spectrum, noticed in the wavelength region below 10 or 15 Å, could tentatively be interpreted as indicating the local existence of coronal temperatures between 5×10^6 and 10^7 °K.

References

De Jager, C.: 1959, *Handbuch der Physik* (ed. by M. Flügge) **52,** p. 80.

Kiepenheuer, K. O.: 1965, *The Solar Spectrum* (ed. by C. de Jager), D. Reidel Publishing Co., Dordrecht, p. 240.

Severny, A.: 1965, *The Solar Spectrum* (ed. by C. de Jager), D. Reidel Publishing Co., Dordrecht, p. 221.

Smith, H. J. and Smith, E. van P.: 1963, *Solar Flares*, Macmillan.

CHAPTER 5

RADIO AND X-RAY EMISSION ASSOCIATED WITH SOLAR FLARES

C. DE JAGER

Sterrewacht 'Sonnenborgh', Utrecht, The Netherlands

The observed emission of solar radio waves, X-rays, and energetic particles indicate that processes involving great energies may develop in certain parts of the sun which we identify as the corona. Although radio quanta from the sun have small energies, the circumstances under which they are emitted indicate that they are produced by processes involving great energies and – often – energetic particles.

1. Radio Emission from the Quiet Sun and the Activity Regions

We can receive radio emission from the sun in the region from about 5 mm to 10 meter which is the atmospheric radio window.

Different wavelengths originate in different levels of the chromosphere and the corona. This is due to the dependence of the absorption coefficient κ on the electron density N_e, temperature T, frequency f, and refraction index n as follows (see also Chapter III, Section 4):

$$x = \zeta N_e^2 / f^2 T_{en}^{3/2} \, (\mathrm{cm}^{-1})$$

In this expression ζ is a slowly varying function which is of the order 0.1. We have for the ordinary ray,

$$n_{\mathrm{ord}} = \left(1 - \frac{f_0^2}{f^2}\right)^{1/2}$$

and for the extraordinary ray

$$n_{\mathrm{extr}} = \left(1 - \frac{f_0^2}{(f - f_H) f}\right)^{1/2}$$

Here f_0 is the plasma frequency:

$$f_0 = (e^2 N_e / mc)^{1/2} \simeq 9.0 \cdot 10^{-3} \sqrt{N_e} \quad (\mathrm{Mc/sec})$$

f_H is the gyrofrequency:

$$f_H = eH / 2\pi mc = 2.80 \, H \qquad (\mathrm{Mc/sec})$$

The magnetic field H is expressed in gauss.

It is easily seen that $\kappa \to \infty$ for $f \to f_0$ and negligible magnetic field; hence radiation with frequency f must certainly originate from regions above the layer were $f = f_0$.

Table I shows the relation between height and critical frequencies in the chromo-

TABLE I

CRITICAL FREQUENCIES f_0 AT SEVERAL HEIGHTS IN THE CHROMOSPHERE

Height (*km*)	*Plasma frequency* (*Mc/sec*)	*Corresponding wavelength* (*cm*)
1000	50.000	0.6
2000	20.000	1.5
4000	8.000	4
6000	450	66

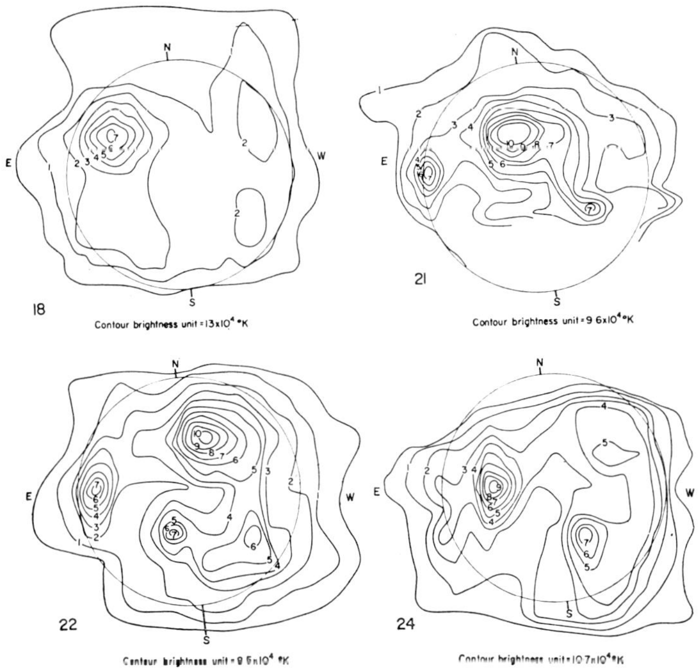

Fig. 1. The radiosun at 21 cm on July 18, 21, 22, and 24, 1957. The maxima correspond with the position of the faculae.

TABLE II

CRITICAL FREQUENCIES AT SOLAR MINIMUM IN THE EQUATORIAL AND POLAR CORONAL REGIONS

Height (*km*)	*Equator* (*frequency in Mc/sec*)	*Pole* (*frequency in Mc/sec*)
7500	280 (1.1 m)	235 (1.3 m)
15000	125 (2.4 m)	100 (3 m)
42000	100 (3 m)	75 (4 m)
140000	56 (5.3 m)	30 (10 m)
280000	30 (10 m)	14 (21.4 m)

sphere and Table II shows the critical frequencies at solar minimum in the equatorial and polar coronal regions. Hence cm-waves originate in the low chromosphere ($h < 5000$ km). The dm-waves are emitted in the transition region to the corona (4000–7500 km), while meter-waves come from the corona ($h > 7500$ km). Since in activity centers the electron density is enhanced by a factor 5 to 10 with respect to the quiet corona the average height of emission increases in coronal activity centers. Also in the chromospheric part of the activity centers the density seems slightly enhanced with respect to that of the quiet chromosphere, but the factor is not well known.

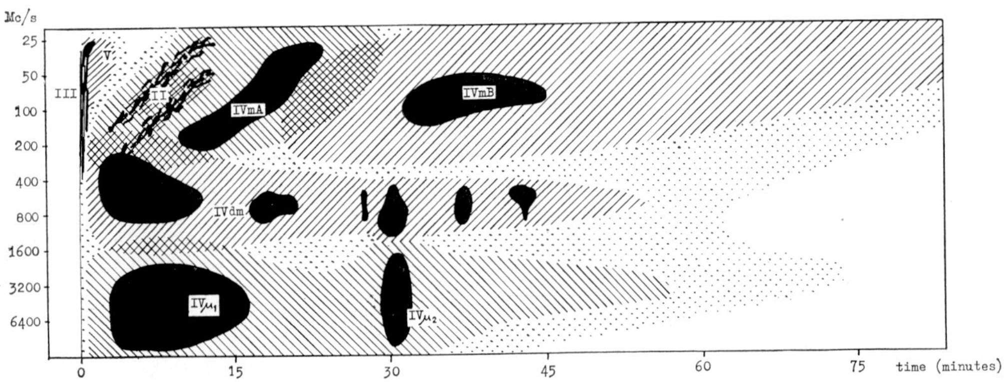

Fig. 2. Schematic review of flare associated radio phenomena (after Fokker).

The greater height of the level of origin of radiowaves emitted from coronal activity regions (as compared with the quiet sun) has the consequence that the decimeter-radiation emitted by activity regions comes from hotter layers than the normal dm radiation. At dm waves the activity centers manifest themselves by a greater radiation intensity, as is clear from a look at radioheliograms (Figure 1).

2. Radio Emission Associated with Solar Flares

There are different types of radio-emissions associated with solar flares, the types are characterized by Roman numbers I to V. Only rarely do these five types occur together with the same flare; in most cases only a few of them appear.

Hence Figure 2, which is a schematic diagram, showing in a wavelength-time diagram the position of the various types of flare-associated bursts, is an idealized case, which in reality hardly occurs. A real dynamic spectrum of radio-phenomena associated with a flare is shown in Figure 3.

It is important that Tables I and II of this chapter should enable us to associate a height to each wavelength. However, in establishing the relation $f_0(h)$ it has not been taken into account that the electron density is five to ten times greater in an activity region than outside it.

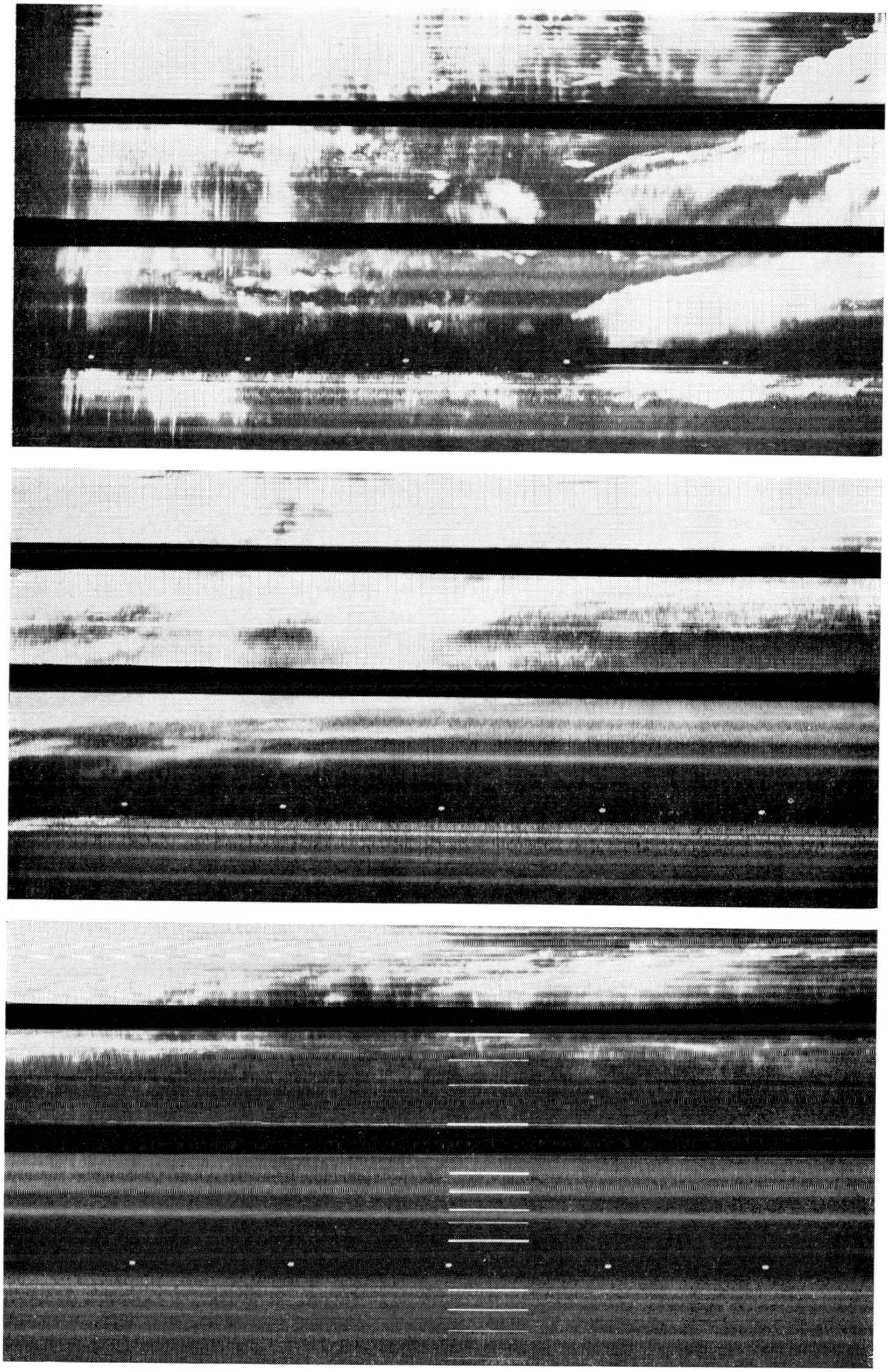

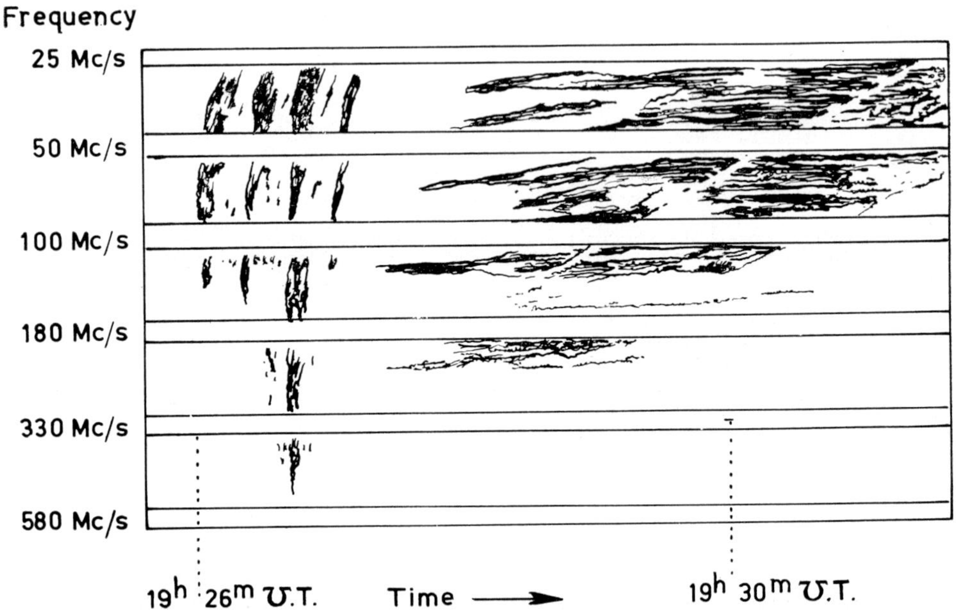

Fig. 3. The great solar radioburst of August 11, 1960, 19.26 UT, as observed with the dynamic spectrograph of Harvard College Observatory, at Ford Davis, Texas. The observations cover the frequency range from 25 to 580 Mc/sec, as shown by the legends to the picture. During the first minute (19.26–19.27 UT) a Type III storm occurred, concurrently with an X-ray burst. After 19.27 UT a Type-II burst occurs (Courtesy: A. Maxwell, Fort Davis, Texas).

The duration, the rate of wavelength chance, polarization, and the proposed mechanism of origin of the five types of radiobursts are summarized in Table III.

The detailed development of a Type III burst is shown in Figure 4.

The observed decrease of the frequency with time for Type-III and II bursts

TABLE 3

CHARACTERISTICS OF RADIO BURSTS

Type	*Name*	*Duration*	*Wave-length*	*Bandwidth*	*Speed*	*Polar-ization*	*Origin*
III		sec	m, dm	some tens of Mc/sec	c/2 ± factor 2	—	plasma-oscillation
II	outburst	some minutes	m, dm	same	10^3 km/sec	—	same
I	storm-burst	< 1 sec	m	2 to 8 Mc/sec	—	—	?
IV		10–60 min	cm, dm	large	10^3 km/sec and later standstill	strong	synchrotron
V		some minutes	long m	some tens of Mc/sec		—	?

corresponds to a propagation of an exciting disturbance upward through the corona. Since each frequency corresponds to a certain height of emission of the waves, we can calculate the propagation velocity from the slope of the emission in the (f, t) diagram. This indicates very clearly the motion of an exciter through the corona. Note that the "paths" of the Type-III and Type-II bursts, when prolongated backward in time, cross each other at a time that could be identified with the start of the flare.

The most probable explanations of the various types of bursts are the following:

Type III burst: jets of electrons accelerated to velocities of the order $c/2$.

Type II burst: the wake of a shock front originating at the flare's collapse which propagates outward with $v \simeq 1000$ km/sec.

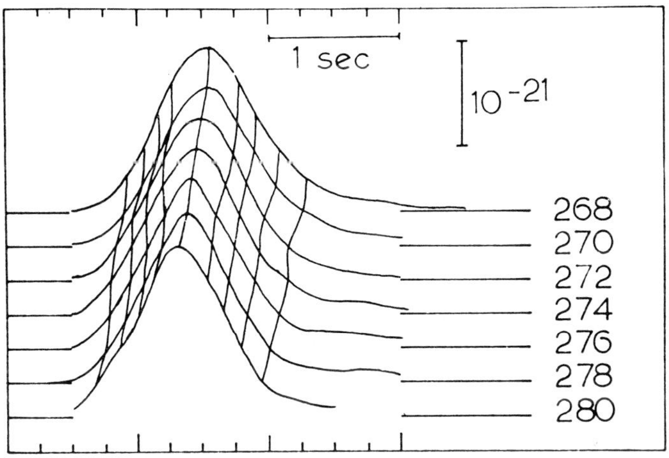

Fig. 4. Fine structure of a Type-III radioburst observed by T. de Groot (Utrecht and Dwingeloo Observatories). The seven recordings were obtained simultaneously at seven frequencies ranging from 268 to 280 Mc/sec. In this case apparently $df/dt = 50$ Mc/sec.

Type IV burst: (in the dm and m range): synchrotron radiation emitted by a cloud of (sub-)relativistic particles, kept together by a magnetic field.

Type V burst: similar explanation as the Type-IV bursts, but they seem to occur in denser solar regions.

Type I burst: the explanation is not yet clear.

3. X-Ray Emission Associated with Solar Flares

The X-ray events correlated in time with flares have been observed partly from rockets, but for the greater part from satellites. Many fine observations were secured with the U.K. Ariel satellite, others were obtained by means of OSO I, launched in April and March 1962, respectively.

A. QUASI-THERMAL FLARE ASSOCIATED X-RAY EMISSION

There appear to be two components in the X-ray emission associated with flares.

The first, treated here, is called *thermal* because the radiation is due to atomic transitions in a gas of a well-defined kinetic temperature; the adjective *quasi* is added because the gas is not in local thermodynamic equilibrium.

The cool, optical flare does not contribute to this radiation, its temperature is too low; the hot coronal source of the radio flare must certainly also be the source for the quasi-thermal X-rays. During some flares hard X-rays have been observed; the emission lasted for a time comparable to the life time of the flare and the emission time of the cm or dm radio waves.

With a scintillation counter Chubb and co-workers observed the short wavelength tail spectrum (20–60 keV) of flare associated emission on August 31, 1959, and found it to be describable by a color temperature of 1.2×10^8 °K. The emission was observable for at least six minutes. The electron temperature corresponding to this high colour temperature is still to be determined, as well as the other physical conditions necessary to have a gas emitting radiation with the observed characteristics.

Yefremev and co-workers observed the solar radiation in the wavelength region between 1.4 and 110 Å by means of open multipliers with interchangeable filters, and found it to consist of two components; one is emitted by the quiet corona and has a color temperature of 0.9×10^6 °K; the other component occurs during flares and has a color temperature of 6.5×10^6 °K.

The U.K. Ariel satellite obtained several flare spectra in the wavelength region 2–15 Å with the aid of a proportional counter, the spectra being obtained with a time resolution of 25 sec to one minute (Pounds and co-workers). The X-ray flare of May 3, 1962 was of the "simple" type. It started at 06.45 UT and had in its maximum phase the following color temperatures:

Wavelengths (Å)	13–11	11–9	9–7	7–5
T_c (°K)	2	2.2	2.7	3.8×10^6

For this flare $\int N_e^2 \, dV = 6 \times 10^{47}$ cm^{-3}.

The general characteristic of the "simple" flare spectra is that of a gradual hardening and intensification of the spectrum during the phase of growth of the flare.

The flare of April 27, 1962 developed slowly with a short "flash" of importance 3 or 3^+; at that moment a short *burst* of X-rays was observed, apparently due to a non-thermal source.

We conclude that X-ray observations confirm the existence of "hot" components of a flare. The location and structure of this component is still unknown.

Theoretically one may expect a gas cloud with a temperature of several 10^6 °K to expand and to leave the sun with near explosive force, shooting a blast wave into interplanetary space, unless it is held near the sun by a magnetic field.

B. NON-THERMAL RADIATION BURSTS FROM FLARES

Bursts of hard X-rays (10^4 to 10^6 eV) with durations ranging from some seconds to one or two minutes at maximum have been observed on several occasions.

These short-lived bursts show the following properties:

(a) they occur mostly during the flash phases of flares. In this connection we may

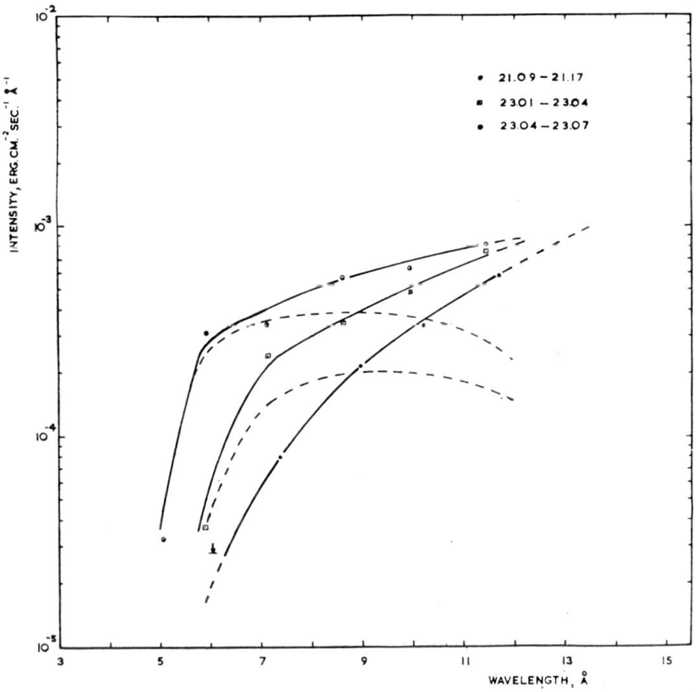

Fig. 5. Ariel spectrum of the X-ray radiation connected to a flare of importance 1 or 1⁻ on April 27, 1962. The visible flare occurred from 23.00 – **23.05** – 23.15 UT. The dashed curve shows the "flare emission" after substraction of the pre-flare spectrum. See *Proc. Roy. Soc.* 1964. By courtesy of K. A. Pounds, Leicester.

remark that the flash phase tends to coincide with the acceleration of particles to high speeds; during a flare-flash, radio-observations show (a) the emission of the Type-III bursts, due to electrons with velocities between 0.2c and 0.9c, and the Type-II bursts excited by a proton-jet or by the wake of a shock front moving with $v \simeq 1000$ km/sec.

(b) The X-ray bursts are correlated with cm radiobursts. These bursts occur in large activity centers, in particular when the flare "touches" the penumbrae of sunspots.

(c) The energies of these bursts are of the order $10^4 - 10^6$ eV, so that they must be due to Bremsstrahlung or magnetic Bremsstrahlung (synchrotron radiation) by electrons moving with velocities of 60000 to 290000 km/sec. Similar velocities are observed for the electrons that produce Type-III radiobursts, so that the acceleration of electrons with energies between 10^4 and 10^6 eV must be quite a common feature in a flare. However, electrons that excite Type-III bursts can generally *not* excite X-ray bursts and *vice-versa*, since X-rays are produced by Bremsstrahlung of electrons shot downward and braked in the dense parts of the solar photosphere, while Type-III

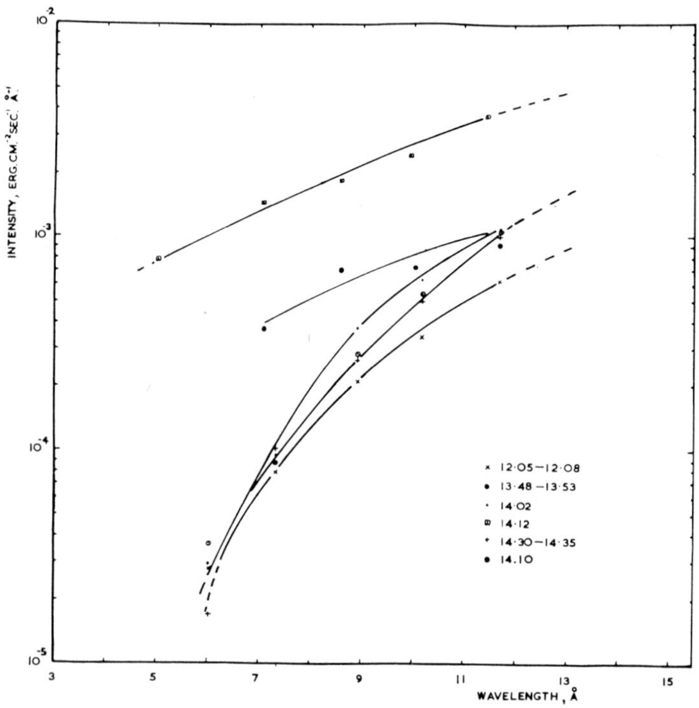

Fig. 6. Ariel spectrum of the X-ray radiation connected with the importance 2 flare of April 27, 1962. The visible flare occurred between 13.50 – **14.30** – 14.40 UT. The X-ray enhancement was small during the first twenty minutes of the flare. Then, at 14.10 UT a rapid burst commenced. This is seen here to have a quite different spectral shape. The onset of the burst was coincident with a 2800 Mc/sec burst and with various types of SID's. See *Proc. Roy. Soc.* 1964. By courtesy of K. A. Pounds, Leicester.

bursts are emitted by plasma oscillations in the corona excited by upward going electrons – these latter are not braked since their mean free path is much longer than the scale height of the corona. It is indeed remarkable that Type-III bursts with $v < 60000$ km/sec do not occur; for velocities around this value the mean free path becomes of the same order as the coronal scale height so that these slower electrons do not reach very far – one should in particular bear in mind that the mean free path is proportional to v^{-4}.

(d) A characteristic and well-studied example is the X-ray burst observed during the flash of the flare of April 27, 1962, at 14.12 UT (Figure 6). The relevant data are given below:

Measured burst intensity 10^{-2} erg cm^{-2} sec^{-1} at earth's distance.

Total X-ray emission from sun (assumed isotropic) 2 to 7×10^{25} erg sec^{-1}.

Energy of electron stream (assume photon production efficiency of 5.10^{-5}) 5×10^{29} erg sec^{-1}.

Total emission (supposed duration 160 sec) 8×10^{31} ergs.

In 160 seconds 10^{39} electrons are involved.

Volume with $N_e = 10^{13}$ cm^{-3} = $(5 \times 10^3$ km$)^3$.

Volume with $N_e = 10^{10}$ cm^{-3} = $(5 \times 10^4$ km$)^3$.

This latter value is of the same order of the chromospheric meshwidth. We thus estimate that in order to emit the X-ray burst on April 27, 1962 a volume of gas with $N_e = 10^{10}$ cm^{-3} and contained in $(5 \times 10^4)^3$ km^3 was involved and accelerated to energies of 50 keV (120000 km/sec).

This is an astonishingly vast volume and a huge amount of energy. Where is the theory that will explain all this?

References

Coutrez, R.: 1960, *Radioastronomia Solare*, Varenna Summer School, Corso XII (ed. by G. Righini, Zanichelli, Bologna, p. 117.

Friedman, H.: *Space Science* (ed. by D. P. Legalley), Wiley, New York.

De Jager, C.: 1959, *Handbuch der Physik* **52** (ed. by M. Flügge), p. 80.

De Jager, C.: 1960, *Radioastronomia Solare*, Varenna Summer School. Corso XII (ed. by G. Righini), Zanichelli, Bologna, p. 296.

De Jager, C.: 1961, *Vistas in Astronomy* 4 (ed. by A. Beer), Pergamon.

Kundu, M. R.: 1965, *The Solar Spectrum* (ed. by C. de Jager), D. Reidel Publishing Co., Dordrecht, p. 408.

Maxwell, A.: 1965, *The Solar Spectrum* (ed. by C. de Jager), D. Reidel Publishing Co., Dordrecht, p. 342.

Minnaert, M.: 1965, *The Solar Spectrum* (ed. by C. de Jager), D. Reidel Publishing Co., Dordrecht, p. 3.

Schatzman, E.: 1965, *The Solar Spectrum* (ed. by C. de Jager), D. Reidel Publishing Co., Dordrecht, p. 313.

Smerd, S. F.: 1965, *The Solar Spectrum* (ed. by C. de Jager), D. Reidel Publishing Co., Dordrecht, p. 398.

CHAPTER 6

PARTICLE EMISSION ASSOCIATED WITH SOLAR FLARES – MODEL OF FLARE EVENTS

C. DE JAGER
Sterrewacht 'Sonnenborgh', Utrecht, The Netherlands

1. Proton Streams

We first discuss the plasma streams which move through interplanetary space with velocities of the order of 1000 to 1500 km/sec. These plasma clouds may appear to be responsible for those geomagnetic storms, which show a sudden commencement, the so-called sc geomagnetic storms. They might also be the same clouds that cause Type-II radio bursts, since examination of the frequency-time characteristics of these bursts shows that their initial velocities are of the order of 1700 km/sec (observations at a frequency around 200 MHz, corresponding with the lowest parts of the corona). When moving upward the velocity seems to decrease to about 700 km/sec at the 50 Mc/sec level (Figure 1).

The observed radiospectra of Type-II bursts can only be interpreted in a consistent way if it is further assumed, that the particle density in those parts of the corona where the Type-II exciters move, is about ten times the normal corona density (Figure 2). Since the Type-II exciters originate from flares, and hence, move in coronal activity centers, this finding of Maxwell and Thompson (Fort Davis Radio Observatory, Texas) nicely confirms results derived from other sources on the electron density in coronal CA's.

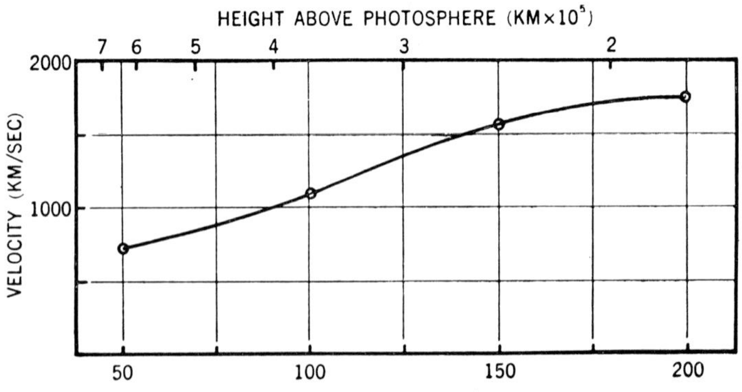

Fig. 1. Average radial velocity of the slow-drift bursts, generally responsible for Type-II radiobursts (Maxwell and Thompson).

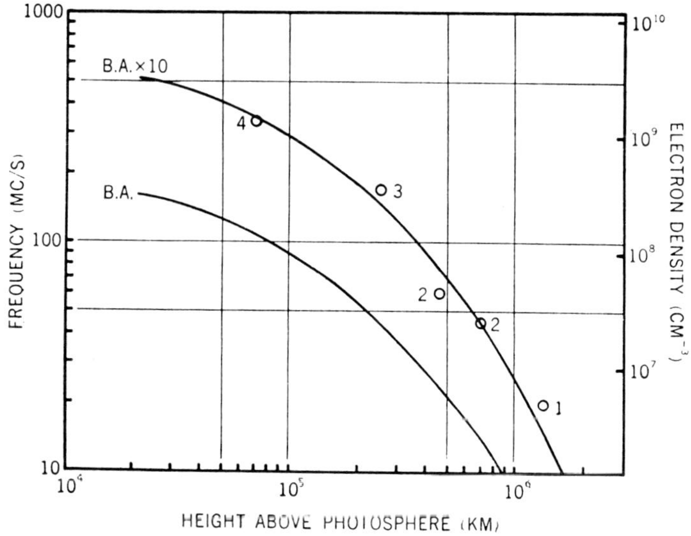

Fig. 2. Electron density and equivalent plasma frequencies in the outer solar atmosphere as derived by Maxwell and Thompson. *BA* is the quiet corona equatorial electron density, as summarized by Baumbach and Allen.

The interpretation of these observations is that a hydromagnetic shock wave is produced as a result of the collapse of the flare region, and that this shock progresses through the corona. Uchida has attempted to predict theoretically the motion of such a shock and found that the velocity in its wake should decrease from about 2000 km/sec at a height of 10^5 km to about 100 km/sec at $3 \cdot 10^5$ km. The corresponding decrease measured in Type-II bursts is in fact from 1700 km/sec to 700 km/sec, but with different assumptions about the magnetic field and about the electron density in the coronal CA, no doubt, theory and experiment could be made to agree satisfactorily.

2. Relativistic Particle Emissions

A. COSMIC RAY PARTICLES

Relativistic particles have been detected only during certain solar flares. In fact, after the first detection of such particles in 1942, only fourteen flares have been found to increase significantly the cosmic ray flux. The emitted particles themselves can only be detected in the earth's atmosphere above heights of about 30 km, but a significant increase may occur in the intensity of secondary neutrons and protons at sea level. For a number of years the standard method for detecting the secondaries has been with the Simpson-type neutron monitor, but fundamental improvements have recently been made in the efficiencies of such instruments (the Carmichael-type monitor has about fifty times the efficiency of the Simpson monitor).

Figure 3 shows how the intensity of cosmic rays increased at the earth during

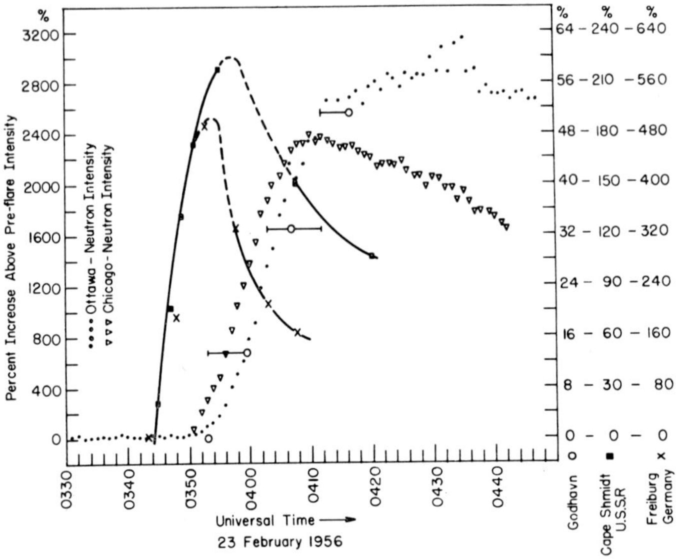

Fig. 3. Intensity of cosmic rays at several stations on the earth following the great solar flare on February 23, 1956, Note the difference between stations situated in and outside the impact zones; in the impact zones the rise is more sudden and occurs some minutes earlier.

and after the flare of February 23, 1956. The neutron intensity is seen to increase very rapidly in regions known as the impact zones, where particles can arrive at the earth along direct trajectories from the sun. In the case of this flare it was some minutes later that particles arrived outside the impact zones. These latter particles impacted on the earth from *all* directions out of space, and this isotropic bombardment kept on for at least fifteen hours, which is much longer than the life time of the flare.

Similar phenomena occurred during other cosmic ray flares and this indicated very clearly that the cosmic ray particles can be kept stored in space in a gigantic "magnetic bottle" which occasionally may surround the earth. They may stay in these "bottles" for times of the order of 24 hours.

Solar cosmic rays appear to be an important means for studying the magnetic field configurations in the interplanetary medium.

The *spectra* of solar cosmic ray particles can be determined from an investigation of the neutron-yield in its dependence on geomagnetic latitude. The energy spectrum of solar particles decreases much faster toward high energies than is the case for the spectrum of the galactic cosmic rays (Figure 4).

B. SUB-COSMIC RAY PARTICLES (PROTON EVENTS)

Particles of energies below about one GeV (10^9 eV) are not able to reach the earth's surface, neither are their disintegration products detectable at sea level. They have been detected in balloon experiments (Winckler and associates) and in rocket and

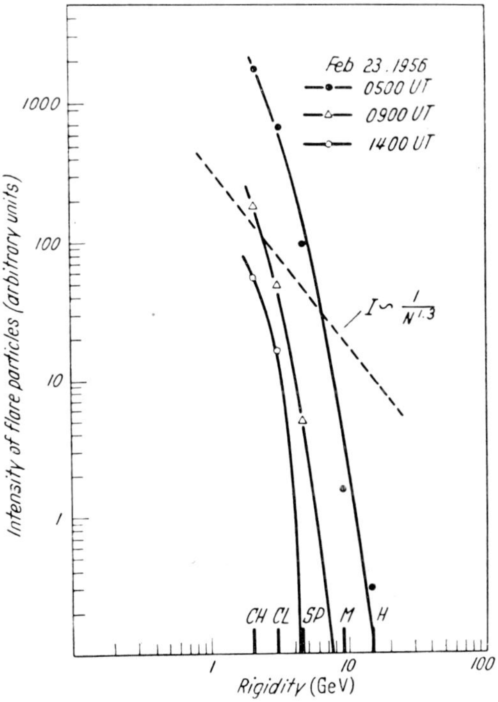

Fig. 4. Spectrum of solar cosmic ray particles observed after the solar flare of February 23, 1956. The spectrum is much weaker than that of the galactic cosmic rays (dashed).

satellite experiments. These particles may be detected in various ways; first by *direct methods:*

(1) With *counters*, brought upward with balloons to above 30 km, or flown in satellites. A rough impression of the energy spectrum of these particles may be obtained from the number of counted particles as a function of height.

(2) *Solid state detectors* enable to determine dE/dx at passage through a thin crystal; and to determine E when the particle is stopped in a thick crystal. Also *scintillation counters* enable a determination of the energy of the particles.

(3) With the *nuclear emulsion* technique. Recovery of the nuclear emulsion plates shows that the particle density may be very great in these sub-cosmic ray bursts.

Indirect methods for detecting these sub-relativistic particles make use of the ionization of the polar low ionosphere by the particles. Their energies are such that they are capable of ionizing the atmospheric regions just below the ionosphere so that their effect is a lowering of the ionospheric base where we have a great damping of plasma oscillations due to the great neutral particle density. *This happens only in the polar regions*, since these particles have too small energies to be deflected to low geomagnetic latitudes. At geomagnetic latitudes below about 60° their effects are hardly discernible. In the polar regions these particles produce:

(4) *Polar Cap absorption* events (PCA): a fade-out of short wave radio communications in polar regions owing to the damping of electromagnetic waves in the dense lower parts of the ionosphere.

(5) *Cosmic noise fade-outs:* these arise because of enhanced absorption of cosmic radiowaves at wavelengths of 10–15 meters. These absorptions are measured with Riometers (Relative Ionospheric Opacity meters).

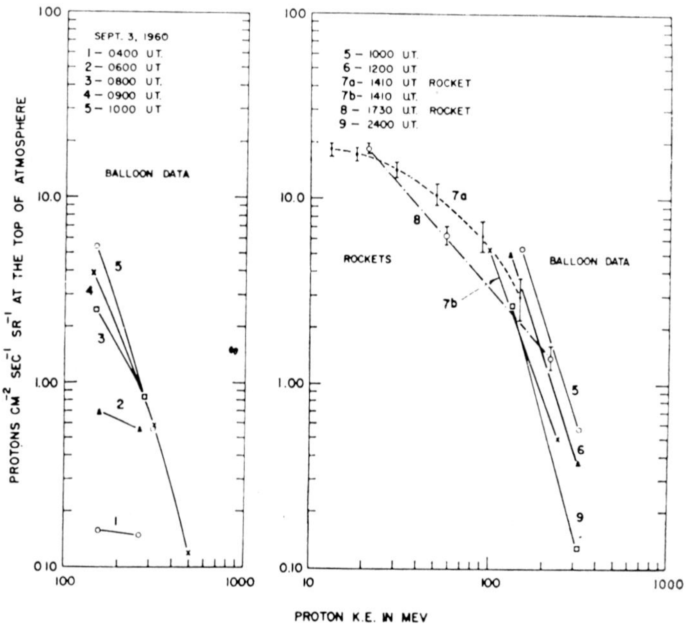

Fig. 5. Integral energy spectrum of solar protons for the event of September 3, 1960. The spectrum begins flat, then increases in steepness and fades away. The low-energy particles are increasing in number when the high-energy part is increasing (J. R. Winckler in J. P. Legalley, *Space Science,* Chapter 11).

During the three years just after the 1957 sunspot maximum about sixty sub-cosmic noise increases have been observed.

The spectrum of this kind of particle is shown in Figure 5.

A comparison of the particle observations with solar phenomena shows:

(1) 50% of all flares of importance 3 or 3^+ are followed by sub-cosmic ray increases.

(2) Of all flares with dm bursts (10 cm) with intensities exceeding 5×10^{-20} MKS and with the radio maximum after the Hα maximum, 90% are followed by a sub-cosmic ray increase.

(3) All solar particle events are associated with a Type-IV burst on the sun.

(4) 40% of all Type-IV bursts are followed by protons events.

(5) A flare with (a) importance 3 or 3^+, (b) type IV radio emission, (c) increase of 10 cm radio emission, (d) partly covering of a spot's umbra, is always followed by a proton event.

3. Acceleration of Solar Protons, a Qualitative Discussion

The solar proton events show the closest correlation with flares presenting Type-IV bursts. These bursts are synchrotron bursts occurring in the first phase of flares and have an average life time of about fifteen minutes. The synchrotron radiation, is due to relativistic electrons captured in a magnetic field existing low in the corona and which has a value of about 200 gauss (Takakura) and apparently a life time of 15 minutes; the size of the field region is of the order of 50000 km. It may be speculated whether the dissolution of this field may not induce electric fields sufficiently large to accelerate protons to the desired energies.

The induction law is written as

$$\oint E_{\text{ind}}\,\mathrm{d}s = \partial/\partial t \int H\,\mathrm{d}\Sigma \times 10^{-8}\,\text{volts} \tag{1}$$

where s is the circumference of the area Σ, H is the magnetic field. We assume that s is of the order λ, then $\Sigma \simeq \lambda^2$. The dissolution takes place in a time τ, then (1) may be approximated to

$$E\lambda \simeq \frac{1}{\tau} H\lambda^2\, 10^8\,\text{volts}$$

With $H = 200$ gauss, $\tau = 10^3$ sec, and $\lambda = 5 \cdot 10^9$ cm the induced field becomes

$$E\lambda \simeq 5 \times 10^{10}\,\text{volts}$$

which is larger by about a factor ten than the average solar cosmic ray particles energy, and five hundred times the energy of the average sub-cosmic ray particle. The agreement could be improved by assuming either $H = 20$ G or $\lambda = 15000$ km for cosmic ray particles; and either $H = 0.4$ G or $\lambda = 2000$ km for sub-cosmic ray particles. These values seem wholly acceptable and certainly not too large, so that the most probable acceleration mechanism for these particles is the *dissolution of a magnetic field system in the low corona*, immediately after the flare's maximum.

4. Summary; Model of Flare Events

In this Section we give in three successive pictures the course of the solar flare events as pictured in the preceding Sections. We shall illustrate the description by three drawings, for the different phases; our basic concept is the one first put forward by Dungey and Sweet of two bipolar spot groups between which, upon mutual approach, a flare may originate.

(a) (Figure 6): Two bipolar spotgroups, each with their own system of field lines, approach each other. The field is strong in the photosphere, much weaker in the corona and chromosphere. Still, even in the corona, the magnetic energy greatly exceeds the kinetic energy of matter, and coronal matter moves with the moving field. The matter in between cannot be penetrated by the field lines and essentially behaves as a diamagnetic body. It is compressed. Some sort of instability arises,

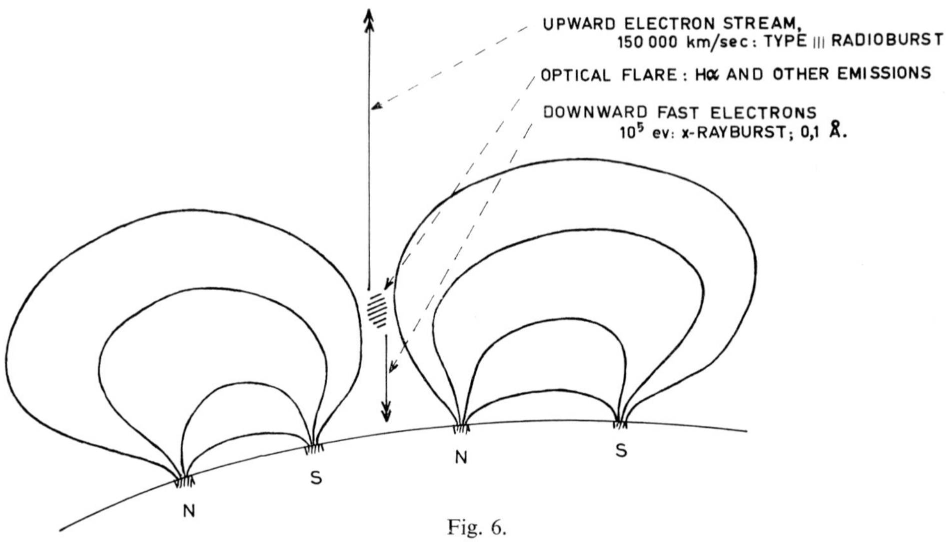

Fig. 6.

perhaps of the kind described by Severny or Kiepenheuer, leading to a sudden collapse-like compression of matter: the optical flare.

In this very early phase local accelerating electric fields may originate; they are not longer than 10 or 100 meters, their field-strengths may amount to 10^5 V of even 10^6 V. These fields directed in one way may accelerate electrons downward, where they are braked in the denser parts of the photosphere and produce X-ray Bremsstrahlung bursts in the energy range 10^5 eV ($\simeq$0.1 Å).

Other fields, in other parts of the flare may be directed upward and produce jets of electrons with velocities around 150000 km/sec; these may excite coronal matter to plasma oscillations: the Type-III bursts.

These clouds are detected up to 2 and 2.5 $R_{\odot}$ above the photosphere; there they have an extension of about one solar radius.

(b) (Figure 7): The processes occurring in the first phase give rise to several secondary phenomena.

(1) The compression which leads to the flare is the source of a hydromagnetic shock wave. Behind the shock front matter propagates outward with an initial speed of the order of 1500 km/sec, decreasing at greater distances to the source, to some hundreds of km/sec.

(2) The electrostatic interaction between the outgoing electrons and the remainder of the sun – at a certain moment slightly positively charged – may cause the Type-V radio-bursts.

(3) As Dungey has shown, the compression phenomenon may lead to a local rearrangement of the field: the outermost lines of force of the two bipolar magnetic systems may break and join with those of the other system. In this way the two, initially independent, systems are partly combined in one. So, systems of lines of

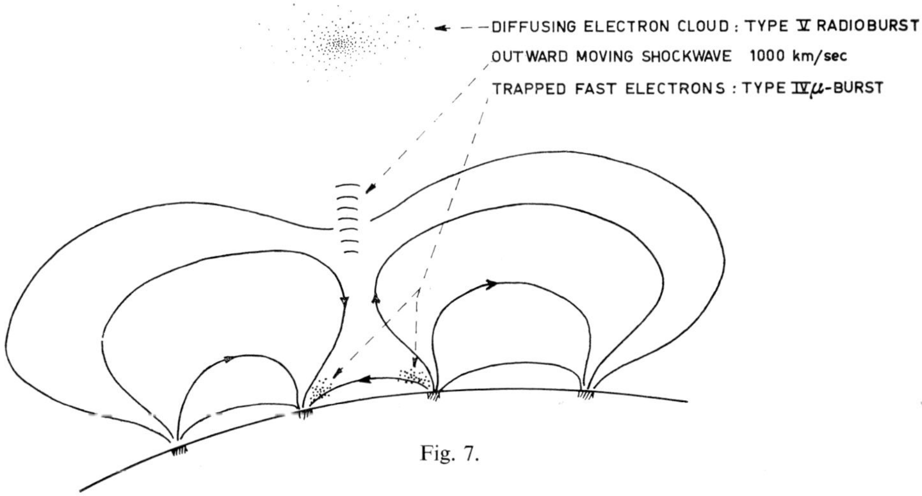

Fig. 7.

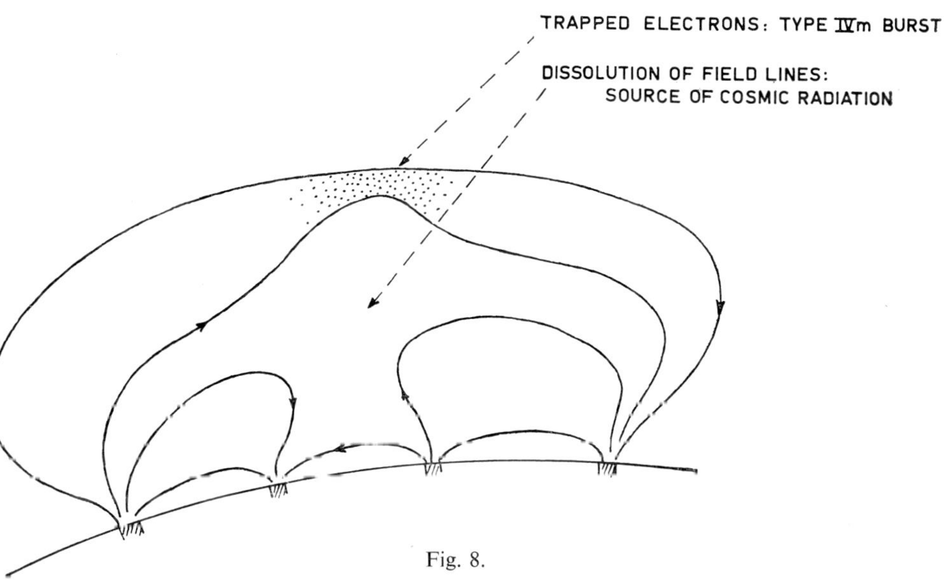

Fig. 8.

force may originate which are capable of trapping a part of outgoing electrons. These electrons bounce back and forward and receive further acceleration during their motions to and fro. The innermost of the trapped electrons may be the source of the Type-IVm or IVdm radiobursts. The diameter of the Type-IV source is still small in this phase. Kundu and Firor found that at 340 Mc/sec the apparent diameter is smaller than 4′, the source is situated lower than 40000 km. Boischot and Pick-

Gutmann found that the first phase of the Type-IV event has a diameter of 6′ to 10′.

(c) (Figure 8): The rearrangement of the field proceeds further. The slow outward going plasma which in the second phase produced the Type-II radio-bursts may now be trapped in the outermost part of the field: the Type-IVm radio-bursts. These may have a larger diameter than the cm and dm Type-IV events. Kundu and Firor found a diameter $\simeq 10'$ and a source height greater than 20000 km. The source moves outward with a velocity $\simeq 1000$ km/sec (Kundu). Boischot and Pick on the other hand, found that the metric event has a small diameter ($\leqslant 3'$) and a great directivity.

These particles and the fields gradually diffuse away so that after one or several hours the situation from before the flare event may be restored. But also even if this were not the case, this would not prevent a second and third flare from originating in the same configuration; in other words: flare recurrence does not premise an exact recurrence of the magnetic configuration.

The dissolution of the field because of the – now no longer prevented – approach of oppositely directed field lines initially belonging to the two independent systems, may give rise to sufficiently strong accelerating electric fields to produce the cosmic ray bursts.

References

De Jager, C.: 1959, *Handbuch der Physik* **52** (ed. by M. Flügge), p. 80.
De Jager, C.: 1963, *Space Sci. Rev.* **1,** 487.
Minnaert, M.: 1965, *The Solar Spectrum* (ed. by C. de Jager), D. Reidel Publishing Co., Dordrecht, p. 3.
Schatzman, E.: 1965, *The Solar Spectrum* (ed. by C. de Jager), D. Reidel Publishing Co., p. 313.
Winckler, J. R.: 1963, *Space Science* (ed. by D. P. Legalley), Wiley, New York.

CHAPTER 7

THE HYDROGEN CONVECTIVE ZONE OF THE SUN

E. SCHATZMAN

Institut d'Astrophysique, Paris

Introduction

The observation of the solar granulation shows that at a certain depth below the photospheric level exists a region in convective equilibrium. We can refer for example to the film obtained by RÖSCH (1960) at the Pic du Midi. Photographs taken from this film have been reproduced in *l'Astronomie*. We shall consider the following:

(1) The instability and the structure in the convective zone;

(2) The velocity field;

(3) The wave generation and propagation in the outher layers of the sun.

1. Stability and Structure

The stability problem can be studied in two different ways. We can either use a phenomenological description of the motions or study the stability of small motions by the known technique of linearized differential equations.

(a) The oldest description of stability problems is by K. SCHWARZSCHILD (1906). Let us consider a bubble with a slightly higher temperature than the surrounding medium; its density being smaller, the buoyancy force pulls it up and, if we assume adiabatic expansion, the bubble will keep going if it remains lighter than the surrounding medium – then the condition for convection will be:

$$\left|\frac{\mathrm{d}T}{\mathrm{d}z}\right|_{\text{adiabatic}} < \left|\frac{\mathrm{d}T}{\mathrm{d}z}\right|_{\text{actual}}. \tag{1}$$

As the change of the pressure with the height is the same everywhere, that relation is more conveniently written

$$\left(\frac{\mathrm{d}\log T}{\mathrm{d}\log P}\right)_{\text{adiabatic}} < \left(\frac{\mathrm{d}\log T}{\mathrm{d}\log P}\right)_{\text{actual}}. \tag{2}$$

(b) The structure of the convective zone can be roughly obtained with this rising and descending bubble model. Let us assume that after a vertical motion over a distance l the bubble stops and accommodates to the surrounding medium, and leaves there the energy which it has carried up. According to ŽEVALKIN (1956) we can introduce a heat capacity per gram C which is intermediate between heat capacity at constant pressure and at constant volume, and hence the flux of mechanical energy can be written

$$F_m = \varrho \bar{V} C l \left[\left(\frac{\mathrm{d}T}{\mathrm{d}z} \right)_{\text{adiabatic}} - \left(\frac{\mathrm{d}T}{\mathrm{d}z} \right)_{\text{actual}} \right]$$

(ϱ is the density). In this expression $\bar{V}$ can be estimated by a simple analysis of the motion under the buoyancy force, writing that the work over a distance l is equal to the kinetic energy of the bubble (SCHATZMAN and PECKER, 1959). The main difficulty in this problem is the choice of the distance l of the convective transport of energy.

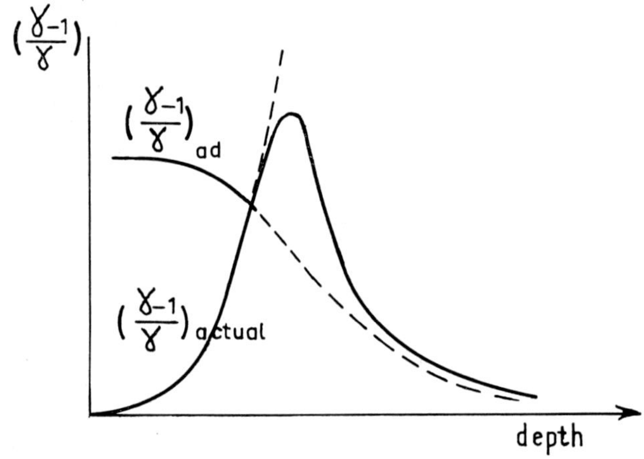

Fig. 1.

Usually it is assumed that l is of the order of the scale height of the pressure of the atmosphere.

The length l has been taken as 1 or 2 scale heights or even greater. The general aspect does not depend upon the value taken for l, but naturally the exact structures and especially the geometrical depth depend on l. The general features can be summarized by the following curve (Figure 1) showing the adiabatic and actual gradients as a function of depth.

From the surface to optical depth about 0.8 the radiative gradient is smaller than the adiabatic one. Below that, the actual gradient overcomes the adiabatic one. The difference has a sharp maximum and then decreases, the actual gradient being very close to the adiabatic gradient at great depth. The most elaborated model is given by BÖHM-VITENSE (1958) and includes the effect of radiative exchange of energy between bubbles.

(c) The method of small motions has been applied for the first time to convective instability by RAYLEIGH (1916) for the incompressible fluid. For a layer with solid boundaries, temperature gradient β, expansion coefficient per degree α, horizontal wave number k and vertical number k_z the rate of growth n of any parameter of the motion is given by

$$n^2 = -\frac{\alpha\beta g k^2}{k^2 + k_z^2} \qquad (g, \text{gravity}).$$

If β is negative the layer is unstable and the convection sets in. For the case of a gas with adiabatic expansion the temperature gradient β has to be changed to

$$\beta^* = -\frac{\mathrm{d}T}{\mathrm{d}z} + \left(\frac{\mathrm{d}T}{\mathrm{d}z}\right)_{\text{adiabatic}}$$

and it can be seen that the instability condition expressed by the sign of β^* is the same as in the bubble theory. The difficulty with the structure of the convective zone is that the linearized theory of the motion cannot be applied. But it is useful to know the modes of the linearized equations in order to analyse the actual motion with the help of these modes.

Using the BÖHM-VITENSE (1958) models, BÖHM (1963) has calculated the modes and has found the rate of growth as a function of the horizontal wave length (Table I).

TABLE I

RATE OF GROWTH OF SEVERAL HORIZONTAL WAVELENGTHS

Horizontal wavelength $\lambda/2$	*Rate of growth* s^{-1}
50 000 km	3×10^{-4}
12 500 km	1.2×10^{-3}
3 125 km	5×10^{-3}
781 km	1.3×10^{-2}

It should be noticed that the rates of growth of the small wavelength and the observed rate of growth on the granulation can very readily be compared, but the latter seems to be too fast.

2. Velocity Field

The non-stationary aspect of the granulation suggests very strongly that the convective zone is turbulent. The difficulty is that very little is known about turbulence in a non-uniform medium. To solve the problem we should consider the following set of questions: (a) knowing the density and temperature distributions, find the properties of the turbulent field; (b) knowing the velocity field of turbulence, determine this convective flux and the structure of the convective zone.

(a) The full problem has not been solved, but an interesting attempt has been made, by LEDOUX, SCHWARZSCHILD and SPIEGEL (1959). They use the so called Boussinesq approximation in which the scale height for density and temperature is large compared to the thickness of the convective zone. They calculate the rate of growth of modes assuming free boundaries and taking into account the viscosity.

Then, the general idea of the treatment of the turbulence is a generalization of Heisenberg's phenomenological theory. In the theory of homogeneous turbulence the average kinetic energy of all modes with total wave numbers between k and $k + dk$ is called $F(k)\,dk$, and Heisenberg considered the energy loss through ordinary viscosity and through turbulent viscosity, this last effect being due to the transfer of energy from big vortices to small vortices. The generalization consists of estimating the energy gain. For a given wave number the linear theory gives an effective growth rate $n(k)$ varying smoothly with k.

The amplitude growing like $e^{n(k)t}$, the energy of a mode grows like $e^{2n(k)t}$ and the rate of growth of the energy of the k-mode is $2n$.

Consequently, the energy gain in the interval k_0, k can be expressed by:

$$E_{\text{gain}} = \int_{k_0}^{k} 2n(k)\,F(k)\,\mathrm{d}k\,.$$

Where k_0 corresponds to the minimum wave number.

On the other hand the energy loss in the same interval is given by Heisenberg as:

$$E_{\text{loss}} = 2\left\{\nu + \eta \int_{k}^{\infty} \left[\frac{F(k')}{k'^3}\right]^{1/2} \mathrm{d}k' \int_{k_0}^{k'} F(k'')\,k''^2\,\mathrm{d}k''\right\}.$$

The first term is the dissipation by ordinary viscosity, the other one is the dissipation by turbulent viscosity and has been derived by Heisenberg by dimensioned considerations. η is a numerical factor of the order of unity which is taken later by LEDOUX, SCHWARZSCHILD and SPIEGEL (1959) as 1/3. For the sake of simplification they take for $n(k)$ the maximum rate

$$n(k) = -\,\nu k^2 + \frac{g\alpha\beta^*}{2x}\frac{1}{k^2}$$

where g is the gravity, $\alpha = 1/T$ is the coefficient of thermal expansion, ν is the kinematic viscosity, x is the thermometric conductivity, β^* is the temperature gradient excess over the adiabatic gradient.

The numerical solution of the equation $E_{\text{loss}} = E_{\text{gain}}$ has been obtained by the authors. The minimum wave number k_0 is related to the geometrical thickness d of the layer by

$$k_0 = \sqrt{2}\,\frac{\pi}{d}\,.$$

With the following values of the physical parameters corresponding to the part of the solar convective zone near the surface $d = 300$ km; $\alpha = 10^{-4}$; $\beta^* = 10^{-4}$; $g = 2.7 \times 10^4$; $\nu = 10^3$ cm^2/sec; $x = 10^{12}$ cm^2/sec. They obtain $\langle v^2\rangle^{1/2} = 2$ km/sec and $\langle \Delta T^2\rangle^{1/2} = 300°$.

These values differ appreciably from the observed values. The root mean square

velocity is about half the observed one, while the rms temperature fluctuation is about three times larger than the observed one.
However, in view of the very simplified picture of the convective zone, it should be considered that the agreement in order of magnitude is a proof of the exactness of the physical picture which has been used to describe the turbulence.

(b) An important step towards the understanding of the nature of turbulence produced by a temperature gradient has been made by MALKUS and VERONIS (1960) who considered a theory of the coupling between modes inspired by the kinetic theory of gases. This theory of the excitations of higher modes by lower modes is very similar to the resonance theory in celestial mechanics. They have obtained a fairly interesting confirmation of their theory in an experiment in which turbulence in water is produced by the density inversion at 4 °C. A layer of water is kept at 8 °C at

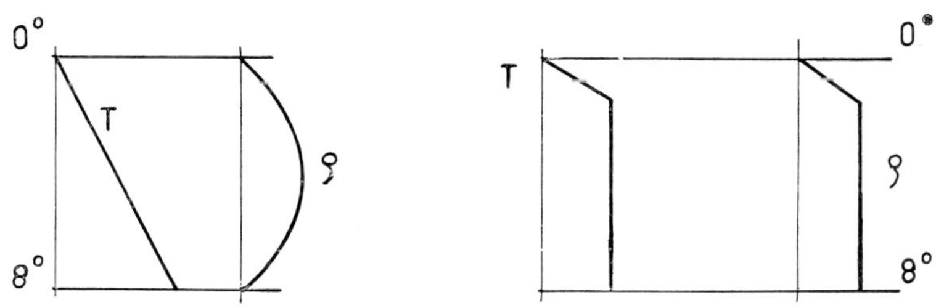

Fig. 2. Radiative equilibrium (on the left); Convective zone (on the right).

its bottom and at 0 °C at the top. In between the two horizontal surfaces a convective zone develops, bounded above by a region in which the heat transfer is due to thermal conductivity, and below by a solid surface. This kind of natural boundary is very similar to the boundary in the convective zone (Cf. MALKUS and VERONIS, 1960).

3. Generation and Propagation of Waves

The problem of propagation of waves in an atmosphere is a classical problem in hydrodynamics and reference should be made to the chapter "atmospheric waves" of Lamb's Hydrodynamics. In the first papers on the heating of the chromosphere only plane waves parallel to the surface have been considered. However the consideration of horizontal waves only is misleading. In order to understand the situation let us consider first adiabatic oscillations in an isothermal atmosphere.

If the motion is analyzed in sine waves it is possible to obtain the dispersion equation. It can be seen that there is a forbidden interval for wave propagation. In the interval of

$$\omega^2 \left[\frac{\gamma g^2}{4c^2}, \frac{(\gamma - 1) g^2}{c^2}\right]$$

only standing waves can exist (c is the sound velocity).

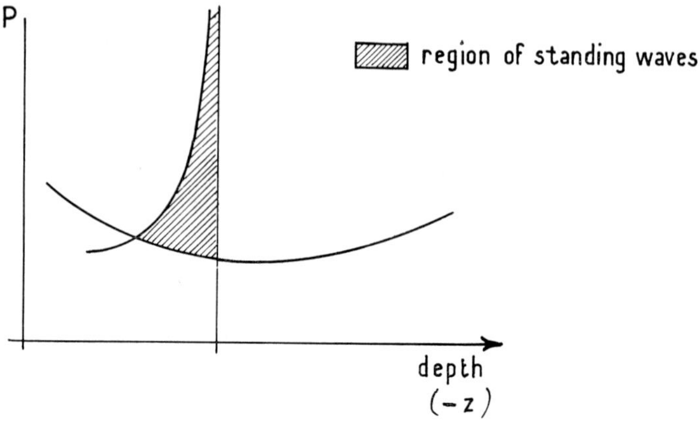

Fig. 3.

The first limit corresponds to the lower frequency limit of horizontal waves, the other one corresponds to gravity waves and can be very easily described in the following way: a bubble of slightly lower density rises in a stable region until its density becomes larger than the density of the surrounding medium. The gravity pulls it down and the period of oscillation is just given by

$$\omega^2 = \frac{(\gamma - 1) g^2}{c^2}.$$

In the convective region the gravity waves cannot exist, as they would correspond to a negative value of ω^2. In a non-isothermal atmosphere similar relations can be

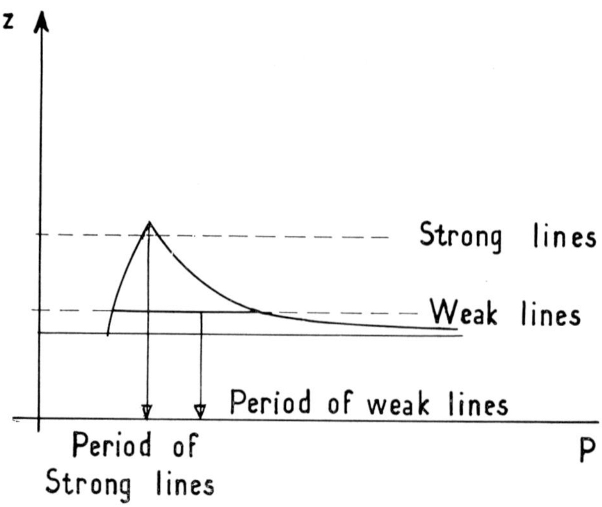

Fig. 4.

found using the BKWJ approximation – with a parabolic representation of velocity of sound $c(z)=(c_0{}^2(1+\varepsilon^2 z^2)$ the forbidden interval is limited by ω_1 and ω_2

$$\omega_1^2 = \frac{\gamma^2 g^2}{4c^2(z)} \qquad \omega_2^2 = \frac{g^2(\gamma-1)}{c^2(z)}\left(1-\frac{\beta}{\beta_a}\right)$$

where β is the gradient of temperature, and β_a is the adiabatic gradient.

In Figure 3 the characteristic periods P_1 and P_2 are plotted as a function of height for

$$\frac{1}{\varepsilon} = 360\,\text{km} \qquad \gamma = 1.1\,.$$

Spiegel and Moore suggest that the waves observed by Evans and Michard have just the frequency of standing waves ($P \simeq 280$ sec). If we plot the boundary region of the standing waves, z as a function of the period, we can interpret the difference between strong and weak lines, strong lines being formed at a greater height than weak lines show a shorter average period (Figure 4).

References

Böhm, K. H.: 1963, *Astrophys. J.* **137**, 881.

Böhm-Vitense, E.: 1958, *Z. Astrophys.* **46**, 108.

Ledoux, P., Schwarzschild, M., and Spiegel, E.: 1961, *Astrophys. J.* **133**, 184.

Malkus, W. V. R. and Veronis, J. A.: 1960, Symposium on Cosmical Gas Dynamics, Varenna

Rayleigh, E.: 1916, *Phil. Mag.* **32**, 529

Rösch, J.: 1960, *L'astronomie* **74**.

Schatzman, E. and Pecker, J.-C.: 1959, *Astrophysique générale*, pp. 292–295.

Schwarzschild, K.: 1906, *Nachr. Akad. Wiss. Göttingen, II Math.-Physik. Kl.* **41**.

Ževalkin, S. A. : 1956, *Astron. Ž.* **33**, 137.

CHAPTER 8

WAVE PROPAGATION IN THE PHOTOSPHERE, CHROMOSPHERE AND CORONA

E. SCHATZMAN
Institut d'Astrophysique, Paris

Introduction

The outer layers of the sun consist of two regions, both at a higher temperature than the surface of the sun. These are the chromosphere and the corona. The upper regions are at a temperature of the order of 10^6 °K, compared with 4800 °K for the solar surface. At first sight, this appears to be thermodynamically impossible, since heat travels from hot regions to cold regions.

SCHWARZSCHILD (1948), BIERMANN (1948), SCHATZMAN (1949) suggested that the heating of the chromosphere and corona could arise from mechanical energy transported from the convective zone. This raises two points, the first is to find the source of energy in the convective zone, and the second is to solve the problem of the transport of energy from this zone to the corona. It should be noticed here that the surface of the sun is not uniform. There are different chromospheric regions, the plages, seen bright in calcium K line and the other parts of the solar surface. As can be seen from Leighton's photographs of the solar magnetic field, the regions where the magnetic field is stronger reproduce exactly the pattern of calcium plages. It can be concluded that the thermal balance above the region of the plages is different from the other regions, and therefore that both the mechanical flux and the mechanism of transport of the mechanical energy are different.

1. Wave Propagation

In order to study the phenomenon of wave propagation in the outer layers of the sun, it is necessary to know the physical conditions of wave propagation:

(a) As long as the mean free path is small compared to the wave length, the waves are ordinary acoustic waves, and the adiabatic relation can be considered as valid between pressure and density.

(b) When the mean free path is large compared to the wave length, no longitudinal waves can propagate. However, ionic waves can propagate if the wave frequency is intermediate between the collision frequency and the plasma ionic frequency:

$$\omega_c \ll \omega \ll \sqrt{2}\,\Omega_p .$$

If we take the following values (SPITZER, 1963; Ch. 5)

$$\omega_c = \frac{2\pi}{t_c} \qquad t_c = \frac{11.4\; T^{3/2}}{N_e \log \Lambda} \qquad \log \Lambda \simeq 10$$

and

$$\sqrt{2}\,\Omega_p = \left(\frac{8\pi n_e e^2}{m_i}\right)^{1/2} = 10^{3.27}\, n_e^{1/2}$$

we find the condition

$$n_e \ll T^3 \cdot 10^{5.06} \tag{1}$$

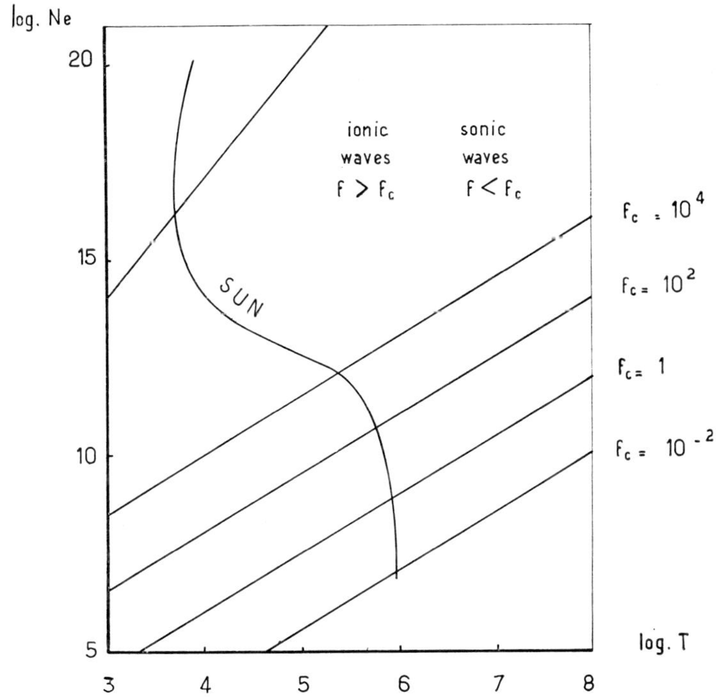

Fig. 1.

In the meantime, the conditions on the frequency read

$$n_e T^{-3/2} \lesssim f \lesssim 3 \times 10^2\, n_e^{1/2} \tag{2}$$

In the diagram with Log T, log n_e (Figure 1), we can plot the condition (1), the curve representing the chromosphere and corona, and the collision frequency.

As long as $f < f_c$, we can speak of sound waves. When we reach the point where the collision frequency becomes smaller than the wave frequency, we are in the region where ionic waves can propagate at sound velocity, but now, the sound velocity is given by

$$V_s^2 = \frac{\gamma k T (Z+1)}{m_i}$$

with $\gamma = 3$ for collisionless plasma (instead of $\gamma = 5/3$ for a plasma within collisions).

However, in the collisionless plasma, another effect becomes important, the so-called Landau damping.

As the velocity of the wave is comparable to the sound velocity, the motion of the ions can smear out the wave very quickly. Following SPITZER (1963, Chapter 3) we can estimate the time of damping of the ionic wave:

$$\sigma \simeq \frac{\Omega_p^2}{\omega^2} \simeq 1.8 \times 10^6 \frac{n_e}{\omega^2}.$$

The ionic wave is damped very quickly as $(\Omega_p/\omega) \gg 1$: we can conclude that, *above a certain level in the solar atmosphere no longitudinal wave can propagate.*

(c) Gravity waves cannot propagate either when the collision frequency is too small; the density of the outher layers of the star is so small that the characteristic frequency corresponding to the change in gravity due to the change in density is extremely low.

(d) However, Alfvén waves, which do not imply any collision effect as they are transverse waves, can propagate even when the density is very low and the mean free path is very long.

2. Generation of Waves in the Convective Zone

We can describe the generation of waves in a compressible atmosphere using the classical definition of monopole, dipole and quadrupole, and interpreting properly the equation of hydrodynamics.

A. SOURCES

If we imagine for example that at a certain point, there is a source of gas, which alternatively blows and whales we can generate expanding sound waves, diverging from a single source or monopole.

If we have two single sources very near each other, such that the amount of matter ejected by one pole is equal to the amount of matter absorbed by the other we have a production of sound waves by a double source or dipole. It is easy to generalize to a quadrupole.

B. EQUATION OF MOTION

We can write the *equation of motion* in the following way

$$\frac{\partial^2 \varrho}{\partial t^2} - \frac{\partial}{\partial x_j} c^2 \frac{\partial}{\partial x_j} \varrho = \frac{\partial q}{\partial t} - \frac{\partial}{\partial x_j}\left[\varrho\left(F_j + \frac{\partial c^2}{\partial x_j}\right)\right] + \frac{\partial^2}{\partial x_i \partial x_j}\left[(p - \varrho c^2)\delta_{ij} + \tau_{ij} + \varrho u_i u_j\right].$$

The left hand side of the equation expresses that the change of density propagates at the sound velocity c. The right hand side represents the various sources of sound waves. The first term, $\partial q/\partial t$ represents the monopole. It is important only when considering mass ejection. The second term represents the dipole source. It is made of two parts; the first part $\partial/\partial x_i(\varrho F_j)$ represents the dipole source due to the density fluctuations in an external field F_j. The second part represents the dipole source due to the density fluctuations in presence of a temperature gradient. The third term represents the quadrupole source. It is made of three parts. The first one represents the source due to the fluctuations of the pressure; the second one is the source due to the viscous forces; the third term represents the source due to the velocity field.

Lighthill has studied the quadrupole source of acoustic waves, produced by isotropic turbulence. The result, he obtained is that the power radiated per unit volume is proportional to the rate of dissipation of kinetic energy multiplied by the Mach number of the turbulence to the power 5:

$$\varepsilon_{\text{acoustic}} = \text{constant}\, \varepsilon_{\text{turbulence}} M^5 .$$

UNNO and KATO (1963) have studied the dipole source of acoustic waves in an atmosphere.

In presence of a magnetic field, two other sources of waves have to be added: the dipole source due to the Lorentz forces (OSTERBROCK, 1961) and the quadrupole source due to magnetic stresses (KULSRUD, 1955).

C. FLUX OF MECHANICAL ENERGY

It is not easy to calculate the flux of mechanical energy produced by these various sources.

The flux in the regions of small magnetic fields is probably not larger than 4×10^7 ergs $\text{cm}^{-2}\,\text{sec}^{-1}$.

The dipole source, in the presence of a magnetic field is fairly important and it is quite possible to explain a flux of about 4×10^8 ergs $\text{cm}^{-2}\,\text{sec}^{-1}$ in the magnetic regions. Finally, it should be noticed that the flux of energy present in solar flares is of the order of 10^9 ergs $\text{cm}^{-2}\,\text{sec}^{-1}$, and that the total flux radiated by the sun is about 6×10^{10} ergs $\text{cm}^{-2}\,\text{sec}^{-1}$.

The flux of mechanical energy which is necessary, in order to explain the chromosphere is less than 1/1000 of the total solar flux.

3. Wave Propagation

Let us consider again the graph giving the critical period as a function of height for the compression waves and the gravity waves (see Figure 3 of the preceding chapter).

Only waves of period shorter than $P_{\min}$ can propagate through the whole atmosphere. In the convective zone, self amplification takes place, as it results from the properties of the convective zone.

There is a certain region where it is impossible for any wave of period $P > P_{\min}$ to propagate. A wave taking place below this zone can induce a quasi stationary wave which radiates into the chromosphere; but the amplitude of the wave which has crossed the region of stationary waves is very small.

When a wave of short period propagates in an atmosphere, its amplitude increases with height. A very simple analogy can be obtained by comparing with surface waves

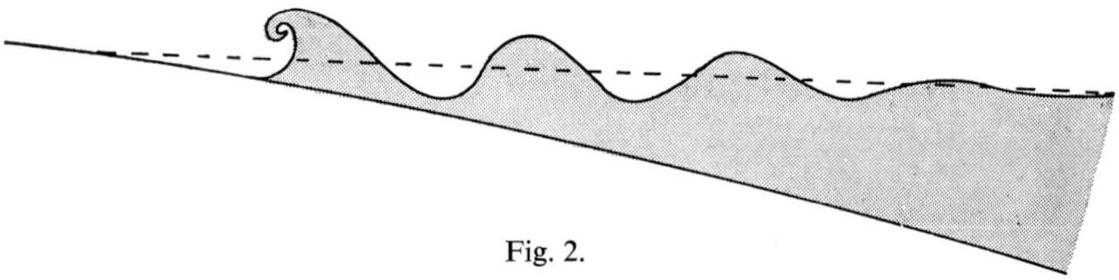

Fig. 2.

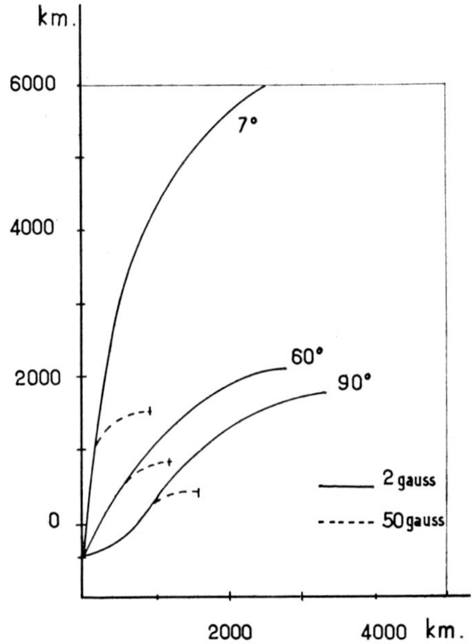

Fig. 3.

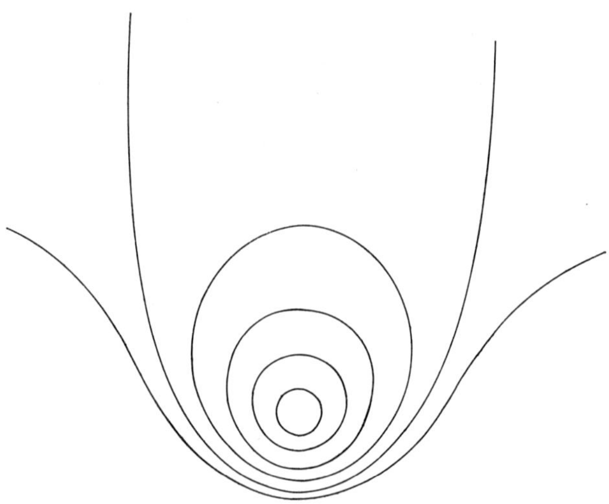

Fig. 4.

in shallow water (COURANT-FRIEDRICH: *Supersonic Flow and Shock Waves*). In the oscillation there is a relation between pressure and mass which is similar to the relation between pressure and density for a gas of adiabatic compressibility $\gamma=2$. When a wave propagates towards a shore or small slope, the amplitude increases, the top of the wave propagating faster than the bottom. It is well known that a kind of shock can develop, and on certain beaches, like the Californian beaches, or like the beaches near Bayonne, big breakers can develop, as a result of the amplification of the wave when the depth of the water decreases (Figure 2). In the same way, a plane wave propagating in an atmosphere increases its amplitude with height. If the density is ϱ, V the velocity of propagation of the wave, v the material velocity, $\varrho V v^2 =$const. expresses the law of conservation of the energy. In an isothermal atmosphere $V=$const. and v increases like $\varrho^{-1/2}$. The velocity of propagation is $V+v$ at the top of the wave, and $V-v$ at its bottom. The skewness of the wave increases with height, and after a few wave lengths, the wave becomes a shock wave.

The situation is not exactly the same with gravity waves; however when considering the motion of the matter, it can be seen that the variation of the pressure is in phase with the horizontal velocity. The propagation of the gravity wave along the surface of the sun is accompanied by the propagation of a pressure wave. Therefore, the condition for the existence of gravity waves is the same as the condition for the existence of pressure waves. The time of relaxation of the gas has to be smaller than the period of the wave.

Due to dissipation, the amplitude of the waves is never very large, and the velocity of propagation remains close to the sound velocity. It is therefore interesting to consider sound rays, and wave surfaces. A recent paper of Weinberg develops the Eikonal method, which can be applied when the wavelength is small compared to the scale of the density gradient.

SCHATZMAN (1949) and OSTERBROCK (1961) have studied the shape of the sound rays for a point source just below the photosphere for acoustic waves (without magnetic field) and, in the presence of a magnetic field, for pseudo-acoustic waves.

One can find on Figure 3 the ray paths for fast mode waves (see Section 4) in the hydrogen atmosphere, from the upper hydrogen convective zone to the photosphere and chromosphere. It is important to notice the refraction of acoustic or pseudo-acoustic waves. The wave which develops from a point source has a finger shape (Figure 4).

In the expression of the energy balance in the chromosphere, as we shall see (Section 5) the refraction of the mechanical wave is very important.

4. Modes of Hydromagnetic Waves

There are three modes of propagation of these waves (see for example DENISSE and DELCROIX (1961). The Alfvén wave of velocity of propagation $V_A \cos \Theta$

$$V_A = B(4\pi\bar{\varrho})^{1/2}$$

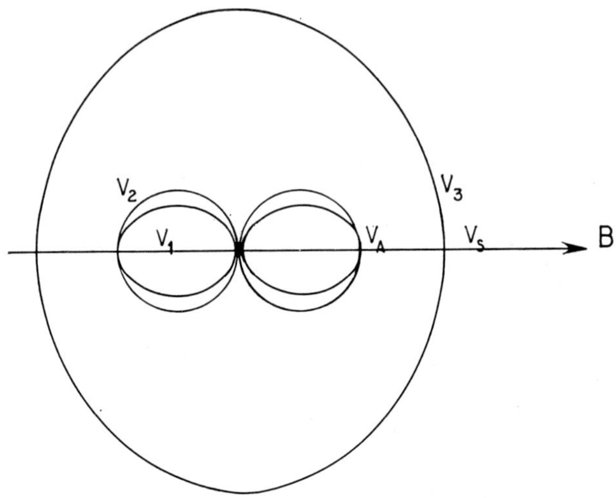

Fig. 5.

and two other modes, the fast and slow modes, where Θ is the angle of the direction of propagation with the magnetic field. The velocity of propagation of the slow and fast mode is given by

$$V_1^2, V_3^2 = \tfrac{1}{2}\left[V_A^2 + V_S^2 \mp \{(V_A^2 + V_S^2)^2 - 4V_A^2 V_S^2 \cos^2 \Theta\}^{1/2}\right]$$

where

$$V_S^2 = \frac{\gamma k T}{\mu}$$

is the sound velocity.

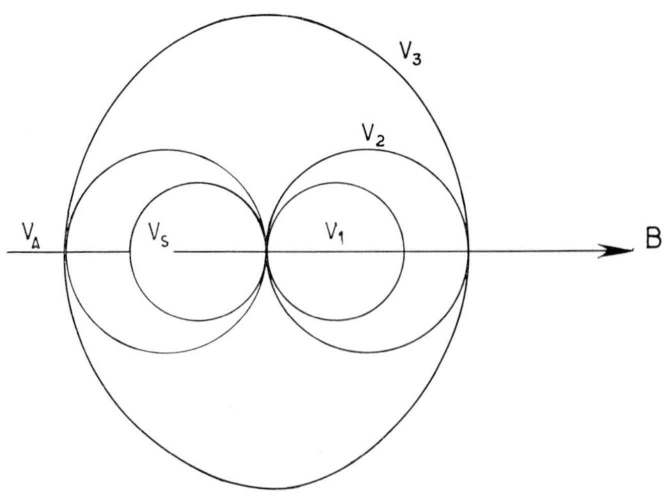

Fig. 6.

It is interesting to use a polar representation (Figures 5 and 6). Two cases have to be considered, $V_A < V_S$ and $V_A > V_S$.

If the velocity of the Alfvén wave, travelling at a velocity $V_A \cos \Theta$ is plotted in the polar diagram, we have the two circles touching at the origin. The Alfvén wave is a transverse wave with the vibration perpendicular to the plane defined by the direction of propagation and the magnetic field. V_1 and V_3 are the pseudo-acoustic modes, the fast mode travels almost at the velocity of sound, and the vibration is almost along the direction of propagation; the slow mode has its direction of vibration in the plane of the paper and is almost perpendicular to the direction of propagation. Another case

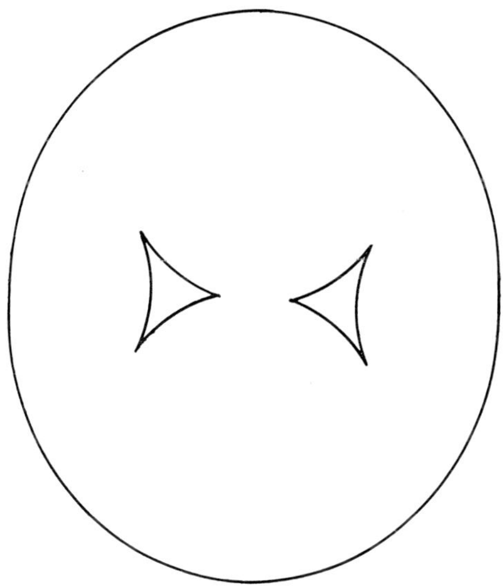

Fig. 7.

can be drawn for $V_A > V_S$; in this case, the slow mode has its vibration almost parallel to the magnetic field, the fast mode now has its vibration in the plane of the paper and almost perpendicular to the magnetic field.

In nature it is not known if the slow mode exists, only the fast mode has been observed in the laboratory.

We can now consider these waves, and their bearing upon the transport of mechanical energy in the chromosphere.

From these polar diagrams, it is possible to draw the shape of the wave surface. Figure 7 gives the shape of the wave surface for the case $V_A > V_S$.

It can be seen (1) that the fast mode propagates almost isotropically and that (2) the Alfvén mode propagates exactly along the lines of force.

It follows that the Alfvén mode can carry energy at a greater distance from the source than the fast mode. One should study the production of these waves, and in particular the coupling of modes (the Eikonal method does not give this).

The only way coupling could take place would be from gradients in the magnetic field.

An oscillation propagating through a region where the direction of the magnetic field is changing generates an oscillation perpendicular to the magnetic field and induces the production of transverse waves.

In the general case of the equations of propagation (Leray), high order differentials occur and solutions are not available. Ferraro has obtained solutions for a vector field in the atmosphere. Jen has studied a field of fixed direction, but of variable magnitude. However, neither of these cases give coupling of the pseudo-acoustic and Alfvén waves. In order to obtain coupling, one must have twisted lines of force. Then any oscillation produces a change in the field, generating an oscillation partly perpendicular to the field, giving an Alfvén wave.

Frisch (1964) has studied the propagation of waves in a stratified medium, that is, dependent only on z the height, and allowing for a magnetic field varying with height. In order for the coupling to be efficient it is seen that it is necessary to have the velocities of the Alfvén, and fast waves, almost equal. This can be easily seen since the two modes must be in phase if the amplitude is to be built up.

If the coupling is done in a region of a twisted field, where typical dimensions are small compared with λ, fast waves can thus give Alfvén waves.

This may well be the case above the "plage" regions.

5. Mechanics of the Dissipation of Waves and Chromospheric and Coronal Heating

The possible ways of coronal and chromospheric heating are first summarized. (1) Dissipation of shock waves; (2) Dissipation of Alfvén waves; (3) Gravity waves (Parker, Whitaker).

(1) The first will now be considered in detail. It has previously been mentioned that the waves propagating upwards into the less dense regions of the sun will have an increasing velocity amplitude.

We have seen (Section 3) how far it takes to produce a shock wave. The velocity at the "top" of the wave is greater than that at the bottom $u \sim \exp(-gz\mu/2kT)$, hence $c+u$ increases very much more rapidly than $c-u$, and quickly a shock wave is produced. The number of waves to reach this state can be obtained by comparing c with u at a given height.

For example starting at u_0 we have a shock wave when

$$\exp \frac{z - z_0}{2H} = 1 + \frac{c}{u_0} \frac{\lambda}{2H}.$$

For the convective region of the sun $u_0 \simeq 1$ to 2 km/sec and $c \simeq 7$ km/sec. Hence, when $\lambda \ll 2H$

$$\frac{z - z_0}{\lambda} \simeq \frac{c}{u_0} \simeq 5.$$

Then $z-z_0$ is of the order of a few scale heights, thus the change from a sine wave to a shock front will take place after 200 to 500 km.

The dissipation in the shock is rapid and it can be shown that the finite change of entropy across the shock is proportional to the (amplitude of the shock)^3. One can say that, but for a numerical constant, the amount of energy in the wave is equal to the amount of kinetic energy in the shock wave. That is

$$\varrho u^2 Vt\,.$$

Here V is the velocity of propagation of the shock, and provided the amplitude is small it equals the velocity of sound. "t" is the time length of the wake of the shock, that is to say the time needed to cover a distance Vt. Now energy is dissipated at a rate proportional to ϱu^3, and for small amplitude u is proportional to the amplitude of the shock.

Then the change of entropy is proportional to (amplitude of the shock)^3. The numerical term mentioned above, is difficult to estimate.

Osterbrock used 1/8 $\varrho u^2 Vt$ and Weyman 1/12. The reason for the difficulty in estimation is that the term depends on the structure of the shock itself. One can attempt to write an expression for the conservation of energy, and obtain the rate of dissipation as a function of height (SCHATZMAN, 1949). Osterbrock has done this for the fast mode in magneto hydrodynamic waves.

To compare the mechanical flux with the total flux we see that the total flux is $10^{10.8}$ ergs/cm^{-2}/sec^{-1}. The mechanical flux is 3 to 4×10^7 ergs/cm^{-2}/sec^{-1} that is approximately 1/1000 of the total flux and at disturbed places it will be 1/100.

(2) The possibility exists also, that heating may occur by the dissipation of Alfvén waves by electron-neutral particle collisions. This is based on the fact that neutral particles and ions do not have the same inertia as electrons (PIDDINGTON, 1956).

(3) WHITAKER (1963) suggests dissipation of gravity waves by normal conduction.

References

BIERMANN, L.: 1948, *Z. Astrophys.* **25**, 161.
DENISSE, J. F. and DELCROIX, J. L.: 1961, *Théories des ondes dans les plasmas*, Dunod, Paris.
FRISCH, U.: 1964, *Ann. Astrophys.* **27**, 224.
KULSRUD, R. M.: 1955, *Astrophys. J.* **121**, 461.
LIGHTHILL, L. J.: 1952, *Proc. Roy. Soc. London* A **211**, 564.
OSTERBROCK, D. E.: 1961, *Astrophys. J.* **134**, 347.
PIDDINGTON, J. H.: 1956, *Mon. Not. Roy. Astron. Soc.* **116**, 314.
SCHATZMAN, E.: 1949, *Ann. Astrophys.* **12**, 203.
SCHWARZSCHILD, M.: 1948, *Astrophys. J.* **107**, 1.
SPITZER, L.: 1963, *Physics of Fully Ionized Gases*, North Holland Publishing Co., Amsterdam.
UNNO, W. and KATO, S.: 1963, *Publ. Astron. Soc. Japan*, in press.
WHITAKER, W. A.: 1963, *Astrophys. J.* **137**, 914.

CHAPTER 9

CORONAL HEATING AND SOLAR WIND

E. SCHATZMAN
Institut d'Astrophysique, Paris

1. Heat Balance

We have now to consider the heat carried mechanically in the chromosphere and corona and hope to explain in this way the heating of these two regions. It can be said that the present theories give a picture of the temperature in the chromosphere and corona which show all the main features of the observed temperature distribution. However the theory fails to be in quantitative agreement with the observations. The situation is the following. De Jager and Kuperus (1961) estimate the mechanical flux leaving the photosphere, and calculate the structure of the chromosphere and corona, assuming hydrostatic equilibrium and heat balance.

$$\operatorname{div}(F_{\text{mech}}) - \operatorname{div}(F_{\text{radiation}}) = 0\,.$$

Their solution (Figure 1) shows a steep rise of the temperature in a few hundred kilometers, from the upper chromospheric temperature of 100000° to the coronal temperature of 650000°. A very important feature of the flux of mechanical energy is the refraction of acoustic waves, which produces a very quick decrease with height of the rate of energy dissipation, ε, per cubic centimeter per second.

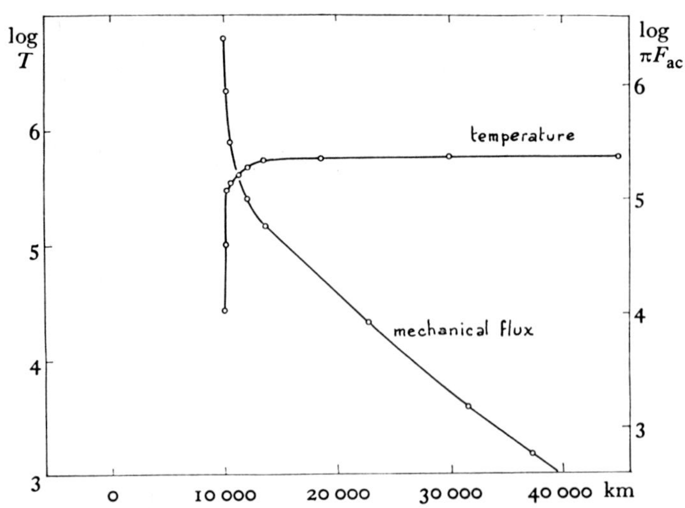

Fig. 1. Log $F_{ac}(h)$ and log $T(h)$ in the assumed coronal model.

Osterbrock has tried to explain the difference between plages and undisturbed regions of the chromosphere. His treatment of the problem is different. Assuming a certain vertical distribution of temperatures and densities, he calculates the rate of energy dissipation as a function of height.

When considering only the fast mode, the main features are the following. The rate of dissipation increases with height, as the amplitude of the shock increases, reaches a maximum, and then, due to refraction, decreases.

In the presence of a magnetic field, the dissipation takes place at a lower level.

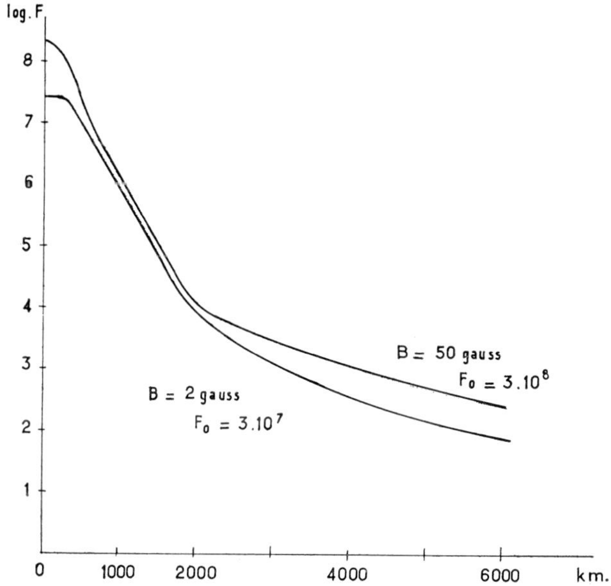

Fig. 2.

When taking into account also the larger mechanical flux above the disturbed regions, it seems possible to explain the different vertical distribution of temperatures. However, there is no quantitative agreement with the theory and the observations, the amount of energy carried in the lower photosphere being insufficient to explain the brightness of the faculae, observed at the limb in white light.

We give Figure 2 the flux of mechanical energy as a function of height in the disturbed and undisturbed regions, and Figure 3 the dissipation ε per gram per second.

An obvious difficulty appears then. The flux of mechanical energy due to fast waves above the plage regions is much too small to explain the emission from the upper chromosphere, and is much too small to provide the coronal heating.

It is necessary to assume that some energy is transferred from the fast mode, which is the only one to be able to carry energy at low heights to other modes, Alfvén and slow mode. As these modes propagate more or less along the lines of force,

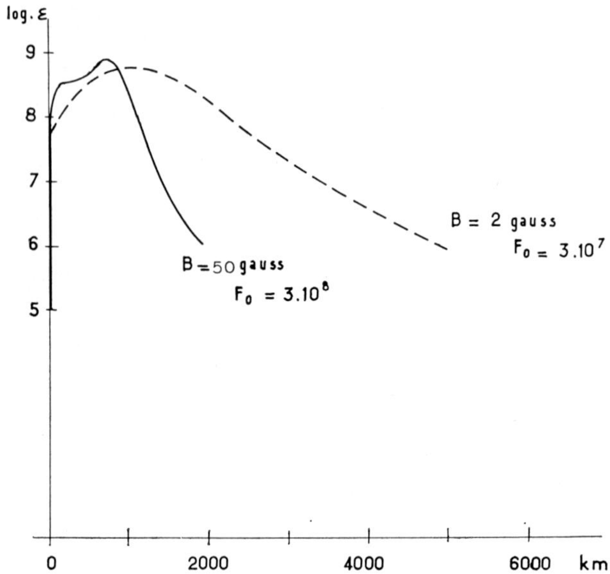

Fig. 3.

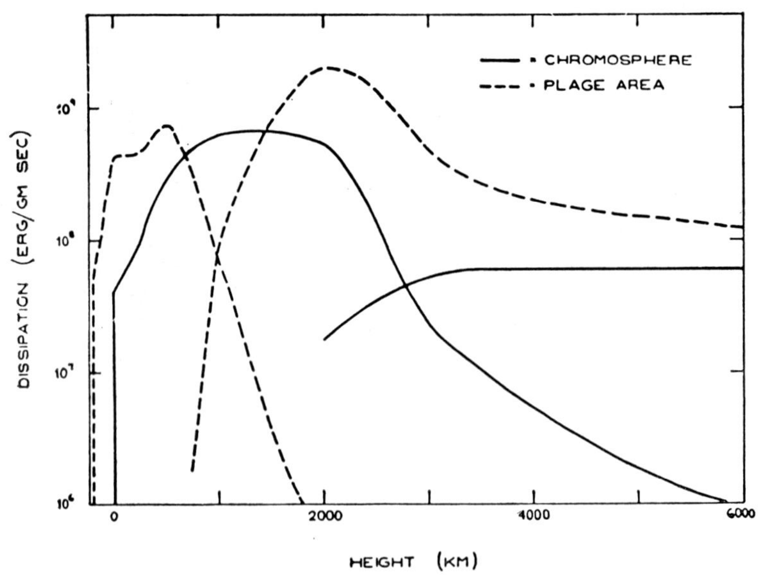

Fig. 4. Computed dissipation in the chromosphere above the undisturbed sun, $B = 2$ gauss; solid line; and in plage area, $B = 50$ gauss; dashed line. In both cases left-hand curve is result of fast-mode disturbances; right-hand curve, of slow-mode disturbances (assumed to cover 10 per cent of the sun's area at any moment and averaged in this drawing over the whole area).

they are not influenced by refraction and can carry mechanical energy at much larger distances. It is the assumption of Osterbrock that some energy is transferred from the fast mode to the Alfvén mode and to the slow mode. It is readily seen that energy dissipation can take place at a much higher level. Osterbrock writes for the equation of conservation of the mechanical energy of slow waves, an expression which is similar to the expression for fast waves but with the difference that there is no refraction to reduce the flux. An example of the solutions obtained by Osterbrock can be found in Figure 4.

It should be noticed that the coupling between electromagnetic modes and gravity has not been studied yet. Considering the importance of gravity waves as shown by Whitaker, this question needs to be clarified.

2. Thermal Equilibrium

In order to understand the situation let us consider the thermal equilibrium of a chromospheric region of temperature T and density of ions, electrons, hydrogen atoms: N_i, N_e, N_H; let us assume that each radiative capture is accompanied by an energy loss $\bar{w}$. We can write two equations:

(1) Dissociative equilibrium

$$\frac{\mathrm{d}N_H}{\mathrm{d}t} = - DN_HN_e + RN_iN_e$$

where it is assumed that we have only collisional dissociation and radiative recombination;

(2) Energy equilibrium

$$C_p\frac{\mathrm{d}T}{\mathrm{d}t} + R\bar{w}N_iN_e = \varepsilon$$

We have naturally the conditions of conservation:

$$N_i + N_H = N_0 \qquad (3)$$
$$N_i = N_e.$$

First of all, we must consider the equilibrium conditions. The solution has been obtained by Thomas and Athay (1961), and we give Figure 5 ε as a function of the temperature. In order to be complete, we give also the radiation loss by free-bound transitions in Helium He II.

Thomas and Athay (1961) have shown that the chromosphere is stable when the radiation losses at equilibrium increase with the temperature:

$$\frac{\mathrm{d}\varepsilon}{\mathrm{d}T} > 0.$$

Beyond a maximum, the chromosphere is unstable: for $T > 12000°$ for hydrogen emission, for $T > 50000°$ for Helium emission.

The study of the differential equations (Souffrin) shows that, along the branch $(\mathrm{d}\varepsilon/\mathrm{d}T)>0$ the solutions are stable; when the temperature and the density are disturbed, the curve $N_e(T_e)$ spirals around the point of equilibrium, and the medium comes back to equilibrium. In other words, the system oscillates with positive damping.

The period of the oscillation is very short, and is essentially given by the time of recombination. For $N_e=10^{13}$, the characteristic time is of the order of 1 second. Therefore, we can say that the chromosphere adjusts immediately to the amount of mechanical energy injected.

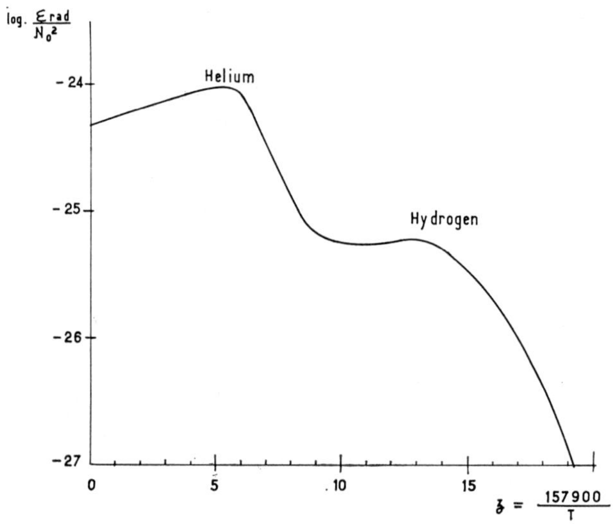

Fig. 5.

If the amount of mechanical energy changes, the chromosphere reaches a new state of equilibrium in a very short interval of time.

It can be shown that an increase in the gradient of ε produces a decrease of the temperature.

A simultaneous increase of ε produces an increase of the density, and consequently the point representing the equilibrium of the log (ε/N_0^2) curve goes down. If the point is on the Helium branch, it can reach the hydrogen unstable zone and finally, the equilibrium can shift from a hot helium zone to a cool hydrogen zone.

Physically, it can be said that a decrease in the scale height of ε produces a decrease in the scale height of the temperature, and then, a decrease in the temperature itself.

This stability problem has a great importance. It is certainly the behaviour of the chromosphere with respect to the thermal balance which explains the calcium plages, the spicules, and, eventually the flares. It should be stressed again that the change in the chromospheric equilibrium is a result both of a change in the mechanical flux put into the chromosphere and of a change in the gradient of the rate of dissipation.

REMARK ON CORONAL HEATING

Assuming that the corona is heated by a succession of shock waves, Billings wonders whether something of this mechanism could be observed.

Two facts have to be considered: (1) that the line broadening corresponds to a temperature higher than the kinetic temperature derived from the ionisation equilibrium, (2) the coronal line never shows any Doppler shift.

From these facts it is possible to derive an order of magnitude of the rms velocity. The absence of any Doppler shift, implies that many clumps of the corona are seen together, and it is possible to derive an upper limit to the size of these clumps. Billings suggests, as a possible fit, clumps of about 200 km, and shock waves following each other with a very low frequency (period $\simeq$500 sec); another possibility is a high frequency (period $\simeq$6 sec), so that the photograph with a much longer exposure time superimposes many waves of very great wave length.

3. The Solar Wind

The problem of the solar wind, or more generally, the problem of the stellar winds, has raised many discussions during the last years. It seems that now the problem is physically understood though the mathematics of the problem remain very intricate (see especially PARKER, 1960, 1963).

In its generality the problem of the solar wind would imply the solution of the following equations:

(A) Equation of conservation of mass;

(B) Equation of motion, or equation of conservation of momentum,

(C) Equation of energy,

(D) Equation of conservation of the mechanical flux.

We are very far from having enough knowledge of the heating mechanism to be able to write exactly the equation of conservation of the mechanical flux; the equation

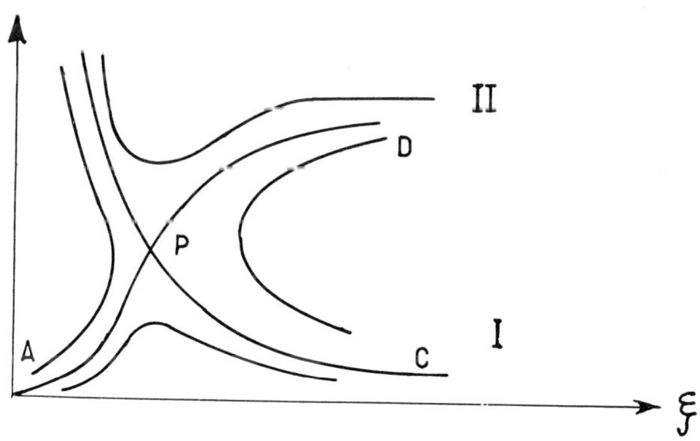

Fig. 6.

of energy implies the knowledge of the energy sources, and the effects of thermal conduction; however, in the outer layers of the corona, the heating mechanism is not important any more, and it is possible to consider the solution of equations (A), (B), and (C). This has recently been done by Parker.

However, even simple assumptions replacing the exact equation of energy (C), give the possibility of understanding the dynamics of the expansion of the solar corona. Various assumptions can be made. If we write

$$T \sim N^{\alpha - 1}$$

we know that for $\alpha = 1$ we have an isothermal corona; for $\alpha = 5/3$, we have an adiabatic

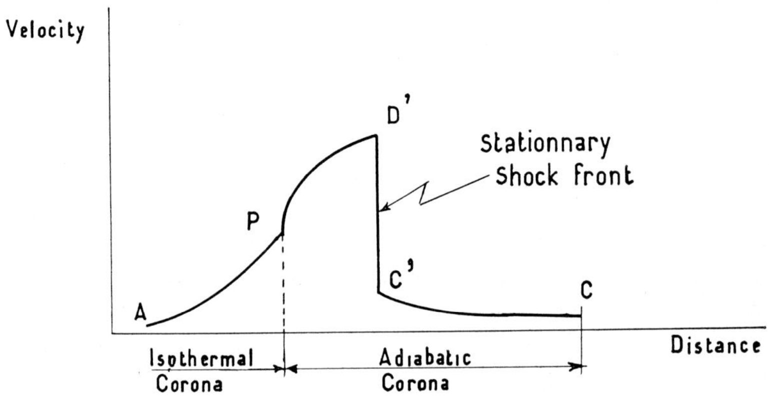

Fig. 7.

corona. For $\alpha < 5/3$ it is possible to give the general behaviour of the solutions. Let us plot the velocity as a function of the distance to the centre of symmetry (Figure 6). Starting from small values of the distance we follow the velocity curve. We find first solutions with zero velocity and finite density at infinity. There is a singular point. The solution crossing the singular point gives a finite velocity to infinity and a vanishing density.

In order to understand the physics of the solar wind, we can follow Parker's argument. Let us consider an isothermal corona extending to distance b, and an adiabatic corona, extending beyond b. The isothermal corona implies that the coronal heating is such that it can balance all radiative and conduction losses. Beyond distance b, no heating is supposed to exist, and we assume that the corona cools as fast as possible and it cannot cool faster than adiabatically. Let us first assume that the corona is not expanding. We have only the hydrostatic equilibrium. The temperature decreases beyond the point b.

$$T = T_0 - \frac{3}{5} \frac{GM}{R} \left(\frac{1}{b} - \frac{1}{r} \right)$$

and we see that the temperature vanishes at a certain point r_0

$$r_0 = \frac{b}{1 - (3/5)(bRT_0/GM)}.$$

When b becomes sufficiently large, $b = b_{max}$

$$b_{max} = \frac{3\ GM}{5\ RT_0} = 13.7\, R_\odot \quad \text{for} \quad T_0 = 10^6 \text{ degrees},$$

the adiabatic corona extends to infinity. There is no acceptable solution for $b > b_{max}$; in other words, when increasing the energy input in the corona, the isothermal corona expands, but the corona remains finite. When a certain value of the energy input is reached, the adiabatic corona extends to infinity. For larger energy flux the excess of energy is transformed into kinetic motion, and the corona expands.

When considering a slow expansion, one finds first the solution corresponding to zero velocity at infinity. For a certain value of the energy input the solution follows a curve which is similar to the singular curve *APC* (Figure 6). It is necessary, then, to jump from curve *APC* to curve *APD*, with finite velocity at infinity. There is a finite jump in the necessary energy input when jumping from curve *APC* (zero velocity at infinity) to curve *APD* (with finite velocity at infinity). It is impossible to fulfil the equations with a continuous series of solutions, but by placing a discontinuity at a certain distance (Figure 7), it is possible to obtain solutions with intermediate values of energy input. The abscissa of the shock front $C'D'$ moves outwards, as the energy input increases, until it has gone to infinity.

We can state in other words the physical conditions for the existence of the coronal wind. If the kinetic kT_0 at the boundary of the isothermal corona is too small compared to the potential energy (GMm_H/r), the particles cannot leave the Sun. On the contrary, if the isothermal corona extends sufficiently far, the velocity of the particles are sufficiently large at the boundary, and the expansion is possible (COWLING, 1959). The conditions for a sufficiently efficient heating of the corona seem to be difficult to fulfil.

4. Effect of the Solar Wind

The solar wind is a very small phenomenon. If the solar wind, at the distance of the earth is characterized by the following values: $N = 10\ \text{cm}^{-3}$, $v = 300\ \text{km/sec}^{-1}$, it corresponds in the chromosphere to the following values: $N = 10^{13}\ \text{cm}^{-3}$, $v = 1\ \text{cm sec}^{-1}$.

If we accept the correspondence between the streamers and the larger values measured by Mariner II, we may have velocities ten times larger in the chromosphere.

In the lower corona, the velocities are larger. When $N \simeq 10^9$, at about 6000 km, the velocity is of the order of 10^4 to $10^5\ \text{cm/sec}^{-1}$, and the gas of these regions is lost in a time of the order of 10 days to 1 day.

TABLE I

TILT ANGLE α AS A FUNCTION OF SOLAR WIND VELOCITY V_m

V_m	α
1000 km/sec	23°
300 km/sec	53°
100 km/sec	76°

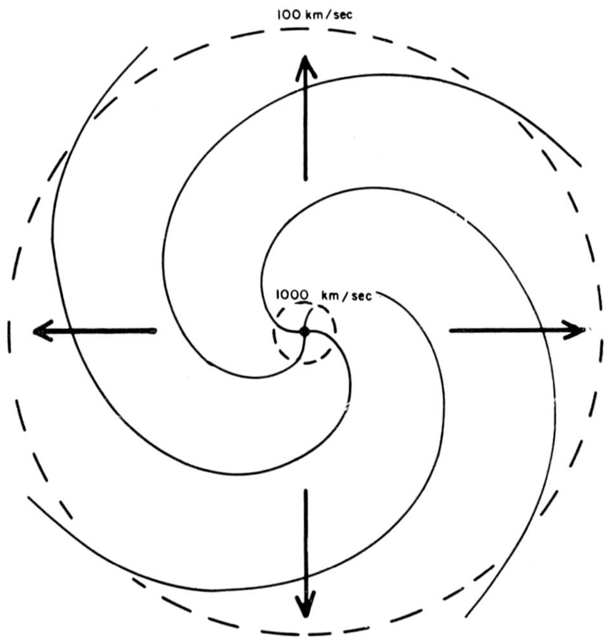

Fig. 8. Lines of force from the rotating sun near the solar equatorial plane, assuming a uniform outward radial gas flow. Broken circles represent the scale of the orbit of the earth relative to the spiraling lines of force for the indicated gas velocities.

The lines of force of the solar magnetic field can be carried away by the solar wind, and we can consider (1) the general magnetic field, (2) the magnetic field above magnetic groups.

If there were no rotation, the magnetic field at large distance, from the sun, would be radial. However, due to rotation the Coriolis force would tilt the lines of force. Assuming that the tangent of the tilt angle is equal to the ratio of radial velocity to the rotational velocity PARKER (1958) gives the following values for the radial and azimuthal field:

$$B_R = B_0 \left(\frac{R}{V}\right)^2 \cos \Theta$$

$$B_e = B_0 \frac{\Omega R}{V} \frac{R}{r} \cos \Theta \sin \Theta \,.$$

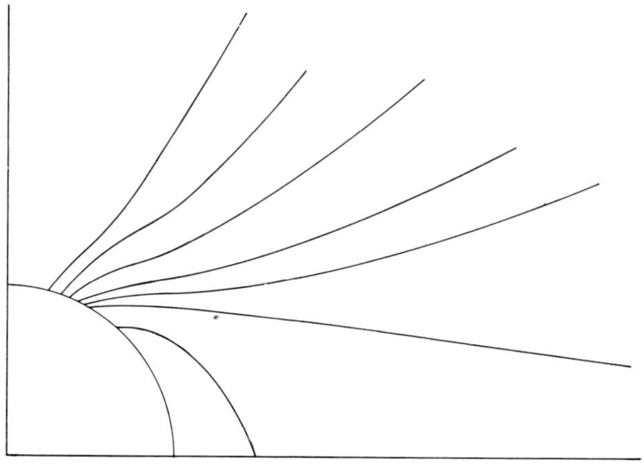

Fig. 9.

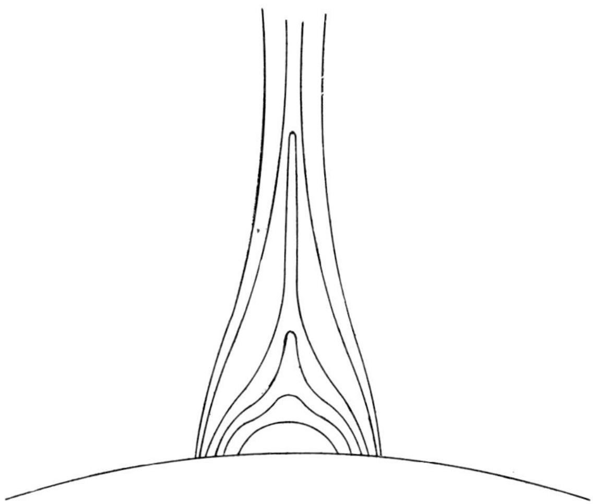

Fig. 10.

The tilt angle at the distance of the earth depends naturally on the velocity V_m of the solar wind. The different values are given in Table I (Figure 8). Near the Sun, where the rotational velocity is very small, we can compute the shape of the lines of force assuming for example, that a dipole field at the surface of the Sun is carried by an isotropic solar wind. As an example, we can take a density decreasing like r^{-6} and we obtain the following expressions for the field

$$B_r = B_0 \left(\frac{R}{r}\right)^2 \cos \Theta$$

$$B = \tfrac{1}{2} B_0 \left(\frac{R}{r}\right)^5 \sin \Theta$$

from which we derive the shape of the lines of force:

$$\frac{r}{R} = \left(1 - \tfrac{3}{2} \log \frac{\cos \Theta}{\cos \Theta_0}\right)^{-1/3}.$$

They have been represented in Figure 9. In the same way it is possible to show that the lines of force of a bipolar field are carried to infinity (Figure 10). These pictures can be compared to the solar corona at minimum (similar to Figure 9) or at maximum (Figure 10). It should be especially noticed that the existence of magnetic streamers results from the effect of the solar wind on the magnetic field, and are confirmed by the observations of the solar wind.

Mariner II, in its flight to Venus has shown great variations in the intensity of the solar wind.

The narrow regions of strong wind can correspond to the magnetic streamers. The greater flux of mechanical energy above the plages could easily provide the necessary energy for production of a strong wind.

References

Cowling, T. G.: 1959, *Modèles d'études et évolution stellaire*, Liège.
De Jager, C. and Kuperus, M.: 1961, *Bull. Astron. Inst. The Netherlands* **16,** 71.
Parker, E.: 1958, *The Plasma in a Magnetic Field* (ed. by Landshoff).
Parker, E.: 1960, *Astrophys. J.* **132,** 821.
Parker, E.: 1963, *Astrophys. J.* (in the press).
Thomas, R. N. and Athay, R. G.: 1961, *Physics of the Solar Chromosphere.*

CHAPTER 10

THE SOLAR MAGNETIC FIELD AND THE SOLAR ACTIVITY

E. SCHATZMAN
Institut d'Astrophysique, Paris

1. Structure of the Sun

The Sun has a mass of 2×10^{33} g, a radius of 7×10^{10} cm, a luminosity of 3.6×10^{33} ergs sec^{-1}. The internal structure of the Sun is determined by the equation of hydrostatic equilibrium, the equation of transfer of energy and the equation of energy generation by thermonuclear reactions.

Our knowledge of the energy transfer in the convective zone is not good, and consequently, the extension of the solar convective zone is not known with certainty. The depth of the solar convective zone is of the order of 100000 km. Inside the outer convective zone we find a radiative zone in which the density increases steadily, and very probably the central region of the Sun, where most of the energy generation takes place, is convective. The central temperature is of the order of 15.10^6 degrees, and the density of the order of 120 g cm^{-3}. The Sun rotates slowly, with a period of about 27 days, and shows a different period for different latitudes: this is the differential rotation.

We do not know how deep is the differential rotation. This might well be a superficial phenomenon associated with the convective zone (KIPPENHAHN, 1963).

2. Generation of the Solar Magnetic Field

It has been known for a long time that finite conductivity produces a decay of the magnetic field (COWLING, 1945). For a dipole field, the time scale of decay is of the order of 10^{10} years. The actual time scale is not known, as we do not know if the solar magnetic field permeates the whole sun. Furthermore, the surface field is carried away by the solar wind, and consequently, a regeneration of the magnetic field is necessary.

The motion of matter across the magnetic field, producing a current which produces a magnetic field, it has be assumed since LARMOR (1919) that the internal motions of a star could regenerate constantly the magnetic field. This kind of dynamo effect has been discussed several times. COWLING (1933) has shown that a stationary axisymmetric velocity field cannot maintain a stationary magnetic field. More generally, BACKUS (1957) has shown that an axisymmetric magnetic field cannot be prevented from decaying by a velocity field.

However, PARKER (1955) has shown that in a rotating object, turbulence and differential rotation can regenerate the magnetic field by a non-stationary dynamo mechanism.

The mechanism of Parker can be described in the following way. (1) Let us consider a poloïdal field, with the lines of force in the meridian plane. The differential rotation bends the lines of force in the direction of the parallels, and generates a toroïdal component. (2) Let us now consider the motions in the convective zone. When matter rises, the Coriolis force deflects the fluid flowing into the rising column and produces a cyclonic motion.

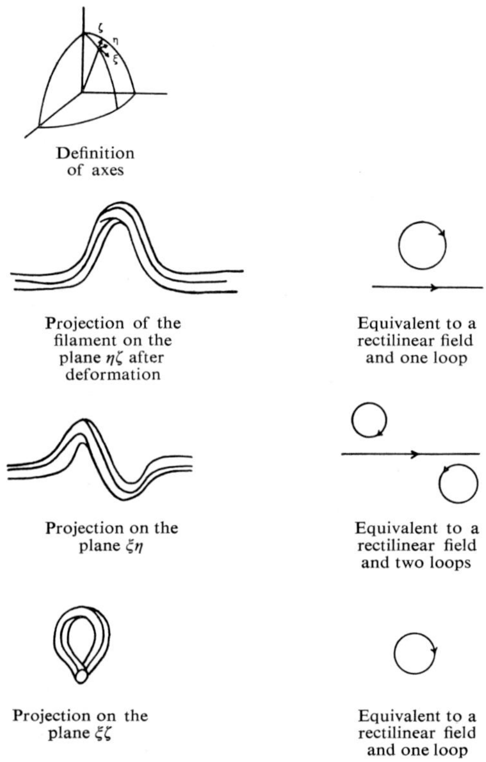

Fig. 1. Plasmas in astrophysics.

A toroïdal field, elongated by the rising flow, is also twisted by the cyclonic motion, and it is possible to see that a twisted tube of force can be analysed as the superimposition of a toroïdal tube of force and four loops of magnetic field. Figure 1 shows the shape of the twisted tube of force and the loops in which the magnetic field can be analysed. Ascending and descending motion produce tubes of force twisted in the same direction.

Parker assumes then that, due to conductivity, all the loops can merge into each other. We can see that the horizontal loops cancel exactly each other. The loops in a vertical plane will not cancel exactly, as the rising and the sinking motions have not exactly the same importance. They will contribute to the regeneration of the poloïdal field, without acting, on the average, on the toroïdal field.

It is then possible to write a system of equations for the toroïdal component and the poloïdal component of the magnetic field, each equation expressing the destruction of the magnetic by conductivity, and regeneration by the motions. The magnetic field which is produced is only statistically axisymmetric, and this explains why the theorem on the dynamo effect for axisymmetric field does not apply there.

The system of equations shows the possibility of a kind of progressive wave,

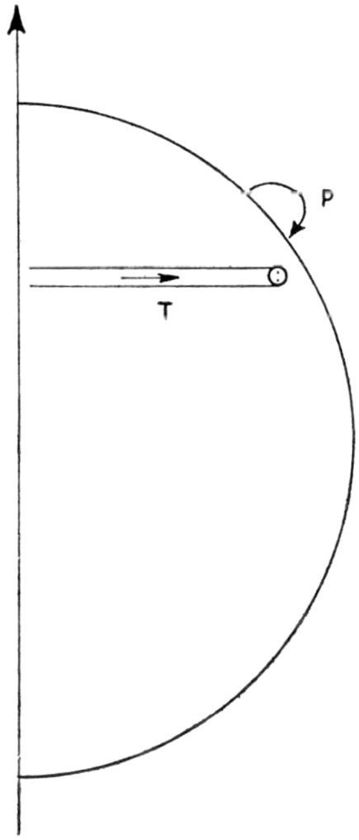

Fig. 2.

where the field is successively predominantly poloïdal and toroïdal. This dynamo wave may prove eventually to be the base of the theory of stellar activity.

On the Sun we find that the field of the tube gives rise to a poloïdal field (p) and a toroïdal field (T) by the differential rotation of the Sun (Figure 2).

3. Buoyancy of Tubes of Force

Let us assume that, underlying the photosphere, can exist a magnetic tube.

Due to conservation of $(p+(B^2/8\pi))$ across the boundary of the tube of magnetic field, the gas pressure inside the tube is smaller than the pressure outside, and the density is then smaller. The tube of force is then unstable with respect to the exchange of stabilities, which results essentially from the buoyancy force due to the lower density of the matter in the tube of force.

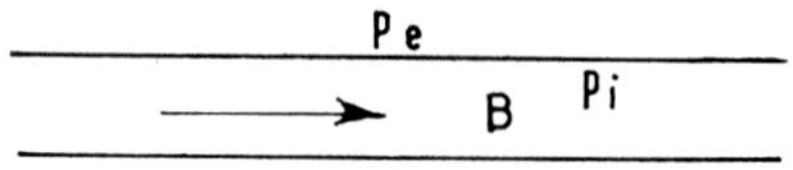

Fig. 3.

Let us follow Parker in his theory of this model. He assumed a tube with pressure p_i, density ϱ_i and with a field B (Figure 3).

If we call the external pressure p_e we have, according to Parker, the equation expressing the pressure balance:

$$p_i + \frac{B^2}{8\pi} = p_e \rightarrow p_i < p_e$$

And if we take the temperature (T) inside and outside being the same we have

$$\varrho_i < \varrho_e$$

and the tube will move upwards. By the upward motion the tube will tend to bend. If the bending wavelength is great enough, the resistance force of the field will not be sufficient to prevent the bending. For this wavelength (L) Parker found the following expression:

$$g\,(\varrho_e - \varrho_i) = \frac{B^2}{4\pi L}$$

and with

$$\frac{R\varrho_i T}{\mu} + \frac{B^2}{8\pi} = \frac{R\varrho_e T}{\mu}$$

we find

$$\varrho_e - \varrho_i = \frac{\mu B^2}{8\pi RT} \rightarrow L \sim \frac{2RT}{g\mu},$$

which is just two times the normal scale height in the atmosphere.

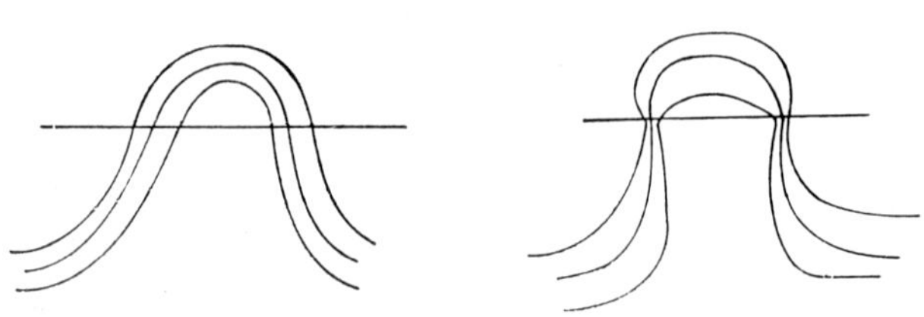

Fig. 4.

If one calculates this effect with the theory of small motions and for a slab of magnetic field instead of the tube, one gets a much greater value for L and also the time scale of development turns out to be greater.

As soon as the field has got the configuration as in Figure 4 the convection, which cannot penetrate into the tubes of field lines, cannot either reach the spot. Therefore the spot receives less energy than the surrounding medium and is therefore much cooler. Also the field lines will be pressed together by the lateral pressure of the gas.

4. The Solar Cycle

Let us assume with ALLEN (1960) that there is a large scale circulation below the solar surface, with two toroïdal magnetic strands, each circulating in an opposite direction (Figure 5). Following COWLING (1946) and PARKER (1955) he assumes that when the strands are near the surface they can float up here and there and break the surface to form bipolar spot groups. Allen includes also in the higher latitudes, a circulation in the opposite direction. This explains why the sunspots tend to reach the equator at the end of a solar cycle.

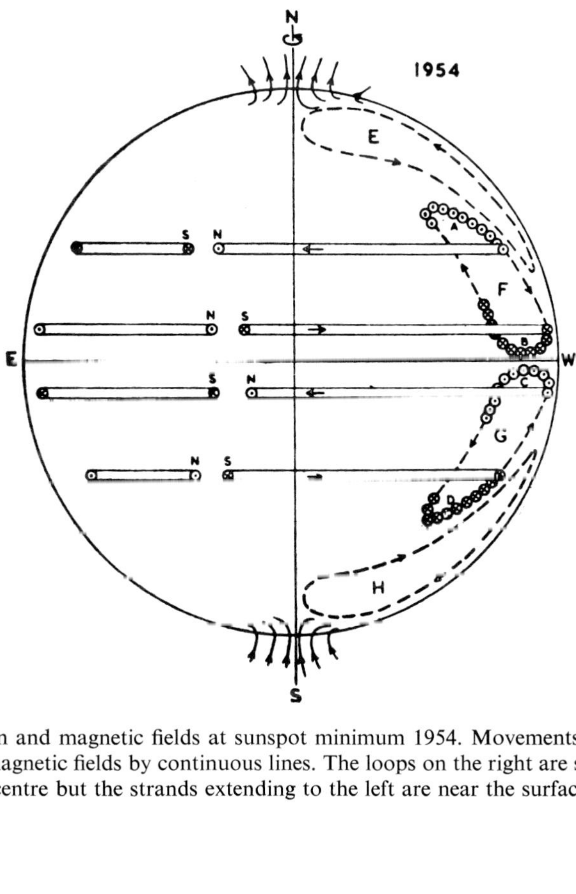

Fig. 5. Circulation and magnetic fields at sunspot minimum 1954. Movements are represented by broken lines and magnetic fields by continuous lines. The loops on the right are sections through the centre but the strands extending to the left are near the surface.

The change of polarity in the sunspots groups from one cycle to the next would then be explained by the opposite polarity of the magnetic strands.

5. Equilibrium of Sunspots

Our knowledge of the sunspots provides us with the following facts: The magnetic fields are strong; the energy density of the field is larger than the energy density of the gas; the temperature is lower than in the surrounding photosphere there exists a velocity field in the spots; the radiative flux depends on the position in the spot, so there is no radiative equilibrium; all parameters change only slowly with time.

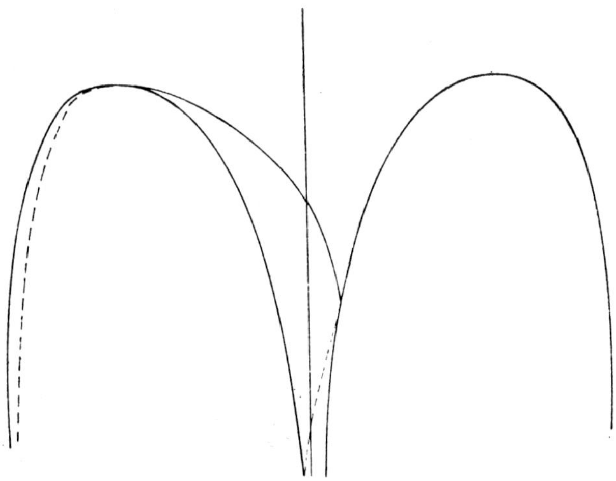

Fig. 6.

If we consider a stationary spot we still have to make one more assumption on the distribution of B.

For instance, the lines of force are in meridian planes and at each altitude there is a characteristic parameter which describes the lateral distribution of B (e.g. $B=\alpha B_0 e^{-\beta^2}$ with α only depending on height and β only depending on the lateral distance).

We can assume a force free field. We have in the equation of equilibrium a term $j \times B$. In the higher layers of the spot we have: $B^2/8\pi \gg P$ and the term $j \times B$ has to be small; thus j and B are almost colinear. In this way we find the condition for a force free field, $\nabla \times B=\alpha B$, where α is a constant. With this condition we obtain twisted lines of force (Figure 6).

In an Hα picture of sunspots we see that the fibers very often show a twisted structure.

Bumba has made measures of B in the umbra and in the penumbra (Figure 7). His picture of the lines of force show clearly their twisted nature. However he did not succeed in connecting together the parts relative to the umbra and to the penumbra

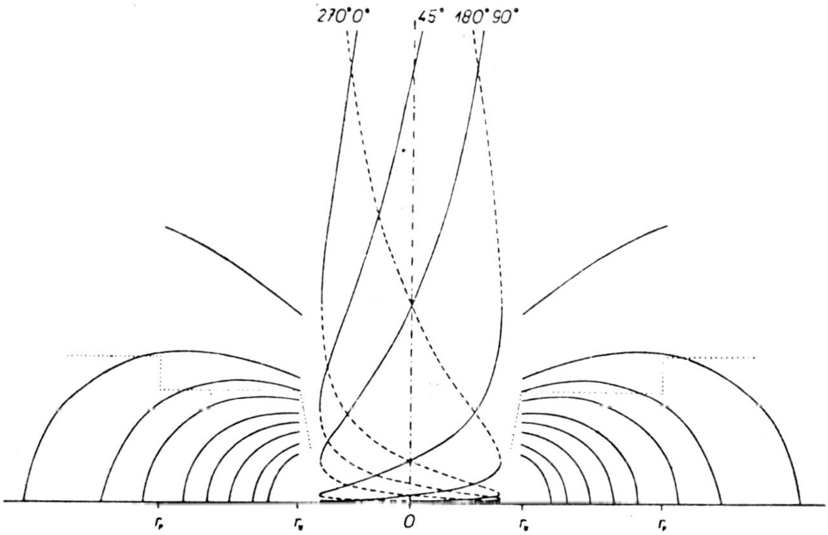

Fig. 7. Schematic layout of the umbral and penumbral system of lines of force in the sunspot. The axis of the spot is given in a dot-and-dash line (O). The height scale is rather arbitrary and considerably enlarged. The probable boundary between the photosphere and chromosphere is given schematically (dotted line). The mutual height of the umbral and penumbral systems is also arbitrary so far. The existing arbitrariness in the coupling of both systems is illustrated by the gap between them.

essentially because it is difficult to find the exact difference of level of the two parts.

The time dependence, and the velocity field have been neglected in the calculations so far. It should be noticed that the term $(1/c)(\partial B/\partial t)$ has the same order of magnitude as $\nabla \times B$.

6. The Need for an Accelerating Mechanism

The need for a mechanism to accelerate particles to energies of the order of a few MeV is to explain several solar phenomena. For example, it is known that solar X-rays are produced by the bremsstrahlung of electrons (Anderson and Winckler) but in order for this to happen it is necessary for these electrons to be highly energetic.

In addition, it has been shown by Anderson and Winckler that there exists a remarkably close association in time between the solar X-rays and the Type-III radio bursts and it is therefore suggested that these radio emissions are directly excited by the fast electrons. These Type-III bursts can be explained by electrons of 20–100 keV causing plasma oscillations within the corona (Wild, Sheridan-Neylan), and Ginzburg and Železniakov suggest that these plasma waves are coupled to electromagnetic waves which are able to propagate away from the region of the oscillating plasma. If the oscillations occur at a sufficiently high level in the solar corona the electromagnetic waves are not appreciably attenuated and may propagate away from the sun.

Solar cosmic rays are also observed on the earth and these particles have energies of a few 100 MeV. An accelerating mechanism is essential for the explanation of these phenomena.

Since the Type-III radio emissions originate at levels of about 10^5 km (Wild, Sheridan and Trent) it is thought that these high energy electrons must also exist at this level. However, if these same electrons are the cause of X-rays then they must exist at a much lower level because the density of nuclei at a height of 10^5 km is too low for X-rays to be produced. It is therefore postulated that these high energy electrons undergo large scale movements in the solar atmosphere and move from the lower chromosphere where there is a sufficient density of nuclei to produce X-rays up to heights of 10^5 km where the Type-III radio bursts are excited.

7. The Mechanisms for Particle Acceleration

The accelerating mechanisms can be classified in two classes: The Fermi mechanism

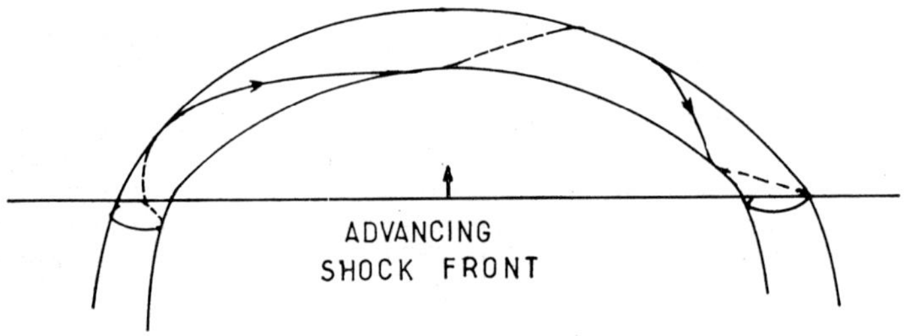

Fig. 8.

and the betatron mechanism. The first mechanism was suggested by Fermi who considered the possibility of accelerating particles in a medium in which the lines of magnetic force are moving and the magnetic moment of the particle is conserved. The particles encounter regions of different magnetic field strength and are either trapped or escape according to their pitch angle. For those particles which are trapped the varying magnetic field is seen with an electric field and the particles are either accelerated or decelerated. The energy change of the particle is therefore proportional to the mean value of $(v/c)^2 B$ where v is the particle velocity, B the magnetic field strength and c the velocity of light. This is a small effect except if the velocity v is already near c (second Fermi process). In case of the sun, this would mean that we have to assume the presence of a turbulent magnetic field at the time of a flare.

In the second mechanism (betatron mechanism) a variation of the magnetic field strength, slow compared to the Larmor period, can give to a particle a considerable acceleration. The ratio of the energies of the particle before and after acceleration is equal to ratio of the magnetic field strengths. In the case of the sun, where the ratio of the magnetic fields is unlikely to exceed a few thousand, the energy increase of the particle is unlikely to exceed this factor. Since the energies required are of the order of 100 MeV for protons, this mechanism would seem insufficient to produce a constant stream of high energy particles.

A mechanism derived from the Fermi mechanism has been considered by Wentzel, who considered the possibility of accelerations being produced by the movement of a hydromagnetic shock front. He assumes that the refraction of the lines of force across the shock front produces a magnetic bottle in which the particles can be trapped. When the shock moves, the particles, being reflected from one bottle neck to the other can be accelerated (Figure 8).

The condition for the trapping of particles in front (or behind) of the shock is:

$$\vartheta > \vartheta_c$$

where ϑ is the pitch angle.

ϑ_c is given by

$$\sin^2 \vartheta_c = \frac{H_0}{H_{sh}}$$

where H_0 is the steady magnetic field strength and H_{sh} is the magnetic field behind the shock front.

For those particles which are trapped in the magnetic bottle, the movement of the shock front causes them to accelerate rapidly until their energies are sufficiently large for them to escape from the bottle before its size decreases to zero. Those particles which are able to cross the advancing shock front will be decelerated by decreasing magnetic field. Wentzel, however, has calculated that this effect of deceleration is small and that even for small shocks in which $(H_{sh}/H_0) < 1.25$ there is a net gain in energy.

However, the main problem, like in Fermi mechanism, remains the problem of the energy of injection.

It is necessary that the particles have large mean-free-paths, so that the gains on the bottle necks can exceed the losses due to collisions. The particles must be injected into the bottle neck with highly super-thermal energies, of perhaps 1 MeV for protons.

Schatzman has considered a mechanism which can be said to derive from both the Fermi and betatron mechanisms. A particle crossing a perpendicular M.H.D. shock can be accelerated, in a way similar to the betatron acceleration. However, in order to have continuous acceleration, the particle has to keep going with the shock front, a result which can derive only from collisions of the particles with irregularities of the magnetic field. At each instant, a particle would have a certain probability of being lost or of continuing to move with the shock front.

The number of particles that are kept near the shock front in this way decreases inversely with time, thus a large number of particles could be kept for a short time and accelerated, whilst only a few would be retained for a considerably longer period and accelerated to much higher velocities. This mechanism would therefore seem able to accelerate a small proportion of the particles to very high velocities. For example, even in the case of a very weak shock in which:

$$\eta = \frac{B_2 - B_1}{B_1} = 10^{-3}$$

where B_2 is the magnetic field strength behind the shock front and B_1 the magnetic field strength in front of the shock and in which $B_1 = 50$ gauss, protons can be accelerated to several hundred MeV, which is the sort of energy required.

8. Theories of Solar Flares

It has been suggested that solar flares are generated near a place of zero magnetic field. This suggestion seemed reasonable from observational evidence of flares in

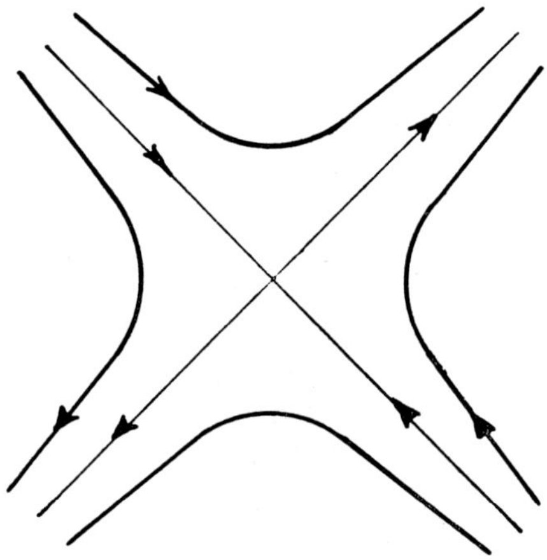

Fig. 9.

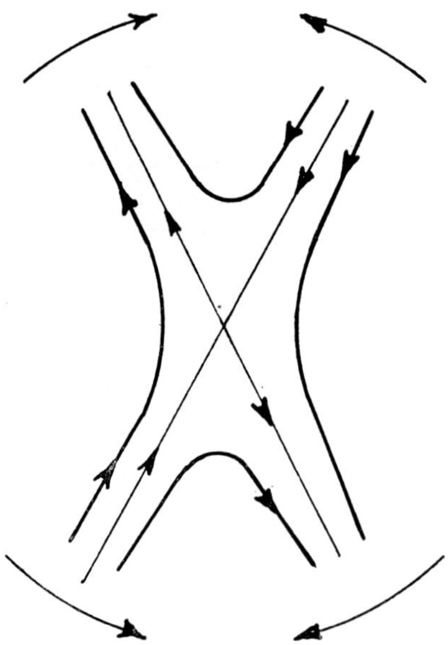

Fig. 10.

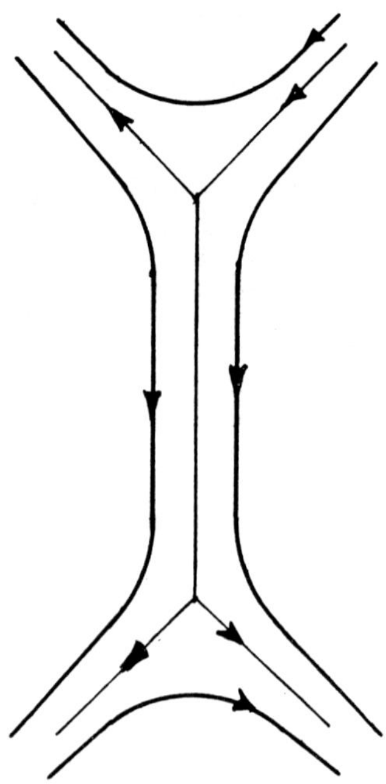

Fig. 11.

which it is always seen that a flare takes place in between two sections of opposite polarity. It was Severny who first suggested that solar flares occur at the neutral points of the magnetic field but although it has been observed that the longitudinal field is zero, Michard, Mouradian and Semel have shown that the field is still present, but is horizontal and parallel to the solar surface. Deubner has given two examples where flares developed in a region in which the magnetic field is horizontal.

Krivsky has considered which type of flares produce cosmic ray particles and concludes that all have a "Y" shape.

Gopasyuk has observed that after a solar flare the hills of the magnetic field surrounding the flare move towards the position where the flare took place. Since the flare must take its energy from the energy of the magnetic field, the matter surrounding the neutral point will collapse and the hills of the magnetic field will be drawn towards the original neutral point as has been observed by Gopasyuk. When this happens we have two magnetic fields approaching each other and shock waves may be produced and particles accelerated.

Around a neutral point of the magnetic field, the field lines may possess many different configurations. Dungey has considered the situation in which the neutral point is surrounded by lines of force which are hyperbolic in shape. (Figure 9). Dungey considers the stability of this pattern by investigating the currents produced in the medium and its stability.

Dungey concludes that the singular lines of force will be drawn together due to the production of a further magnetic field from the currents induced in the medium (Figure 10).

Sweet has also considered this problem and investigated the analogous situation to that of Dungey with the configuration given in (Figure 11). Sweet investigates the possibility of this system being unstable by energy considerations.

Sweet concludes that if the magnetic pressure is smaller than the gas pressure, stability is maintained and that the Dungey process can take place if the magnetic energy is large enough. In the Dungey process the lines of force approach each other and the particles can therefore be accelerated; the variable magnetic fields necessary for large accelerations can be easily produced near sunspots.

9. Solar Cosmic Rays

The total number of cosmic ray particles produced in a flare is believed to be about 10^{34} to 10^{36} and since the region is not likely to be greater than 10^{18} cm^2 in area there must be about 10^{16} 10^{18} cosmic ray particles per square centimeter.

Since the total number of particles above the photosphere is of order 10^{24} there must be about 1 particle in 10^6 which is accelerated to an energy of about 100 MeV. This fraction of high energy particles is extremely large. It is possible that a shock wave will be produced in the lower chromosphere and that the trapped particles are accelerated and swept out into a region from which they can escape.

Cosmic rays are observed from the earth after a short delay and so it seems that these particles must be temporarily trapped in the magnetic field of the flare. However, since the lines of magnetic force near a flare are likely to be highly curved, (Fig. 12a) there exists a centrifugal force acting on the trapped particles which would cause them to drift away from the center of the flare and cross the lines of magnetic force. They would then enter regions from which they would be able to escape and to travel towards the earth (Fig. 12b).

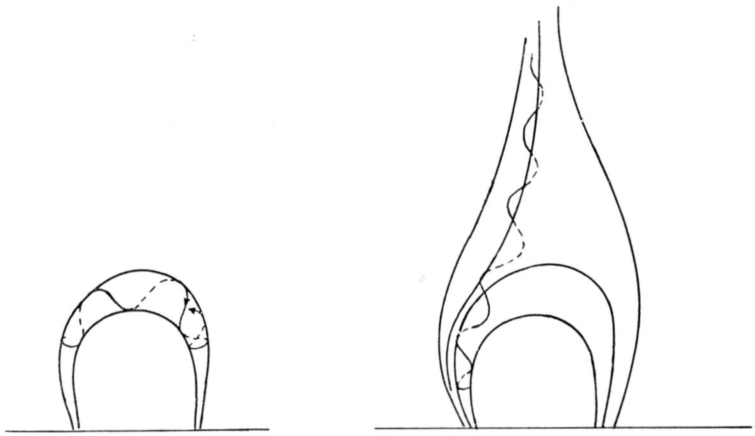

Fig. 12.

This region in which the cosmic ray particles are trapped has been considered by Stepanjan and Vladimirski and they suggest that a region of 10^{31} cm^3 is required for trapping, with a drift velocity of the order of

$$\frac{|\nabla H|\, \beta^2 c^2}{H\omega}$$

they obtain a trapping time of about 10^3 to 10^4 sec which seems to be in reasonable agreement with the observations of cosmic rays since they are observed about 10 minutes after the start of the flare.

References

Solar Magnetic Field

Kippenhahn, R.: 1963, *Astrophys. J.,* **137,** 664.

Magnetic Field (Decay)

Cowling, T. G.: 1945, *Mon. Not. Roy. Astron. Soc.* **105,** 166.
Backus, G.: 1957, *Astrophys. J.* **125,** 500.

Dynamo

Parker, E. N.: 1955, *Astrophys. J.* **122,** 293.

Generation of Sunspots

Cowling, T. G.: 1946, *Mon. Not. Roy. Astron. Soc.* **106,** 218.
Parker, E. N.: 1955, *Astrophys. J.* **121,** 491.

Solar Cycle

ALLEN, C. W.: 1960, *Observatory* **80**, 94.

Equilibrium and Sunspots

SCHLUTER, A. and TEMESVARY, S.: 1957, *Trans. I.A.U. Symposium No. 6.*

Shock Waves

ANDERSON, J.E.: 1963, *Magnetohydrodynamic Shock Waves*, M.I.T. Press, Cambridge, Mass., U.S.A.

PART II

INTERPLANETARY MEDIUM

CHAPTER 11

INTERPLANETARY MEDIUM

H. ELSÄSSER
Landessternwarte, Königsstuhl–Heidelberg

In this lecture, problems and results on interplanetary matter will be discussed. As we know today, the space between the planets is not empty. It contains diffuse matter in the form of ionized gas – the interplanetary plasma – and in the form of small particles – the so-called interplanetary dust or smoke. The questions concerning the interplanetary plasma will be discussed in Dr. Lüst's lecture whereas my topic will be mainly the interplanetary dust particles. But we shall see that it is impossible to separate distinctly these two aspects of the interplanetary medium from each other; therefore we cannot avoid touching on the problems of gas, but from a quite different point of view from that which Dr. Lüst will discuss.

1. Main Sources of Information on Interplanetary Medium

(I) The *solar corona* ("white corona") and the *zodiacal light* are due to scattering of sunlight in interplanetary space. By measuring their brightness distribution, polarization and spectral features, information concerning the composition of the interplanetary medium, its spatial density, the sizes and the nature of the dust particles can be obtained and models of the interplanetary medium can be derived. In this way we get information on the conditions in those regions of the planetary system near and within the earth's orbit.

(II) A *meteor* appears when an interplanetary particle enters the earth's atmosphere. Optical and radar observations of the meteor trails give the influx rate and thus again information on the density. These observations apply to particles near the earth's orbit and with diameters mainly in the range $a > 10^{-2}$ cm, whereas the solar corona and the zodiacal light inform us on particles with $a < 10^{-2}$ cm.

(III) *Direct measurement of interplanetary dust from space crafts* is a new and important source of information. We have in principle two kinds of measurements: (a) Flux measurements with microphone systems and similar devices which can give densities and the particles' size-distribution, and (b) collection of dust by means of recoverable sounding rockets which gives us the possibility to see and to investigate the interplanetary dust in its original form. Comparison of the results of the indirect methods (I and II) and the direct method (III) is of great interest.

2. Solar Corona and Zodiacal Light

A. SOLAR CORONA

In connection with problems of the interplanetary medium we are interested in the outer parts of the corona which can be observed during a solar eclipse as the "white corona".

(a) *Brightness distribution.* The distribution of surface brightness in the solar corona has been studied at many eclipses and is well known up to about $\varrho=5$ solar radii distance from the sun. The mean photographic brightness averaged over all position angles was derived by Baumbach and expressed by the formula

$$I_c(\varrho) = \frac{2.56}{\varrho^{17}} + \frac{1.42}{\varrho^7} + \frac{0.053}{\varrho^{2.5}} \quad (1.2 < \varrho < 5) \tag{1}$$

The intensity-unit is 10^{-6} times the brightness at the center of the solar disk, ϱ is measured in units of the solar radius from the center of the sun. The mean brightness varies during the solar activity cycle by about $\pm 20\%$. The enormous contrast in brightness between the photosphere and the corona demonstrates the difficulties of corona observations.

Measurements from the earth's surface beyond $\varrho=5$ are severely hampered by the intense sky light due to scattering in the earth's atmosphere. This is much better higher up in the atmosphere. Some years ago Rense *et al.* and Blackwell made successful observations of the outer corona during eclipses from high flying aircrafts. Blackwell was able to measure the brightness as far as $13°.5(\varrho=50)$ from the sun.

(b) *Spectrum.* Grotrian has demonstrated that the spectrum of the outer corona consists of two components. The coronal spectrum contains the same absorption lines as the photospheric spectrum but the line depths are reduced in the inner corona and increase with increasing ϱ. If we denote by t_c and t_{ph} the line depths of coronal and photospheric lines, Grotrian derived the following numbers

$$\begin{aligned} t_c/t_{ph} &= 0.05 \quad \text{at} \quad \varrho = 1.1 \\ &= 0.73 \quad \text{at} \quad = 2.6 \\ &= 1.0 \quad \text{at} \quad > 4\ . \end{aligned}$$

At $\varrho>4$ the coronal and photospheric spectra are identical. This phenomenon was explained by Grotrian by the assumption of two components: (1) *K-corona* ("kontinuierliche" corona): Shows complete continuum without any lines. Arises from scattering of sun light by free electrons. Mean thermal velocity of electrons in the corona about 5000 km/sec ($T=10^6$ °), the resulting Doppler broadening obliterates absorption lines completely. (2) *F-corona* (Fraunhofer corona): Shows Fraunhofer lines in full depth. Due to scattering by interplanetary dust particles far away from the sun. Spurious corona which is projected on the real (solar) electron corona.

Since I_F/I_K (I_F intensity of F-corona, I_K of K-corona) is increasing with increasing ϱ, t_c becomes larger in the outer corona. Since

$$I_c = I_F + I_K \quad \text{and} \quad I_F = t_c/t_{ph} \times I_c$$

I_F and I_K can be obtained if t_c/t_{ph} is known. In this way the brightness of the F-corona has been derived by Van de Hulst as

$$I_F(\varrho) = \frac{0.30}{\varrho^7} + \frac{0.053}{\varrho^{2.5}} \tag{2}$$

(compare with (1)).

(c) *Polarization*: Another possibility to separate F- and K-corona are polarization-

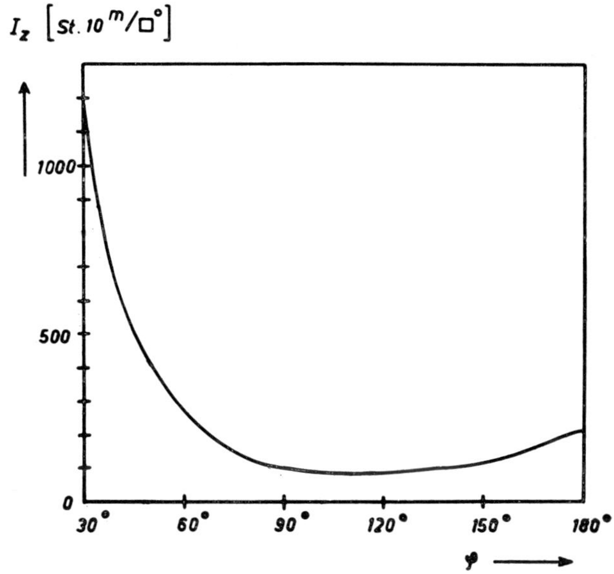

Fig. 1. Brightness I_z of the zodiacal light (visual spectral region).

measurements. Light scattered by electrons show linear polarization (Thomson scattering). The polarization of light scattered by dust particles at small scattering angles as in the case of the F-corona is expected to be very small.

The polarization p in the corona has a maximum near $\varrho = 1.5$ and is then decreasing with increasing distance from the sun's limb. The electric vector of the coronal light vibrates preferentially parallel to the solar limb. The background sky again affects the measurements from the ground in the outer corona but Blackwell has made observations from an aircraft up to $\varrho = 20$.

B. ZODIACAL LIGHT

The zodiacal light is a phenomenon only visible in the night sky. After twilight has disappeared and before dawn the diffuse luminous cone of the zodiacal light becomes visible. The axis of the cone coincides with the ecliptic. The brightness of the zodiacal light decreases with increasing distance from the sun, near the horizon at the end of twilight it is brighter than the brightest parts of the milky way.

(a) *Photometry of the zodiacal light.* Since the zodiacal light is a phenomenon which is best observed close to the horizon at low altitudes it is difficult but important to remove the effects arising from extinction and atmospheric scattering.

Another problem is to separate the different components which contribute to the observed brightness: the zodiacal light, the light from background stars and the air-glow light of ionospheric origin.

Figure 1 shows the variation of the zodiacal light brightness I_z with increasing elongation φ. The unit of I_z is 1 Star $10^m/\square^\circ$

Near the anti-solar point ($\varphi=180°$) a brightening of the zodiacal band, the "Gegenschein", is recorded. The brightness distribution in the Gegenschein is shown in Figure 2.

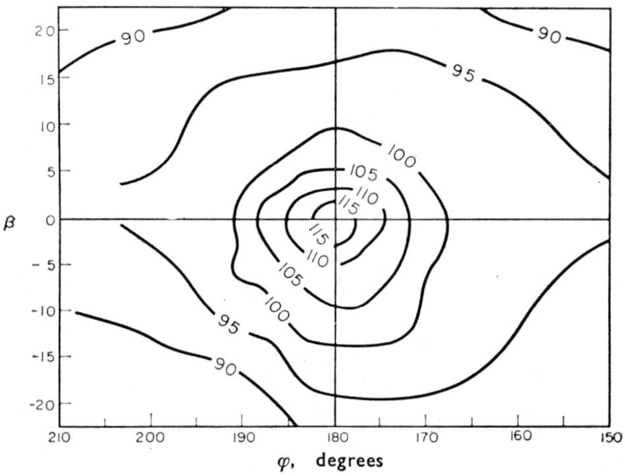

Fig. 2. Isophotal map of the Gegenschein (blue spectral region), φ elongation, β ecliptical latitude.

(b) The *polarization* of the zodiacal light was detected by Dufay (1925) and the first measurements with modern equipment were produced in 1953 by Behr and Siedentopf. At $\varphi=30°$ p is found in the order of 20% and is increasing according to more recent observations by Elsässer and Blackwell and Ingham with increasing φ in the range 30°–90° (compare Figure 4). The plane of polarization is located in the ecliptic (electric vector perpendicular to the ecliptic).

(c) The *spectrum*: Since the intensities of the zodiacal light are very low, it is difficult to obtain information on the spectral characteristics by ordinary spectroscopy. By colour-measurements, i.e. the comparison of the intensities in different broad spectral regions, it is found that the colour of the zodiacal light is closely similar to that of the solar light. Spectra have been obtained by Hoffmeister and Blackwell and Ingham. They came to the conclusion that the spectrum of the zodiacal light is nearly identical with the photospheric spectrum. But one must bear in mind that observations of this kind are again strongly affected by the contributions of the airglow and the stellar background sky.

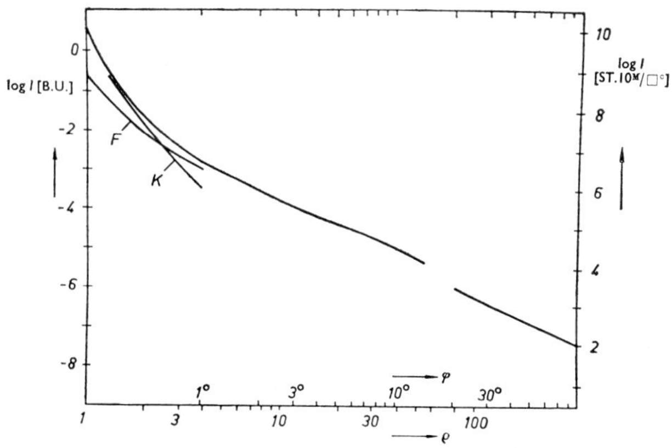

Fig. 3. Brightness of the corona and the zodiacal light. F = F-corona, K = K-corona. (B. U.) Baumbach Units = 10^{-6} times the surface brightness at the centre of the sun's disk.

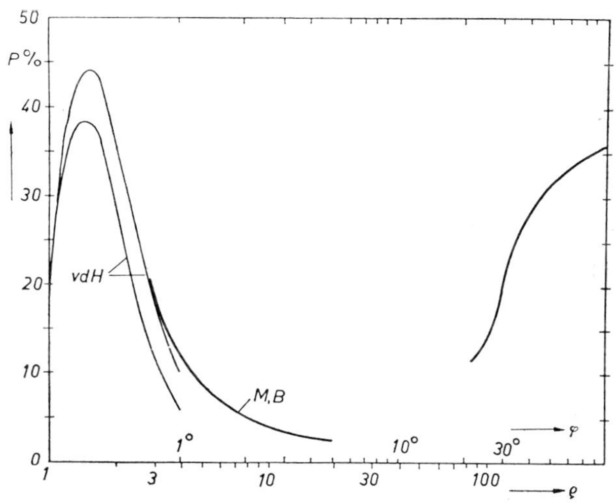

Fig. 4. Polarization p of the corona and the zodiacal light vdH = Van de Hulst's curves for maximum and minimum corona. M, B = Michard's *et al.* and Blackwell's results.

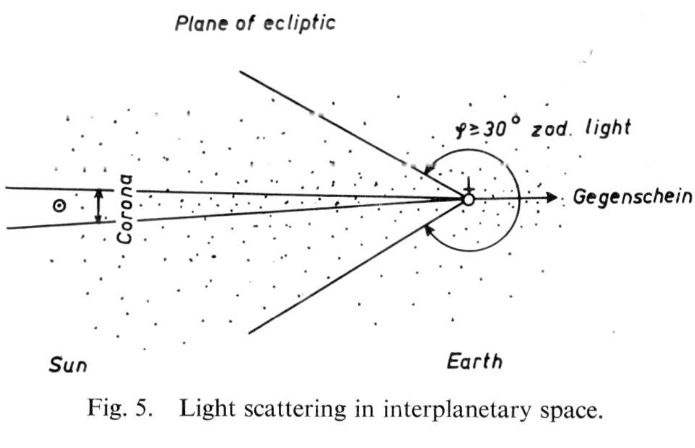

Fig. 5. Light scattering in interplanetary space.

C. CONNECTION BETWEEN ZODIACAL LIGHT AND CORONA

In 1934 it was pointed out by Grotrian that the corona and the zodiacal light are connected to each other by one continuous brightness curve. At that time the gap between the corona and the zodiacal light not accessible to observations extended from 1° to 30° elongation. Nowadays we have the results from high flying aeroplanes (Blackwell *et al.*). Figure 3 shows the brightness distribution in the corona and in the zodiacal light region together in one diagram.

A smooth connection between the corona and the zodiacal light is also indicated by the polarization data (Figure 4). There seems to be a marked minimum of p near $\varphi = 5°$.

These results made it obvious that the zodiacal light at elongations $\varphi \geqslant 30°$ is only one source of information on the interplanetary medium; one has also to take into account the observations in the outer corona which is simply the zodiacal light at small elongations. Figure 5 demonstrates that the corona, especially the F-corona, and the zodiacal light are identical phenomena seen from different aspects.

3. Particles in Interplanetary Space. Interpretation of Corona and Zodiacal Light Data

Up to now two types of models of interplanetary material have been discussed. The first is based on the assumption that the zodiacal light is caused exclusively by scattering of sunlight by dust particles. After the detection of considerable polarization a second type was examined involving the assumption that, in addition to the dust particles, interplanetary electrons contribute to the zodiacal light. The opinion was that high polarization could not be explained by solid particles alone. A contribution by electrons would indicate a real solar component in the zodiacal light, therefore this idea found wide interest and was discussed extensively during recent years.

A. LIGHT SCATTERING BY SMALL PARTICLES

(a) *Size of dust particles.* Only when the radiation pressure P, exerted by the solar light on a particle, is smaller than the gravitational attraction S by the sun will the particle stay in the planetary system. P/S depends mainly on the particle size. The cross-section for light pressure follows from the theory of Mie for small spheres. The critical radius is of the order 2×10^{-5} cm for metallic and 1×10^{-4} cm for dielectric particles, for smaller sizes is $P/S > 1$. In the case of dielectric spheres P/S becomes again < 1 for very small radii a, namely $a < 4 \times 10^{-7}$ cm.

From the similarity in colour of the zodiacal light and the sun it follows too that the particles mainly have radii above $a \simeq 10^{-4}$ cm. If $a < \lambda$ (λ wavelength of scattered light) the scattered intensity depends strongly on λ and then the zodiacal light should be bluer than the sun. This is not observed. Here we find the same order of magnitude of a as above.

We know of no upper limit for a. It seems reasonable to assume that the meteorites and asteroids with diameters in the order of meters and kilometers represent the upper end of the size distribution function.

(b) *Scattering functions.* In order to obtain the number $n(r)$ of particles per cm^3 at distance r from the sun we must know the intensity of light scattered by a single particle as function of the particle size a and the scattering angle ϑ.

The geometry of light scattering in interplanetary space is shown by Figure 6: Δ distance particle P – earth E, r distance sun $S - P$. φ elongation, ϑ scattering angle,

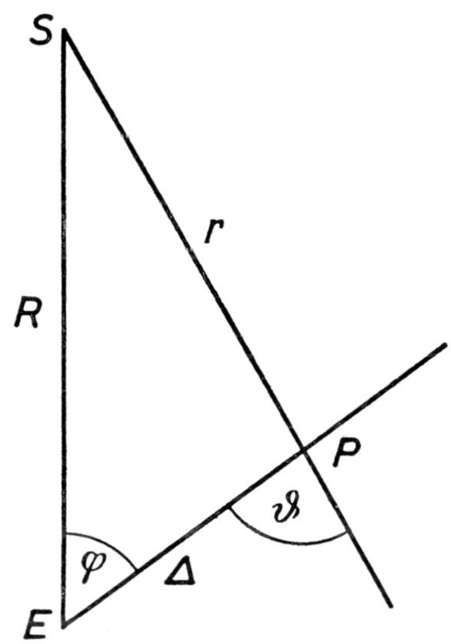

Fig. 6. Geometry of scattering.

$R = 1$ AU. The minimum distance of the line of sight from the sun is $r_{\min} = R \sin \varphi$. Information on interplanetary matter can therefore be obtained from zodiacal light observations ($\varphi > 30°$) only for $\geqslant 0.5$ AU. At smaller distances from the sun the data of the outer corona have to be used.

Let I_0 denote the intensity of the incident unpolarised wave and $I(\vartheta)$ the intensity of the scattered radiation at a distance Δ from P. The scattering function $S(\vartheta)$ is defined by

$$I_k(\vartheta) = \frac{I_0 \cdot S_k(\vartheta)}{\Delta^2} \qquad (k = 1, 2) \; . \tag{3}$$

The index 1 is valid for the component with the electric vector perpendicular, index 2 for the vector parallel to the plane of vision. S_1, S_2 depend on the material, size, form and orientation of the scattering body, on the wavelength and on ϑ. For spherical particles of homogeneous material the scattering functions can be computed according to the Mie-theory. During the last years results of extensive machine computations for

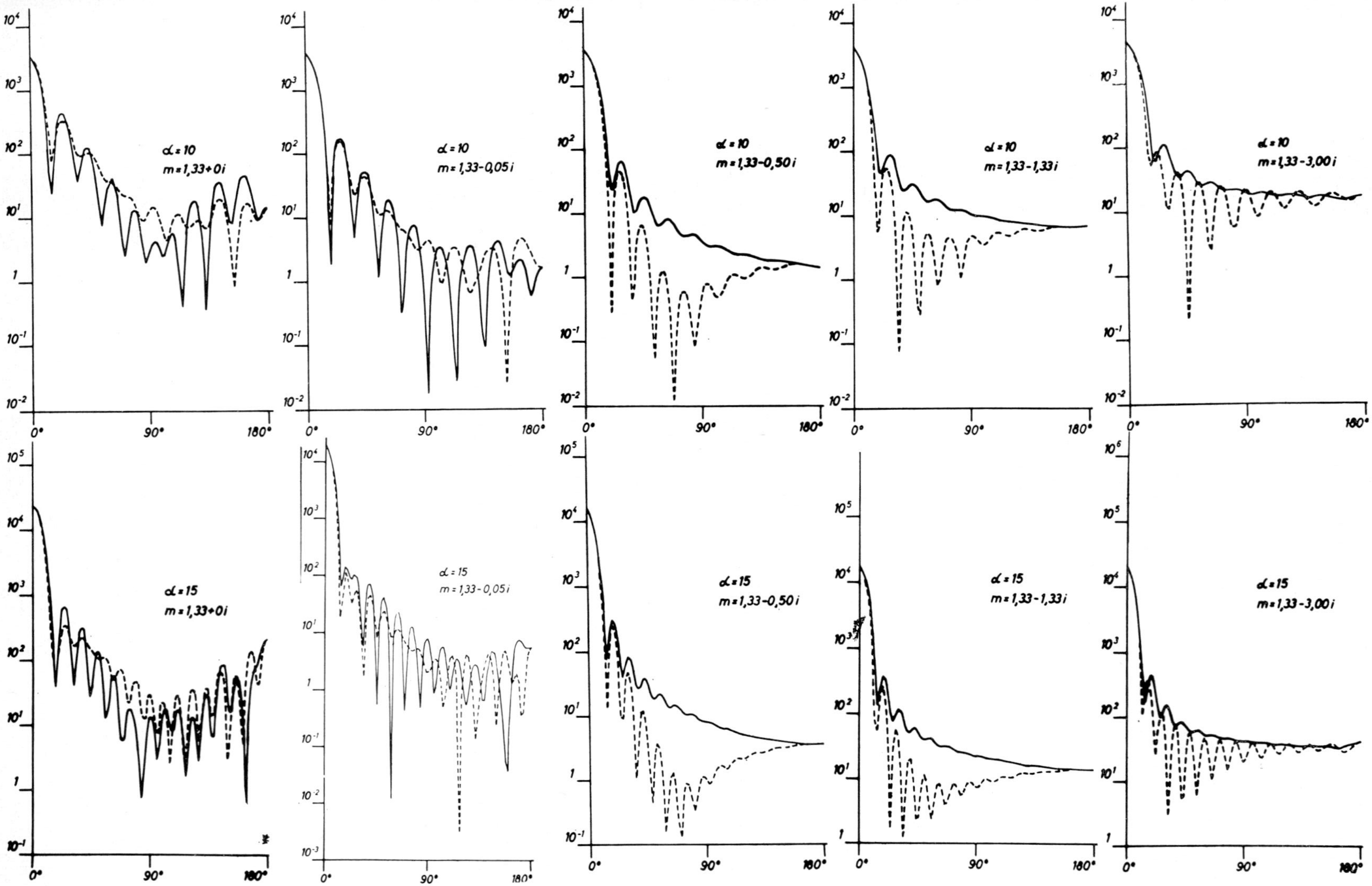

Fig. 7. Scattering functions of spherical particles according to calculations by Giese. Abscissae: Scattering angle ϑ. Ordinates: Intensities $i_k = (4\pi^2/\lambda^2)\, S_k\ (k = 1, 2)$; i_2 dotted, m refractive index.

different materials and sizes have been published (see list of references). Typical examples of scattering functions derived in this way are presented in Figure 7.

Mie's theory applies only to spheres. No analogous theory is available for nonspherical particles apart from a few very special cases. It is therefore of great interest to measure directly scattering functions in the laboratory. Experiments of this kind with samples of meteoritic dust have been performed by Richter during the last years. In another type of experiment, instead of small particles and visible light a single test body and microwaves are used so that the ratio a/λ is of the same order. But up to now no large differences in $S(\vartheta)$ for spherical and nonspherical particles have been observed.

For first order approximations the scattering functions of larger particles can be represented by

$$S(\vartheta) = S_d(\vartheta) + S_r(\vartheta) \tag{4}$$

$S_d(\vartheta)$ describes the scattering due to diffraction and is independent of the refractive index whereas $S_r(\vartheta)$ is due to reflection and refraction of the incident light. The former predominates at small scattering angles and consequently at small elongations; it is unpolarized. The width of the diffraction pattern depends on the size of the particle. For a sphere with radius $a \simeq 10^{-4}$ cm the halfwidth of this pattern is of the order of 10°, for $a \simeq 10^{-3}$ cm it is about 1°. As Allen and Van de Hulst have shown, diffraction by interplanetary particles gives rise to the phenomenon of the F-corona and the steep decline in brightness between the outer corona and the zodiacal light.

At large scattering angles and thus at the elongations of the zodiacal light, the light scattered by reflection and refraction is essential. Often a simple form is used, for instance (written as the sum of components 1 and 2 of polarization)

$$S_r(\vartheta) = \frac{a^2}{4} A \tag{5}$$

which is the function for an isotropic scatterer with albedo $A(S_1 = S_2)$. That there is some physical justification for this simplification is shown by the laboratory experiments and the calculations according to the Mie-theory.

The functions for light scattering by free electrons are given by

$$S_1(\vartheta) = \tfrac{1}{2}\left(\frac{e^2}{mc^2}\right)^2, \qquad S_2(\vartheta) = \tfrac{1}{2}\left(\frac{e^2}{mc^2}\right)^2 \cos^2\vartheta \tag{6}$$

(e charge, m mass of the electron, c velocity of light). The scattered light is unpolarized for $\vartheta = 0°$ and 180°, otherwise the plane of vibration lies perpendicular to the plane of vision.

B. DENSITY OF INTERPLANETARY MEDIUM

(a) *Equation connecting brightness distribution and number of particles.* If we denote by $n_d(r, a)$ the number of dust particles per cm^3 with radius a at P (Figure 6) and by $n_e(r)$ the corresponding number of electrons the surface brightness $I_{1,2}(\varphi)$ of the scattered sun light at elongation φ follows by integration over the line of sight as

$$I_k(\varphi) = LR^2 \left\{ \int_0^\infty \frac{1}{r^2} \int_{a_0}^{a_1} n_d(r, a)\, S_k^d(a, \vartheta)\, \mathrm{d}a\, \mathrm{d}\Delta + \int_0^\infty \frac{n_e(r)}{r^2} S_k^e(\vartheta)\, \mathrm{d}\Delta \right\} \qquad (k = 1, 2) \tag{7}$$

L is the luminosity of the sun. It is taken into account that the particle size may have values between a_0 and a_1. S^d and S^e are the scattering functions for dust particles and electrons respectively. It seems plausible to assume, as we have done, that the interplanetary medium is distributed symmetrically to an axis which passes through the sun and lies perpendicular to the plane of the ecliptic. – The observational results mostly are represented by the total brightness $I(\varphi)$ and the polarisation p, it is

$$I = I_1 + I_2\,, \qquad p = I_1 - I_2/I_1 + I_2\,. \tag{8}$$

As was mentioned already one has to consider the functions $I_1(\varphi)$ and $I_2(\varphi)$ in the whole range of elongations for which photometric observations are available. This is of importance because the brightness distribution in the corona region ($\varphi < 10°$) depends strongly on parameters which are inessential in the range of the "classical" zodiacal light ($\varphi \geqslant 30°$).

(b) *Analysis of F-corona.* As we have seen already, the steep increase in the brightness of the F-corona towards the sun is due to diffracted sunlight. The gradient depends strongly on the width of the diffraction pattern and thus on the size of the particles. A comparison of the observed brightness as a function of elongation with the dependence of the diffracted light on the scattering angle ϑ makes obvious that the former can be produced only on the assumption of a mixture of particle sizes with a mean particle diameter near

$$a = 10^{-3}\,\mathrm{cm}.$$

This is an important result because in older investigations much larger sizes have been assumed. The exponent β of an assumed particle distribution

$$n(a) = n(a_0)(a/a_0)^{-\beta} \tag{9}$$

follows in the range of 2.0–3.0 on the assumption of $a_0 \simeq 10^{-4}$ cm. So from an analysis of the F-corona it follows conclusively that the interplanetary dust consists mainly of small particles, their frequency diminishing with increasing size.

The particle number per cm^3 corresponding to the observed brightness of the dust corona is deduced from Equation (7) as

$$n_d \simeq 1 \times 10^{-14}\,\text{particles/cm}^3 \qquad (a_0 = 10^{-4}\,\mathrm{cm}, a_1 = 10^{-1}\,\mathrm{cm}) \tag{10}$$

and the density as

$$d_d \simeq 10^{-20}\ \mathrm{g/cm}^3$$

assuming a specific weight $s = 7.5$ g/cm^3. The quoted values should be characteristic for distances $r \simeq 1$ AU, since the main contribution to the F-corona comes from particles seen under small scattering angles in the immediate neighborhood of the earth. The variation of n_d with r cannot be derived in this way since the range of φ in question is not wide enough.

(c) *Analysis of zodiacal light*. The main difficulty here arises from the fact that $S_r(\vartheta)$ depends on the material, i.e. on the refractive index of the particles, which is unknown. The older investigations mostly worked with simple forms of the scattering function, for instance the one of Equation (5). Then from the observed gradient in brightness a flat increase of $n_d(r)$ towards the sun follows. An analysis according to Equation (7) actually means the determination of $n_d(r)$ multiplied by the unknown albedo A. We already had information on n_d from the dust corona, so a comparison of the two results allows a rough estimate of A and consequently a statement on the nature of the particles. The albedo A derived in this way is of the order of

$$A = 0.01 .$$

According to measurements on meteoritic material one would expect rather an albedo of 0.10, but the calculated value may be fairly uncertain. The result indicates that the interplanetary dust probably consists mainly of absorbing particles.

After the first reliable polarization measurements in the zodiacal light have been made, models of the interplanetary medium consisting of a mixture of dust particles and electrons were discussed on the assumption the contribution by the dust to the observed polarization would be zero. By solving Equation (7) then electron densities of several hundred electrons per cm^3 at $r = 1$ AU have been derived. The contribution by these electrons to the brightness I_z of the zodiacal light would be less than 50% at $\varphi = 30°$ and decreases with increasing φ. So the quoted conclusions, concerning the interplanetary dust, are practically not affected by the hypothesis on interplanetary electrons.

During the past few years an extensive investigation of possible models of interplanetary matter satisfying the observations of the zodiacal light has been undertaken by Giese and Siedentopf. They made use of exact scattering functions according to the Mie-theory. Their results clearly show the possibilities of causing polarization by spherical particles. Their main conclusions concerning the polarization of the zodiacal light are:

(1) If one assumes the interplanetary matter to be consisted of dielectric dust only, the calculated polarization of the zodiacal light is much less than the observed one, for some models even a negative polarization is to be expected.

(2) Assuming metallic particles at $\varphi - 40°$ the calculated polarization p goes as high as 50%, but with increasing elongation p decreases.

(3) Only on the assumption of additional interplanetary electrons the calculated results could be brought into agreement with the observations. But then a variety of combinations dust + electrons are possible solutions. Giese and Siedentopf in this way found as a minimum electron density near the earth's orbit $n_e = 300$ electrons per cm^3.

In a paper published a few months ago Giese discussed one more different model, assuming as the main constituent very small Rayleigh-like scattering particles. In this case for the explanation of the polarization of the zodiacal light, electrons are unnecessary. But models of this kind are not realistic since they cannot reproduce the F-corona. Furthermore the color of the zodiacal light deviates strongly from the sun's color for a model of this type. There must be a fair amount of particles with diameters near 10^{-3} cm in interplanetary space. In addition there might be submicron particles present responsible for the observed polarization. Models of this kind have not been discussed yet.*

These results show that unique solutions for the interplanetary density and its variation with the distance from the sun are not yet available.

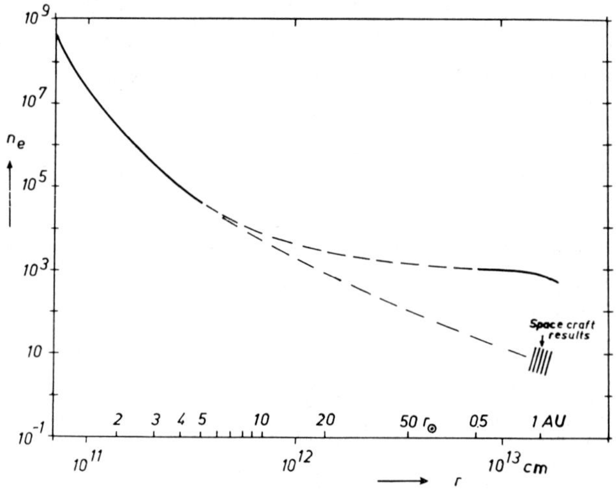

Fig. 8. Electron densities in interplanetary space.

(d) *Interpretation of the Gegenschein.* The most simple explanation is the following: The scattering functions of dielectric spheres show an increase at $\vartheta \simeq 180°$ which depends on the particle size (Figure 7). If a relatively small part of the interplanetary dust (about 10%) is dielectric, the Gegenschein phenomenon is to be expected, as Walter has shown. The analysis along this line is very similar to the analysis of the F-corona. Walter has derived spatial densities and a size distribution of the particles in excellent agreement with the results deduced from the dust corona.

(e) *Interplanetary gas.* The correct number of interspatial free electrons per cm^3 near $r = 1$ AU (= number of protons) would be of great interest because there are certainly intimate connections between the interplanetary gas and the solar corona. The concentration of free electrons in the corona follows from an analysis of the K-corona as about 10^8 electrons/cm^3 in the lower corona. The variation with increasing distance from the sun is shown in Figure 8. Electron densities of a few hundred per cm^3 at $r = 1$ AU would be very well in line with the concentration in the corona and its outward decrease. But this is also true for the smaller gas densities measured by space crafts (Figure 8).

* Cf. Herlofson's lecture. See footnote on page V.

An unambiguous analysis of the corona and the zodiacal light at elongations $\varphi = 10° - 30°$ would be very helpful in this respect, but at present we know too little about the properties of interplanetary dust, and therefore we meet the same difficulties here as in the range $\varphi \geqslant 30°$.

Considering the fact that the methods so far discussed gave only unsatisfying results concerning the interplanetary gas, one may look for other possibilities which are free of doubtful hypotheses. One approach to obtain more direct information by observations from the earth's surface could be a careful analysis of the spectrum of the zodiacal light, in the same way as the separation of the corona-spectrum in the electron and dust components was performed. On account of their random velocities, which are determined by the kinetic temperature of the gas, the interplanetary electrons scatter a spectrum in which the Fraunhofer lines have been almost obliterated, whereas the interplanetary dust scatters the solar spectrum almost unchanged. Therefore a reduction of the line depths in the composed spectrum of the zodiacal light is to be expected. The effect depends on the one hand on the contribution by the electron component to the observed brightness and consequently on the electron density, and on the other hand the temperature of the gas is important.

Should a few hundred electrons near $r = 1$ AU be present, a detectable reduction of the line depths in the zodiacal light spectrum is to be expected, as calculations by Schmidt and Elsässer have shown. But the accuracy of the spectroscopic observations performed so far is not high enough to allow definite conclusions on interplanetary gas densities. An upper limit of the electron concentration at $r = 1$ AU compatible with spectroscopic observations by Blackwell and Ingham amounts to $n_e \leq 400$ electrons/cm^3.

4. Meteors and Interplanetary Densities

A meteor appears in the sky when an interplanetary particle, a so-called meteorite, enters the earth's atmosphere. Those particles are revolving around the sun in different directions, and their geocentric velocities are not larger than 70 km/sec near the earth's orbit. Heating of the meteorite through collisions with the air molecules results in an ablation of atoms from its surface and gives rise to the luminuous trail.

By observation with the naked eye one is able to detect meteors brighter than 5^m. On a clear moonless night a few sporadic meteors up to about 60 per hour during a meteor shower are recorded in this way. By telescopic and radar observations meteors of magnitude 10 and fainter can be detected. Using radar techniques the ionized trail of a meteor serves as reflector.

To derive fluxes and densities is a difficult problem because the observed meteor brightness depends on different, not well known, parameters such as the size or mass of the meteorite, its composition, its velocity, the angle of incidence, etc. In addition the recorded number of meteors might not be free from considerable uncertainties; so Millman estimates that his team of six observers would record all meteors of 0^m but only 0.5% of 5^m.

The following figures are rough estimates of particle sizes for meteors of different brightness

Brightness	$-1^m.5$	5^m	10^m
Size	0.3 cm	0.03 cm	0.01 cm

They show that meteor observations inform us mainly on interplanetary particles with sizes above $a=10^{-2}$ cm.

According to Watson and Van de Hulst the particle density above the earth's atmosphere derived from meteor observations is given by

$$n(a) = 10^{-28} a^{-5} \text{ (particles/cm}^3\text{)} \tag{11}$$

The exponent is quite different from what we found in the case of the F-corona (smaller sizes). This indicates that the size distribution function really does not follow an exponential law $a^{-\beta}$ with β=const, β rather seems to be increasing with increasing a. (This effect is already indicated by the F-corona itself, assuming β in the range of 2–3 one has to cut the distribution function near 10^{-2}–10^{-1} cm). The high value of β follows also from radar observations of meteors. On the other hand the frequency of meteorites with $a=10^{-2}$ cm is lower by several orders of magnitude according to this formula than follows from an analysis of the F-corona. But the error of the factor 10^{-28} might be considerable (compare the paper of Kaiser).

From the data on interplanetary dust as they are deduced from an interpretation of the F-corona, the zodiacal light and the meteor observations, one comes to the conclusion that several thousand tons of interplanetary matter per day are falling on to the earth.

5. Space Craft Results on Interplanetary Dust

A. MEASUREMENTS BY MEANS OF MICROPHONE SYSTEMS

Microphone systems for recording impacts of interplanetary particles have been flown during the last years in many space crafts. The most extensive measurements have been performed by Explorer VIII; the results from other space crafts agree well with the Explorer VIII data.

As experiments in the laboratory have shown a microphone system is sensitive to the momentum of the impacting particle. The recorded pulse heights can therefore be expressed in terms of the particle mass if a proper average speed for the particle relative to the space craft is assumed (20–30 km/sec).

The sensitive area of the microphone system on Explorer VIII was 7×10^{-2} m^2

TABLE I

Sensitivity	*Impacts*	*Influx rate (particles/m²/sec)*
10^{-9} g–10^{-8} g	$\simeq$ 3650	3.0×10^{-2}
10^{-8} g–10^{-7} g	$\simeq$ 75	6.2×10^{-4}
$>10^{-7}$ g	1 or 2	$\simeq 1\times10^{-5}$

big, its mass sensitivity was in the range of 10^{-7}–10^{-9}grams (this means it was sensitive to particle sizes in the range of about 10^{-2}–10^{-4} cm). Table I gives the total number of impacts (second column) recorded during an exposure time of about one month in three different ranges of sensitivity (first column). The resulting cumulative influx rates (third column) follow from the number of impacts after applying a cor-

Venus Flytrap nose cone. *Left to right:* fully closed, fully open, and partly open.

Fig. 9. Venus Flytrap nosecone.

rection factor 2 in order to convert the observed influx rates to omnidirectional values. During the flight large variations of the influx rate have been recorded.

From this mass distribution function a size distribution follows which can be represented by $n(a) \sim a^{-\beta}$ with $\beta \simeq 6$, a result which agrees quite well with the result (11) on meteors.

Assuming an average speed of the particles relative to the space craft of 30 km/sec from the influx rate of 3×10^{-2} particles/m^2/sec a particle density

$$n_d = 10^{-12}\ \text{particles/cm}^3 \tag{12}$$

follows (particle sizes $a > 10^{-4}$ cm).

The particle number per cm^3 derived from the F-corona (Equation (10)) is 10^2 times lower. This discrepancy led to the concept of a dust cloud about the earth with spatial densities 10^2–10^3 times bigger than the densities in free interplanetary space.

Little information about the extension of this cloud is available. From older satellite maesurements Whipple derived the diameter of the cloud to be in the order of 200000 km. If a cloud of this kind really is present we cannot be sure that the data derived from meteor observations and space craft measurements performed in the vicinity of the earth are representative of interplanetary space.

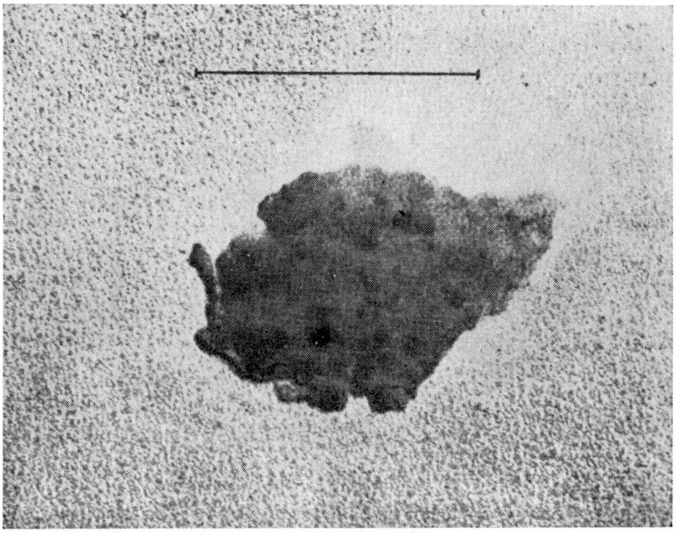

Fig. 10a. Irregular micrometeorite (scale = 1 μ).

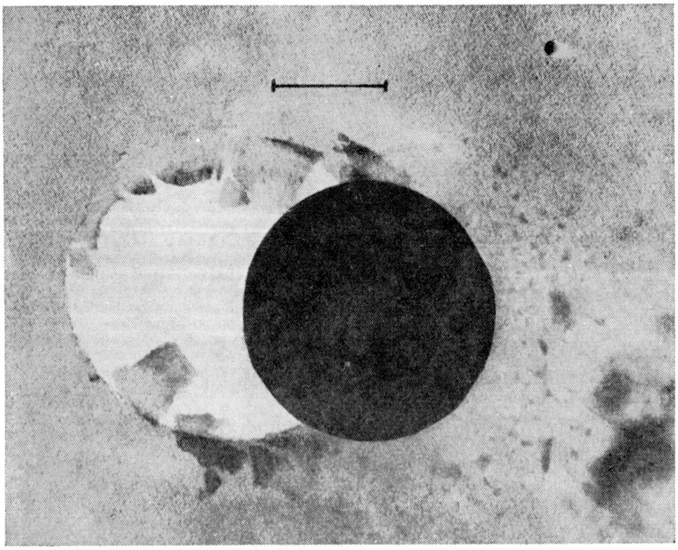

Fig. 10b. Spherical particle (scale = 1 μ).

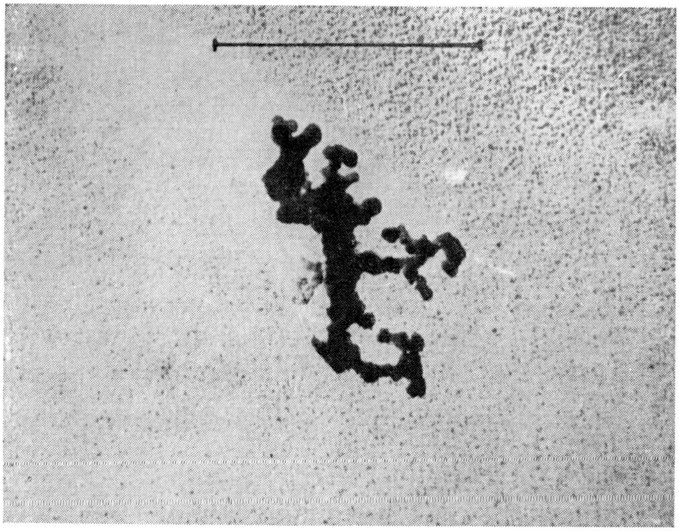

Fig. 10c. "Fluffy" particle (scale = 1μ).

B. THE "VENUS FLYTRAP" EXPERIMENT

The "Venus Flytrap" experiment was performed in June 1961 with a recoverable Aerobee rocket by Soberman, Hemenway and co-workers. They intended to collect interplanetary particles at altitudes above 100 km for investigations in the laboratory. At an altitude of 88 km the lower part of the rocket's nose cone could be opened, then eight "leaves" carrying collectors for particles were extended outward (Figure 9). By an ingenious technique particles of terrestrial origin could be distinguished from the dust collected at high altitudes. In this experiment 133 particles have been picked up which are believed to be of extraterrestrial origin. From this number again spatial densities above the value of Equation (10) follow; the dimensions of those particles were mainly in the range $a=0.1-1\mu$. Three different types of particles have been found: 72% were of an irregular shape with sharp edges (Figure 10a), 16% were spherical particles (Figure 10b) and 12% showed a fluffy structure (Figure 10c). Investigations concerning the nature of these particles are being performed at the moment in the laboratory.

References

Review Articles

ALEXANDER, W. M. *et al.*: 1963, *Space Research* **3**, North Holland Publishing Co., Amsterdam, p. 891.
ELSÄSSER, H.: 1957, *Mitt. Astron. Ges.* **2**, 61.
ELSÄSSER, H.: 1963, *Plan. Space Sci.* **11**, 1015.
GIESE, R. H.: 1963, *Space Sci. Rev.* **1**, 589.

KAISER, T. R.: 1963, *Space Sci. Rev.* **1,** 554.
MINNAERT, M.: 1955, *Mém. Roy. Soc. Sci. Liège* (International Astrophysical Symposium) **15,** 15.
SOBERMAN, R. K. *et al.*: 1961, *GRD Research Notes* No. 71.
WHIPPLE, F. L.: 1961, *Nature* **189,** 127.

Textbooks

KUIPER, G. P. and MIDDLEHURST, B. (eds.): 1953, *The Solar System*, Vol. I, 'The Sun', The University of Chicago Press, Chicago, Ill.
KUIPER, G. P. and MIDDLEHURST, B. (eds.): 1963, *The Solar System*, Vol. IV, 'The Moon, Meteorites and Comets', The University of Chicago Press, Chicago, Ill.
ALLER, L. H.: 1953, *Astrophysics I and II*, Donald Press Company, New York.
FLÜGGE, S.: 1959, 'The Solar System', *Handbuch der Physik* **52.**
UNSÖLD, A.: 1955, *Physik der Sternatmosphären*, Springer Verlag, Berlin.
VAN DE HULST, H. C.: 1957, *Light Scattering by Small Particles*, John Wiley and Sons, Inc., New York.

CHAPTER 12

INTRODUCTION TO PLASMA PHYSICS

R. LÜST

COPERS and Max-Planck-Institut für Physik und Astrophysik, München

1. Introduction

The ionized gas is called a plasma if it contains charged particles in such a large number that its properties are influenced in an essential way and determined by their presence.

In most cases we can regard the plasma as quasineutral, which means that the number of positive and negative charges is equal. The special properties of a plasma depend on the electromagnetic interaction of the particles themselves and with the external fields. The hydrodynamical properties are furthermore, important, hence we have a coupling between hydrodynamical and electrodynamical phenomena. The collective behaviour of a normal fluid is due to the short range forces of the colliding particles. In the plasma, the long range Coulomb forces among the particles also give rise to the collective behaviour, and these are often more important than the effect of the short range forces.

The description of a plasma can be made in several different ways. One is the microscopic approach, based on the distribution functions of positive and negative charged particles. These distribution functions are determined by the respective Boltzmann equations. Certain effects, as for example the micro-instabilities, can be described only by this method. The description of a plasma based on this procedure is the most rigorous and complete one.

For a large number of phenomena, however, it is sufficient to represent the plasma from a macroscopic point of view, and describe it in terms of average quantities, like density, pressure, temperature and macroscopic velocity. This is equivalent to using the fluid theory.

We know from hydrodynamics that the fluid description holds only if a certain number of particles stay together for some time. The mechanism to provide this behaviour is the collisions among particles, and this means that the mean free path of the particles must be small compared with the characteristic length scale of the fluid.

In a number of practical cases this might not be the case, but if magnetic fields are present, the fluid model can still be used, since the charged particles are spiralling around the lines of force and are kept together at least perpendicularly to the magnetic field. The condition which now replaces the requirement of short mean free path for the validity of the fluid model is replaced by the requirement that the gyration radius of the particles be small compared with the dimensions of the system. In this case,

Ortner & Maseland (eds.), Solar Terrestrial Relations.

however, one can no longer assume the pressure to be isotropic, but a tensor pressure has to be used.

In this introduction to plasma physics we will limit ourselves to the discussion of the macroscopic equations. In these two lectures we will assume them as given and shall analyse their structure, the importance of different terms, and certain applications of these equations. We can point out that the rigorous way of deriving these equations is to take moments of Boltzmann equations for ions and electrons and performing averages on them. We will accomplish this in three steps. First, the so-called one-fluid theory will be investigated, which is also known as magnetohydrodynamic approximation. In this approximation ions and electrons are not taken explicitly into account. Second, we will consider the two-fluid theory and some of its complications. Third, we will discuss a model where the collisions are very infrequent and the pressure consequently is no longer isotropic.

2. Magnetohydrodynamics

(a) As already pointed out we may describe the plasma by combining ten equations of electrodynamics and of hydrodynamics. In the following we will use gaussian units and will not distinguish between **B** and **H** or **E** and **D**, since electrons and protons do not possess magnetic and electric susceptibilities, $\mu=\varepsilon=1$. Even complex ions would not give any noticeable effect. The only contribution from matter will be currents and space charges. The Maxwell equations will therefore be written as

$$\operatorname{curl}\mathbf{B} = \frac{4\pi}{c}\mathbf{j} + \frac{1}{c}\dot{\mathbf{E}} \tag{1}$$

$$\operatorname{curl}\mathbf{E} = -\frac{1}{c}\dot{\mathbf{B}} \tag{2}$$

$$\operatorname{div}\mathbf{E} = 4\pi\lambda \tag{3}$$

$$\operatorname{div}\mathbf{B} = 0 \tag{4}$$

where $\mathbf{B}, \mathbf{E}, \mathbf{j}, \lambda$ are the magnetic field, electric field, electric current density and charge density respectively.

Maxwell equations are Lorentz-Invariant, while the normal hydrodynamical equations are only Galilei-invariant. To have a consistent formulation, and in order to avoid fictitious effects arising from inconsistent transformation rules we will make the Maxwell equations also Galilei-invariant.

Let us have two reference systems Σ and Σ', the latter moving with a constant velocity $\mathbf{w}$ with respect to the first. Assuming $E^2 \ll B^2$ and neglecting the displacement current we apply the following non relativistic transformation rules:

$$\mathbf{B}' = \mathbf{B} \tag{5}$$

$$\mathbf{E}' = \mathbf{E} + \frac{\mathbf{w}}{c} \times \mathbf{B} \tag{6}$$

$$\mathbf{j}' = j \tag{7}$$

$$\lambda' = \lambda - \frac{\mathbf{wj}}{c^2} \tag{8}$$

$$\nabla' = \nabla \tag{9}$$

$$\frac{\partial}{\partial t'} = \frac{\partial}{\partial t} + \mathbf{w}\nabla . \tag{10}$$

It can now easily be shown that Maxwell equations are Galilei-invariant according to the above transformation rules. The displacement current can be neglected if the electric fields are small compared with magnetic fields and if the characteristic velocity in a plasma (for example, ten phase velocities of waves) are small compared with the velocity of light.

(b) Maxwell equations have to be supplemented by a relation between the current density and the electric field. If $\mathbf{v}$ is the velocity of the plasma, this relation, Ohm's law for a moving fluid, has the form

$$\mathbf{j} = \sigma\left(\mathbf{E} + \frac{v}{c} \times \mathbf{B}\right) \tag{11}$$

where the electric conductivity, σ, is a constant of the material. This equation also fulfils the above transformation rules.

(c) The macroscopic velocity of the plasma is determined by the momentum equation of hydrodynamics, to which however, we have to add a term to take into account the force exerted by the magnetic field on the plasma.

One way of deriving this force is through the energy equation for the electromagnetic field. This can be derived by multiplying (11) by $\mathbf{j}$, and using (1) neglecting displacement current to obtain

$$\frac{\mathbf{j}^2}{\sigma} = \frac{c}{4\pi}(\mathrm{curl}\,\mathbf{B}\cdot\mathbf{E}) + \frac{1}{c}(\mathbf{vB} \times \mathbf{j}) \tag{12}$$

and combining this equation with the following vector identity:

$$\frac{c}{4\pi}\mathrm{div}\,(\mathbf{E} \times \mathbf{B}) = \frac{c}{4\pi}\mathbf{B}\cdot\mathrm{curl}\,\mathbf{E} - \frac{c}{4\pi}\mathbf{E}\,\mathrm{curl}\,\mathbf{B}\,. \tag{13}$$

Applying Maxwell equations (2) (1) and adding, we obtain the energy equation

$$-\frac{1}{8\pi}\frac{\partial}{\partial t}B^2 = \frac{\mathbf{j}^2}{\sigma} + \frac{1}{4\pi}\mathrm{div}\,(\mathbf{E} \times \mathbf{B}) + \frac{1}{c}(\mathbf{v}\cdot\mathbf{j} \times \mathbf{B})\,. \tag{14}$$

On the left hand side we have the change in time of electromagnetic energy ($E^2 \ll B^2$) and the terms on the right hand side are the dissipation of foule heat, the divergency of the Poynting energy flux and the work alone by the force of the magnetic field on the plasma respectively.

This force is

$$\mathbf{K} = \frac{1}{c}\mathbf{j} \times \mathbf{B}\,. \tag{15}$$

This force can also be written, making use of as

$$\mathbf{K} = \frac{1}{4\pi}(\operatorname{curl}\mathbf{B} \times \mathbf{B}) \tag{16}$$

and using the expansion of the curl of a vector product

$$\mathbf{K} = -\frac{1}{8\pi}\operatorname{grad}\mathbf{B}^2 + \frac{1}{4\pi}(\mathbf{B}\operatorname{grad})\mathbf{B}. \tag{17}$$

We may notice incidentally at this point that the magnetic force could have been derived from the Maxwell stress tensor **T**

$$\mathbf{K} = \operatorname{div}\mathbf{T} \tag{18}$$

where **T**, in a system aligned with the magnetic lines of force, is

$$\mathbf{T} = \begin{vmatrix} -\frac{1}{8\pi}B^2 & 0 & 0 \\ 0 & \frac{1}{8\pi}B^2 & 0 \\ 0 & 0 & \frac{1}{8\pi}B^2 \end{vmatrix} \tag{19}$$

with a generic element given by

$$T_{ik} = -\frac{1}{4\pi}(B_iB_k - \tfrac{1}{2}B^2\cdot\delta_{ik}) \tag{20}$$

(δ_{ik} is the Kronecker delta).

The equation for conservation of momentum then becomes, neglecting viscosity,

$$\varrho\frac{\mathrm{d}\mathbf{v}}{\mathrm{d}t} = -\operatorname{grad}p + \frac{1}{c}[\mathbf{j}\times\mathbf{B}] \tag{21}$$

where ϱ is the mass density and p is the pressure.

(d) The mass density can be obtained from the equation of continuity for mass

$$\frac{\partial\varrho}{\partial t} + \operatorname{div}(\varrho\mathbf{v}) = 0 \tag{22}$$

and the energy equation is the equation connecting density and pressure, which, neglecting viscosity and heat conduction, is

$$\frac{\mathrm{d}p}{\mathrm{d}t} = \gamma\frac{p}{\varrho}\frac{\mathrm{d}\varrho}{\mathrm{d}t} + \frac{\gamma - 1}{\varrho}\frac{j^2}{\sigma} \tag{23}$$

with γ representing as usual the ratio of specific heat. We have seen at the outset that we have assumed the electric field and the characteristic velocities to be small. This fact allows us to neglect the displacement current.

(e) The quasineutrality of the plasma implies through (3) the smallness of the electric field. This means that N_p, the number density of protons is almost equal to N_e, the number density of the electrons, so that at any point the relationship

$$\frac{\lambda}{e} = |N_p - N_e| \ll N_p + N_e \tag{24}$$

holds.

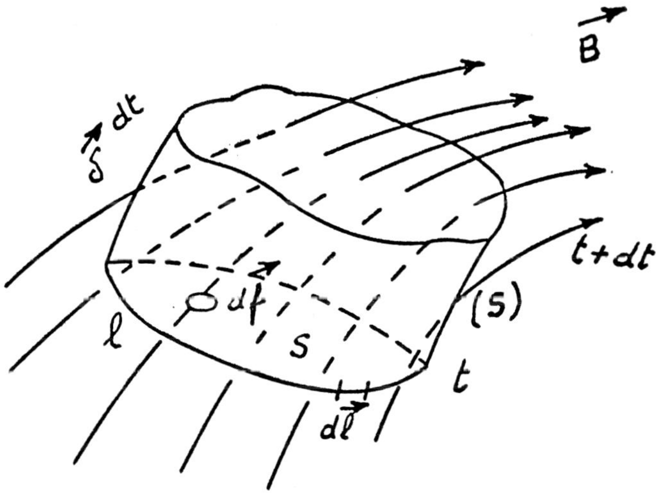

Fig. 1

This condition will be violated for a plasma at thermal equilibrium at distances which are small compared with the Debye length.

$$h = \sqrt{\left(\frac{kT}{4\pi N e^2}\right)} = 12 \sqrt{\left(\frac{T\,(^\circ\mathrm{K})}{N\,(\mathrm{cm}^{-3})}\right)}. \tag{25}$$

We can therefore redefine a plasma as a gas where the characteristic length is larger than the Debye length. If deviation from quasineutrality would occur over a volume larger than the Debye sphere, the resulting electrostatic forces would yield a potential energy per particle enormously greater than the mean thermal energy. The Debye length is also of importance in the neighbourhood of walls surrounding the plasma. Deviation from quasineutrality will occur in the transition layer.

If inertia effect of electrons is taken into consideration care should be taken; oscillation will be set up. These oscillations are oscillations of electrons around the ions, where the restoring force is supplied by the electric field between ions and electrons.

The plasma oscillation frequency is

$$\omega_p = \sqrt{\frac{4\pi N e^2}{m_e}} \tag{26}$$

where m_e is the electron mass.

Deviation from quasineutrality should be expected for changes in time close to the plasma frequency.

It should finally be noted that deviation from quasineutrality could arise due to the presence of external forces (for example in a gravitational field). These effects are, however, very small.

(f) As a first application of these equations, we shall consider in which way the magnetic flux through a surface varies when the surface is moved and deformed by the motion of the plasma (cf. Figure 1).

At time t the flux is given by

$$\Phi = \int_s \mathbf{B}\,\mathrm{d}\mathbf{f} \tag{27}$$

where $\mathbf{df}$ is a surface element of the surface s surrounded by a line l of elementary length $\mathbf{dl}$.

The flux may change for the following reasons: (a) There might be a time variation of B; (b) Some lines of force might no longer traverse the surface since the surface moves with velocity and is also deformed. These lines of force are therefore last through the lateral surface (S). We will therefore have:

$$\frac{\mathrm{d}\Phi}{\mathrm{d}t} = \int_s \frac{\partial \mathbf{B}}{\partial t}\cdot \mathrm{d}\mathbf{f} + \int_s \mathbf{B}\cdot(\mathbf{v}\times \mathrm{d}l) \tag{28}$$

which applying Stokes theorem can be written

$$\frac{\mathrm{d}\Phi}{\mathrm{d}t} = \int \frac{\partial \mathbf{B}}{\partial t}\cdot \mathrm{d}\mathbf{f} + \int \operatorname{curl}(\mathbf{B}\times \mathbf{v})\cdot \mathrm{d}\mathbf{f} \tag{29}$$

taking the curl of (11)

$$\operatorname{curl}\mathbf{E} = -\operatorname{curl}\left(\frac{\mathbf{v}}{c}\times \mathbf{B}\right) + \operatorname{curl}\frac{\mathbf{j}}{\sigma} \tag{30}$$

and using (2)

$$\frac{\partial \mathbf{B}}{\partial t} = \operatorname{curl}(\mathbf{v}\times \mathbf{B}) - c\operatorname{curl}\frac{\mathbf{j}}{\sigma} \tag{31}$$

we obtain

$$\frac{\mathrm{d}\Phi}{\mathrm{d}t} = \int \operatorname{curl}(\mathbf{v}\times \mathbf{B}) - c\int \operatorname{curl}\frac{\mathbf{j}}{\sigma} - \int \operatorname{curl}(\mathbf{v}\times \mathbf{B}) \tag{32}$$

and if we make the assumption of infinite conductivity $\sigma \to \infty$ we finally get

$$\frac{\mathrm{d}\Phi}{\mathrm{d}t} = 0\,. \tag{33}$$

This means that the magnetic flux through a surface moving with and deformed by the fluid is not changed.

The magnetic field is transported by the velocity field in such a way that every part of the line of force moves with the local velocity $\mathbf{v}$.

One says in this case that the magnetic lines of force are frozen into the fluid.

3. Two-fluid Model

(a) We will now try to take into account the presence of the ions and of the electrons in an explicit way. We will assume the existence of two fluids, one composed by the electrons, the other by the ions.

We will now make the following assumptions: (1) There are only protons and electrons. No neutral particles are present; (2) The plasma is quasi neutral: $N_p = N_e = N$; (3) The displacement current is neglected; (4) The pressure is assumed to be isotropic. This of course implies that the collision frequency is high enough; (5) Viscosity is neglected.

The two fluid equations for the protons and electrons are

$$m_p \frac{\mathrm{d}_p \mathbf{v}_p}{\mathrm{d}t} = e\mathbf{E} + \frac{e}{c} \mathbf{v}_p \times \mathbf{B} + \mathbf{K}_p + \nu \frac{m_p m_e}{m_p + m_e} (\mathbf{v}_e \quad \mathbf{v}_p) \tag{34}$$

and

$$m_e \frac{\mathrm{d}_e \mathbf{v}_e}{\mathrm{d}t} = - e\mathbf{E} - \frac{e}{c} \mathbf{v}_p \times \mathbf{B} + \mathbf{K}_e + \nu \frac{m_p m_e}{m_p + m_e} (\mathbf{v}_e - \mathbf{v}_p) \tag{35}$$

where K_e and K_p represent external forces of non electromagnetic origin; ν is the collision frequency for momentum transfer, m_e and m_p are the electron and proton masses, v_e and v_p are the average electron and proton fluid velocities, and the total differential operators are given by

$$\frac{\mathrm{d}p}{\mathrm{d}t} = \frac{\partial}{\partial t} + (\mathbf{v}_p \operatorname{grad}) \tag{36}$$

and

$$\frac{\mathrm{d}_e}{\mathrm{d}t} = \frac{\partial}{\partial t} + (\mathbf{v}_e \operatorname{grad}) . \tag{37}$$

The collision frequency is approximately

$$\nu = 10^{1.4} N T^{-3/2} \,(\mathrm{sec}^{-1}) . \tag{38}$$

The mean velocity of the plasma is

$$\mathbf{v} = \frac{m_p \mathbf{v}_p + m_p \mathbf{v}_e}{m_p + m_e} \tag{39}$$

and the current density is

$$\mathbf{j} = Ne(\mathbf{v}_p - \mathbf{v}_e) \tag{40}$$

By adding (34) and (35) we have

$$\varrho \frac{\mathrm{d}_1 \mathbf{v}}{\mathrm{d}t} = \frac{1}{c} \mathbf{j} \times B + N(\mathbf{K}_p + \mathbf{K}_e) \tag{41}$$

which is the equation for the mean mass velocity, where $\mathrm{d}_1\mathbf{v}/\mathrm{d}t$ is defined by

$$\frac{\mathrm{d}_1\mathbf{v}}{\mathrm{d}t}=\frac{\partial\mathbf{v}}{\partial t}+(\mathbf{v}\,\mathrm{grad})\,\mathbf{v}+\frac{m_e m_p}{(m_e+m_p)^2}\frac{1}{Ne}(\mathbf{j}\,\mathrm{grad})\frac{\mathbf{j}}{Ne} \tag{42}$$

and

$$\varrho=N(m_p+m_e). \tag{43}$$

The external forces $K_{e,p}$ could be for example

$$NK_p=-\,\mathrm{grad}\,p_p \tag{44}$$

and

$$NK_e=-\,\mathrm{grad}\,p_e \tag{45}$$

respectively, with the total pressure p

$$p=p_e+p_p \tag{46}$$

Dividing (34) by m_p and (35) by m_e and subtracting we obtain the equation for electrical current density, which we also call the generalized Ohm's law.

$$\frac{1}{e^2N}\frac{m_p m_e}{m_p+m_e}\frac{\mathrm{d}_2\mathbf{j}}{\mathrm{d}t}+\frac{1}{e^2N}\frac{m_p m_e}{m_p+m_e}\gamma\mathbf{j}=$$
$$=\mathbf{E}+\frac{1}{c}\mathbf{v}\times\mathbf{B}+\frac{1}{e\varrho}\left\{(m_p-m_e)\frac{1}{c}\mathbf{j}\times\mathbf{B}+Nm_eK_p-Nm_pK_e\right\}. \tag{47}$$

The first term in the parenthesis is the Hall term and the sum of the other two is the pressure diffusion term if $K_{p,e}$ is grad $p_{p,e}$

The total derivative $\mathrm{d}_2\mathbf{j}/\mathrm{d}t$ is defined by

$$\frac{\mathrm{d}_2(\mathbf{j}/N_e)}{\mathrm{d}t}=\frac{\partial}{\partial t}\left(\frac{\mathbf{j}}{N_e}\right)+(\mathbf{v}\,\mathrm{grad})\frac{\mathbf{j}}{N_e}+\left(\frac{\mathbf{j}}{N_e}\mathrm{grad}\right)\mathbf{v}-\frac{m_p m_e}{m_p+m_e}\left(\frac{\mathbf{j}}{N_e}\mathrm{grad}\right)\frac{\mathbf{j}}{N_e}. \tag{48}$$

Normally it is assumed that the additional terms in the substantial derivatives (42) and (48) of $\mathbf{v}$ and $\mathbf{j}$ are small. This is true if one of the following conditions is fulfilled:

(a) v_p and v_e are small. In this case we can neglect all additional terms in the total derivative ($\mathrm{d}/\mathrm{d}t=\partial/\partial t$).

(b) Only creeping diffusion is present

$$\frac{\mathrm{d}_p\mathbf{v}_p}{\mathrm{d}t}=\frac{\mathrm{d}_e\mathbf{v}_e}{\mathrm{d}t}\rightarrow\frac{\mathrm{d}_1\mathbf{v}}{\mathrm{d}t}=\frac{\mathrm{d}\mathbf{v}}{\mathrm{d}t}\quad\text{and}\quad\frac{\mathrm{d}_2}{\mathrm{d}t}\left(\frac{\mathbf{j}}{N_e}\right)=0. \tag{49}$$

(c) Small diffusion velocity

$$|\mathbf{v}_p-\mathbf{v}_e|\ll\mathbf{v}\rightarrow\frac{\mathrm{d}_1}{\mathrm{d}t}\mathbf{v}=\frac{\mathrm{d}}{\mathrm{d}t}\mathbf{v} \tag{50}$$

and

$$\frac{\mathrm{d}_2}{\mathrm{d}t}\left(\frac{\mathbf{j}}{N_e}\right)=\frac{\mathrm{d}}{\mathrm{d}t}\left(\frac{\mathbf{j}}{N_e}\right)+\left(\frac{\mathbf{j}}{N_e}\mathrm{grad}\right)\mathbf{v}. \tag{51}$$

Furthermore since $m_e/m_p \ll 1$

$$\omega_p = \frac{4\pi e^2 \varrho}{m_p m_e} \simeq \frac{4\pi e^2 N}{m_e} \tag{52}$$

and the electrical conductivity is

$$\sigma = \frac{e^2 \varrho}{m_p m_e} \frac{1}{\gamma} = \frac{\omega_p^2}{4\pi} \frac{1}{\gamma} \approx 10^7 T^{3/2} \tag{53}$$

which is almost independent of the density and is not affected very much by a magnetic field. (The difference between the value parallel and perpendicular to the magnetic field is not more than a factor of 2).

Finally we may introduce in (47) the electric field $\mathbf{E}^c$ measured in a co-moving system

$$\mathbf{E}^c = \mathbf{E} + \frac{1}{c}\mathbf{v} \times \mathbf{B} \tag{54}$$

and treat the other terms on the right hand side of (47) as an impressed electric field $\mathbf{E}^i$

$$\begin{aligned} \mathbf{E}^i &= \frac{1}{e\varrho}\left\{(m_e - m_p)\frac{1}{c}\mathbf{j} \times \mathbf{B} + m_e N K_p - m_p N K_e\right\} = \\ &= \frac{1}{e}\left\{(m_e - m_p)\frac{\mathrm{d}_1 \mathbf{v}}{\mathrm{d}t} + \frac{m_p K_p - m_e K_e}{m_p + m_e}\right\}. \end{aligned} \tag{55}$$

In (55) we have eliminated the so-called Hall-term by using (41). Therefore in this formation the magnetic field can be eliminated from Ohm's law (47) with the exception of the term in E^c. Equation (47) now has the form

$$\frac{4\pi}{\omega_p^2} \frac{\mathrm{d}_2 \mathbf{j}}{\mathrm{d}t} + \frac{1}{\sigma}\mathbf{j} = \mathbf{E}^c + \mathbf{E}^i. \tag{56}$$

(b) Let us now discuss some conclusions of these equations.

If we neglect the inertial term of the current, we obtain the one fluid equation if curl $\mathbf{E}^i = 0$. Very often this assumption is made without stating it.

Sometimes one has introduced a conductivity tensor for the relation between current density and the electric field $\mathbf{E}^c$. But very often one has neglected the impressed electromotive force and obtained in this way a strong reduction of the electrical conductivity in the direction perpendicular to the magnetic field.
For instance, the following equation

$$\mathbf{E}^c = \frac{\mathbf{j}}{\sigma} + \frac{m_p - m_e}{\varrho e c}\mathbf{j} \times \mathbf{B} \tag{57}$$

can be solved for $\mathbf{j}$

$$\mathbf{j} = \frac{\sigma}{1 + \alpha^2}\left(\mathbf{E}^c - \alpha \mathbf{E}^c \times \frac{\mathbf{B}}{B} + \alpha_2 E^c \frac{\mathbf{B}}{B}\right) \tag{58}$$

with

$$\alpha = \frac{\omega_{gr}}{\gamma} \quad \text{and} \quad \omega_{gr} = \frac{eB}{m_e c}. \tag{59}$$

But this solution is not in agreement with the equation of motion. Either a pressure gradient is needed to balance $(1/c)\mathbf{g} \times \mathbf{B}$ or a stationary solution is only possible if

$$\mathbf{j} \times \mathbf{B} = 0. \tag{60}$$

But from Equation (57) in this case follows

$$\mathbf{E}^c \times \mathbf{B} = 0. \tag{61}$$

This equation determines a velocity to the plasma from (54)

$$\mathbf{v} = c\frac{\mathbf{E} \times \mathbf{B}}{B^2} + \text{const}\,\mathbf{B}. \tag{62}$$

The plasma therefore has to move, or, if such a motion is not possible (e.g. for a confined plasma) a pressure gradient will develop, which must be taken into account.

Let us assume that $p_p = p_e = p/2$. The impressed electromotive force

$$\mathbf{E}^i \cong \frac{1}{2e\varrho} m_p \operatorname{grad} p \tag{63}$$

gives rise to an additional current at the beginning in the direction of grad p which will build up a space charge. This space charge will give rise to an additional electrical field $\mathbf{E}^s$ just compensating the pressure term. Therefore the first equation will be

$$\mathbf{E} = \mathbf{E}^a + \mathbf{E}^s \tag{64}$$

where $\mathbf{E}^a$ is the applied electrical field, and

$$\mathbf{E}^s = \frac{1}{2e\varrho} m_p \operatorname{grad} p = - \mathbf{E}^i \tag{65}$$

Therefore

$$\mathbf{j} = \sigma(\mathbf{E} + \mathbf{E}^i) = \sigma\mathbf{E}^a \tag{66}$$

$$\operatorname{grad} p = \frac{1}{c}\mathbf{j} \times \mathbf{B} \tag{67}$$

and

$$\mathbf{v} = 0 \tag{68}$$

there is no reduction of the conductivity and no Hall current at all. The condition for this is of course that curl $\mathbf{E}^i$ must vanish, since otherwise $\mathbf{E}^i$ cannot be compensated by a space charge. If this is not the case, one has to solve the whole set of equations.

(c) We have seen that one can say that the magnetic field is frozen into the matter. We may also invert the argument by saying that the plasma is coupled to the magnetic field. This property is of special importance for the experiments in connection with the problem of thermonuclear fission. The plasma might be confined only for a finite time since the collisions of the particles tend to dissipation and to diffusion of the particles across the magnetic lines of force.

If the confining magnetic field is produced by electric currents flowing through the plasma the time of confinement is given by the time of decay of the magnetic field due to ohmic losses. If the confining magnetic field is produced by external currents, the time of confinement is given by the time of diffusion of the plasma across the magnetic field due to the collisions between ions and electrons.

Let us first estimate the decay time of a magnetic field. Elimination of **j** and **E** from (1) and (2) gives

$$\frac{\partial \mathbf{B}}{\partial t} = \operatorname{curl}(\mathbf{v} \times \mathbf{B}) - \frac{c^2}{4\pi}\operatorname{curl}\left(\frac{1}{\sigma}\operatorname{curl}\mathbf{B}\right). \tag{69}$$

If one replaces the curl by $1/L$ where L is a characteristic length, the second term gives a decay time due to ohmic losses of

$$\tau_0 = \frac{4\pi L^2 \sigma}{c^2}. \tag{70}$$

Let us now assume that the plasma is confined by a magnetic field produced by external currents. In the steady state we are left with the two equations (67)

$$\operatorname{grad} p = \frac{1}{c}\,\mathbf{j} \times \mathbf{B} \tag{67}$$

and

$$\frac{1}{\sigma}\mathbf{j} = \mathbf{E} + \frac{\mathbf{v}}{c} \times \mathbf{B} - \frac{m_p - m_e}{2e\varrho}\operatorname{grad} p \tag{71}$$

(with $p_p = p_e = p/2$).

Eliminating **j** between (67) and (71) we have

$$\operatorname{grad} p = \frac{\sigma}{c}\left\{\mathbf{E} \times \mathbf{B} + \mathbf{B}\left(\frac{\mathbf{v}}{c}\mathbf{B}\right) - \frac{\mathbf{v}}{c}\mathbf{B}^2 - \frac{m_p - m_e}{2e\varrho}\operatorname{grad} p \times \mathbf{B}\right\}. \tag{72}$$

From this equation we can determine the plasma velocity **v**. For the question of confinement we are only interested in the velocity component perpendicular to the magnetic field. If $\sigma = \infty$ we obtain

$$v_\perp = \left\{\frac{\mathbf{E} \times \mathbf{B}}{B^2} + \frac{m_p - m_e}{2e\varrho}\operatorname{grad} p \times \mathbf{B}\right\} c. \tag{73}$$

The first term gives rise to the normal drift motion, while the second term is a motion parallel to the isobars which is not dangerous as far as confinement is concerned. If σ is finite, however, we get an additional contribution $v_\perp^2$ which is given by

$$v_\perp^2 = -\frac{c^2}{\sigma}\frac{1}{B^2}\operatorname{grad}_\perp p. \tag{74}$$

This component will lead to a loss of particles and with a characteristic length L one gets for the diffusion time

$$\tau = \frac{4\pi\sigma L^2}{c^2}\frac{2}{\beta} \tag{75}$$

where

$$\beta = \frac{p}{B^2/8\pi}. \tag{76}$$

(d) Before closing this point, a few words should be added about the importance of the different additional terms in the generalized Ohm's law (47) in comparison to the one fluid theory (11). We shall do this by comparing all the terms with the $\mathbf{v}(c)\mathbf{B}$ term. We assume that the different terms in the equation of motion are of the same order of magnitude. Furthermore it is convenient to introduce the gyration frequency of the protons and electrons

$$\omega_i = \frac{eB}{m_p c} \tag{77}$$

$$\omega_e = \frac{eB}{m_e c} \tag{78}$$

hence

$$\frac{1}{\omega_i^2}\frac{\partial j}{\partial t} \approx \frac{\omega}{\omega_p^2} j \tag{79}$$

$$\frac{1}{\sigma} j \approx \frac{\gamma}{\omega_p^2} j \tag{80}$$

$$\frac{1}{c}|\mathbf{v} \times \mathbf{B}| \approx \frac{\omega_e \omega_i}{\omega_p^2 \omega} j \tag{81}$$

$$\frac{m_i}{ce\varrho}|\mathbf{j} \times \mathbf{B}| \approx \frac{\omega_e}{\omega_p^2} j \tag{82}$$

and

$$\frac{m_i}{e\varrho}|\text{grad } p_e| = \frac{\omega_e}{\omega_p^2} j. \tag{83}$$

From these estimates we may see that the last two terms in (47) are small compared to the term $1/c|\mathbf{v} \times \mathbf{B}|$ if $\omega < \omega_i$. The inertial term is small compared to this term if $\omega \ll \sqrt{\omega_i \omega_e}$ which is fulfilled when $\omega \ll \omega_i$. Finally the term $(1/\sigma)\mathbf{j}$ is small compared to $1/c|\mathbf{v} \times \mathbf{B}|$ if $\gamma\omega \ll \omega_i \omega_e$. Therefore the electric field in a plasma is determined by $(\mathbf{v}/c)\mathbf{B}$ if we are only interested in small frequencies and large electrical conductivity. In this limit we return to the equations of magnetohydrodynamics.

4. Fluid Description with Anisotropic Pressure

As already pointed out in many practical cases the collisions are very infrequent. If a magnetic field however is present we may still apply a fluid picture for the description of the plasma, since the particles are gyrating around the magnetic lines of force. The particles cannot disperse in the two directions perpendicular to the magnetic field, but only in a direction to the field lines. In this case we cannot except however the pressure to be isotropic, for which collisions are needed.

In the most general case we would expect that the pressure is completely anisotropic and we are dealing with the full stress tensor. But for quite a number of applications it is still sufficient to assume that the pressure is only different for the directions parallel and perpendicular to the magnetic field.

In the momentum equation and in the generalized Ohm's law (47) we must replace now the scalar pressure p by a pressure tensor $\mathbf{p}$. If we assume that the magnetic field is very strong (its change over the ion Larmor radius is small), then the stress tensor is approximately diagonal in a local cartesian co-ordinate system, one axis of which is directed along $\mathbf{B}$ and is invariant under rotation about $\mathbf{B}$. If $\mathbf{e}$ is a unit vector parallel to $\mathbf{B}$ we finally get for the components of $\mathbf{p}$

$$p_{ik} = p_{\parallel} e_i e_k + p_{\perp}(\delta_{ik} - e_i e_k). \tag{84}$$

We now have two new quantities, the pressure parallel to the magnetic field $p_{\parallel}$ and the pressure perpendicular to the magnetic field $p_{\perp}$. We therefore need two new equations to determine these quantities. These new equations will replace the adiabatic relation (23) between the pressure and the density (assuming infinite conductivity since the collisions are very infrequent).

We may obtain these equations by applying the first law of thermodynamics. We can obtain the internal energy per unit mass u noticing that it is one half of the terms of the stress tensor, and is given by

$$u = u_{\parallel} + u_{\perp} \tag{85}$$

where

$$u_{\parallel} = \frac{p_{\parallel}}{2\varrho} \tag{86}$$

and

$$u_{\perp} = \frac{p_{\perp}}{\varrho}. \tag{87}$$

If collisions are infrequent $u_{\parallel}$ and $u_{\perp}$ are independent. If we further assume that there is no heat flow, we can calculate the change of $u_{\parallel}$ and $u_{\perp}$ by applying again the first law of thermodynamics.

$$\mathrm{d}u = -\, p\,\mathrm{d}V \tag{88}$$

(with $V = 1/\varrho$ the specific volume), separately for $u_{\parallel}$ and $u_{\perp}$. For this purpose we have to calculate the change of the volume V due to a small displacement $\mathbf{v}\mathrm{d}t$. Let us consider a specific volume element $V = L \cdot S$ where L is an element of length along $\mathbf{B}$ and S an element of area perpendicular to $\mathbf{B}$. The change of length $\mathrm{d}L$ due to the displacement $\mathbf{v}\mathrm{d}t$ is given by

$$\begin{aligned}\mathrm{d}L &= \mathbf{ev}(\mathbf{r} + L\mathbf{e})\,\mathrm{d}t - \mathbf{evr}\,\mathrm{d}t \\ &= (\mathbf{e}\ \mathrm{grad})\,\mathbf{ve}L\,\mathrm{d}t\,.\end{aligned} \tag{89}$$

Therefore the change in volume associated with this change in length is

$$(\mathrm{d}V)_{\|} = S\,\mathrm{d}L = V(\mathbf{e}\,\mathrm{grad})\,\mathbf{ve}\,\mathrm{d}t\,. \tag{90}$$

Furthermore, the total change in volume is given by

$$\mathrm{d}V = V\,\mathrm{div}\,\mathbf{v}\,\mathrm{d}t\,. \tag{91}$$

Subtracting (91) and (90) we can compute the change in volume associated with a change in area dS perpendicular to **B**:

$$(\mathrm{d}V)_{\perp} = V\,\mathrm{div}\,\mathbf{v}\,\mathrm{d}t - V(\mathbf{e}\,\mathrm{grad})\,\mathbf{ve}\,\mathrm{d}t\,. \tag{92}$$

If we now apply the first law of thermodynamics for $u_{\|}$ and $p_{\|}$ we finally get

$$\frac{\mathrm{d}p_{\|}}{\mathrm{d}t} = -\,p_{\|}\,\mathrm{div}\,\mathbf{v} - 2p_{\|}\,(\mathbf{e}\,\mathrm{grad})\,\mathbf{ve} \tag{93}$$

$$\frac{\mathrm{d}p_{\perp}}{\mathrm{d}t} = -\,2p_{\perp}\,\mathrm{div}\,\mathbf{v} + p_{\perp}(\mathbf{e}\,\mathrm{grad})\,\mathbf{ve}\,. \tag{94}$$

These are two equations for $p_{\|}$ and $p_{\perp}$. They can be written in a simpler form by using

$$\frac{\partial \mathbf{B}}{\partial t} = \mathrm{curl}\,(\mathbf{v} \times \mathbf{B}) \tag{95}$$

so that

$$\frac{\mathrm{d}}{\mathrm{d}t}\left(\frac{p_{\|}B^2}{\varrho^3}\right) = 0 \tag{96}$$

and

$$\frac{\mathrm{d}}{\mathrm{d}t}\left(\frac{p_{\perp}}{\varrho B}\right) = 0\,. \tag{97}$$

Equations (96) and (97) replace the adiabatic relation (23) in case of anisotropic pressure.

CHAPTER 13

INTERPLANETARY PLASMA

R. LÜST

COPERS and Max-Planck-Institut für Physik und Astrophysik, München

Introduction

The space, in which the sun and the planets are imbedded is usually called the interplanetary space. From astronomical observations and other astrophysical evidence we know that the space between the stars, the so-called interstellar space, is not empty but filled with gas and dust of low density. Should we expect to find the same conditions in the near neighborhood of the sun, e.g. at the orbit of the earth? In fact a number of indications point out that there are pronounced differences between the interstellar and interplanetary space, so that it is justified to attribute two different names to these spaces which do not just differ in their location with respect to the solar system. It is not easy to define the boundaries for the zone of the interplanetary medium, and it may turn out that a definition of the outer boundary, that is the region where it emerges into the interstellar space, is more distinct than its inner boundary which separates it from the outermost layers of the solar atmosphere. We have good reasons to assume that the interplanetary matter is in one way or another just the smooth and physical extension of the solar corona itself. In the following exposition we use the term "interplanetary medium" for that material which is located in the planetary system some solar radii away from the sun as well as some distance away from each planet (e.g. the atmosphere of the earch is very probably separated from the interplanetary space by the magnetic field of the earth).

The special interest and attention which is given to the interplanetary medium results from several reasons. A number of effects observed in the earth's atmosphere like aurora or geomagnetic storms is caused or influenced by the interplanetary medium. Furthermore, the latter might modulate the cosmic radiation which is coming from outer space. Finally it is the nearest cosmical object with exception of our earth itself and its atmosphere which can now directly be investigated by means of space probes.

Observation of this medium from the surface of the earth is very difficult, actually more difficult than observations of the interstellar medium. The main reason for this is that its total mass is not very large at a given moment, and therefore its direct effects (e.g. its brightness in light intensity) or more indirect indications are not too easily detectable. For instance, we cannot expect interplanetary absorption lines superposed on the solar spectrum analogous to the interstellar absorption lines.

1. Evidence for the Interplanetary Medium and its Properties

A. OBSERVATIONS OF THE ZODIACAL LIGHT

The first indications about the existence of an interplanetary medium probably stem from observations of the zodiacal light. The zodiacal light is produced by the scattering of sunlight by the dust particles and electrons in interplanetary space. The densities of the dust and of the electron component can be determined, in principle, by the observation of the brightness and of the central intensities of the Fraunhofer lines of the scattered sunlight. These will be described in more detail in the chapter by Prof. Elsässer.

B. SCATTERING OF RADIO-WAVES DURING OCCULTATIONS OF RADIO SOURCES

It had been suggested as early as 1951 by MACHIN and SMITH (1951) as well as by VITKEVIČ (1958) to use the occultations of radio sources by the gas in the neighborhood of the sun for obtaining information about the material there. Since then, extensive measurements in this field have been made in different countries (Australia, England, USA, and USSR).

Mainly the occultation of the Crab Nebula (a radio source originated from a supernova) has been observed since this is a very strong source. Recently SLEE (1961) in Australia detected the occultations of a number of other sources and extended the observations out to larger solar distances (in some cases up to 120 $R_{\odot}$).

The radio waves emitted from a source behind and passing through the medium in the neighborhood of the sun are influenced by the electrons in two ways. Firstly they are scattered by random variations in the electron density and this leads to an increase of the apparent angular dimensions of the radio source. Furthermore the refraction due to a density gradient may change the apparent position of the source. The first effect has been mainly used in interpreting the observations, which show clearly the presence of electrons in the interplanetary space. These electrons are not uniformly distributed. HEWISH (1958) finds a fine structure in radial directions which might be of the order of 10000 km and which might be connected with lines of force of the magnetic field. But since the size of these irregularities is not well determined, the value for the electron density is rather uncertain. The observations by SLEE further indicate, that the density distribution is flattened with respect to the solar poles. But further measurements are needed to see if this flattening is changing with time.

In summary, this method has been very valuable to demonstrate the existence of the interplanetary medium, but it was not yet possible to derive quantitative data. In the future a combination with other observations and further increase in sensitivity and spatial resolution might lead to more direct quantitative data.

C. IONIZED COMET TAILS

BIERMANN (1951, 1952, 1953) pointed out in 1951 that the comet tails of Type I indicate the presence of an interplanetary medium. But they also show that the interplanetary gas is streaming radially outwards from the sun with a high velocity. Therefore the interplanetary medium should be identical with the corpuscular radiation coming from the sun (BIERMANN 1957, 1961), which is now, following Parker, very often called the "solar wind".

Comet tails of Type I are straight, long (up to 10^8 km), and narrow, and show very often fine structure. They are composed of ionized gas in contrast to the tails of Type II and III, which are broader and curved. The latter are diffuse and in most cases do not show any structure. From the observed spectrum it is clear that they contain dust particles and neutral molecules. The observed ionized molecules of Type I tails are mainly CO^+, N_2^+, CO_2^+, and CH^+.

The best direct evidence from the ionized tails for the existence of the corpuscular radiation is their almost radial direction in contrast to the direction of the dust tails. This almost radial direction should be due to the interaction with the corpuscular stream coming almost radially from the sun. The direction of the tails should differ slightly from the solar wind direction, since the comet is moving too and the comet tail should lag behind the radius vector of the sun, an effect which has actually been observed. The angle between the radius vector and the tail should give the ratio of the tangential component of the orbital velocity of the comet and the solar wind. The observed angles correspond to a velocity of about several hundred km/sec up to 1000 km/sec for the solar wind.

Other strong evidences of the existence of a solar wind are the observed high acceleration in Type I tails, which are much larger than observed in the tails of Type II and III. The accelerations in the ionized tails amount to at least 100 times the gravitational acceleration of the sun and must be due to a momentum transfer from the solar wind to the gas particles in the comet tails. But this interaction mechanism cannot be explained by direct collisions of the particles, and magnetic fields will probably play a role. That magnetic fields are of importance is also indicated by the fine structures which have been observed in ionized comet tails in contrast to the non-ionized tails (e.g. RH. LÜST, 1962). But until now no detailed mechanism has been developed, and therefore it is not yet possible to obtain an estimate for the particle density from the observed acceleration. It is planned however to create an artificial ionized cloud in the interplanetary space (BIERMANN, R. LÜST, RH. LÜST, and SCHMIDT, 1962), which might help to get a better understanding of the mechanism of the interaction with the solar corpuscular radiation and also to determine the density of the solar wind.

Finally the ionization of the molecules in the tails might be connected with the solar corpuscular radiation. BIERMANN proposed charge exchange between the solar protons and the non-ionized molecules as a possible process for the ionization, since ionization by the solar ultraviolet radiation would give a time scale much larger than that observed. But if one accepts the solar particle flux observed by spacecrafts, the time scale for ionization by charge exchange is also too large. At present no mechanism is known which can explain the observed short time scale for the ionization.

Further studies showed that the solar wind is also present during times of low solar activity. Four comets have been found with ionized tails during the time of sunspot minima 1900 and 1955 and during a period of very low geomagnetic activity (RH. LÜST, 1961). In one of these comet tails certain changes were observed to be related to geomagnetic disturbances. The time lag between the geomagnetic disturbance and the event on the comet indicated a velocity of the solar particles of about 300 to 400 km/sec^{-1}. Other investigations also using comet data indicated that the emission of corpuscular radiation seems to be present in all heliographic latitudes, but depends somewhat on the solar latitude in such a way that the intensity is somewhat lower at higher latitudes (STUMPFF, 1961).

Summarizing this part, we may say that it is also not yet possible by this method to deduce reliable estimates of the density of the interplanetary medium. However, the observed features in the ionized comet tails and especially their correlations with the solar and geomagnetic activity closely indicate the presence of the solar wind and give some values for the velocity during perturbed periods.

D. BOUNDARY OF THE GEOMAGNETIC FIELD

We cannot expect to find the interplanetary plasma at the surface of the earth, since the plasma interacts with the geomagnetic field. A certain volume surrounding the earth is probably very efficiently shielded from the solar wind, which is identical with the interplanetary medium (BIERMANN, 1957). The geomagnetic field should be completely confined within a cavity formed in the interplanetary plasma. The boundary of this cavity should be determined by the interaction of the interplanetary plasma with the geomagnetic field. The part of the boundary facing the sun (the upstream side) will depend mainly on the kinetic pressure of the solar wind, since the gas pressure is probably smaller than the kinetic pressure, while the downstream side of this interface and its extension will be influenced by the temperature of the interplanetary medium (JOHNSON, 1960). Therefore observations about the shape of its surface would allow us to draw certain conclusions about the properties of the interplanetary plasma. Especially the distance of this interface from the surface of the earth in the solar direction should enable us to determine the kinetic pressure of the solar wind. Of course it is necessary for the interpretation of such observations that we have some kind of a theory about this boundary.

We shall not discuss in detail here all the theoretical considerations which have been carried out already in connection with this problem. It was treated first by CHAPMAN and FERRARO (1930). More details and literature may be found in a review paper by PARKER (1962). The theoretical problem is very difficult since it is a so-called freeboundary problem, where the boundary is not given, but must be determined. Into the boundary conditions enter the values of the functions at the boundary which are not known beforehand.

If one regards the solar wind as composed of independently moving protons and electrons and assumes that the particles are elastically reflected from the sharp outer boundary of the geomagnetic field, the boundary condition is that the magnetic pressure inside the boundary must equal the impact pressure of the solar wind:

$$\frac{B_s^2}{8\pi} = 2Nmv^2 \cos\vartheta ,$$

if we neglect the hydrostatic pressure of the solar wind. Here N is the number of particles per cm^3, m is the mass of the particles, v is velocity of the particle and ϑ the angle between the incident velocity and the normal of the surface. B_s is the value of the magnetic field at the boundary.

Different kinds of approximations have been carried out in treating this problem in two and three dimensions (ŽINGULEV and ROMIŠEVSKIJ, 1959; DUNGEY, 1961; HURLEY, 1961; BEARD, 1960b; FERRARO, 1960; SPREITER and BRIGGS, 1962; MIDGLEY and DAVIS 1962; SLUTZ, 1962). As far as the actual shape of the boundary is concerned the problem in three dimensions has not yet been really solved. But the distance of this boundary from the earth in the equatorial plane on the sunward side might give a first estimate for the strength of the solar wind. This distance can be calculated approximately by replacing the impact pressure of the solar wind by a constant outside pressure. For this case MIDGLEY and DAVIS (1962) as well as SLUTZ (1962) have determined the shape of the cavity surface separating the earth's magnetic field from the interplanetary plasma. Different methods have been applied in these two papers, and their results are in good agreement. The magnetic field is compressed by the outside pressure, and its value is higher (2.6 times) at the boundary compared to the undisturbed dipole field. Therefore according to these calculations the boundary at the equator is 37 per cent farther out than the point where the undisturbed dipole field would balance the assumed external pressure.

The distance of the boundary at the equator is given by

$$R_b = 1.37\left[\frac{M}{\sqrt{8\pi p}}\right]^{1/3} = 1.37\left(\frac{B_0^2}{8\pi p}\right)^{1/6} R_e$$

where M is the dipole moment of the earth, B_0 the magnetic field at the surface of the earth, R_e the radius of the earth and p the outside pressure exerted by the plasma.

If we take the measurements obtained with Explorer X (see page 187) ($N = 3.10^8$ Protons/cm^3 and $v = 300$ km/sec), this boundary should be located at 9.6 R_e in the equatorial plane assuming that the particles are reflected at this surface and that they exert a pressure of 2 Nmv^2. If the particles would be brought to rest at this point, the pressure would be just half as large, and the equatorial radius of the boundary would increase to 10.8 R_e.

E. MEASUREMENTS FROM SPACE VEHICLES

During the last three years it has been possible to obtain the first direct evidence about the interplanetary medium by measurements from space vehicles. While the observations of the scattered sunlight and of scattered radiowaves give indications of the electron component of the interplanetary plasma, the space vehicles carried plasma probes for measuring the ion component. The expected mass motions should be

TABLE I

INTERPLANETARY PLASMA DATA GAINED WITH SATELLITES AND SPACE PROBES

Vehicle	*Distance from the earth's centre* km	*Ion flux* $cm^{-2} sec^{-1}$	*Energy* eV	*Velocity* km sec^{-1}	*Density* cm^{-3}	*Temperature* 10^5 °K	*Kp-index*
Lunik I	$> 10^5$	1–2 10^8	> 15	–	–	–	0–1
Lunik II	2.5–3.7 10^5	2 10^8	> 15	–	–	–	2–3
Lunik III	1.26 10^5	4 10^8	> 30	–	–	–	4–6
Lunik III	4.5 10^5	$\ll$ 4 10^8	> 30	–	–	–	2
Sov. Venus probe	1.9 10^6	1 10^9	> 50	–	–	–	5–7
Explorer X	1.4–3.0 10^5	1–4 10^8	250–800	200–400	6–20	$\simeq 1$	2–3
Mariner II	7 10^5–8.7 10^7	10^8	500–3000	360–700	0.3–10	0.6–5	

more easily detectable from the ions, since the thermal motion of the ions is much smaller than that of the electrons assuming equal temperatures for both components.

These plasma probes were essentially Faraday caps which measured the received ion current on a collector plate. The collector plate had been shielded by two or more grids to suppress photoelectric effects from the collector plate and to measure the energy of the incoming particles by changing the potential of one of the grids with respect to the body of the space vehicle. (Nothing is known at present about the potential of the vehicle with respect to its surrounding).

Such plasma probes were carried by a number of Russian space crafts (GRINGAUZ *et al.*, 1950, 1961) and by the American satellite Explorer X (BRIDGE *et al.*, 1962) and by the American space probe Mariner II (NEUGEBAUER and SNYDER, 1962, 1963). The results are summarized in Table I.

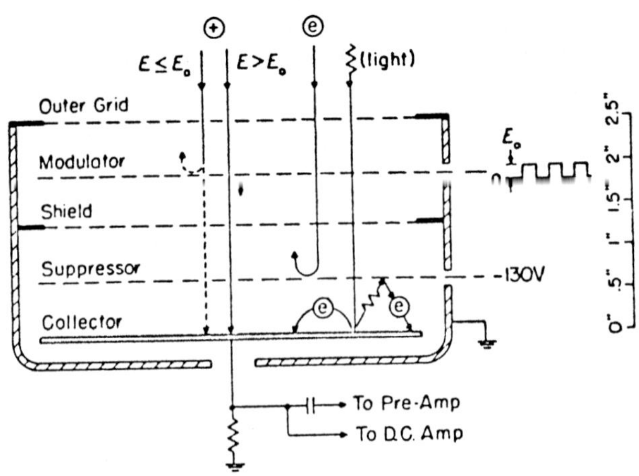

Fig. 1. A schematic drawing of the plasma cup showing the modulation of protons with energies less than E_0 eV. Protons with larger energies reach the collector while *dc* currents due to plasma electrons and photoelectrons are reduced by a suppressor grid.

These observations show that the interplanetary plasma is only observed beyond a certain distance from the earth, since the earth's magnetic field shields a certain region around the earth from the solar wind.

In Figure 1 the method used in Explorer X and in Figure 2 the apparatus used in Mariner II are shown respectively.

The orbit of Explorer X was in the downstream direction (140° to the earth-sun line), and it crossed the boundary apparently at 22 R_e. This shows that the distance of this boundary from the surface of the earth is larger in the direction away from the

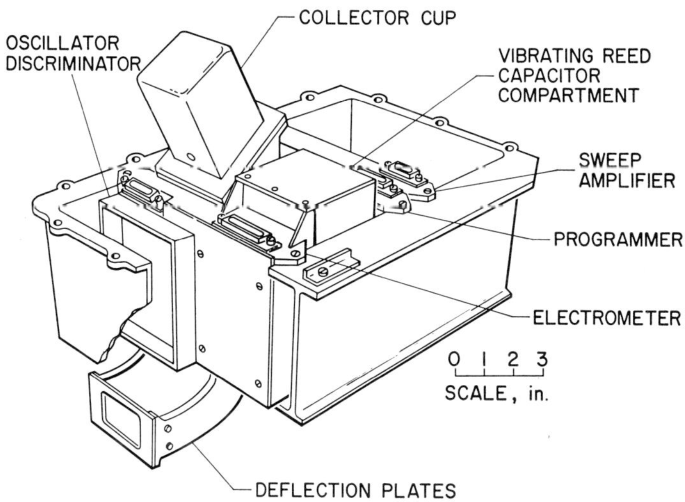

Fig. 2. Plasma spectrometer, mounted in a spacecraft instrument case (shown partially cut away).

sun compared to the direction towards the sun (e.g. the measurements obtained by Explorer XII). But the observed plasma flow fluctuated outward from this distance; it disappeared and reappeared on several occasions. When the plasma was observed, the magnetic field was weak, fluctuating in magnitude and direction, while the absence of a noticeable plasma was associated with a relatively strong steady field of almost radial direction away from the earth. The disappearance of the plasma together with the change of the magnetic field might be due to the fact that the satellite was moving almost along the boundary of the geomagnetic cavity and that this surface was changing its position in space. The motion of the boundary might be caused by fluctuations of the intensity of the solar wind. Since Explorer X was so close to this boundary all the time and since the geomagnetic cavity might be surrounded by a kind of "bow wave" as OLBERT (1962) pointed out, the observed properties of the plasma flow might not be the same as in the free interplanetary space.

The flux observed by Explorer X fluctuated around a mean value of about 3×10^8 particles/cm^2 sec with a mean kinetic energy of 500 eV. This gives a mean mass velocity for protons of about 300 km/sec. From the observed energy and angular spread, a temperature between 10^5–10^6 °K might be concluded, but this high temperature might be due to the fact that the plasma passed through the bow wave mentioned above. The plasma arrived from the general direction of the sun, since it has been observed within a "window" of about 20° × 60° aperture, which included the sun.

Finally it should be mentioned that an increase of the particle flux and the mean kinetic energy of the particles was observed at about 40 R_e simultaneously with a sudden commencement of a magnetic storm recorded at the earth.

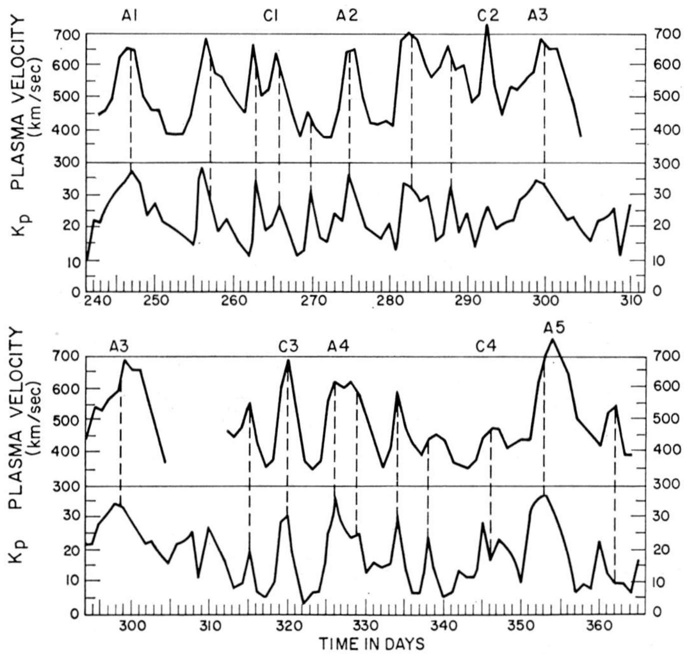

Fig. 3. Daily "average" plasma velocity compared with planetary magnetic activity index K_p (See text for definition of average velocity).

The most detailed data concerning the interplanetary plasma are from Mariner II. It covered a period of more than 100 days in measuring the particle flux and magnetic fields in the interplanetary space out to a distance of about 87 million km from the earth and 0.708 AU from the sun. For the reduction of the measurements it was assumed that the bulk velocity of the plasma is always radially outward from the sun.

The measurements showed that there existed a measurable flow of plasma from the direction of the sun at all times during the Mariner mission. The velocity had values between 330 km/sec and about 800 km/sec, the particle density ranged from about 0.3 particles/cm^3 to about 30 particles/cm^3. A temperature was derived from the energy distribution in the radial direction and 90% of the temperature was between 6×10^4 and 5×10^5 °K. The plasma appeared to contain protons as well as alpha particles. The velocity was seldom constant for more than a few hours and it showed a series of maxima which were correlated with the geomagnetic activity and which showed a clear tendency to recur after 27 days (cf. Figure 3).

Several attempts have been made to measure the magnetic fields in the interplanetary space by magnetometers carried in space vehicles (COLEMAN *et al.*, 1960 and 1962; HEPPNER *et al.*, 1962). The results and the consequent picture are still not too clear, but at least it is clear that there exist magnetic fields of the order of some 10^{-5} gauss under quiet conditions. More details will be discussed in the seminar by Prof. L. Davis.

F. INFLUENCE ON GALACTIC AND SOLAR COSMIC RAYS

It would go beyond the scope of this article to discuss in detail the behavior of the galactic and the solar cosmic rays in the interplanetary space. It is clear that these energetic particles may be regarded as probes which can give us some important information about the state of the interplanetary medium, in particular about the strength and the shape of the magnetic fields present there. There are a number of effects, which must be attributed to the interaction between cosmic rays and interplanetary magnetic fields, but care is needed in applying their consequences to the general conditions in the interplanetary space.

In the behavior of the cosmic rays coming from outer space, two effects have been observed which are not of geocentric origin, but which must be caused in the interplanetary space. The first effect is the variation of the intensity of the cosmic radiation over an eleven-year solar cycle (MEYER and SIMPSON, 1955, 1957; NEHER, 1956). Different models have been proposed to explain this variation (MORRISON, 1956; PARKER, 1958). Their main feature is that the sun controls and modulates the galactic flux of cosmic rays through magnetic fields transported by the solar or interplanetary plasma. For the understanding of this effect, the effects at the interface between the interplanetary and the galactic plasma are very probably also of importance. That the plasma from the sun might make a large cavity or a bubble in the galactic magnetic field was suggested first by DAVIS (1955). (Some of the features of this active boundary of the interplanetary medium shall be discussed on page 203.) As DAVIS (1962) has pointed out, this interface might be unstable since its surface is concave towards the plasma. This instability might allow bubbles of gas to escape into the galactic field and exite Alfvén waves to run along the galactic field away from the cavity. Thus galactic cosmic rays trapped inside the cavity are pumped out, and the diffusion in it is impeded. The theories mentioned here offer an explanation of the eleven-year intensity variation, but it is impossible to deduce at present some more or less precise figures for the strength of the magnetic field, its shape and the size of its irregularities.

The second effect is the so-called Forbush decrease. There the galactic cosmic ray intensity is decreased within a few hours, while its recovery to its normal value takes place over a much longer period (several days to weeks). It has been demonstrated clearly by the measurements with the space probe Pioneer V (FAN, MEYER and SIMPSON, 1961; COLEMAN *et al.*, 1961) that the Forbush decrease is also a phenomenon which occurs in the interplanetary medium. But since this phenomenon is related to the perturbed state of the interplanetary medium we shall not discuss it further (For references see CARMICHAEL, 1962; PARKER, 1962; GOLD, 1962). It has been estimated that magnetic fields of a strength of the order of 10^{-4} gauss would be able to produce these events, and the observations with Pioneer V indicated a field strength of about 4.10^{-4} gauss.

Of particular importance are the solar cosmic ray intensity increases in connection with solar flares. The high-energy solar particle events have been reviewed by CARMICHAEL (1962) and those of lower particle energy causing the polar cap absorption effect by ROSE and ZIAUDDIN (1962). Therefore, only the results related to the interplanetary space shall be stated here. Historically the solar cosmic ray effect was the first evidence for the existence of magnetic fields in the interplanetary space (SCHLÜTER 1953).

There are a number of solar cosmic ray effects that indicate the presence of magnetic fields between the sun and the earth. (1) The arrival of energetic solar protons at high geomagnetic latitudes, since these regions are forbidden for direct arrival from the sun. (2) Arrival from a direction differing from that to the sun. (3) The time of arrival with respect to the onset of the flare event on the sun. (4) The isotropy after the initial increase and the slow decay of the increase. Furthermore it should be mentioned in this connection that the detection of flare-produced cosmic rays in the vicinity of the earth is more likely if the flare is in the western portion of the solar disc and that the rise time of the intensity is shorter for flares near the western limb of the sun compared to those events where the flare is near the center of the disk.

Trajectories of the solar cosmic ray particles in the geomagnetic field have been calculated to analyze these effects, (e.g. LÜST, SCHLÜTER and KATTERBACH, 1956; JORY, 1956; LÜST, 1957) and models have been proposed by several authors (MEYER, PARKER and SIMPSON, 1956; COCCONI *et al.*, 1957; LÜST and SIMPSON, 1957; GOLD, 1959).

A very detailed analysis of the different events has been carried out by MCCRACKEN (1962). His general conclusion was that the cosmic ray flux has been the greatest from a direction 50° to the west of the sun and about 10° north of the plane of the ecliptic. This indicates that during the times of the flare events the magnetic lines of force have an angle of about 50° with respect to the earth-sun line. This angle – the so-called "garden-hose" angle – together with the other facts mentioned above suggests that the magnetic field has a spiral structure between the sun and the earth during the time of the flare event connecting in the plane of the ecliptic the western limb of the sun with the earth. One would expect such a spiral structure of the magnetic field connected to the sun due to the rotation of the sun. The spread in the arrival directions of the solar cosmic ray particles indicates small-scale irregularities. GOLD (1959) assumes that both ends of the magnetic lines of force are connected to the same active group at the sun and form a bottle which is set up by the plasma ejected by a solar flare. PARKER assumes that the magnetic lines of force are stretched out by the solar wind all the time and that a solar flare causes a blast wave running outwards and shearing the lines of forces. Both models describe the main features of the solar flare event and of a Forbush decrease. But it should be emphasized again that the evidence about the magnetic fields from these cosmic ray data refers to the fields existing during substantial geophysical disturbance. Therefore during undisturbed periods the field might be different in character (see page 206).

Summarizing this part, it can be said that the cosmic rays have given already very valuable information about the magnetic fields in the interplanetary space but related to mere exceptional periods. Since the cosmic ray particles probe a large volume – either the region between the sun and the earth or the entire interplanetary space – at one instant of time, cosmic ray observations will always be very useful and will supplement direct measurements of the field using magnetometers carried in space probes which give data from different points at different times. But the cosmic ray data provide information only in an indirect way and their interpretation depends, therefore, quite seriously on theoretical assumptions.

G. GEOMAGNETIC EVIDENCE

Since we tried to restrict ourselves in this discussion to the undisturbed conditions in the interplanetary space, we shall not go into any detail about the geophysical effects and their correlations with solar activity. It is clear that the different geomagnetic effects are directly related to the perturbed conditions in the interplanetary space caused by the sun and they have given the first indications of the corpuscular radiation from the sun. It was even concluded from the perturbations in the geomagnetic field, e.g. measured by the three-hour planetary index K_p (BARTELS, 1951, 1958), that the corpuscular radiation must arrive almost all the time since quiet intervals ($K_p=0$) are extremely rare (even rarer than large magnetic storms) (BARTELS, 1945).

The small geomagnetic disturbances show a tendency of recurrence at regular intervals of 27 days, which is the rotational period of the sun as seen from the earth. It is generally assumed that from certain regions on the rotating sun an enhanced corpuscular radiation is emitted continuously during a certain period of time. These regions on the sun have been called M-regions by Bartels. The particles will form a spiral at a given moment since the solar rotation is superposed to the radial outward motion. More detailed studies have shown that the corpuscular emission from the sun is connected with local magnetic fields at the solar surface and with a brightening of certain chromospheric spectral lines. In this way MUSTEL and his co-workers (1961) derived velocities of the order of 200–300 km/sec for the corpuscular stream and about 400 km/sec during periods of higher activity. The large geomagnetic storms are caused by plasma clouds which are moving with a velocity of the order of about 1000 km/sec. This velocity has been derived from the delay time between the occurrence of the solar flare responsible for the ejection of the plasma and the onset of the geomagnetic storm. Such a large geomagnetic storm is accompanied by a number of other geophysical effects such as aurora, radio blackouts, perturbations and changes in the radiation belts and sometimes, as mentioned above, Forbush-decreases of the cosmic ray intensity.

TABLE II

SOME CHARACTERISTICS OF THE INTERPLANETARY PLASMA

			Interplanetary space ($B = 10^{-5}$ gauss)				Corona ($B = 1$ gauss)
	Symbol	*Unit*	$n[cm^3]$	$T[°K]$: 10^4	10^5	10^6	$n = 10^8\ cm^{-3}$, $T = 10^6$ °K
Plasma frequency	ω_p	sec^{-1}	1		5.64×10^4		5.64×10^8
			10		17.85×10^4		
			10^2		56.4×10^4		
Cyclotron frequency for	ω_{gi}	sec^{-1}			9.58×10^{-2}		9.58×10^3
protons and electrons	ω_{ge}				1.76×10^2		1.76×10^7
Radius of gyration	R_{gi}	cm		1.64×10^7	5.18×10^7	1.64×10^8	1.64×10^3
	R_{ge}			3.83×10^5	1.21×10^6	3.83×10^6	3.83×10^1
Debye length	h_D	cm	1	6.9×10^2	2.18×10^3	6.9×10^3	0.69
			10	2.18×10^2	6.9×10^2	2.18×10^3	
			10^2	6.9×10^1	2.18×10^2	6.9×10^2	
Scale height	H	cm		2.79×10^{12}	2.79×10^{13}	2.79×10^{14}	6.04×10^9
Mean free path	λ	cm	1	6.9×10^{11}	6.0×10^{13}	5.4×10^{15}	8×10^7
			10	7.3×10^{10}	6.3×10^{12}	5.7×10^{14}	
			10^2	7.6×10^9	6.6×10^{11}	5.9×10^{13}	
Electrical conductivity	σ	esu		6.4×10^{12}	1.75×10^{14}	4.9×10^{15}	6.9×10^{15}
conductivity		$(\Omega m)^{-1}$		7.1×10^2	1.95×10^4	5.45×10^5	7.7×10^5
Alfvén velocity	V_A	cm. sec^{-1}	1		2.18×10^6		2.18×10^8
			10		6.9×10^5		
			10^2		2.18×10^5		
Sound velocity	v_s	cm. sec^{-1}		1.66×10^6	5.25×10^6	1.66×10^7	1.66×10^7

From the change of the geomagnetic field (about 100 γ) during the initial phase of the storm it was possible to get an estimate for the particle flux of the enhanced corpuscular emission from the sun. But this depends also on the theoretical model. FERRARO (1955) has suggested an upper limit of 100 protons/cm^3 for the particle density, while PARKER (1962) states a value of 1500 km/sec and 30 protons/cm^3. BLACKWELL and INGHAM (1961b) derived 300 particles/cm^3 from an increase in the brightness of the zodiacal light after a solar flare.

2. Theoretical Models

A. GENERAL CONSIDERATIONS

Before attempting to find some theoretical models describing the state of the interplanetary plasma, we shall discuss some general properties of the medium. Especially we must consider in what way we can describe the phenomena in the interplanetary space. Since the interplanetary gas is very tenuous, it is not obvious that the hydrodynamical or magneto-hydrodynamical picture can be applied. The condition for this is that the mean free path λ of the particles is small compared to the dimensions of the system L or to relevant lengths which are of importance, such as the scale height H (that is, the height interval in which the density decreases by a factor e). In the presence of a magnetic field it might be sufficient for the validity of the fluid picture that the radius of gyration of the particles is small compared to the scales mentioned above.

In the following discussion we shall assume for simplification a particle density of $N=10$ particles/cm^3, a temperature of $T=10^5$ °K, a velocity of $v=300$ km/sec, and a magnetic field strength $B=10^{-5}$ gauss near the orbit of the earth. But, as mentioned above, we should keep in mind the large uncertainties about these values; other parameter values are given in Table II.

But first of all we must see whether the gas is ionized or not and if we are dealing with a plasma. Even if we knew nothing about the origin of the gas in the interplanetary space, it would be certain that the gas would be ionized out to a considerable distance from the sun. Due to the ionization by the solar radiation, the sun will in any case be surrounded by an HII region – a so-called Strömgren sphere (STRÖMGREN, 1939) – and according to this picture the interplanetary gas would be ionized till beyond the orbit of Jupiter (ÖPIK, 1954). The time of photo-ionization for a hydrogen atom may be estimated by using the rocket observations of HINTEREGGER (1961). With a flux of $\sim 10^{10}$ quanta/cm^{2}sec (beyond the Lyman limit) and an absorption coefficient of 6×10^{-18} cm^2 for an H-atom (in the ground level), a mean lifetime of about 1.5×10^7 sec $=1.7\times10^2$ days follows.

Since, as we have seen from the observations, the gas in the interplanetary space is coming from the solar corona, it is already ionized. In the corona, the ionization is caused by the collisions of the electrons. Since the recombination rate is very long compared to the travel time through the corona and the interplanetary space, the gas will remain ionized (KAHN, 1961). The recombination time is given roughly by

$\tau = 2 \times 10^{10}\ T^{1/2} N_e^{-1}$ (sec), where N_e is the electron density and T the temperature in degrees Kelvin. With $T = 10^5$ °K and $N_e = 10$ electrons/cm^3 we obtain a recombination time $\tau = 6 \times 10^{11}$ sec $= 1.9 \times 10^4$ years. With a velocity of 300 km/sec the gas travels over one astronomical unit (AU) in 5×10^5 sec $= 5.8$ days. The ionized gas may be described as a plasma if the Debye length

$$h_D \approx 6.9 \left(\frac{T}{N_e} \right)^{1/2} \text{[cm]}$$

is small compared to the relevant scales of the system. With $T = 10^5$ °K and $N_e = 10$ electrons/cm^3 we obtain $h_D = 6.9 \times 10^2$ cm. This shows that the Debye length is certainly small with respect to all relevant dimensions like the scale height or the mean free path. But this figure indicates also that the Debye length might be comparable to the scale of the space probes and its instruments measuring the properties of the plasma. We see further that the Debye volume is large enough to contain a sufficient number of particles.

The mean free path of the protons and electrons is roughly given by

$$\lambda \approx 1.6 \times 10^5 \frac{T}{N \ln \Lambda} \text{[cm]}$$

where $\ln \Lambda \approx 20$ for the density and temperature range in question. This gives in the corona with $T = 10^6$ °K and $N = 10^8$ particles/cm^3: $\lambda = 8 \times 10^7$ cm and near the orbit of the earth with $T = 10^5$ °K and $N = 10$ particles/cm^3: $\lambda = 6.3 \times 10^{12}$ cm. Therefore, for this example the mean free path in the corona is small compared to the radius of the sun ($\approx 7 \times 10^{10}$ cm), but near the earth it is almost comparable to the mean radius of the orbit of the earth (1 AU $= 1.5 \times 10^{13}$ cm). But for higher temperatures of the interplanetary plasma the mean free path would be of the same order of magnitude as the dimensions of the solar system. The scale height of an isothermal atmosphere

$$H = 6 \times 10^3\, T \left(\frac{r}{r_\odot} \right)^2$$

(r = distance from the center of the sun, $r_\odot$ = solar radius) is in the solar corona $H \approx 6 \times 10^9$ cm and near the orbit of the earth $H \approx 28 \times 10^{13}$ cm.

Finally we should state the radius of gyration for the protons and electrons in a magnetic field of 10^{-5} gauss. For the protons we have $R_{gp} \approx 1.64\ T^{1/2}\ B^{-1}$ cm and for the electrons $R_{ge} \approx 3.83 \times 10^{-2}\ T^{1/2}\ B^{-1}$ cm. This gives $R_{gp} \approx 5.2 \times 10^7$ cm and $R_{ge} \approx 1.2 \times 10^6$ cm. Therefore the radius of gyration is always extremely small compared to the scale of the system.

From this comparison we see that a fluid picture can be applied for describing the plasma in the interplanetary space. But if the temperature is too high (about 10^6 °K), the collisions might not be frequent enough to establish an isotropic pressure. This might have some consequences as far as stability is concerned in certain regions. The magnetic fields present in the interplanetary space are frozen into the plasma since the electrical conductivity σ is very large. The electrical conductivity is given by

$$\sigma \approx 1.38 \times 10^8 \frac{T^{3/2}}{1n} [\text{sec}^{-1}]$$

which leads to $\sigma \approx 1.75 \times 10^{14}$ sec^{-1} near the orbit of the earth. The decay time of a magnetic field due to Ohmic dissipation is approximately

$$\tau = \frac{4\pi L^2 \sigma}{c^2} [\text{sec}]$$

with L as a characteristic length and c the velocity of light. Taking $L = 1$ AU, we obtain $\tau = 5.5 \times 10^{20}$ sec $= 1.75 \times 10^{13}$ years. Therefore the magnetic field will be carried by the plasma. The sound velocity for $T = 10^5$ °K is about $V_s \approx 5.3 \times 10^6$ cm/sec, and the Alfvén velocity about $V_A \approx 6.9 \times 10^5$ cm/sec assuming $N = 10$ particles/cm^3 and $B = 10^{-5}$ gauss. This means that the observed velocities of about 300 km/sec are supersonic.

In Table II the different quantities mentioned above and some others are summarized for the solar corona and for the interplanetary space (near the orbit of the earth) by covering a density range from 1 to 10^2 particles/cm^3 and a temperature range from 10^4 to 10^6 °K.

Before we apply the fluid equations, we have to consider whether one can treat the interplanetary medium by a static model, by a stationary model, or by a nonstationary model. Whether a static model (CHAPMAN, 1957) is appropriate is mainly a question of whether the boundary conditions can be fulfilled. We shall see that a static model seems almost impossible. The choice between a stationary and instationary model depends on the possibility that there is really an unperturbed interplanetary medium at all and on the time scales of the perturbations.

Since the mathematical difficulties are so much greater for a nonstationary model, it is natural to try stationary models first, (PARKER, 1958, 1960, 1961b), and we shall see that these models will give us already a good description of the behaviour of the interplanetary medium. A certain nonstationary model has been treated too for the perturbed state of the interplanetary medium by using certain time-dependent similarity solutions (PARKER, 1961a). But since we restrict ourselves in this article to the unperturbed state of the interplanetary medium we shall only discuss the static and stationary models. The influence of magnitic fields will be neglected completely in these two models.

B. STATIC MODEL

Static models of the interplanetary medium have been assumed by CHAPMAN (1957). He treats the interplanetary medium as an extension of the solar corona and assumes a number of idealized conditions: (1) No magnetic fields, (2) Spherical symmetry, (3)

Material at rest, (4) Steady state. In this case, the pressure gradient must be balanced by the gravitational force:

$$\frac{dp}{dr} = -\varrho g\,, \tag{1}$$

where p is the pressure, ϱ the density, g the gravitational acceleration, and r the radial distance. Furthermore assumption (4) requires the heat flux F be constant:

$$F = -4\pi r^2 K \frac{dT}{dr} = \text{const}\,, \tag{2}$$

where T is the temperature and K the thermal conductivity. (The heat loss by radiation is small compared to the conductive loss in the corona and therefore has been neglected by Chapman up to the orbit of Jupiter (CHAPMAN, 1959 and POTTASCH, 1960). If one takes into account only the thermal molecular conductivity K_m which for a fully ionized gas is given by

$$K_m = 5 \times 10^{-7} T^{5/2} \text{ ergs cm}^{-1} \text{ deg}^{-1} \text{ sec}^{-1}\,,$$

Equation (2) can be integrated directly and gives

$$\frac{T}{T_0} = \left(\frac{r_0}{r}\right)^{2/7} + \frac{T_\infty}{T_0}. \tag{3}$$

Here T_0 is the temperature at the level r_0 (the level of the corona). The heat flux is given by

$$F = \frac{40\pi}{7} r_0 T^{7/2}\,, \tag{4}$$

and T_∞ is the temperature at infinity which can be chosen freely. Since T_0, the temperature of the coronal level, is of the order of a million degrees and T_∞ should not be higher than 10^4 degrees, the last term in Equation (3) is not of significance for fixing the temperature near the earth's orbit ($r = r_e$).
By introducing

$$p = 2nkT \qquad \text{and} \qquad \varrho = n\bar{m}\,, \tag{5),(6}$$

where n denotes the ion number density, $\bar{m}$ the mean ion mass and k the Boltzmann constant, and using Equation (3) with $T_\infty = 0$ one gets:

$$\frac{n}{n_0} = \left(\frac{r}{r_0}\right)^{2/7} \exp\left[-\frac{7r_0}{5H_0}\left\{1 - \frac{r^{-5/7}}{r_0}\right\}\right]. \tag{7}$$

Here H_0 is the scale height of the corona at the level $r = r_0$

$$H_0 = \frac{2kT_0}{\bar{m}g_0} \tag{8}$$

with g_0 the value of the solar gravity at this level, namely $GM_\odot/r_0^2$, where G is the gravitational constant and $M_\odot$ the solar mass.

The pressure is then given by

$$\frac{p}{p_0} = \exp\left[-\frac{7r_0}{5H_0}\left\{1 - \frac{r^{-5/7}}{r_0}\right\}\right]. \tag{9}$$

According to Equation (7) the density reaches a minimum value $n_{\min}$ with

$$\frac{n_{\min}}{n_0} = \left(\frac{7r_0}{H_0 2}\right)^{2/5} \exp\left[-\frac{7r_0}{5H_0} + \frac{2}{5}\right] \tag{10}$$

at the distance

$$r_{\min} = r_0\left(\frac{7r_0}{2H_0}\right)^{7/5} \tag{11}$$

while the density at infinity becomes infinite but in such a way that the pressure remains finite there, namely:

$$p_\infty = p_0 \exp\left[-\frac{7r_0}{5H_0}\right]. \tag{12}$$

The temperature decreases, according to this model, only very slowly with increasing distances from the sun. It might be compared with the observations out to some 20 solar radii by using electron density data based on eclipse observations and assuming again hydrostatic equilibrium as has been done by CHAPMAN (1959) and POTTASCH (1960). Both authors find that the temperature decreases more rapidly than predicted by Equation (3) assuming only molecular heat conduction as the determining factor for the temperature distribution. Chapman suggests that the conduction due to turbulence contributes to the heat flow, and in this way the temperature gradient should be increased. Pottasch points out the possibility that cooling by radiation could be of importance due to the emission from metallic ions.

PARKER has stressed another difficulty with this solution, namely that the final pressure remains finite at infinity (Equation (12)) due to the slow temperature decrease. The corresponding values of p_∞ are much higher than those to be expected in the interstellar space; therefore, this idealized solution cannot be fitted to the interstellar space. But the observations already give a steeper decrease of temperature. The limit would be that the temperature falls off with the adiabatic gradient, since a faster decrease would cause a convective instability and this would restore the adiabatic gradient.

Therefore, let us assume now that the density and the pressure obey the adiabatic relation

$$\frac{p}{p_1} = \left(\frac{n}{n_1}\right)^\gamma \tag{13}$$

where γ is the ratio of the specific heats (for a monatomic gas $\gamma = \frac{5}{3}$), p_1 and n_1 are the pressure and the density at a certain level $r=r_1$. With Equation (13) we get from the hydrostatic equilibrium Equation (1)

$$\frac{n}{n_1}=\left\{\frac{\gamma-1}{\gamma}\frac{r_1^2}{H_1}\left(\frac{1}{r}-\frac{1}{r_1}\right)+1\right\}^{1/\gamma-1} \tag{14}$$

$$\frac{p}{p_1}=\left\{\frac{\gamma-1}{\gamma}\frac{r_1^2}{H_1}\left(\frac{1}{r}-\frac{1}{r_1}\right)+1\right\}^{\gamma/\gamma-1} \tag{15}$$

and

$$\frac{T}{T_1}=\frac{\gamma-1}{\gamma}\frac{r_1^2}{H_1}\left(\frac{1}{r}-\frac{1}{r_1}\right)+1 \tag{16}$$

where

$$H_1=\frac{2kT_1}{mGM_\odot}r_1^2=\frac{kT_1}{mg_1}. \tag{17}$$

Equation (15) shows that in principle the pressure may become zero if the temperature gradient is adiabatic from a certain level. The condition for this to happen is that

$$\frac{\gamma-1}{\gamma}\frac{r_1}{H_1}\geqslant 1$$

or

$$\frac{\gamma-1}{\gamma}\frac{mGM_\odot}{EkT_1r_1}=\frac{\gamma-1}{\gamma}\frac{7.94\times 10^{17}}{T_1r_1}\geqslant 1\,. \tag{18}$$

If this condition is not fulfilled, the pressure at infinity is

$$p_\infty=p_0\left\{1-\frac{\gamma-1}{\gamma}\frac{r_1}{H_1}\right\}^{\gamma/\gamma-1}. \tag{19}$$

The observations mentioned above show that inside of $20r_\odot$ the temperature gradient is certainly not the adiabatic one (according to Chapman about half the adiabatic), and outside this region no good data are available. With a temperature $T=2.3\times 10^5$ °K at $r=20r_\odot$ as given by Pottasch one finds a value of

$$\frac{r_1}{H_1}=2.32$$

and

$$\frac{\gamma-1}{\gamma}\frac{r_1}{H_1}=0.928$$

with $\gamma=\frac{5}{3}$. Therefore if the temperature fell off from this level adiabatically, the pressure at infinity could drop almost to zero. Chapman extrapolates from the observation that the adiabatic temperature gradient is reached at $r=50\,r_\odot$ and that the temperature at this distance should be $T=1.57\times 10^5$ °K. In this case we would obtain $r_1/H_1=1.46$ and

$$\frac{\gamma-1}{\gamma}\frac{r_1}{H_1}=0.56$$

for $\gamma=\frac{5}{3}$. Since the uncertainties of this value are quite high and since also γ might be somewhat higher if radial magnetic fields are present, it cannot be excluded at present that the pressure falls off sufficiently rapidly and that it can be balanced in this way by the pressure of the interstellar medium.

But on the other hand it is also not unreasonable to investigate nonstatic solutions which might then be regarded as the natural consequence of the unbalanced pressure of the corona. Before this is done, it should however, be remarked that the static solutions as presented above will be directly applicable to observations only on the rotational axis of the sun, since the solar rotation has not been included. Furthermore, a possible influence of magnetic fields has not been taken into account.

C. STATIONARY MODEL

1. *Basic equations*

It will now be assumed that the matter is not at rest, but streaming with the velocity v. In the stationary case we then have the following equations:

The equation of continuity

$$\operatorname{div}(\varrho v) = 0, \tag{20}$$

the momentum equation

$$\varrho (v \operatorname{grad}) v = - \operatorname{grad} p - \varrho g, \tag{21}$$

and the energy equation

$$(v \operatorname{grad} e) = - p\left(v \operatorname{grad}\left(\frac{1}{\varrho}\right)\right) + \operatorname{div}(\kappa \operatorname{grad} T), \tag{22}$$

where e is the specific internal energy, which for an ideal gas is

$$e = \frac{1}{\gamma - 1} \frac{p}{\varrho} \tag{23}$$

In the case of spherical symmetry these equations can be simplified. Equation (20) can be integrated and gives

$$nr^2 v - C, \tag{24}$$

where $4\pi C$ – constant is the total number of particles leaving the sun per unit of time. Some dimensionless variables are useful for further discussion. With the radius a for the base of the corona, we define:

$$\xi = \frac{r}{a}. \tag{25}$$

The ratio of the kinetic energy to the thermal energy at $r=a$ shall be

$$\psi(\xi) = \frac{1}{4} \frac{mv^2(\xi)}{kT_0}, \tag{26}$$

and that of the gravitational potential energy to the thermal energy

$$\lambda(\xi) = \frac{GM_{\odot}m}{2kT_0 r}. \tag{27}$$

Using these relations in Equations (21) and (22), these can be written:

$$\frac{\mathrm{d}}{\mathrm{d}\xi}\left(\psi + \frac{T}{T_0} - \lambda\right) = \frac{T1}{T_0\xi^2\psi^{1/2}}\frac{\mathrm{d}}{\mathrm{d}\xi}(\xi^2\psi^{1/2}), \tag{28}$$

and

$$\frac{\mathrm{d}}{\mathrm{d}\xi}\left(\tfrac{1}{2}\psi + \frac{\gamma}{\gamma-1}\frac{T}{T_0} - \lambda\right) = A\frac{\mathrm{d}}{\mathrm{d}\xi}\left\{\kappa_0\left(\frac{T}{T_0}\right)^{5/2}\xi^2\frac{\mathrm{d}}{\mathrm{d}\xi}\left(\frac{T}{T_0}\right)\right\} \tag{29}$$

where

$$A = \frac{a\kappa_0}{Ck}. \tag{30}$$

For the heat conduction it has been assumed that:

$$\kappa = \kappa_0\left(\frac{T}{T_0}\right)^{5/2} \tag{31}$$

which should be valid for an ionized gas.

Equation (29) can be integrated directly and gives

$$\psi + \frac{\gamma}{\gamma-1}\frac{T}{T_0} - \lambda - A\kappa_0\left(\frac{T}{T_0}\right)^{5/2}\xi\frac{\mathrm{d}}{\mathrm{d}\xi}\left(\frac{T}{T_0}\right) = E \tag{32}$$

where E is an integration constant and proportional to the energy flux. The character of the solutions of these equations depends on whether or not E is zero. Chamberlain (1960, 1961) argued that E should be zero in contrast to Parker. If the observations were accurate enough (e.g. in the solar corona or at large distances), this question could be solved easily. At present the observations do not allow the exclusion of the possibility that E is zero, but the observed velocities near the earth's orbit indicate that an E different from zero is much more probable. Clauser (1961) and Germain (1961) have pointed out that the solutions of the above equations are analogous to the solutions of a flow through a Laval nozzle, with the gravitational field playing the role of the throat of the nozzle. In these cases solutions with E not equal to zero have been observed.

With the heat conduction term the equation can be solved only numerically. This has been done by Chamberlain for the case $E=0$. But he had to reduce the thermal conductivity in an arbitrary way to produce an agreement with the observed coronal densities. The reduction of the heat conduction could be due to magnetic fields or to deviations of the Maxwellian distribution according to Chamberlain. For the general case including the heat conduction the equation has been solved by De Jager (1963).

2. *Polytropic solutions*

For a more general discussion of the properties of the solutions we now follow Parker and treat polytropic models instead of taking the heat conduction term explicitely into account (See also DEUTSCH, 1961; WEYMANN, 1960; and BARDEEN, 1962). In this case analytical solutions can be obtained. The energy equation (22) will be replaced by the relation between the temperature and the density

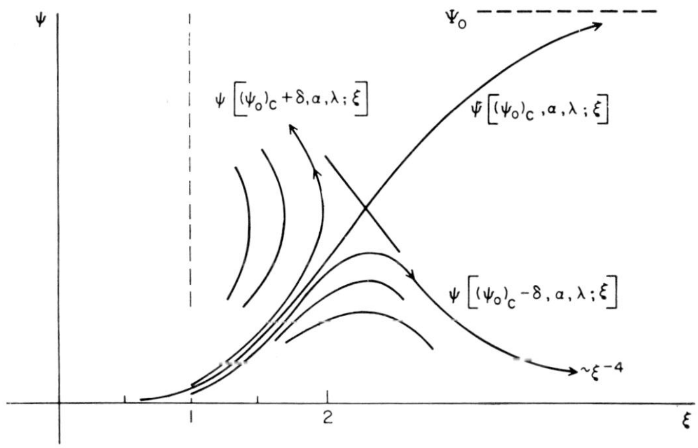

Fig. 4. The general form of the radial velocity function (ψ_0, α, λ, ξ) for ψ_0 equal to the critical value $(\psi_0)_c$, and also slightly less and greater. It is supposed that

$$\alpha < \tfrac{3}{2} \text{ and } 1 \ll \lambda \ll \frac{2}{\alpha - 1}$$

ψ is proportional to v(v = velocity of the plasma), is proportional to the radial distance from the sun (PARKER, 1961).

$$T = T_0 \left(\frac{\varrho}{\varrho_0}\right)^{\alpha - 1}, \tag{33}$$

where α is a polytropic index which is bounded between the isothermal and the adiabatic case with

$$1 \leqslant \alpha \leqslant \gamma . \tag{34}$$

Using Equations (33) and (34) we can eliminate the temperature in the momentum Equation (28) and get

$$\frac{d\psi}{d\xi}\left[1 - \frac{\alpha\psi_0 e^{(\alpha-1)/2}}{\psi e^{(\alpha+1)/2} \xi e^{2(\alpha-1)}}\right] = \frac{2\alpha\psi_0 e^{(\alpha-1)/2}}{\xi e^{2(\alpha-1)} \psi e^{(\alpha-1)/2}} - \frac{\lambda}{\xi} \tag{35}$$

where

$$\psi_0 = \frac{1}{4} \frac{mv^2(\xi - 1)}{kT_0} . \tag{36}$$

Integration of Equation (35) gives the Bernoulli Equation that corresponds to the energy Equation (32):

$$\psi + \frac{\alpha}{\alpha - 1}\left(\frac{\psi_0}{\psi\xi^4}\right)^{(\alpha-1)/2} - \lambda = E_1 , \tag{37}$$

where E_1 again is an integration constant and proportional to the energy flux. It is given by

$$\psi + \frac{\alpha}{\alpha - 1}\left(\frac{\psi_0}{\psi \xi^4}\right) e^{(\alpha-1)/2} - \lambda = E_1 \tag{38}$$

with

$$\lambda_0 = \frac{GM_\odot m}{2akT_0}. \tag{39}$$

In this way Equation (37) gives in an implicit form the solution ψ.

Let us first assume that $E_1 \neq 0$. For a given value of α and λ_0 the solution ψ depends on the parameter ψ_0, the ratio between the kinetic energy and the thermal energy at the base level a. The general form is plotted in Figure 4 for different values of ψ_0. There is a critical value $(\psi_0)c$, which is determined by the condition that the right hand side of Equation (35) and the coefficient of $d\psi/d\xi$ on the left hand side of Equation (35) vanish simultaneously. This gives an equation for

$$(\psi_0)c + \frac{\alpha}{\alpha - 1} - \lambda_0 - \frac{5 - 3\alpha}{\alpha - 1}\left[\left(\frac{4}{\lambda_0}\right)^{4(\alpha-1)}\left(\frac{\alpha}{2}\right)^2 (\psi_0)_c^{\alpha-1}\right]\exp\left[\frac{1}{5 - 3\alpha}\right] = 0. \tag{40}$$

As long as $\psi_0 < (\psi_0)c$, the value of ψ increases first with increasing ξ, passes through a maximum, and decreases finally as ξ^{-4}. From Equation (24) we can see that the density remains finite for $\xi \to \infty$ as well as the pressure (from Equation (33)). For all these solutions the velocity remains subsonic. If $\psi_0 = (\psi_0)_c$, the value of ψ increases monotonically with increasing ξ. At the critical level ξ_c the velocity reaches the sound velocity, and for $\xi \to \infty$, ψ approaches E_1. Therefore for this solution the density and the pressure go to zero at large distances. It is this solution that gives high velocities at large distances from the sun and thus describes a solar wind.

Let us now see why it seems plausible that this solution will be relevant under certain conditions for the phenomena in the interplanetary medium. The following conditions must be fulfilled for the existence of this critical solution, which is the only one allowing a transition from the subsonic region to the supersonic one.

(1) The behavior of ψ for small values of ξ shows that only if

$$\alpha < \tfrac{3}{2} \tag{41}$$

does ψ go to zero for vanishing ξ. But it might be that this condition need not be taken too seriously since for small ξ the physical conditions do not correspond to the assumptions made here.

(2) To have a solution which goes to infinity it is necessary that $E_1 > 0$. This means that the total energy per particle must be positive. From this it follows that

$$\lambda_0 < \psi_0 + \frac{\alpha}{\alpha - 1} \tag{42}$$

or if we assume $\psi_0 \ll 1$

$$\lambda_0 < \frac{\alpha}{\alpha - 1}. \tag{43}$$

This condition shows that the temperature must either be high enough for given radii a, or that for a given temperature the radius unto which this temperature is maintained should be large enough. If the conditions (42) or (43) are not fulfilled, the corona will not extend to infinity but will have a finite radius ξ_f of polytropic atmosphere:

$$\xi_f = \lambda_0 \bigg/ \left(\lambda_0 - \frac{\alpha}{\alpha - 1} \right). \tag{44}$$

(3) Finally since the velocity should increase outward from the base level $\xi = 1$, the slope of ψ must be positive. This requires that the right hand side of Equation (35) must be negative, since the coefficient on the left hand side is negative. Therefore

$$\lambda_0 > 2\alpha. \tag{45}$$

This means that the temperature of the corona should not be too high or that the region which is heated up to a certain temperature is not large in radius. If this condition is not fulfilled, stationary solutions will no longer be possible according to this simplified theory, and an expansion of a more explosive character should occur.

The expression (43) and (45) give the following restriction for a slowly expanding corona:

$$\frac{\alpha - 1}{\alpha} < \frac{2akT_0}{GM_{\odot}m} < \frac{1}{2\alpha}. \tag{46}$$

3. *Boundary conditions at large distances*

We might finally ask why these solutions which lead from subsonic velocities to supersonic ones are the ones which are relevant here and not the kind which remains subsonic. A solution is determined uniquely by applying boundary conditions. Until now only the conditions at one boundary, namely at the solar corona, have been considered. In addition, one has to take into account the conditions at large distances, where the transition must occur into the interstellar medium. There the pressure is quite low compared to the gas pressure in the corona. Therefore one must have a solution where the gas pressure at large distances is almost negligible compared to the pressure in the solar corona. Furthermore at large distances the velocity should be small too if we assume stationary conditions with respect to the interstellar medium. For the subsonic solutions the velocity would become small at large distances, but for large ξ the pressure is

$$p(\infty)(\sim) E_1 e^{\alpha/(\alpha - 1)} \tag{47}$$

and would therefore require that E_1 has to be rather small. According to Equation (38) and the discussion before (Equation 43) this would mean that λ_0 must be almost equal to $\alpha/(\alpha - 1)$. In this way this kind of solution would depend extremely sensitively on small temperature variations in the solar corona. The same would be true for the solution with $E_1 = 0$ which was required by Chamberlain.

But the boundary conditions can be fulfilled with the supersonic solution similar to the case which is realized in a flow through a Laval nozzle. In this case the supersonic solution will be terminated at some distance from the sun by a stationary shockwave, which is followed by a subsonic solution. The radial distance of the shockwave is determined by the required drop of the pressure.

This means that the solar wind should extend outwards to a distance where the kinetic pressure has decreased to the pressure in the interstellar medium. As DAVIS (1955) has pointed out, the interstellar magnetic fields of the order of 10^{-5} Gauss will lead to an effective pressure of about 10^{-12} dyn/cm^2 in the interstellar space. With the solution given by Parker, this will give a radius of 50 to 100 AU for the extension of the solar wind.

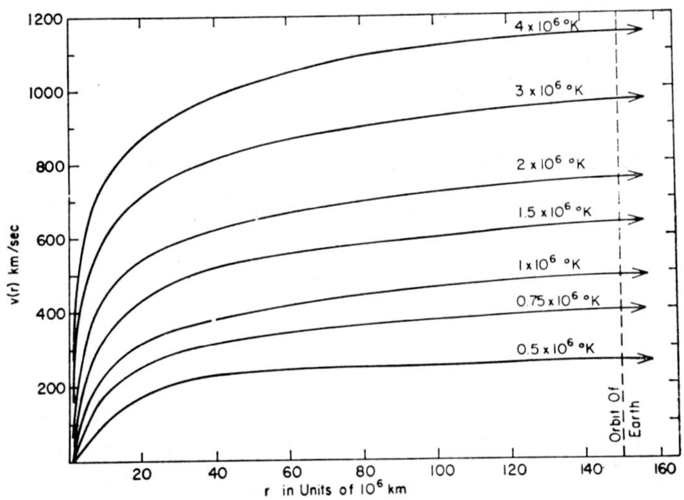

Fig. 5. Expansion velocity v as a function of the distance from the sun for various coronal temperatures assuming an isothermal interplanetary plasma ($\alpha = 1$) (PARKER, 1958).

BRANDT (1962) tried to show from comet tail observations that the wind would extend only to 2 AU. But his conclusion seems not to be justified since from the comet data it is not clear whether these comets had ionized or dust tails.

4. *General discussion of the stationary model*

In applying the stationary model discussed above the main uncertainties are due to our limited knowledge about the temperature of the solar corona and near the sun. As far as the restrictions of Equation (46) are concerned, three unknowns enter: the polytropic index, the temperature T_0 at the base level and the radius a at the base level, i.e. the distance to which the temperature T_0 is maintained by heating the corona. As long as these values are not better known it will be impossible to predict the properties of the solar wind near the orbit of the earth. But on the other hand the range of uncertainty is such that the unequalities of (41) are fulfilled very probably under normal conditions at the solar corona. According to PARKER (1960) it follows from the observed temperature gradient of 3°/km that $\alpha \approx 1.1$ to 1.2. With $a = 10^{11}$ cm the possible range in temperature would be of the order of some millions of degrees Kelvin. Parker found that the gas velocity would be of the order of 300 km/sec and the particle density about 10 particles/cm^3 at 1 AU if one assumes $T_0 = 1.2 \times 10^6$ °K, which is not unreasonable for the corona. As a consequence the whole coronal mass would be lost in about one day. Figure 5 demonstrates the dependence of the expansion velocity on the coronal temperature for an isothermal model ($\alpha = 1$). DE JAGER (1963) took into account the heat conduction and solved the equations numerically. The values do not differ too much from those derived with the isothermal model.

Therefore the hydrodynamical description of the interplanetary plasma gives a model which does not contradict the present observations and which looks very reasonable. This stationary model produces a solar wind and not a solar "breeze". This description should be applicable as long as the temperature is not too high or the density too low, giving a mean free path large to the dimensions of the system. In the latter case exospheric models would be more appropriate (CHAMBERLAIN, 1961a; see also ÖPIK and SINGER, 1960).

D. THE STRUCTURE OF MAGNETIC FIELDS

As we have seen, not too much is known about the magnetic fields in the interplanetary space. We know with certainty that magnetic fields are present. Under normal conditions the field strength is very probably of the order of 10^{-5} gauss. But it is not clear whether these fields have a large scale structure or whether they are fairly disordered and variable under undisturbed conditions. Our theoretical picture is even more scanty.

ELLIOT (1960) proposed a large scale magnetic field with a fair degree of axial symmetry where at least a considerable part of the lines of force would be connected to the solar surface. But the extension of such a field out to 1 AU would cause serious difficulties since such a field should enforce co-rotation out to this distance. But this has not been observed (see Section 2c). As ALFVÉN (1942), LÜST and SCHLÜTER (1955) and GOLD (1962) have pointed out, one should expect that the inner part of the solar system must co-rotate with the sun where the magnetic field is strong enough and where it is connected to the solar surface. All observations are consistent with the picture that the solar corona co-rotates with the sun. But where does the transition take place between the co-rotating part and the outer non-co-rotating part of the interplanetary medium? LÜST and SCHLÜTER (1955) have estimated that the co-rotating sphere should extend to about 50 to 80 solar radii. Inside of this sphere the magnetic field should be essentially force-free with an azimuthal (toroïdal) component. Outside of this region the magnetic field would be disordered and not connected with the sun. Such a field structure inside of the sphere could cause the observed effect that there is a higher probability for the detection, at the earth, of flare-produced cosmic rays when the flare is in the western portion of the solar disc. This picture would imply that at least during certain periods there are no magnetic lines of force that connect the sun with the orbit of the earth.

As has been stated before, GOLD (1959) assumes that during periods of higher activity after solar flares such connection exists, and that the magnetic lines of force are drawn out by the ejected plasma. But also he stressed the idea of co-rotation near the sun. PARKER (1958) developed the picture that the magnetic lines of force are stretched out at least out to the orbit of the earth by the solar wind all the time. This would mean that there should permanently exist a large scale field with a spiral structure.

E. SOME REMARKS TO THE PERTURBED STATE

As pointed out at the beginning, we do not intend to discuss in any detail in this article the perturbed state of the interplanetary plasma. Therefore only some remarks will be added which are more directly related to the discussion of the former pages. From the conditions for the stationary solution it follows that such a solution is no longer possible if the temperature near the sun exceeds a certain value. PARKER (1961a) has stressed these ideas and developed a model for the "enhanced solar corpuscular radiation" following a large solar flare. According to his picture this phenomenon should be a blast wave from the hydrodynamical explosion of the enhanced corona. The blast wave is composed of the quiet-day interplanetary plasma swept up by the outward momentum of the explosion. The quiet-day magnetic field embedded in the quiet-day material is sheared, and the galactic cosmic rays are swept back by the compressed fields in the blast wave leading to fractional cosmic ray decreases (Forbush decrease).

In contrast to this, GOLD (1959) is advocating the magnetic-tongue picture for the perturbed state of the interplanetary plasma (see Section 2F). But since no detailed calculations have been carried out for this model it is difficult to compare it with Parker's model in a more quantitative way.

It should be emphasized that different ideas have been discussed by a number of other authors. Here the work of Alfvén and his group should be mentioned. Also SINGER (1962) dealt with the theory of the Forbush decrease.

F. CONCLUDING REMARKS

In concluding this review about the interplanetary plasma it should be said that there is a strong probability that the solar wind and therefore the interplanetary plasma is not merely a phenomenon of our solar system. There are strong theoretical reasons in connection with the evolution of stars (SCHWARZSCHILD, 1961) and also in connection with the theory of the hydrogen convection zone (BIERMANN and LÜST, 1960) which indicate that stars of later spectral types should have extended coronas and a much larger flux of corpuscular radiation (stellar wind) than the sun. The observations by DEUTSCH (1956) on M-giant stars seem to be an evidence for this.

References

ALFVÉN, H.: 1942, *Arch. Mat. Astr. och Fys.* A **28,** No. 6.
VAN ALLEN, J. A.: 1962, Contribution to the Varenna Summer School.
BARBIER, D.: 1955, *Mém. Soc. Roy. Sci. Liège* **15,** 55.
BARDEEN, J.: 1962, Unpublished manuscript.
BARTELS, J.: 1945, *FIAT Review of German Science* **17,** Part I, Geophysics.
BARTELS, J.: 1951, *Abh. Akad. Wiss. Göttingen, Math.-Phys. Kl.* Sonderheft.
BARTELS, J.: 1958, *Naturwiss.* **45,** 181.
BEARD, D. B.: 1960, *Phys. Rev. Letters* **5,** 89; *J. Geophys. Res.* **65,** 3559.
BIERMANN, L.: 1951, *Z. Astrophys.* **29,** 274.
BIERMANN, L.: 1952, *Z. Naturforsch.* 7a, 127.
BIERMANN, L.: 1953, *La physique des comètes*, p. 251.
BIERMANN, L.: 1957, *Observatory* **77,** 109.
BIERMANN, L. and LÜST, R.: 1960, *Compendium of Astronomy and Astrophysics* (ed. by J. L. GREENSTEIN), Chicago, p. 260.
BIERMANN, L., LÜST, R., LÜST, RH., and SCHMIDT, H. U.: 1961, *Z. Astrophys.* **53,** 226.
BIERMANN, L. and LÜST, RH.: 1962, *The Solar System* (ed. by G. KUIPER and B. MIDDLEHURST), University of Chicago Press, Chicago, Vol. IV, Chapter 18.
BRANDT, J. C.: 1962, *Icarus* **1,** 1.
BRIDGE, H. S., DILWORTH, C., LAZARUS, A. J., LYON, E. F., ROSSI, B., and SCHERB, F.: 1962, *J. Phys. Soc. Japan* **17,** Suppl. A II, 553.
CHAMBERLAIN, J. W.: 1960, *Astrophys. J.* **131,** 47.
CHAMBERLAIN, J. W.: 1961, *Astrophys. J.* **133,** 675.
CARMICHAEL, H.: 1962, *Space Sci. Rev.* **1,** 28.
CHAPMAN, S. and FERRARO, V. C. A.: 1930, *Nature* **126,** 129.
CHAPMAN, S.: 1954, *Astrophys. J.* **120,** 151.
CHAPMAN, S.: 1957, *Smithsonian Contributions Astrophysics* **2,** 1.
CHAPMAN, S.: 1959, *Proc. Roy. Soc.* A **253,** 450.
CHAPMAN, S.: 1961, *Space Astrophysics* (ed. by W. LILLER), McGraw-Hill Book Company, London-New York, p. 133.
CHAPMAN, S. and BARTELS, J.: 1940, *Geomagnetism*, Oxford University Press.
CLAUSER, F. H.: 1961, Fourth Symposium on Gas Dynamics. *IAU Symposium* **12,** 26.
COCCONI, G. P., MORRISON, P., GOLD, T., and HAYAHAWA, S.: 1958, *Nuovo cimento* **8** (16) 61.
COLEMAN, Jr, P. J., SONETT, C. P., JUDGE, D. L., and SMITH, E. J.: 1960a, *J. Geophys. Res.* **65,** 1856.
COLEMAN, Jr. P. J., DAVIS, L., and SONETT, C. P.: 1960b, *Phys. Rev. Letters* **5,** 43.
COLEMAN, Jr., P. J., SONETT, C. P., and DAVIS, Jr., L.: 1961, *J. Geophys. Res.* **66,** 2043.
COLEMAN, Jr., P. J., DAVIS, Jr., L., SMITH, E. J., and SONETT, C. P.: 1962, *Astrophys. J.* **138,** 1099.
DAVIS, L.: 1955, *Phys. Rev.* **100,** 1440.
DAVIS, L.: 1962, *J. Phys. Soc. Japan* **17,** Suppl. A. II, 543.
DEUTSCH, A.: 1956, *Astrophys. J.* **123,** 210.
DEUTSCH, A.: 1961, Fourth Symposium on Cosmical Gas Dynamics, IAU Symposium **12,** 238.
DUNGEY, J. M.: 1961, *J. Geophys. Res.* **66,** 1043.
ELLIOT, H.: 1960, *Phil. Mag.* **5,** 601.
ELSÄSSER, H.: 1958, *Die Sterne* **34,** 166.
FAN, C. Y., MEYER, P., and SIMPSON, J. A.: 1960, *Phys. Rev. Letters* **5,** 269.
FERRARO, V. C. A.: 1955, *Ann. Géophys.* **11,** 284.
FERRARO, V. C. A.: 1960, *J. Geophys. Res.* **65,** 3951; *Rev. Mod. Phys.* **32,** 934.
FORBUSH, S. E.: 1954, *J. Geophys. Res.* **59,** 525.
GERMAIN, P.: 1961, Fourth Symposium on Cosmical Gas Dynamics, IAU Symposium **12,** 296.
GOLD, T.: 1959, *J. Geophys. Res.* **64,** 1665.
GOLD, T.: 1962, *Space Sci. Rev.* **1,** 100.
GOLD, T.: 1962, *J. Phys. Soc. Japan* **17,** Suppl. A I, 187.
GRINGAUZ, K. I., BEZROKIH, V. V., OZEROV, V. D., and RYBCINSKIJ, R. E.: 1960, *Doklady AN SSSR* **131,** 1301; *Soviet Phys.-Doklady* **5,** 361.

Gringauz, K. I.: 1961, *Space Research* **2,** 539.
Heppner, J. P., Ness, N. F., Skillman, T. L., and Scearce, C. S.: 1962, *J. Phys. Soc. Japan* **17,** Suppl. A II, 546.
Hewish, A.: 1958, *Mon. Not. Roy. Astron. Soc.* **118,** 534.
Hinteregger, H. E.: 1961, *J. Geophys. Res.* **66,** 2367.
Hinteregger, H. E. and Watanabe, K.: 1962, *J. Geophys. Res.* **67,** 3373.
Hurley, J.: 1961, *Phys. Fluids* **4,** 854.
De Jager, C.: 1963, Proc. COSPAR Symposium 3.
Johnson, H. M.: 1960, *J. Geophys. Res.* **65,** 3049.
Jory, 1956, *Phys. Rev.* **103,** 1068.
Kahn, F. D.: 1961, Fourth Symposium on Cosmical Gas Dynamics, IAU Symposium **12,** 310.
Lüst, Rh.: 1961, *Z. Astrophys.* **51,** 163.
Lüst, Rh.: 1962, *Z. Astrophys.* **54,** 67.
Lüst, R. and Schlüter, A.: 1955, *Z. Astrophys.* **38,** 190.
Lüst, R.: 1957, *Phys. Rev.* **105,** 1827.
Lüst, R., Schlüter, A., and Katterbach, K.: 1956, *Nachr. Akad. Wiss. Göttingen, II Math.-Physik. Kl.* **8,** 127.
Lüst, R. and Simpson, J. A.: 1957, *Phys. Rev.* **108,** 1563.
Machin and Smith, 1951, *Nature* **168,** 599.
McCracken, K. G.: 1962, *J. Geophys. Res.* **67,** 423, 435 and 447.
Meyer, P. and Simpson, J. A.: 1955. *Phys. Rev.* **99,** 1517.
Midgley, J. E. and Davis, L.: 1962, *J. Geophys. Res.* **67,** 499.
Morrison, P.: 1956, *Phys. Rev.* **101,** 1397.
Mustel', E. L.: 1961, *Astron. Ž.* **38,** 28; *Soviet Astron.-AJ* **5,** 19.
Neher, H. V.: 1956, *Phys. Rev.* **103,** 228.
Neugebauer, M. and Snyder, L. W.: 1962, *Astrophys. J.* **138,** 1095.
Öpik, E. J. and Singer, S. F.: 1961, *Phys. Fluids* **4,** 221.
Olbert, S.: 1962, Summer School Varenna.
Parker, E. N.: 1958, *Phys. Rev.* **110,** 1445.
Parker, E. N.: 1958, *Astrophys. J.* **128,** 664.
Parker, E. N.: 1958, *Astrophys. J.* **128,** 677.
Parker, E. N.: 1962, *Space Sci. Rev.* **1,** 62.
Parker, E. N.: 1960, *Astrophys. J.* **132,** 821.
Parker, E. N.: 1961a, *Astrophys. J.* **133,** 1014.
Parker, E. N.: 1961b, *Astrophys. J.* **134,** 20.
Peterson, A. W.: 1961, *Astrophys. J.* **133,** 668.
Pottasch, S. R.: 1960, *Astrophys. J.* **131,** 68.
Rose, D. C. and Ziauddin, S.: 1962, *Space Sci. Rev.* **1,** 115.
Rossi, B.: 1962, Summer School Varenna.
Schlüter, A.: 1951, *Z. Naturforsch.* **6**a, 613.
Schwarzschild, M.; 1961, General Assembly IAU, Berkeley.
Singer, S. F., Laster, H., and Lenchek, A. M.: 1962, *J. Phys. Soc. Japan* **17,** Suppl. A II, 583.
Slee, O. B.: 1961, *Mon. Not. Roy. Astron. Soc.* **123,** 223.
Slutz, R. J.: 1962, *J. Geophys. Res.* **67,** 505.
Snyder, L. W. and Neugebauer, M.: 1963, *Space Research* **4.**
Spreiter, J. R. and Briggs, B. R.: 1962, *J. Geophys. Res.* **67,** 37.
Strömgren, B.: 1939, *Astrophys. J.* **89,** 526.
Stumpff, P.: 1961, *Astron. Nachr.* **286,** 87.
Sonett, P., Davis, Jr., L., and Coleman, Jr., P. J.: 1962, *J. Phys. Soc. Japan* **17,** Suppl. A II, 525.
Vitkevič, V. V.: 1958, *Astron. Ž.* **35,** 52; *Soviet Astron.-AJ* **2,** 45.
Weymann, R.: 1960, *Astrophys. J.* **132,** 380.
Whipple, F. L. and Gossner, J. L., 1949, *Astrophys. J.* **109,** 380.
Žingulev, V. N. and Romiševskij, E. A.: 1959, *Doklady AN SSSR* **127,** 1001: *Soviet Phys.-Doklady* **4,** 859.
Žingulev, V. N.: 1959, *Doklady AN SSSR* **126,** 521; *Soviet Physics-Doklady* **4,** 514.

General References

Interplanetary Medium

ELSÄSSER, H.: 1958, *Mitt. Astron. Ges.* Hamburg, 1. Aufl., 1951, 2. Aufl., 1958, p. 61.

Gas component of the interplanetary medium

BIERMANN, L.: 1961, *Space Astrophysics* (ed. by W. LILLER), McGraw-Hill Book Company, London-New York, p. 150.

CHAPMAN, S.: 1961, *Space Astrophysics* (ed. by W. LILLER), McGraw-Hill Book Company, London-New York, p. 133.

PARKER, E. N.: 1961, *Space Astrophysics* (ed. by W. LILLER), McGraw-Hill Book Company, London-New York, p. 157.

PARKER, E. N.: 1962, *Plan. Space Sci.* **9**, 461.

ŠKLOVSKIJ, J. S.: 1962, *Physics of the Solar Corona*, Moscow.

Dust component of the interplanetary medium

HOFFMEISTER, C.: 1951, *Naturwiss.* **28,** 227.

MINNAERT, M.: 1955, *Mém. Soc. Roy. Sci. Liège* **15,** 15.

MANRING, E.: 1961, *Adv. Space Sci. Techn.* **3,** 273.

PART III

IONOSPHERE AND HIGH LATITUDE PHENOMENA

CHAPTER 14

RADIO AURORA

B. HULTQVIST
Kiruna Geophysical Observatory, Kiruna C, Sweden

1. Introduction

Reflections of radio waves from ionization associated with aurora – radio aurora – were first investigated in 1938 by Harang and Stoffregen. That radio wave propagation via auroral ionization is possible was, however, discovered earlier in the 1930's by amateur radio operators. After the second world war and particularly during the last ten years radio aurora has been studied by a number of researchers in North America, Britain, Norway, Sweden, Germany, Russia, New Zealand, and the Antarctic.

2. Experimental Results

In the following, results obtained from various observations are listed in order to give a general impression on the behaviour of radio aurora.

A. TIME CHARACTERISTICS

1. *Diurnal variation*

Records taken in Kiruna (lat. 67.8 °N; geomagnetic lat. 65.3°) over 4.5 years show a pronounced maximum of echo occurrence frequency in the early afternoon (Figure 1). No such maximum can be found for visual aurora or magnetic activity. It is typical of the auroral zone.
Other maxima occur in the night and can be seen at all latitudes (Figure 2), although the time of maximum occurrence frequency varies with latitude.

2. *Seasonal variation*

According to observations in Kiruna the echo activity is in most years greatest in September, October, February and March. In addition there is a fairly strong variation from year to year (Fig. 3).

3. *Long term variation*

In Figure 3 a long term variation in the occurrence frequency of radio aurora can be seen. The yearly maximum occurred in 1960, about two years after the maximum sunspot number, as has been found also for visual auroral and magnetic disturbances.

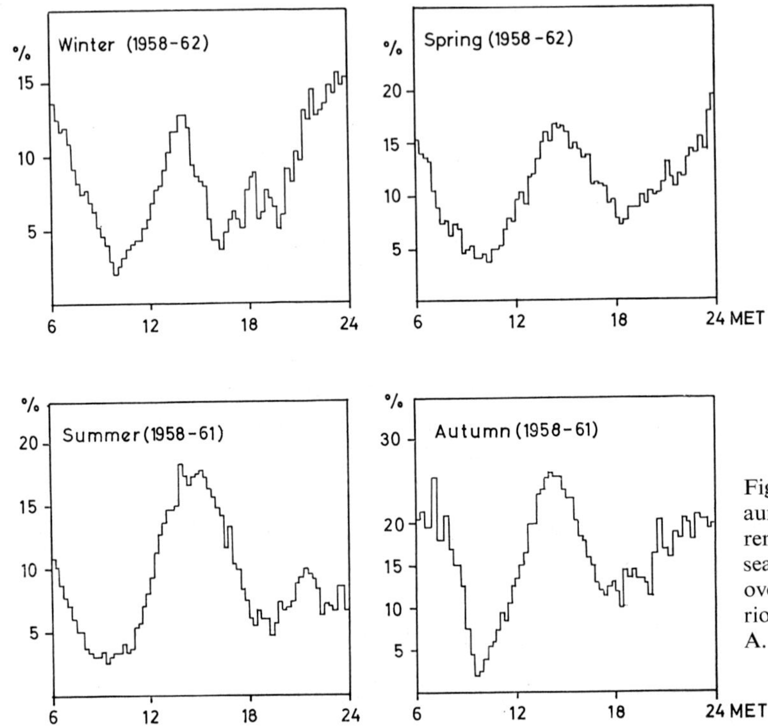

Fig. 1. Percentage of auroral echo occurrence for the various seasons, averaged over the recording period 1958–1962. After A. EGELAND (1962a).

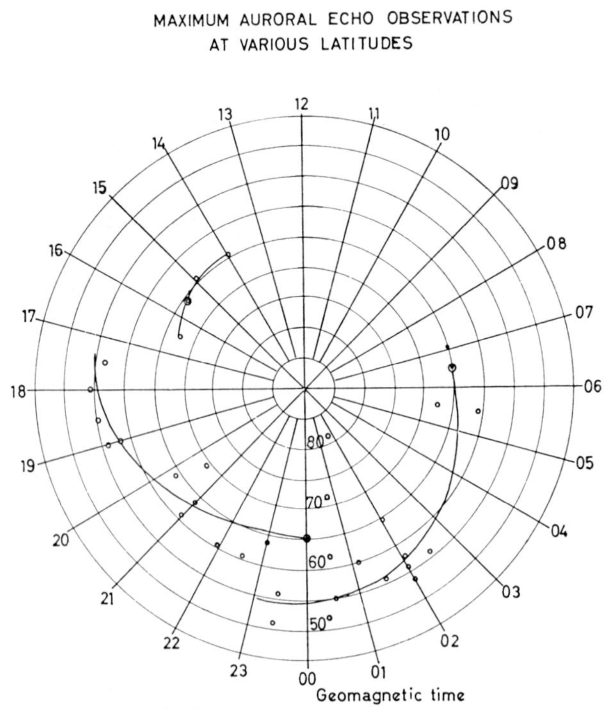

Fig. 2. Times of peak auroral echo activity as observed at various geomagnetic latitudes. After A. EGELAND (1962a).

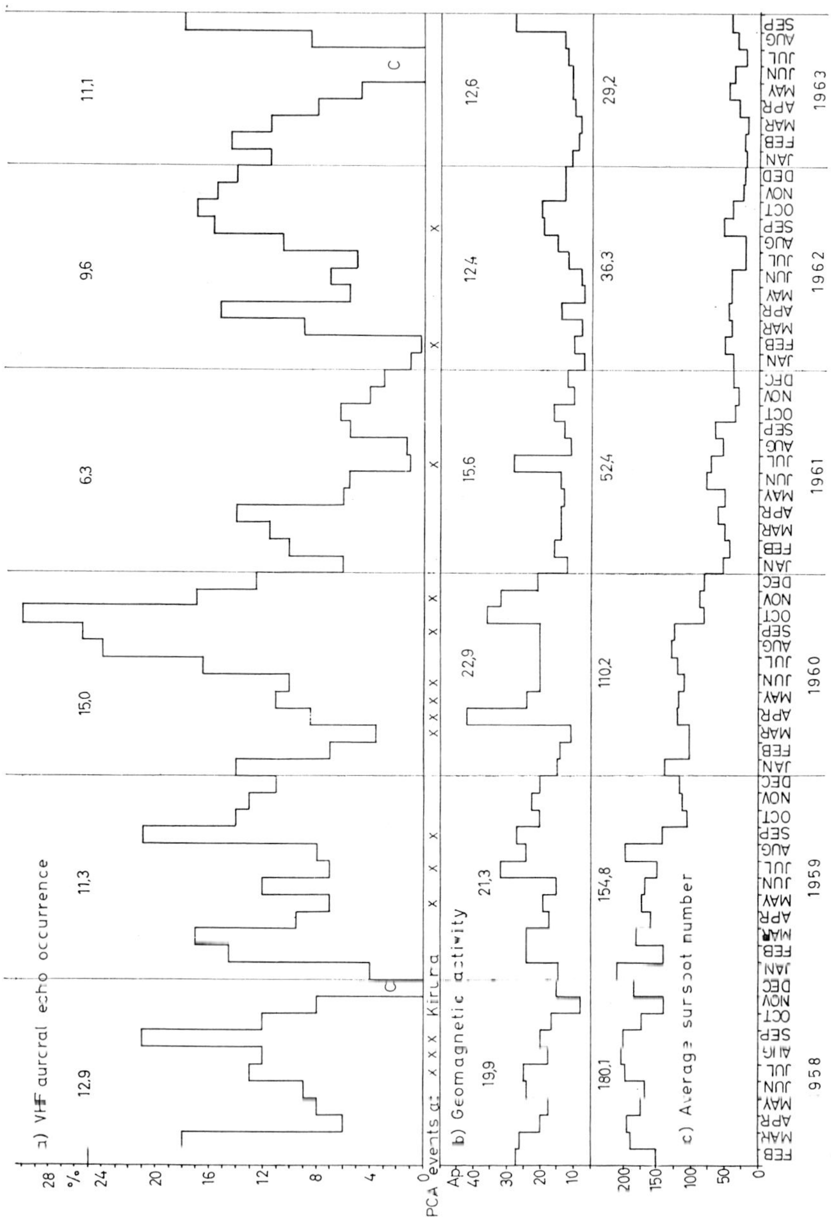

Fig. 3. Long term variation of auroral echo occurrence (February, 1958–September, 1963) as observed at Kiruna close to the auroral zone.

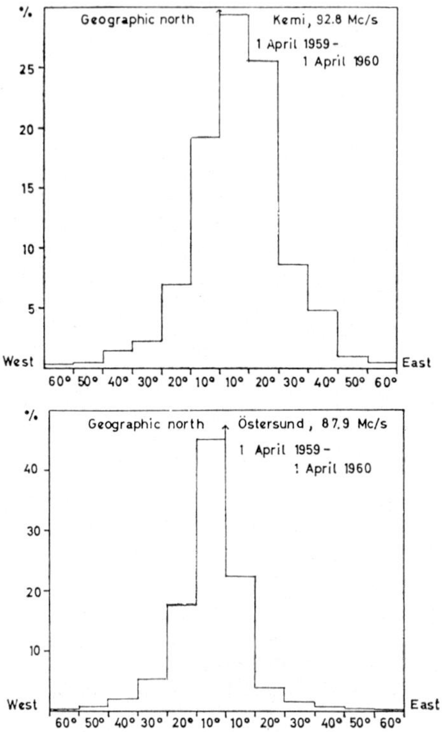

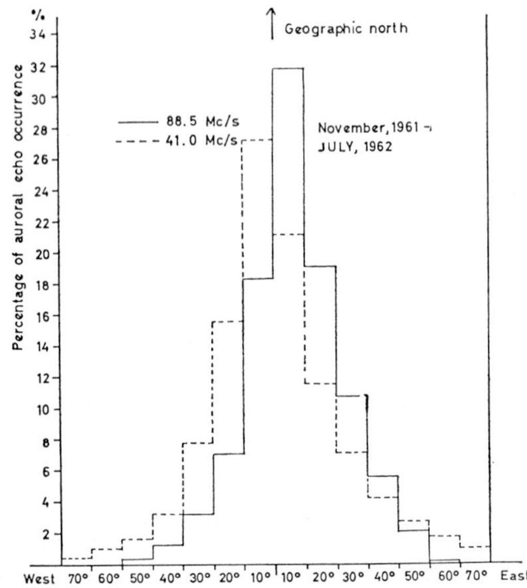

Fig. 4. Histograms of oblique auroral echoes received at Kiruna showing azimuthal distribution between 60 degrees east and west of geographic north. After A. EGELAND (1962a).

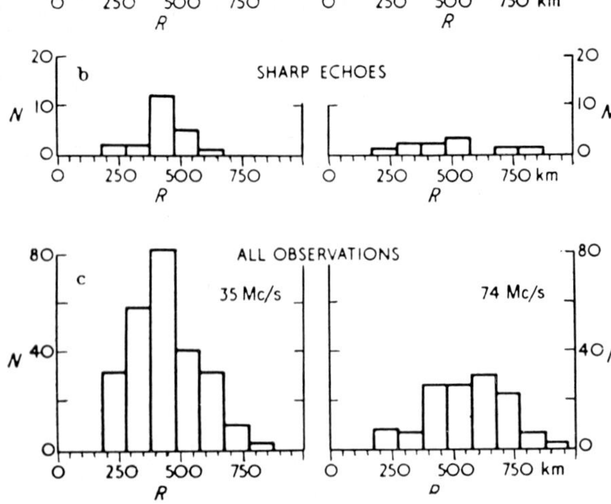

Fig. 5. Frequency of occurrence of echoes on 35 and 74 Mc/sec as a function of range, obtained in Norway. The columns indicate the number, N, observed for each 50-km increase in range, R. From HARANG and LANDMARK, 1953.

4. *Duration of echoes*

The majority of echoes last less than 10 minutes, with some durations extending from a few seconds to 12 hours. The duration depends of course very much on the power of the transmitter.

B. SPATIAL CHARACTERISTICS

1. *Geographic regions of observation*

Radio aurora is observed in the same geographic regions as the visual aurora but with very high power radars also at low latitudes.

2. *Directional distribution*

Radio aurora is observed almost exclusively at low elevation angles in a sector centered somewhere around geographic or geomagnetic north. The distribution of auroral echoes has been studied with the use of radar by many workers. For oblique auroral reflections the distributions shown in Figure 4, were found. The width of the azimuthal distribution decreases rapidly with increasing frequency.

3. *Range and height distribution*

Typical observational results are shown in Figure 5 and Table I.

TABLE I

Frequency	*Height range km*	*Mean height km*
40	75–140	90 – 110
55		110.9±3.1
73	100–125	
73	102–109	106 – 107
78	100–120	
106		108
398	75–135	100 – 110
400	90–160	
448		100 – 105
550	100–120	110±2
1300	80–105	90

For references see EGELAND (1962a).

C. AURORAL ECHO STRENGTH

The daytime echoes were of lower intensity than those recorded during the night and early morning.

It is not clear from experiments whether the amplitude follows an R^{-2} or R^{-4} law or something in between.

D. APPARENT MOTION OF RADIO AURORAE

A smooth changeover from westerly to easterly drift around midnight was observed by e.g. Harang and Tröim at Kjeller but not at Tromsö. The velocities were of the order of one km/sec with a wide dispersion. Good correlation exists between these motions and the variation of the geomagnetic **H**-vector. (Figure 6.)

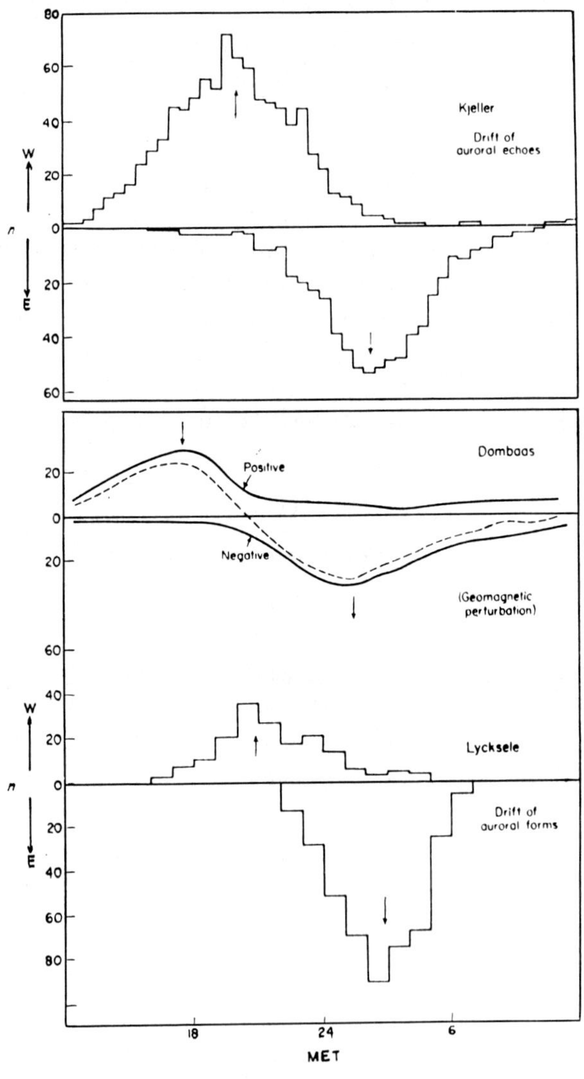

Fig. 6. Mean diurnal variation of (a) E-W drift of ionized areas determined from echoes on 40 Mc/sec at Kjeller, (b) the horizontal component of the geomagnetic perturbing vector at Dombaas (c) E-W drift of auroral forms determined from all-sky photos at Lycksele. After Stoffregen (Harang and Tröim, 1961).

E. AMPLITUDE DISTRIBUTION

Amplitude distributions for two auroral reflection periods observed at 92.8 Mc/sec are shown in Figure 7. The proximity of the observed points to a Rayleigh distribution shows that the aurorally propagating wave is the aggregate of a large number of scattered wavelets uncorrelated in phase.

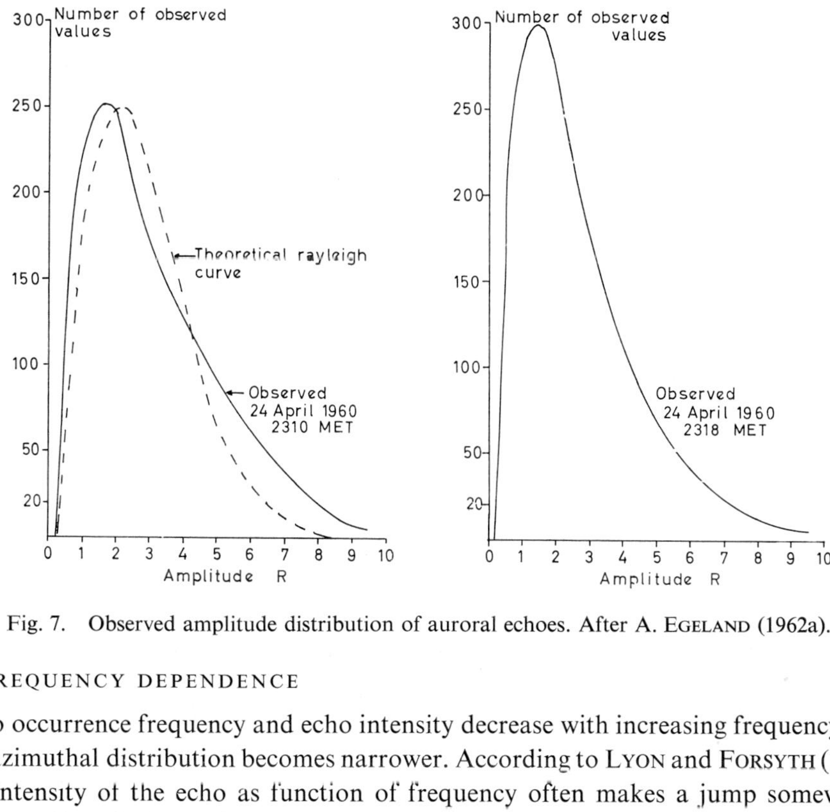

Fig. 7. Observed amplitude distribution of auroral echoes. After A. EGELAND (1962a).

F. FREQUENCY DEPENDENCE

Echo occurrence frequency and echo intensity decrease with increasing frequency and the azimuthal distribution becomes narrower. According to LYON and FORSYTH (1962) the intensity of the echo as function of frequency often makes a jump somewhere between 40 and 100 Mc/sec, from which they conclude that critical reflection sometimes occurs up towards 100 Mc/sec, indicating electron densities in the irregularities of the order of 10^8 cm^{-3}.

G. POLARIZATION CHARACTERISTICS

The polarization of auroral echoes is mostly identical with the transmitted polarization but some exceptions have been reported in the literature.

H. CORRELATION OF RADIO AURORA WITH OTHER PARAMETERS

1. *Visual aurora*

Figure 8 shows that a significant, but not perfect correlation exists between the visible

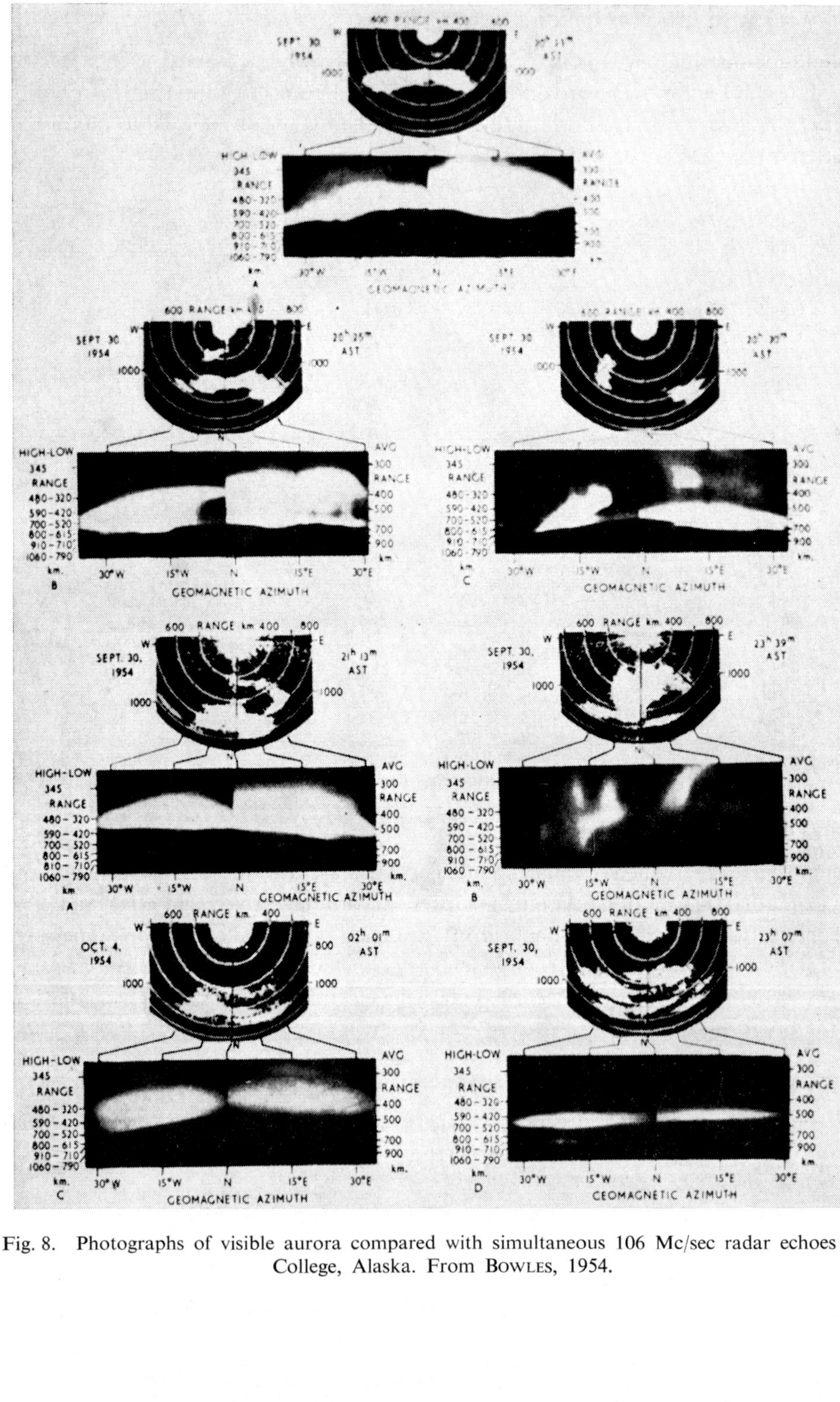

Fig. 8. Photographs of visible aurora compared with simultaneous 106 Mc/sec radar echoes at College, Alaska. From Bowles, 1954.

auroral forms and radar echoes. Visual aurorae do not invariably act as good reflectors.

2. *Geomagnetic disturbances*

The correlation is best when the magnetic disturbances are observed close to the region of auroral reflections. The probability of strong radio aurora is proportional to the geomagnetic activity at middle latitudes, but a maximum occurs in the auroral

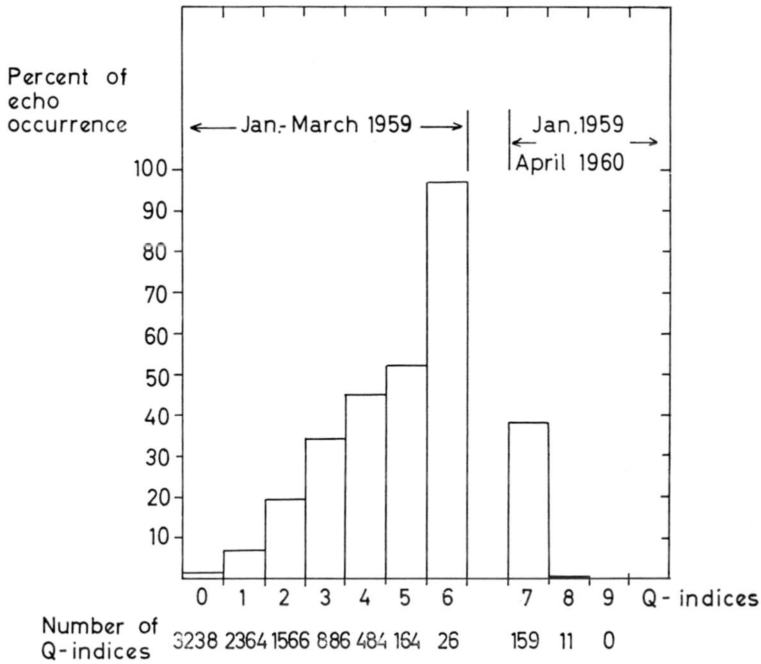

Fig. 9. Correlation between auroral echo occurrence and geomagnetic activity at Kiruna. After A. EGELAND (1962a).

zone at a high disturbance level and then the radio aurora occurrence frequency decreases with increasing magnetic disturbance (cf. Figure 9).

3. *Ionospheric absorption*

The correlation coefficient between ionospheric absorption and auroral echo occurrence observed from the same place is low. At Kiruna it amounts to only 0.14, so the two ionospheric parameters appear to be unrelated. This is certainly due to the fact that the echoes are produced 400–800 km to the north, while absorption is measured overhead.

4. *Polar cap absorption* (*PCA*)

Auroral echoes are in general absent in the auroral zone during magnetic storms associated with PCA-events.

3. Theories for the Auroral Reflection Process

The mechanisms of auroral reflections are not well understood in spite of the long period of time they have been studied. Here some of the proposed mechanisms, which are still under discussion, are reviewed.

A. ASPECT SENSITIVITY

An important characteristic of the auroral echoes is a pronounced aspect sensitivity: radar reflections occur predominantly from a direction perpendicular to the magnetic field lines at the reflection point. Really conclusive quantitative investigations of these effects have not been made because of the difficulties involved.

Chapman proposed in 1952 that auroral ionization might be aligned along the

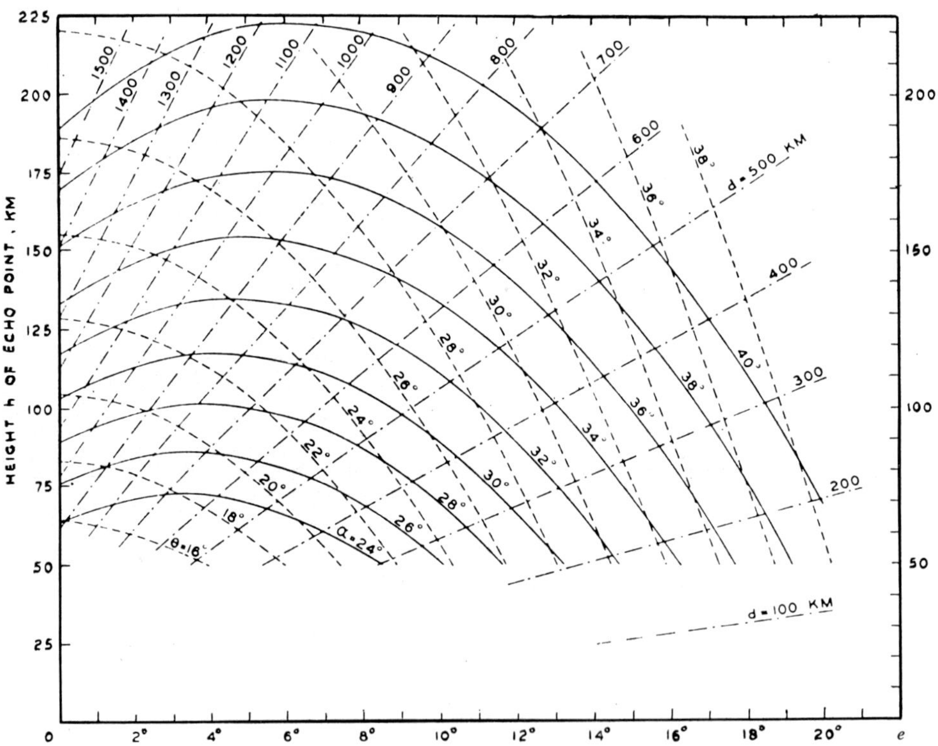

Fig. 10. Each full curve refers to a radar location in the geomagnetic colatitude α marked thereon, and gives the height h (km) of the echo point for a radar beam to the reflecting aurora, at beam elevation e. By interpolation between these curves h can be inferred for a beam of elevation e from a station in any intermediate colatitude. The broken curves (----) similarly show (if necessary, by interpolation) the geomagnetic colatitude θ of the echo point P at a chosen height h for a beam of elevation e, or alternatively for an echo point P corresponding to a beam of elevation e from a radio transmitter in geomagnetic colatitude α. The chain curves (— . — . —) similarly give the distance of the echo point P from the transmitter. After CHAPMAN (1952).

earth's magnetic field. He worked out the geometry in the meridian plane of the case in which the aurora is viewed perpendicular to the lines of the earth's dipole field for various locations. Some of his results are shown in Figure 10.

It follows from Chapman's calculations that auroral echoes should be most easily obtained by radars located in regions where the magnetic dip is of the order of 70 to 75 degrees, because a perpendicularity is then possible from the transmitter on to the lines of the earth's magnetic field with the height of its foot near 100 km.

B. THEORIES OF THE REFLECTION MECHANISM

The first proposals for the reflection mechanism were very simple ones: critical (LOVELL *et al.*, 1947) and partial (HERLOFSON, 1947) reflection from an ionisation configuration similar to the configuration of the auroral form, respectively. The fading of the auroral echoes made it evident that specular reflection, either critical or partial, from large surfaces could not be the relevant process. Several workers have therefore suggested that critical reflection from ionisation blobs are responsible for auroral echoes at least on lower frequencies. Other workers have modified HERLOFSON'S (1947) partial reflection mechanism by introducing irregularities of low electron density in the normal ionosphere, the mechanism of reflection being partial reflection (weak scattering) from these irregularities.

Auroral ionisation can be described conveniently in terms of the distribution of sizes of the individual irregularities, their shape and electron density, the sharpness of their boundaries and the number of scatterers per unit volume. There is considerable disagreement upon many of these aspects of the model. The estimates of size obtained from the various theoretical models range from a fraction of a meter to kilometers. Some types of auroral ionization have been suggested to consist of isotropic scatterers, while estimates of the elongation of anisotropic scatterers occurring in literature range from an axial ratio of 4 : 1 to about 40 : 1. The most fundamental and important difference relates to the strengths of the scatterers. From the time when Herlofson suggested that it would be physically unreasonable to interpret the auroral reflections, obtained by LOVELL *et al.* (1947), as due to critical reflections, the role of critical reflection in auroral echoes has been widely discussed. Booker has suggested weak scattering being the reflection mechanism down to 30 Mc/sec, while Kaiser has interpreted the echoes as due to critical reflections up to frequencies as high as 100 Mc/sec. The Canadian group at Saskatoon, on the basis of multiple-frequency measurements, have concluded that for frequencies between 30 and 100 Mc/sec there are times when the frequency dependence is such that a transition from critical reflection to weak scattering takes place within this frequency range.

The critical reflection has been thought to take place at irregularities much larger than the wavelength by Kaiser and at irregularities of a size comparable with one meter by Moorcroft. The aspect sensitivity can be understood in terms of critical frequency, if the irregularities have a suitable form; two-axial ellipsoids with an axial ratio of about five has been proposed.

The basic physical mechanism in the reflection process is always the same: an incident wave sets electrons in motion and their oscillation in turn re-radiates the electromagnetic wave.

If the auroral ionization is characterized by irregularities of the order of the radio wavelength or smaller, reflection occurs in a manner somewhat different from that from a large surface. When the reflection is critical, nothing is transmitted but all is reflected. When there is an abrupt change in the index of refraction, a portion, determined by the boundary conditions, may be reflected by so-called partial reflection and the remainder transmitted. For a more gradual change in the refraction index, interference between the reflected amplitudes from varying depths reduces the reflected component. In the weak scatter theory it is assumed that the sacattered radiation is negligible as compared with the transmitted component (Born approximation). It is also always assumed that no interaction exists between the scattering processes of the various irregularities. The weak scatter theory has been worked out in considerable detail by BOOKER (1956) for the backscatter case and has been extended to oblique auroral reflections by EGELAND (1962b). The strength and aspect sensitivity of auroral echoes can be interpreted in terms of weak scattering if: (1) the mean electron density is around 10^6 cm^{-3}, (2) the electron distribution has a Gaussian autocorrelation function elongated along the field lines, (3) the correlation distance is about 10 m along the field lines, (4) the correlation distance perpendicularly to the field lines is less than 1 m, and (5) the mean square fractional deviation in the electron density distribution is of the order of 10^{-3}.

Since the weak scatter theories presuppose that the medium is almost completely transparent to the radio waves, it can only be employed for frequencies much higher than the critical frequency. MOORCROFT (1961) has worked out a theory encompassing both critical reflection and weak scattering and also the intermittent frequency range. He has included several different distributions of size and strength of scatterers. All models used by Moorcroft consist of assemblies of scatterers having the electron density distribution $N = N_0 \exp\{-(\xi^2 + \eta^2)/a^2 + \xi^2/c^2\}$, so called gaussoids.

If the irregularities are small compared with the incident wavelength, and the electron density is large, interactions between the electrons can modify the scattered field, giving rise, on occasion, to strong electrostatic fields and to resonances for certain polarizations. For a gaussian cylindrical distribution the scattered field may be twice the value predicted by weak scatter theory for polarization transverse to the cylinder, but no increase occurs for parallel polarization.

C. COMPARISON OF THEORETICAL MODELS WITH EXPERIMENTAL DATA

The experimental *aspect sensitivity* results can be interpreted in terms of any of the processes, described above, provided the shape of the reflecting or scattering irregularities is adjusted to fit the observations. Various authors have found the sizes and axis ratios mentioned earlier. Specular reflection from large, more or less plane surfaces can be eliminated on the basis of the experimental aspect sensitivity results.

The amount of aspect sensitivity involved in the auroral reflection mechanism is still under discussion.

The *azimuthal distribution* of radar echoes is thought to be due to the aspect sensitivity.

EGELAND (1962b) has studied the azimuthal distribution of radar and oblique auroral reflections at 41 Mc/sec and about 90 Mc/sec, respectively, at Kiruna. He found that the observed distribution does not fit well with the one to be expected from weak scatterers, circularly symmetric around the magnetic field lines. The observed distributions were narrower at the top and broader at the bottom than the calculated ones. By assuming three-axial ellipsoidal shape for the scatterers he could improve very much the agreement between observations and theory. At 41 Mc/sec the following average dimensions of the scatterers fitted best: along field lines $L=14$ m, along the isomagnetic latitude lines, $a=0.5$ m, and perpendicularly to the other two axes, $b=0.2$ m. At 90 Mc/sec the corresponding values were found to be: $L=8$ m, $a=0.5$ m, and $b=0.2$ m. The result of Egeland's work thus is that the azimuthal distribution of oblique auroral reflections is consistent with weak single scattering from three-axial irregularities. Studies of the *spatial characteristics* of auroral reflections using two 49.7 Mc/sec receiver antennae, separated by 10 wavelengths, revealed distinctly different echo structure in the two receivers, thereby indicating that the correlation distance of the echoes is less than 30–60 m. It has not been possible to obtain information about absolute *echo strength* for selecting between the various proposed mechanisms, since the number density of irregularities is not known, and therefore any average value of the reflection coefficient can be due to a small number of strong scatterers or a large number of weak scatterers.

Most useful information has been obtained from studies of *amplitude distribution and fading* of the auroral echoes. The experimental results clearly show that a large number of scatterers are involved in the reflection process and there is no measurable specularly reflected component in the echoes. Whether the irregularities are over-dense or under-dense is not clear from these observations.

The investigation by the Saskatoon group of the *frequency dependence* for the auroral echo process in the frequency range 40–100 Mc/sec, seems to show that in the lower part of the frequency band mentioned, critical reflections sometimes play an important part, thereby supporting the theoretical model worked out by MOORCROFT (1961). They found that weak scattering dominates in the upper part of the band and is often the responsible mechanism down to 40 Mc/sec.

The *polarization* studies that have been carried out have not given any support for the existence of a resonance effect perpendicularly to the long axis of the scattering centers and, thus, do not support the proposed resonance scattering mechanism. When *comparing with other upper atmosphere parameters* the following conclusions can be drawn:

Against critical reflection speak the following facts (they seem, however, not completely to rule out the possibility of existence of critical reflections in radio aurora): Sporadic *E*-echoes seen on ionosondes during auroral displays usually do not involve cut-off frequencies much higher than 10 Mc/sec, corresponding to 10^6 el/cm^3. Omholt has deduced maximum electron densities of $2\times10^5-10^6$ el/cm^3 in the ionisation produced by the particles causing visual aurora. Rocket measurements have disclosed no electron density above 10^6 electrons cm^3 at auroral heights.

Against weak scattering from irregularities only a fraction of a meter wide, the following principal objection has been raised: The mean free path and Larmor radius electron gyro radius have about the same values as the irregularity at auroral altitudes. Such irregularities have a duration which is comparable with the collision time (about 10^{-6} second) and it is not clearly understood how such small irregularities can play an important part in the reflection mechanism.

D. CONCLUSIONS ABOUT REFLECTION MECHANISMS

From the review above it seems probable that critical reflection from small irregularities plays a part in the production of auroral echoes in the lowest part of the VHF band, but that weak scattering is the dominant process at VHF and UHF frequencies. Resonance scattering is probably not important. Although the theoretical models proposed in general are very crude, there is good observational support for the existence of a whole size spectrum for the auroral irregularities. Very little is as yet definitely known about its shape.

E. CAUSE OF THE IRREGULARITIES IN AURORAL IONISATION

There are great difficulties in understanding how the auroral electron density irregularities are produced.

Booker (1956) has proposed that turbulence is the relevant production mechanism. Against this speaks the experimental observations by means of rocket technique, according to which no turbulence has been observed above approximately 100 km. In any case turbulence cannot be the only process of importance in auroral ionization, since the high fading rate cannot be produced by turbulence or wind-speed, the velocities involved being one power of ten too large.

Another mechanism that has been proposed is that the irregularities are produced directly by the primary particle radiation. The production of the extremely thin irregularities required by weak-scatter theory is not understood, the auto-correlation function being comparable in width to the spiral of one single incoming electron and therefore probably less than the dimensions of the cross-section of a shower of electrons. Even though the dimensions of the irregularities may differ from those of the autocorrelation function by a factor somewhere between 1 and 10, this objection seems relevant. The large irregularities required by the critical reflection theories can easily be understood as produced by the primary auroral particles.

The two-stream plasma instability, mentioned at the end of the lecture on irregularities producing scintillation, as a possible cause of those irregularities, may very well be of importance for the auroral irregularities too; although it seems improbable that this mechanism can produce the very thin irregularities required by the weak scatter theory. A recent development of this instability model, in which the instability is assumed to produce sound waves, which are thought to be the immediate cause of the electron density irregularities, has recently been proposed.

References

BOOKER, H. G.: 1956, *J. Atm. Terr. Phys.* **8,** 204–221.
BOWLES, K. L.: 1954, *J. Geophys. Res.* **59,** 553.
CHAPMAN, S.: 1952, *J. Atm. Terr. Phys.* **3,** 1–29.
EGELAND, A.: 1962a, *Arkiv för Geofysik* **4,** 103–169.
EGELAND, A.: 1962b, *Arkiv för Geofysik* **4,** 171–209.
HARANG, L. and LANDMARK, V.: 1953, *J. Atm. Terr. Phys.* **4,** 322.
HARANG, L. and TRÖIM, J.: 1961, *Plan Space Sci.* **5,** 33–45.
HERLOFSON, N.: 1947, *Nature* **160,** 867–868.
LOVELL, A. C. B., CLEGG, J. A., and ELLYETT, C. D., 1947: *Nature* **160,** 372–373.
LYON, G. F. and FORSYTH, P. A.: 1962, *Can. J. Phys.* **40,** 749–757.
MOORCROFT, D. R.: 1961, *Can. J. Phys.* **39,** 677–715.

CHAPTER 15

DISTURBANCE EFFECTS IN THE LOWEST IONOSPHERE

B. HULTQVIST

Kiruna Geophysical Observatory, Kiruna C, Sweden

1. Introduction

Storm effects in the lowest ionosphere depend on production of anomalous ionization in and below the D-layer. This ionization can be studied by measuring its absorbing effect on radio waves and also by observing some other influences on propagation of VLF and LF radio waves, which are normally reflected at D-layer heights.

The ionospheric attenuation of radio waves is a phenomenon of great practical and scientific interest. The latter stems mainly from the fact that absorption studies provide information concerning the physical properties of the lowest regions of the ionosphere. In the cosmic noise method of absorption measurement, the cosmic noise power is measured and from its amplitude the absorption of the radio waves in passing through the ionosphere can be evaluated, if the sideral time variation of power on undisturbed days is taken into account. In general these measurements are made by means of riometers – relative ionospheric opacity meters – operating at frequencies between 15 and 60 Mc/sec.

At the frequencies mentioned the refractive index of all parts of the ionosphere is very nearly unity and therefore only nondeviative absorption is of importance. Only that type of absorption will be dealt with in this review.

As early as 1937 it was noted that during the day the intensity of the cosmic noise was lower than expected and this was attributed to ionospheric absorption. In 1949 a series of observations of cosmic noise at 18.3 Mc/sec was undertaken by Shain in Australia and it was apparent that, although the ionospheric attenuation was comparatively small, under certain conditions it was appreciable and could be measured without difficulty. Riometers (cf. LITTLE and LEINBACH, 1959) were first used a few years before IGY and the cosmic noise method of measuring absorption has since become widely used, especially for studies of disturbance effects in the lowest ionosphere.

2. Theory

In the ionosphere the dissipation of the energy contained in radio waves is due almost entirely to collisions of electrons with other particles. Collisionless dissipation by Landau damping of plasma waves excited by the radio waves is negligible. The particles with which the electrons collide when oscillating in the electric field of the radio wave extract momentum from the electrons, thereby transforming radio wave energy into heat. Therefore the absorption coefficient should be proportional to the collision frequency. But if the collision frequency, ν, is much greater than the wave frequency, ω, the collisions will prevent the electrons from moving in the electric field and thereby absorbing energy from it. For such high values of ν the absorption coefficient may be expected to decrease with increasing ν. The transition occurs around 50 km altitude for the riometer frequencies.

Appleton and, independently, Goldstein and Hartree generalized Lorentz's theory for the propagation of electromagnetic waves through a weakly ionized gas – i.e. a gas where the electron-ion and electron-electron interaction can be neglected – and derived a relation between the macroscopic parameter, the absorption coefficient k,

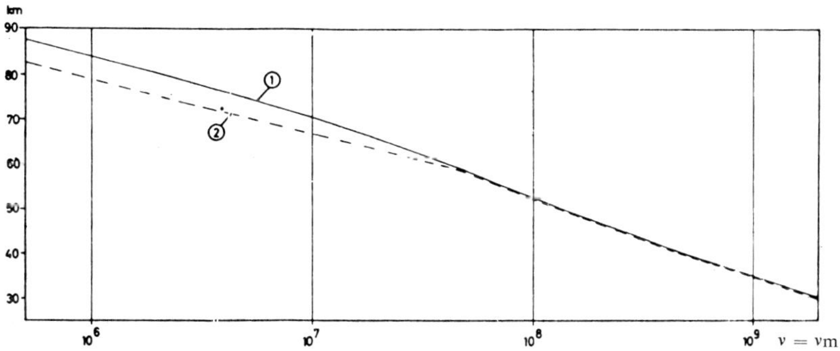

Fig. 1. High distribution of the frequency of collisions between electrons and neutral molecules. Curve 1 is after NICOLET (1959) and Curve 2 has been adjusted to the re-evaluated collision frequency values of KANE (1962) in the upper half of the height interval.

and the microscopic N (electron density) and ν. In their derivation the collisions were taken into account by introducing a friction term, $g.v$, (v is the electron velocity) in the force equation. The coefficient g, which can be formally identified with $m\nu$ (m is the electron mass), was assumed to be independent of the electron velocity.

From the magneto-ionic theory of Appleton and others – it will be called the MI-theory below – the following expression can be obtained for k for the non-deviative case ($\mu \approx 1$) in the quasilongitudinal (QL) approximation:

$$k = \frac{2\pi e^2}{mc} \cdot \frac{N\nu}{(\omega \pm \omega_L)^2 + \nu^2}. \tag{1}$$

Gaussian (cgs) units are used. The QL approximation is valid at the riometer frequencies at all heights of interest for vertical or galactic-pole directed antennae in high and temperate latitudes. In (1) $\omega_L = \omega_H \cdot \cos\alpha$ with $\omega_H = eH/mc$. The letter α indicates the angle between the direction of the geomagnetic field lines and the direction of propagation. H is the geomagnetic field strength. The quantities e and m denote charge and mass of electrons, c is the velocity of light in vacuum and + and − refer to the ordinary and extraordinary rays, respectively.

At the commonly used riometer frequency, 27.6 Mc/sec, ω_L is only about 5% of ω in high and medium latitudes. Well above the transition level in which $\nu \approx \omega$, (about 50 km altitude) the absorption coefficient for the extraordinary component is about

20% greater than that of the ordinary ray, while well below this transition level there is no difference.

In Figure 1 is shown the height distribution for ν due to NICOLET (1959). Evidence has accumulated in recent years, particularly through investigations of propagation of microwaves in various gases, that the velocity dependence of the collision frequency varies from gas to gas. Phelps and Pack measured this dependence for nitrogen and found ν to be proportional to the square of the electron velocity (i.e. to the kinetic energy) instead of constant, as assumed in the MI theory. Huxley showed that the properties of oxygen in this respect are not significantly different from those of nitrogen.

SEN and WYLLER (1960) derived an expression for the absorption coefficient for

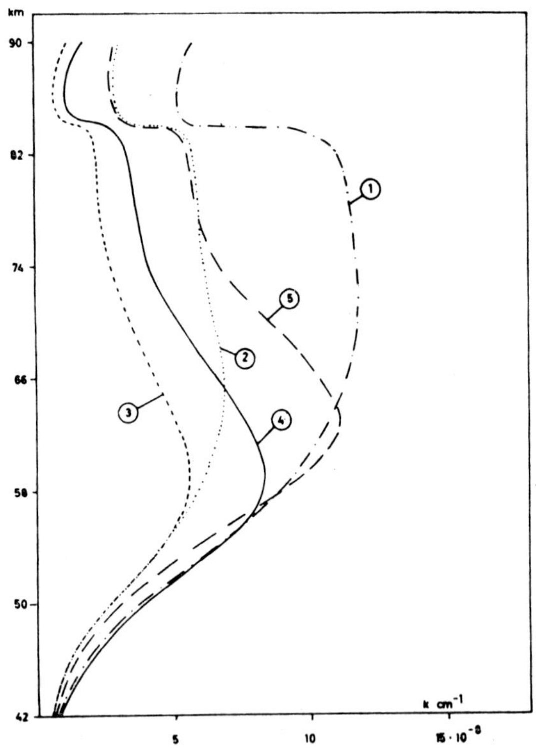

Fig. 2. Height distribution of the absorption coefficient for propagation of 27.6 Mc/sec radiowaves along the lines of force of the geomagnetic field. All distributions are for the ordinary wave and are based on the same electron profile. Curves 1–4 have been calculated by means of magneto-ionic theory (Formula 1), while Curve 5 has been obtained by means of Expression (2) of the generalized theory. For Curve 1 the ν distribution of NICOLET (1953) has been used and for Curve 2 that of NICOLET (1959). Curves 3–5 are based on the re-evaluated ν_m profile of Kane, i.e., Curve 2 in Figure 1. In calculating curve 3, ν in the magneto-ionic formula was taken equal to ν_m in the generalized formula. For curve 4 $3\nu_m/2$ was used as ν in Formula (1). Curve 5, finally, was obtained with the use of generalized theory (Expression (2)) and ν_m profile 2 in Figure 1.

propagation along the geomagnetic field lines. For riometer frequencies their expression can be simplified to

$$k = \frac{5\pi N e^2}{mc \cdot \nu_m} \cdot C_{5/2}\left(\frac{\omega \pm \omega_H}{\nu_m}\right). \tag{2}$$

The function $C_p(x)$ can be written

$$C_p(x) = \frac{1}{p!}\int_0^\infty \frac{\varepsilon^p e^{-\varepsilon}}{\varepsilon^2 + x^2}\, d\varepsilon . \tag{3}$$

For a direction of propagation different from that of the geomagnetic field lines the general expression for the complex refractive index, given by Sen and Wyller – to be referred to by SW below – has to be used. The calculations of k are then perfectly rigorous with no simplifying assumptions involved in their deduction from the generalized theory. However, according to Barrington and Thrane Equation (2) is valid also in the quasilongitudinal case, i.e. for zenith- and poledirected riometers in medium and high latitudes, if ω_H is replaced by ω_L.

In general the effect of the velocity dependence of ν in air is a decrease in the birefringent properties of the medium. The difference from the MI theory is of minor interest as long as ν is negligible compared with ω. For $\nu = 0.1\ \omega$ significant differences between MI and SW theories begin to appear.

If one takes $\nu = \nu_m$ in the MI formula (1), differences varying from 30% to 100% between MI and SW values are found. In deriving (1) one is free to choose any type of mean for identification with the frictional coefficient, g. If one associated ν with the mean square speed, the collision frequency in the MI formula, ν, will be related to ν_m by $\nu = \frac{3}{2}\nu_m$. With such ν values the difference between the MI and the SW values of k for longitudinal propagation is reduced considerably for the ordinary ray and to almost zero for the extraordinary one. For transverse propagation the reductions of the differences are less.

In the asymptotic case $\nu_m \ll \nu$, which is approximately reached at $\nu_m = 1/10\ \omega$, good agreement between the MI and SW theories is obtained if $\nu - \frac{5}{2}\nu_m$. The disparity is not greater than 5%. In the other asymptotic case, $\nu \gg \omega$, the MI formulas can be retained if $\frac{3}{2}\nu_m$ is used as ν value, according to SEN and WYLLER (1960).

Figure 2 illustrates the large disparity between absorption coefficients calculated with the use of one and the same electron density distribution, but with different collision frequency profiles and different theories.

3. Quiet Day Absorption

In order to obtain quantitative absorption values the unattenuated cosmic-noise level over a complete sidereal day has to be deduced. Basically the CNA method cannot provide information about any absorption component which is constant in time. One has to assume that the highest cosmic noise power observed in periods when no sources of measurable absorption are known to exist is completely unattenuated.

Most abnormal absorption associated with ionospheric disturbances of various kinds is thought to take place in or below the height interval of the normal D-layer.

4. Sudden Ionospheric Disturbance (SID)

Fading of short wave signals propagated over paths well separated from the auroral zones was first noted by MÖGEL (1930). He proposed on the basis of having observed

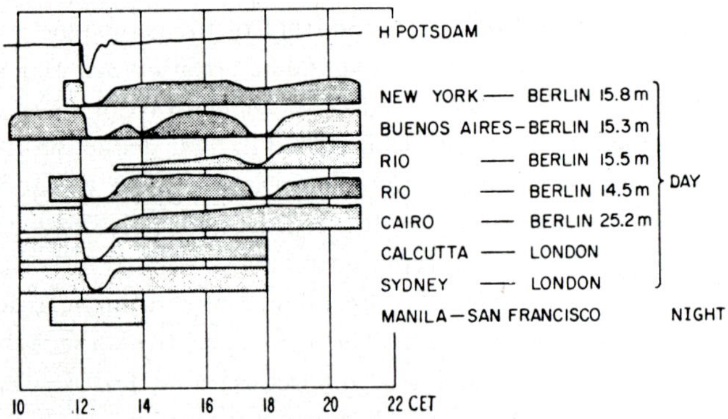

Fig. 3. The disturbance of October 10, 1958, one of the examples of fadeout and geomagnetic crochet presented by Mögel. Note the sudden and simultaneous fadeout at 12^h on all the sunlit paths, and the absence of any effect on the dark path (Manila to San Francisco). The crochet begins at the same time as the fadeout. The second fadeout at 1640 is so late in the day that no crochet is seen at Potsdam, and the fadeout appears in Europe only on western paths that are still sunlit (after WARWICK, 1963).

it only on the day side of the earth, that the effect was produced by X-rays ionizing the atmosphere below the E-layer. A typical example of short-wave fade-out is shown in Figure 3. His discovery was later re-established by Dellinger and by Newton and Barten, who found the connection between solar flares and SIDs.

Various kinds of sudden ionospheric disturbances have been studied in the three decades that have passed.

Dellinger was one of the pioneers in investigating the fade-out of HF radio communication over long distances (called short-wave fade out, SWF). Bureau described the sudden enhancement of atmospherics (SEA), Jansky observed the suppression of the cosmic radio noise level during SIDs (nowadays called sudden cosmic noise absorption, SCNA). Budden and Ratcliffe discovered the sudden phase anomaly (SPA), which is one of the most pronounced SID effects, Berkner and Wells and Martyn studied the SID effect on reflections of vertical transmissions, and McNish was the first to describe the magnetic effect associated with the SID, the magnetic crochet or solar flare effect (SFE).

The name SID contains the most important characteristic of the effect, the sudden onset. The absorption at 30 Mc/sec may increase to values of several decibels in a few minutes. The decay of the absorption is slower; the total duration varies from a fraction of an hour to more than one hour.

SCNA is observed over the entire sunlit hemisphere of the earth except in a zone where the solar zenith angle has values between 90 degrees and a few (at least 2) degrees less than 90.

From simultaneous riometer measurements at four different latitudes a solar zenith angle dependence for the SCNA effect has been found. The measured dependence agrees well with the models for X-ray production of SIDs that have been

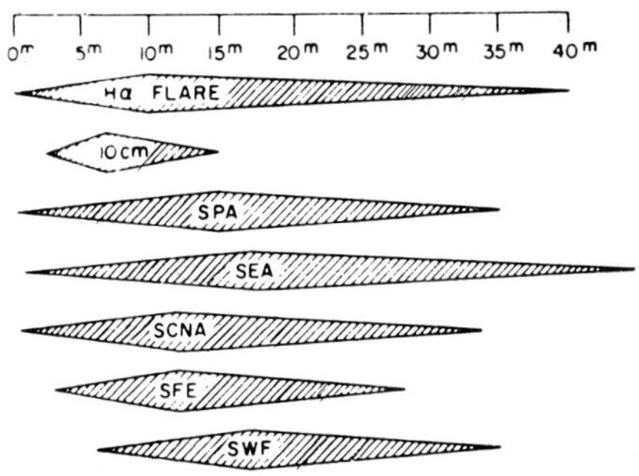

Fig. 4. Average time relations during SID (after WARWICK, 1963).

proposed in the last few years. Reflection is improved during SIDs for wavelengths of about 10 km (30 kc/sec), and the noise from distant thunderstorms increases in a sudden enhancement of atmospherics (SEA). An SEA usually begins about 10 minutes after flare start, i.e., on the average, 2 minutes before flare maximum, and has its maximum about 20 minutes after the flare start. Between the intensity of the SEA and the maximum Hα intensity a better correlation has been found than for any other flare parameter.

The longest waves, 16 km (19 kc/sec) and longer, normally reflected in the D-region, show little change in amplitude during SIDs but a change in phase. This is the sudden phase anomaly (SPA). The SPA method of identifying solar flare effects in the ionosphere is the most sensitive of all existing. Four times as many flares show up as SPAs than as SWF, SEA or SCNA.

At a point located a few hundred kilometers from the receiver, the changing pattern of interference between the reflected wave and the ground wave produces amplitude variations, the sudden field anomaly (SFA). The increased electron density also leads to an enhancement of the current system that flows in the lower E-region. This increase in flow of current is recorded by magnetometers as a small hook, the geomagnetic crochet, also called solar flare effect (SFE). Contrary to what has long been believed, Volland and Taubenheim showed that the SFE current system is not only an intensification of the Sq-system that causes the geomagnetic diurnal variation on quiet days. They found the flare current system to be at a lower level and its eddy center is closer to the subsolar point than that of the Sq-system. Most solar flares produce no observable SIDs but practically every SID observed has been associated with some unusual brightning of the Hα-line.

The time variation of various SID effects have been found to follow very closely the variation of the linewidth and intensity of Hα. The average time relations during an SID is shown in Figure 4 (after WARWICK, 1963). In fact, occurrence and timing of SIDs are more closely related to cm radio bursts than to the Hα flares.

No dependence of the SID intensity on the position of the flare on the sun has been found.

The rocket and satellite measurements of solar X-rays carried out by Friedman's group at US Naval Research Laboratory have definitely shown that X-rays of λ less than 10 Å cause the D-layer disturbance effects described above.

5. Disturbance Effects in the Lowest Ionosphere Associated with Aurora and Geomagnetic Storms

The most important way in which the disturbance effects in the lowest ionosphere, associated with aurora and magnetic storms, show up, is as absorption of radio waves, and the discussion below will be concentrated on this phenomenon.

A. CORRELATION WITH VISUAL AURORA

Studies of the correlation between auroral activity and CNA were recently made by Holt and Omholt in Norway and by Gustafsson and Ortner in Sweden. In both investigations recordings taken with a 5577 Å zenith photometer of 5 degrees field of view and 27.6 Mc/sec riometer recordings obtained with a vertical three element yagi antenna, directed towards zenith, were analysed.

When the opening angles for the radio and photometer equipment are different – – as mostly has been the case – and when recordings of distinct and isolated auroral forms are included in the material, the correlation between the auroral luminosity and CNA is in general poor. It is very much improved by selecting auroral events in which the aurora covers most of the sky and therefore most of the antenna lobe of the riometer.

The correlation is especially good between absorption of radio waves and pulsating auroras.

Even bright coronas may, however, appear without being associated with any measurable amount of absorption.

B. SPATIAL DISTRIBUTION

The auroral type absorption is a fairly local phenomenon, at least compared with sudden cosmic noise absorption or polar cap absorption. The dimensions of a single region of auroral type absorption have been found to be of the order of 100 km or more; such a region may have quite a sharp boundary.

Detailed studies of the latitudinal variation have shown a pronounced maximum at about 67 degrees geomagnetic latitude, i.e. in the auroral zone (see Figure 5). Riometer measurements at conjugate points have revealed many cases in which the absorption events in the two hemispheres were closely related in magnitude and duration, but also some cases where absorption was observed only in one hemisphere. The absorption is in general strongest in the winter hemisphere.

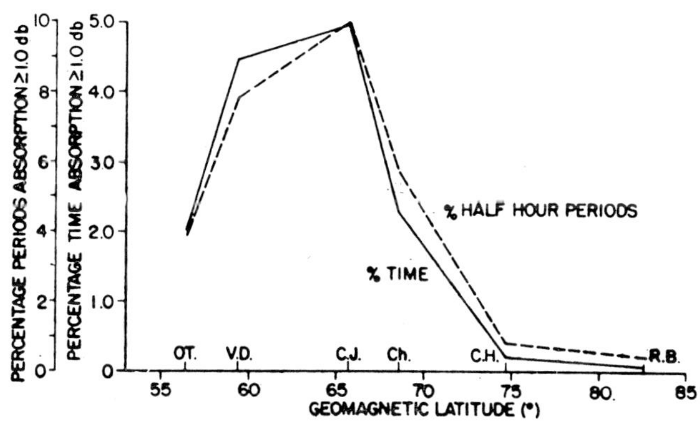

Fig. 5. The percentage of the time (solid line) and percentage of half-hour periods (dashed line) for which auroral absorption equalled or exceeded 1.0 db, as a function of geomagnetic latitude (after HARTZ, 1963).

C. HEIGHT DISTRIBUTION

HEIKKILA and PENSTONE (1961) measured the height distribution of the specific absorption at Churchill by rocket borne equipment during auroral type absorption. They found a pronounced maximum for dA/dh at about 74 km altitude.

The equivalent height of the absorbing layer has been measured by means of several riometers tuned to different frequencies. As the specific absorption is frequency dependent and the frequency dependence varies with height, the ratio of the absorption values for two different frequencies gives an equivalent height (which is closely related to a real physical height only when the absorption is limited to a height interval comparable with the scale height). A big majority of the riometer-measured equivalent height values fall above 65 km. For weak auroral events, altitudes below 75 km have not been found. In a small number of cases the absorption has been found to be independent of frequency (indicating heights below 50 km). It is not yet clear how significant these values are.

D. TEMPORAL CHARACTERISTICS

The absorption periods have been found to be somewhat shorter in the auroral zone than at subauroral latitudes. Periods of several hours' duration sometimes occur although the most common duration is between 20 and 30 minutes.

The diurnal variation of the auroral type CNA in auroral zone latitudes is illustrated in Figure 6. As can be seen from Figure 6 there is a maximum around 06–09 hours in the morning and a secondary maximum in the evening. It is quite different from the corresponding curves for visual aurora and magnetic disturbances. The diurnal variation has been found to be latitude dependent.

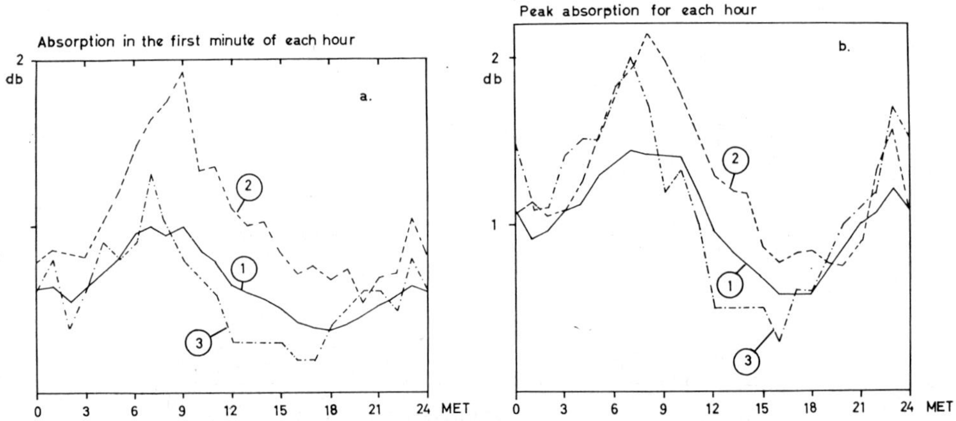

Fig. 6. Diurnal variation of the average intensity of 27.6 Mc/sec cosmic noise absorption for the equinoxes at Kiruna. Months in which strong PCA's occurred are excluded. Curve 1 represents the average for the following months: April 1959, September 1959, October 1959, March 1960, October 1960, March 1961, April 1961, September 1961 and October 1961. Curve 2 is for September 1961 alone and Curve 3 for September 1959.

E. CORRELATION WITH GEOMAGNETIC DISTURBANCES

Even if the correlation between general magnetic activity and cosmic noise absorption is high in auroral and subauroral latitudes, a geomagnetic disturbance is not invariably associated with CNA. Examples of intense isolated geomagnetic bays observed on all three components and not accompanied by cosmic noise absorption at all have been observed.

While a strong positive correlation between geomagnetic disturbance and radio wave absorption exists in auroral and subauroral latitudes – as was reported already by APPLETON in 1937 – the correlation is negative in the equatorial belt (latitude less than 30 degrees), where the main part of the cosmic noise absorption at VHF is thought to take place in the F-region.

F. SUDDEN COMMENCEMENT ABSORPTION (SCA)

All kinds of cosmic noise absorption associated with aurora and magnetic disturbances which have been discussed above are fairly local and irregularly distributed in longitude. There is, however, one type of cosmic noise absorption associated with magnetic activity which often seems to occur simultaneously all along two complete zones centered on the auroral zones, namely absorption starting at the sudden commencement of a worldwide geomagnetic storm.

The character of SCA is similar to that of SCNA discussed earlier. There is a sudden onset – the absorption sometimes increases to 3–4 db within a minute – and the duration is short, only a fraction of an hour. SCA has been found to be associated with X-ray bursts.

G. VARIATION IN THE TWILIGHT PERIODS

It is known that in the lowest ionosphere negative ions, besides free electrons, are carriers of negative electric charges. This has been shown by direct rocket measurements as well as by much indirect evidence. The ratio of the densities of negative ions and free electrons, λ, is lower by day than by night due to the existence of a photodetachment reaction in daytime but not in the night. This means that for a constant production rate for electrons there will be more absorption in the day than in the night. The ratio of day to night absorption, A_D/A_N, is for a given electron density profile determined by the height distributions of λ for day and night. Experimental A_D/A_N values are not easily obtained for the auroral-type absorption, which is irregular in time and space. Because there are no instances of a constant electron production rate throughout twilight intervals, in which absorption changes can result only from alteration of the attachment-detachment equilibrium for the electrons, a statistical approach is the only possible one.

The experimental studies made have shown that the ratio of pre-twilight and post-twilight absorption intensities does not amount to as much as a factor of 2 (it is probably considerably less). This value for the sunrise and sunset effect on auroral absorption is much smaller than predicted on the basis of the λ-distributions proposed in the last few years. This disagreement between observations in the ionosphere and parameter values computed from laboratory measurements is of great significance. Even very drastic changes of the measured electron distribution used could not entail such small A_D/A_N values as those measured.

6. Polar Cap Absorption

A. INTRODUCTION

Polar cap absorption (PCA) seems to have been first recognized by BAILEY (1957) as a special, previously unknown phenomenon after its appearance as a result of the big solar eruption on 23 February 1956.

That PCA is caused by protons of energies much higher than those associated with auroral effects, emitted by the sun in the course of the solar flare, was soon realized, and protons were also observed in balloon flights of cosmic-ray equipment. An example of a strong PCA is that of 7–11 July 1958. No disturbance could be seen on any record other than the cosmic noise one in the first 30 hours after the 3^+-flare of 7 July, except for the ionosounder, which showed blackout from a few hours after the cosmic noise power started to diminish. The geomagnetic conditions were unusually quiet, VHF auroral reflections were almost nonexistent, and no effect could be seen on the cosmic ray records. The riometer record stayed off scale for about 12 hours. Then, after the sudden commencement of a strong magnetic storm some 30 hours after the start of decline of the cosmic noise power, indications other than absorption of disturbed ionospheric conditions were observed. The abnormal absorption decayed slowly during several days.

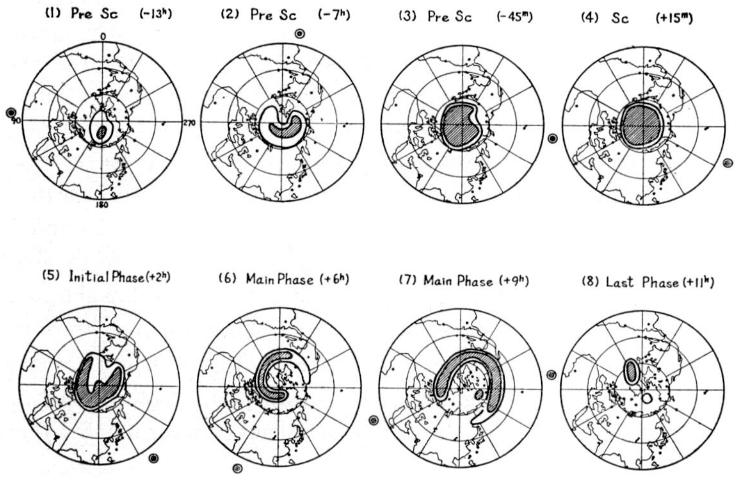

Fig. 7. The world wide patterns of f_{min} on September 15, 1957, showing typical stages of the development of a PCA event (after HAKAKURA *et al.*, 1958).

For references to papers including major lists of PCA events cf. HULTQVIST (1963).

PCA was observed on 80 days during the IGY, i.e. on one day in seven. In June–July 1957 blackout continued for 17 days at Resolute Bay due to overlapping PCA events. It is thus fairly common in periods of high solar activity.

B. GEOGRAPHICAL DISTRIBUTION

The first detailed study of the geographical distribution and its time variation seems to have been made by HAKURA *et al.* (1958) with the use of f_{min} data. Figure 7, given by them, is a good illustration. One can see the fundamental difference between the distribution of PCA – over the entire polar cap – and of auroral absorption – a spiral shaped figure on the night and morning side, approximately at the auroral zone latitude.

PCAs are not always followed by auroral zone absorption.

The lower latitude boundary of polar cap absorption is generally situated between 61 and 65 degrees. On very few occasions it has reached the geomagnetic latitude of King Salmon (57 degrees geomagnetic north), when detected by means of riometers. By balloon borne instruments solar protons have often been observed as far south as at Minneapolis (geomagnetic latitude 59 degrees). However, protons at latitudes below

60° have been found only after the start of the main phase of the associated geomagnetic storm.

C. TIME CHARACTERISTICS AND TIME VARIATIONS

The first observed major PCA events had durations of several days, but also events with a duration as short as three hours have been classified as PCA by several investigators. Protons have been observed to arrive in the earth's atmosphere as long as nine days after the solar flare. It is generally believed that these protons originate at the time of the flare because radio noise and other effects, which are expected to be associated with their production, are only observed during the time that the flare

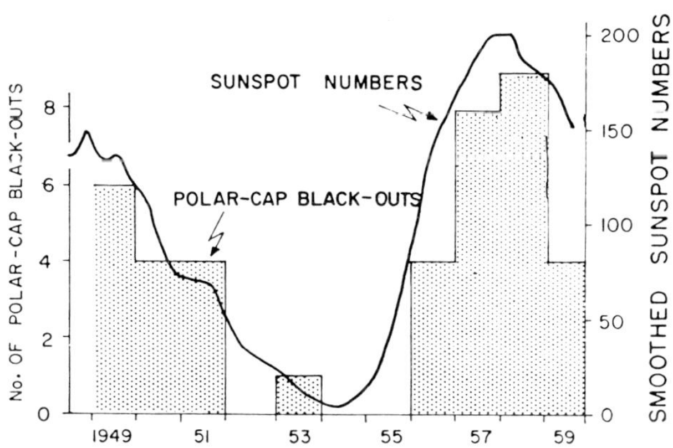

Fig. 8. Distribution of polar-cap blackouts during the sunspot cycle 1949–1959 (after COLLINS *et al.*, 1961).

is visible. Some storage mechanism must therefore exist. There is a fair amount of evidence to show that the slow plasma cloud from the sun is the trapping region.

Variation over a very wide range has also been found for the delay between the start of the associated flare and of the PCA. This variation has been from a fraction of an hour to more than 12 hours.

The rise-time of the absorption also varies very much – from a couple of hours to 24 hours or more.

A close correlation exists between polar cap blackouts and the sunspot numbers, as can be seen from Figure 8. A seasonal variation of the PCA occurrence frequency has been found.

One of the most significant characteristics of polar cap absorption is the strong diurnal variation, illustrated in Figure 9. This variation is discussed in some detail in subsections G and H below.

D. RELATIONS TO DISTURBANCES ON THE SUN

That the major polar cap absorption events are associated with major solar flares was evident from the very beginning of the study of the PCA phenomena.

The most extensive study made hitherto of relations between solar activity and polar cap absorption is the one of WARWICK and HAURWITZ (1962). They made a systematic examination of all unusual solar radio and optical emissions before the onset of 43 polar cap absorption events. They found that in only 3 of 40 cases a high importance solar flare could not be associated with a PCA. The intensity of the absorption event was found to increase with magnitude of the solar event and these relations tended to become stronger with increasing wave length, being weakest for optical flares, and strongest for "decametric" radio bursts.

Warwick and Haurwitz also investigated a possible relation of PCA events to solar activity on the far side of the sun, but they found that the most active region on the sun was located on the invisible side only in 4 of 43 cases. From this they draw the conclusion that mechanisms which provide for transit to the earth from the far side of the sun cannot be generally operative.

E. RELATIONS OF PCA EVENTS TO OTHER DISTURBANCE EFFECTS IN THE EARTH'S UPPER ATMOSPHERE

Because the absorbing ionization is situated exceptionally low in the upper atmos-

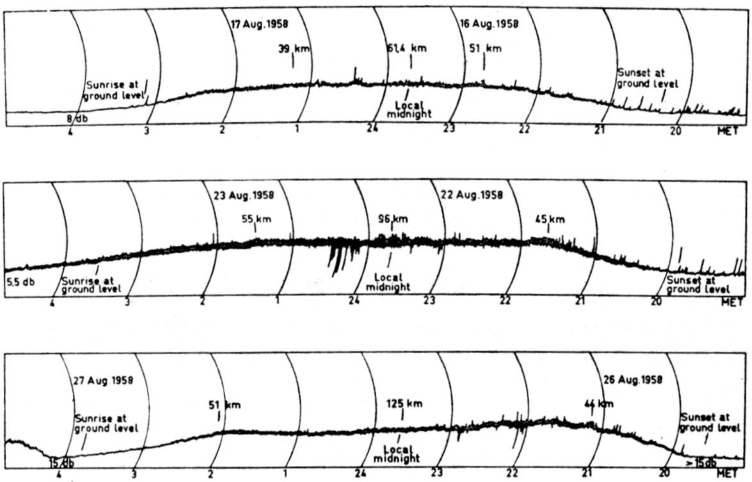

Fig. 9. Examples of records of cosmic-noise power taken at Kiruna showing the diurnal variation of CNA. The figures in kilometers above the curves give the upper limit of the solid earth's shadow vertically above Kiruna at midnight and the estimated beginnings and ends of the periods of night-time absorption.

phere during PCA events, it can be recognized almost only from VLF-propagation data, apart from attenuation of HF and VHF radio waves. The effect on the scatter links described by Bailey is also an absorption effect.

A profound influence of PCA events on the propagation of VLF and LF radio waves, other than attenuation, has been observed. At VLF the normal diurnal variation of the signal strength of the Rugby transmissions received in Kiruna was found to disappear completely. The signal level during PCAs stayed constant approximately at the day level all through day and night.

The interpretation proposed was that the VLF radio waves during strong PCAs are reflected from an ionized layer situated below those which normally are responsible for the propagation, and the electron density of this layer is thought to be sufficient for reflections also in the night.

At LF frequencies a more complicated influence of PCA has been found.

An interesting effect of a different kind associated with PCA has recently been observed. Soon after the start of the PCA on 10 May 1959 the intensity of the 3914 Å band of nitrogen began to increase steadily, reaching a maximum, about one hundred times above minimum night time level some 24 hours later, in coincidence with the peak activity of the associated major geomagnetic storm. The intensity then fell steadily to zero, 48 hours after maximum. Not only the first negative but also the first and second positive bands of nitrogen showed the same behaviour. The intensity of the glow agrees well with what is to be expected on the basis of recent rocket measurements of proton flux and energy spectrum.

7. Electron Reactions in the Lowest Ionosphere

The reactions listed in Table I are those considered to be the most important in the ionosphere below 100 km altitude.

An extensive list of possible reactions in the upper atmosphere was discussed in the seminar by Nicolet.

The numerical values used by different authors for the listed reaction constants cover very wide ranges. Most of them have been determined in laboratory experiments. Representative values used are also given in Table I.

TABLE I

ELECTRON REACTIONS IN THE IONOSPHERE BELOW 100 km ALTITUDE

Reaction	*Reaction constant*	*Value*
Attachment		
$O_2 + O_2 + e \rightarrow O_2^- + O_2 + h\nu$	β	$1.5 \times 10^{-30}\, n(O_2)$ cm^3 sec^{-1}
Photo detachment		
$O_2^- + h\nu \rightarrow O_2 + e$	ρ	0.44 sec^{-1}
Collision detachment		
$O_2 + O_2^- \rightarrow O_2 + O_2 + e$	κ	2×10^{-17} cm^3 sec^{-1}
Dissociative recombination		
$N_2^+ + e \rightarrow N + N + h\nu$	α_{d1}	5×10^{-7} cm^3 sec^{-1}
$O_2^+ + e \rightarrow O + O + h\nu$	α_{d2}	3×10^{-8} cm^3 sec^{-1}
Mutual neutralization		
$O_2^- + N_2^+ \rightarrow O_2 + N_2 + h\nu$	α_{n12}	10^{-7} cm^3 sec^{-1} (both)
$O_2^- + O_2^+ \rightarrow O_2 + O_2 + h\nu$	α_{n11}	

The attachment and detachment reactions are much faster than the recombination reactions. Considering only time changes, which are so slow that attachment detachment equilibrium can be assumed to exist – which is true for variations with

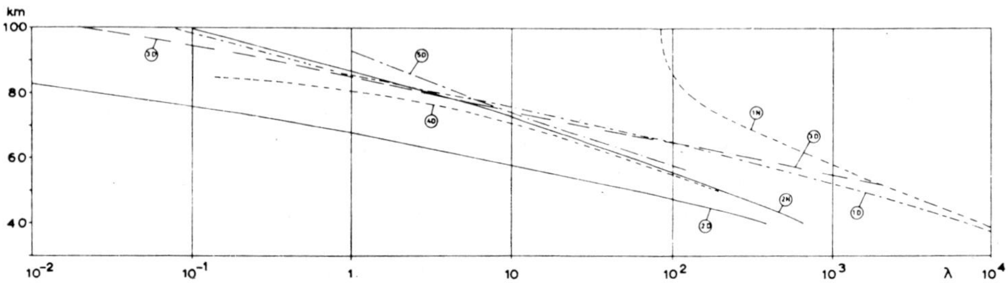

Fig. 10. Height distributions for the ratio between negative-ion and electron densities used in the last few years. The curves designated *D* refer to full daylight conditions and those designated *N* to nocturnal conditions. The different curves have been used by the following authors: 1D-1N: Bailey; 2D–2N; Nicolet and Aikin, Mitra, Brown and Weir, Holt *et al.*, and Aikin; 3D: Moler; 4D: Crain; 5D: Mitra.

time constants of a few minutes or more – it can be shown that the continuity equation for the electron density, N, can be written

$$\frac{\partial N}{\partial t} = \frac{q}{1+\lambda} - N^2 \alpha_{\text{eff}} \tag{4}$$

where

$$\lambda = \frac{\sum_i N_i^-}{N} = \frac{\sum_i \beta_i n_i}{\sum_i \varrho_i C_i^- + \sum_{ij} \kappa_{ij} n_i C_j^-} = \text{const for each height} \tag{5}$$

and

$$\alpha_{\text{eff}} = \lambda \sum_{ij} \alpha_{nij} C_i^- C_j^+ + \sum_i \alpha_{di} C_i^+ \tag{6}$$

in which expressions indices i and j refer to the different constituents – ionized or neutral – which may participate in the reactions in the general case. q is the production rate for electrons. C_i^- and C_i^+ are the relative number densities for negative and positive ions respectively:

$$C_i^- = \frac{N_i^-}{\sum_i N_i^-}, \qquad C_i^+ = \frac{N_i^+}{\sum_i N_i^+},$$

and n_i is the number density of the neutral constituent i.

8. Discussion of Electron Reaction Parameters on the Basis of Experimental Absorption and VLF Propagation Data

A. INTRODUCTION

The ratio of the densities of negative ions and free electrons (λ) is a very important parameter in the lowest ionosphere. Despite its significance no good λ values directly measured in the ionosphere exist. Values have, it is true, been computed indirectly from time constants in radio data but they are not very accurate.

By rocket technique some information about λ has been obtained. These values are, however, very high, in fact so high that they are in disagreement with a number of ground observations, and they must therefore be considered as doubtful.

The λ values used in the literature of the last few years are based on laboratory measurements of the most important reaction constants defining λ. These values diverge widely from those based on radio data.

It should be emphasized that our knowledge of electron reactions in the lowest ionosphere is still in the qualitative stage and that the quantitative values used for the different reaction constants are, in general, very crude.

Height distributions of λ used by a number of different authors in the last few years are shown in Figure 10. As can be seen in the figure, the differences between the various λ profiles are very great for both day and night – amounting to as much as a factor of several hundred in the upper half of the height interval. The lowest λ-distributions are the widely accepted ones of Nicolet and others. They have the value unity at an altitude of 68 km by day and at 87 km by night.

The λ profile is the most important height function determining the form and height of an ionized layer produced by radiation which penetrates to great depths and undergoes maximum interaction with the atmosphere well below the altitude at which $\lambda = 1$. This applies to ionization produced by the ordinary cosmic radiation as well as to solar proton-produced ionization which gives rise to polar cap absorption (PCA). Investigation of such ionized layers may, therefore, provide information on the λ height distribution.

B. THE DAY TO NIGHT RATIO OF ABSORPTION ASSOCIATED WITH AURORA AND GEOMAGNETIC STORMS

Using two different rocket-measured height distributions for the specific absorption recorded under geomagnetically disturbed conditions, a ratio of about 3 for A_D/A_N can be evaluated with Nicolet's λ profiles. With other pairs of λ curves in Figure 10 even greater values are obtained.

Even with extreme assumptions about the amount of absorption in the E-layer, a value of about 3 for A_D/A_N was found.

The experimental results show that A_D/A_N has values between 1 and 2 for auroral type absorption.

C. VARIATION OF POLAR CAP ABSORPTION WITH SOLAR ZENITH ANGLE AT SUNRISE AND SUNSET

In a small number of exceptionally undisturbed riometer recordings of polar-cap absorption it has been observed that absorption decreases at sunset during the interval in which the shadow of the solid earth moves from approximately ground level to about 50 km altitude, and increases during the corresponding interval at sunrise.

The records thus disclosed that the exact position of the earth's shadow above 50 km was immaterial to the nondeviative radiowave absorption. In view of the very low detachment energy of O_2^- – probably the most plentiful negative ion in the lowest ionosphere – it was concluded that the whole solar spectrum may actuate electron detachment from O_2^- and that, in consequence, practically all of the absorbing ionisation must be situated below 50 km altitude. An alternative hypothesis, which has been suggested by several authors, is that the detachment mechanism involves ultraviolet radiation whereby the effective shadow height is increased by the height of the ozone layer, say 30 km. The altitude of the upper boundary of the absorbing layer would thus be situated at about 80 km. The hypothesis that UV light is the main causative factor in electron detachment presupposes that the energy necessary for detachment of electrons from the negative ions is of the order of 4 eV and, consequently, that negative ions other than O_2^- must be the dominant ones in the lowest ionosphere. In the laboratory a large cross-section for formation of NO_2^- by charge transfer has been found. A lower boundary for the electron affinity of NO_2^- is 3.82 eV. In any case NO_2^- must, however, be a trace constituent and it is not understood how it can eliminate the effect of the O_2^- ions.

Indeed, a sharply defined ionized layer below, say, 55 km with no measurable ionization between it and the D-layer, is physically implausible. The above mentioned interpretation of the measurements of sunrise and sunset effects on PCA in terms of O_2^- was based on the assumption that λ may exceed unity up to altitudes much higher than the proposed upper boundary of the layer, as is the case for the curves in Figure 5. It is evident, however, that if λ has values well below 1 in the height interval above 60 km, no sunrise or sunset effect will be manifest there, however high the electron content, since the electron-density change resulting from alteration of the detachment-attachment equilibrium will be negligible.

The inference is that the relation between the variation of the nondeviative absorption and the variation of the solar zenith angle at sunrise and sunset requires a reduction of λ values to less than 1 above approximately 60 km altitude, if the observations are to be accommodated in terms of O_2^- as the principal negative ion.

D. VARIATION OF RADIO PROPAGATION CONDITIONS DURING TWILIGHT PERIODS

If there are any electron detachment effects in the 60–90 km height interval, it should be possible to detect them by studying the propagation conditions for radio waves – especially long and very long waves in the twilight periods.

For 16 kc/sec transmissions over fairly short ground distances the amplitude does not, on the average, start to change until the upper boundary of the solid earth's shadow, moving downward, has reached an altitude of about 60 km at sunrise. The phase starts to change a few minutes after ground sunrise. This might be thought to imply that the electron detachment in the lowermost, cosmic-ray-produced part of the ionosphere increases the attenuation of the radio waves; but for a steep incidence even the post-sunrise electron density is not sufficient to change the refractive index there from very nearly 1.

In fact the VLF effects described above, and others, are fairly consistent with the observed twilight variation of PCA. The VLF and LF radio propagation data seem, therefore, not to provide any real proof that the propagation conditions are affected by photodetachment in the 60–100 km height interval.

E. CONCLUSION

It has been found possible to accommodate all the mentioned effects without postulation of negative ions requiring ultraviolet radiation for electron detachment, by modifying the day and night λ height distributions so that λ is less than 1 down to 60 km. It can be shown that one and the same pair of λ distributions can meet all the requirements of the various observational data in the preceding sections.

The only observational data, mentioned above, which permit a choice between the two alternatives – low λ distribution on one hand and negative ions requiring UV radiation for detachment on the other – is the lack of a strong twilight effect on the auroral type absorption. If most of the auroral absorption is located below the E-layer, for which there is much experimental evidence, then it follows that the ultraviolet detachment hypothesis is incorrect.

If on the other hand large amounts of free electrons are produced by photodetachment above 70 km, then ultraviolet radiation must be the detaching agent and all but a small part of the auroral type absorption must be located above 90 km, contrary to what has been found from the rocket measurements and other observations. These seem to be the two most probable alternatives. At present it is not possible to give a final judgement.

More experimental data are needed for solving the λ puzzle.

References

BAILEY, D. K.: 1957, *J. Geophys. Res.* **62,** 431–463.

COLLINS, C., JELLY, D. H., and MATTHEWS, A. G.: 1961. *Can. J. Phys.* **39,** 35–52.

HAKURA, Y., TAKENOSHITA, Y. and OTSUKI, T.: 1958, *Rep. Ionos. Res. Japan* **12,** 459–468.

HARTZ, T. R.: 1963, 'Multi-station riometer observations', *Radioastronomical and Satellite Studies of the Atmosphere* (ed. by J. AARONS), North Holland Publishing Co., pp. 220–237.

HEIKKILA, W. J. and PENSTONE, S. R.: 1961, *Can. J. Phys.* **39,** 1875–1881.

HOLT, O.: 1963, 'Multistation riometer observations of sudden ionospheric disturbances', *Radioastronomical and Satellite Studies of the Atmosphere* (ed. by J. AARONS), North Holland Publishing Co., pp. 502–524.

HULTQVIST, B.: 1963, 'Studies of ionospheric absorption with the cosmic noise method', *Radioastronomical and Satellite Studies of the Atmosphere* (ed. by J. AARONS), North Holland Publishing Co., pp. 163–219.

KANE, J. A.: 1962, *J. Atm. Terr. Phys.* **23,** 338.

LITTLE, C. G. and LEINBACH, H.: 1959, *Proc. IRE* 47, 315–320.

MÖGEL, H.: 1930, *Telefunken-Zeitung* **11,** 14.

NAWROCKI, P. J. and PAPA, R.: 1961, 'Atmospheric Processes', Geophysics Corporation of America Bedford, Mass., Contract No. AF 19 (604)–7405.

NICOLET, M.: 1953, *J. Atm. Terr. Phys.* **3,** 200.

NICOLET, M.: 1959, *Phys. Fluids* **2,** 95.

Sen, H. K. and Wyler, A. A.: 1960, *J. Geophys. Res.* **65**, 3931–3950.
Warwick, C. S.: 1963, 'The sudden ionospheric disturbance', *Radioastronomical and Satellite Studies of the Atmosphere* (ed. by J. Aarons), North-Holland Publishing Co., pp. 457–475.
Warwick, C. S. and Haurwitz, M. W., 1962: *J. Geophys. Res.* **67**, 1317–1332.
Webber, W.: 1962, *J. Geophys. Res.* **67**, 5091–5106.

CHAPTER 16

IRREGULARITIES IN THE IONOSPHERIC ELECTRON DENSITY DISTRIBUTION CAUSING RADIO WAVE SCINTILLATION

B. HULTQVIST
Kiruna Geophysical Observatory, Kiruna C, Sweden

Introduction

Shortly after the discovery of discrete sources of cosmic radio noise it was found that the strength of the sources was variable. First, it was believed that the sources varied, but Smith, Little and Maxwell found that there was no correlation between the fluctuations recorded at two points on the ground separated by a distance of the order at 100 km. From this observation it could be concluded that the fluctuations originate in the medium through which the radio waves propagate. The effect is similar to the scintillation of stars in the visible frequencies. Scintillations have been observed at radio wavelengths varying between 20 m and 8 mm in the radiation coming from the sun, Cygnus, Cassiopeia, Taurus, Virgo and Centaurus and also in transmissions from artifical satellites.

In analogy with the visible radiation the radio wave scintillations are due to spatial variation in the refractive index of the atmosphere for the waves. The irregularities are strongly inforced in periods of magnetic disturbance. They are also much more pronounced in high latitudes than in medium and low ones.

There seems to exist a very wide size spectrum of irregularities in the ionosphere; from the small ones associated with aurora to very large ones having dimensions of several hundred kilometres. Various parts of the size spectrum influence the propagation of radio waves in various wavelength regions. Very little is as yet known about the shape of this size spectrum and still less about how it varies under various conditions.

Radio waves with frequencies above the maximal usable frequency scattered from the irregularities in the ionosphere can reach far beyond the radio horizon of the transmitter and some small fraction of the radio energy is also scattered back towards the transmitter. This is used for communication in high latitudes during disturbed conditions. In connection with aurora a type of irregularity with characteristics different from those of the more ordinarily existing irregularities in the ionosphere appear in the height range where visible aurora occurs. These irregularities are dealt with in the special chapter on radio aurora. The irregular distribution of the refractive index can also be seen as a fading in records of ionospherically reflected radio waves. In this chapter we will concentrate on scintillations, but some results obtained from drift measurements using manmade transmissions from the ground will also be reviewed.

The characteristics of scintillation will first be described, its time variation and correlations with upper atmosphere parameters. Some relevant results of the theory of diffraction patterns will next be reviewed. The use of these theoretical results for determination of size and shape of the irregularities and of their drift velocity and height will be described. Some physical mechanisms, possibly responsible for the production of the irregularities will finally be reviewed.

Observational Methods, Measures of Scintillation Activity, and General Properties

The scintillations can be observed by means of one antenna and receiver as amplitude fluctuations in reception on the ground of a discrete radio source (see Figure 1). They also appear in interferometer records as irregular variations in the phase difference of the waves received by means of two separated antennae, corresponding to variations in the apparent angle of arrival (cf. Figure 2). The scintillation phenomenon can be due to fluctuations in both amplitude and phase of the wave. A single receiver with antenna is usually sensitive mainly to amplitude scintillations. On the records obtained by means of phaseswitching interferometers the two types of scintillation can be separated to some extent by the fact that amplitude scintillation never can make the curve cross the zero line. The scintillations have great amplitude at the maxima and minima in the interference pattern but are very small when the record crosses the zero line. Phase scintillations, on the other hand, can have large amplitudes also at the zero crossings of the pattern.

For statistical studies the intensity of the scintillation is often given as a simple scintillation index, because the compution of an accurate measure of the irregular variation of the record, like $\overline{(\Delta A/\bar{A})^2} = \overline{((A-\bar{A})/\bar{A})^2}$, requires too much work. For scintillations in transmissions from satellites (see Figure 3) with linearly polarized antennae the following scintillation index has been used by several workers:

S = 0, regular Faraday rotation of polarisation plane visible;

S = 1, Faraday rotation with superimposed scintillations;

S = 2, Faraday rotations completely obscured by scintillation.

The intermediate values 0.5 and 1.5 are also often used. The ratio of the maximal amplitude of the fluctuations in a given time interval to the average amplitude of the received wave, $(A_{max} - A_{min}/A_{max} + A_{min})$ expressed in percentage, is also commonly given as a measure of the scintillation intensity both for cosmic noise and satellite transmissions.

The amplitude scintillation often amount to 100% of the average signal level in high latitudes. The fluctuations required in the ionospheric electron density for production of the scintillations on the earth's surface are only a few percent of the normal electron density. Phase scintillation observed by means of interferometers usually correspond to 2–3 minutes of arc during quiet ionospheric conditions but can amount to half a degree during strong ionospheric disturbances.

The amplitude has Gauss distribution for small scintillations but in the most intense scintillation periods the amplitude distribution is often of the Rayleigh type.

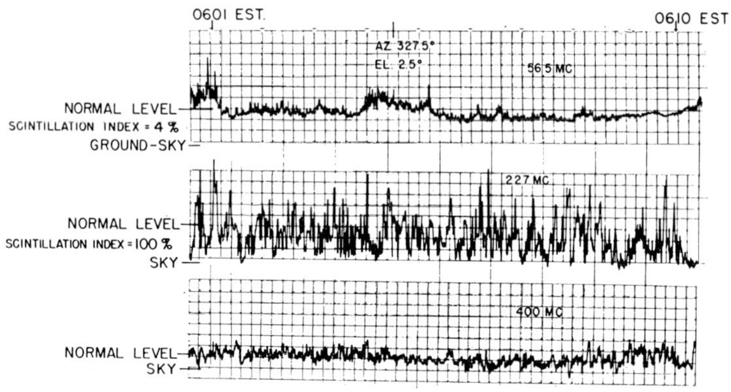

Fig. 1. Two different levels of scintillation of Cygnus at 227 Mc/sec (after AARONS, 1963).

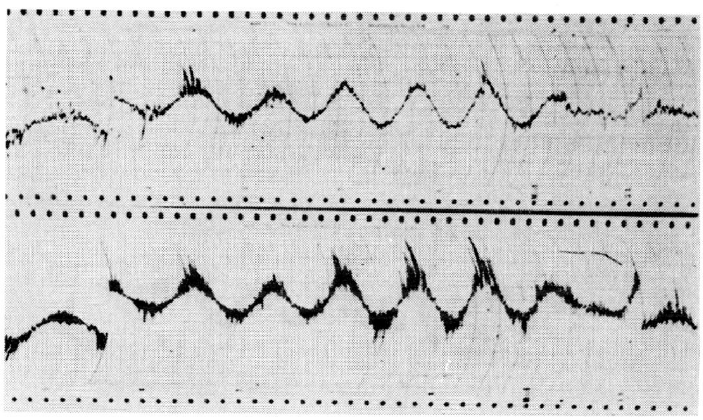

Fig. 2. Typical recording of a radio star obtained by means of a phase-switch interferometer (after LITTLE *et al.*, 1962).

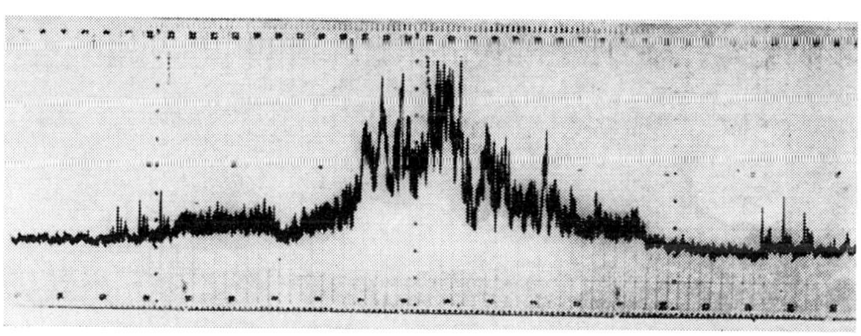

Fig. 3. A typical amplitude record of 1958 $\delta 2$ (Sputnik III) showing amplitude scintillations.

The phase scintillation is generally Gauss distributed and has about the same scintillation rate (half the number of crossings of the average level per time unit) as the amplitude one: that is, in temperate latitudes from a few scintillations per minute to one scintillation in about five minutes. In high latitudes there may occur several scintillations per second. For not too intense disturbances the scintillation rate is independent of the radio frequency of the wave.

Since the departure of the refractive index from unity increases with λ one would expect the linear scintillation index, $\Delta A/A$, to increase in the same way. That has also been found to be the case in general, but sometimes an inversion of the effect is observed with the scintillation index increasing with frequency.

Time and azimuth variations of the scintillation characteristics suggest strongly that the processes responsible for the irregularities are under geomagnetic control.

Spatial Characteristics

The scintillation index increases rapidly at middle latitudes, in the average, with increasing zenith angle of the source; but in the auroral zone the zenith angle dependence is low.

An increase of scintillation rate and amplitude with latitude has been found. LITTLE *et al.* (1962) observed that the intensity of radiostar scintillation is at times more than an order of magnitude greater at College, Alaska than at temperate latitudes and that the intensity is to a large extent connected with the presence of aurora and other disturbance effects in the ionosphere.

Time Variations of Scintillation Occurrence

Radio-star scintillation is observed most often during the night with the maximum generally occurring shortly after local midnight. A secondary maximum at midday has been observed in Australia and Canada but not in England. The disagreement between measuring results for various parameters in different parts of the World is remarkable.

Also in the auroral zone the scintillation index has a maximum around magnetic midnight, but the difference between the average maximal scintillation index in the night and the minimum in the day is less than a factor two (cf. Figure 4), i.e. much less than in middle latitudes.

Correlations with Upper Atmosphere Parameters

The rate of radio star scintillations is generally several times higher during magnetic storms than during quiet conditions, and there is also a definite correlation between radio star scintillation indices and the magnetic disturbance level in auroral latitudes.

Correlations have also been found between the scintillation activity and spread F and sporadic E, but results of different investigators do not agree.

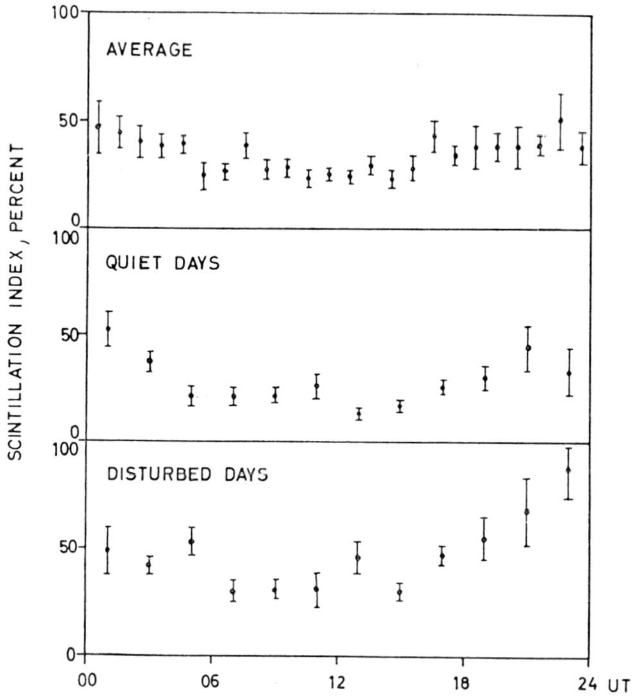

Fig. 4. One-hour averages of the scintillation index for the whole period of recording (a), and two-hour averages of scintillation index for quiet days (b) and for disturbed days (c) as a function of UT (after LISZKA, 1963).

Theory of the Diffraction Pattern

When a plane wave passes through an irregular ionospheric layer the wave front becomes irregular. The wave is no longer plane on the other side of the layer but may be considered as composed of a large number of wavelets of different direction. This irregular wavefront produces a diffraction pattern on the earth's surface. The pattern is in general not stationary; it may change in a random way or it may drift without any change of structure or it may do both simultaneously. A receiver on the ground will see amplitude scintillation.

When recording the fluctuating field strength at two places separated by one or a few hundred meters one often finds that the records are similar but displaced in time. From this can be concluded that the diffraction pattern is fairly stable, but moves as a whole over the earth's surface. In this case the radiostar scintillation is therefore due to drift of the fairly stable configuration of irregularities in the ionospheric electron density distribution.

The drift pattern can be produced not only by the drift of the diffraction screen but also by the motion of the source of radio waves. Examples of the first case is the radio star scintillation and the fading of radiowaves reflected from the ionosphere. The second case is that of satellite scintillation. The way of reasoning is similar for all cases.

The theory of ionospheric diffraction has been developed by Booker, Ratcliffe and Shinn and others and it has been reviewed by RATCLIFFE (1956). The pattern produced by a satellite is especially suitable for measurements of its structure, when the orbit is known, as the time variation of the pattern during the passage in general can be neglected.

For an extensive discussion of the diffraction theory with its applications to the ionosphere the reader is referred to RATCLIFFE (1956).

The size of the ionospheric irregularities, as given by $\varrho_{\Delta N}(\xi)$ – the autocorrelation function of the deviations, $\Delta N(x)$, of the electron density from the mean – is related to the size of the irregularities in the emerging wave front as given by its autocorrelation function, $\varrho_f(\xi)$. The relationship depends on the standard deviation, A_0, of the amplitude, measured in nepers, and of the phase measured in radians, ϕ_0,

$$\phi_m = (A_0^2 + \phi_0^2)^{1/2}. \tag{1}$$

If $\phi_m < 1$, $\varrho_{\Delta N}(\xi) = \varrho_f(\xi)$. Hence the size of the irregularities in the wave field close to the diffracting screen is the same as the size of the ionospheric irregularities. The pattern on the ground is then simply a projection of ionospheric irregularities.

If $\phi_m > 1$ the autocorrelation function, $\varrho_f(\xi)$, of the wave field has a scale which is $1/\phi_m$ times that of the distribution of the electron density.

ϕ_m can be determined from the amplitude distribution on the ground. If the distribution is of the Rayleigh type, $\phi_m^2 \gg 1$, and if it is a displaced Gaussian distribution, then $\phi_m^2 \ll 1$.

In the case of the satellite transmissions the pattern is not produced by a plane wave but by a spherical one. For that reason the scale of the pattern on the ground will be greater than that just below the diffracting screen. If the satellite is at height H and the diffracting layer at height h above the ground, the scale size ξ_{0g} of the autocorrelation function observed on the ground $\varrho_g(\xi)$, is $H/(H-h)$ times larger than the scale size of $\varrho_f(\xi)$, viz. ξ_{of}, i.e.

$$\xi_{of} = \frac{H-h}{H} \cdot \xi_{og}. \tag{2}$$

If the satellite moves along the x-axis with velocity V, the diffraction pattern on the ground moves in the opposite direction with velocity

$$v = \frac{h}{H-h} V. \tag{3}$$

If the pattern velocity is considerably larger than the velocity of ionospheric drift – which in general is true – temporal and spatial variations of the amplitude are related by the expression $x = vt$. Therefore the pattern velocity, when directed along the x-axis may be determined by observing the amplitude, $A(t)$, at two points x_1 and x_2. The time interval $(t_2 - t_1)$ between observation of a particular detail of $A(t)$ at x_2 and x_1, respectively, gives simply the pattern velocity.

The reasoning above may be extended to two dimensions. The autocorrelation function of the radio wave field produced by irregular ionospheric layers is not isometric. The contours of $\varrho_g(\xi, \eta)=\text{const.}$ can then be approximated as concentric ellipses (PHILLIPS and SPENCER, 1955). When the pattern is anisometric the shape of the autocorrelation ellipse – often called characteristic ellipse – must be taken into account in determining the pattern velocity and the height of the irregularities from measurements of the amplitude at two different points (if these points are not situated on a line parallel to the velocity vector of the satellite during the passage).

Determination of the Size and Shape of the Characteristic Ellipse

The dimensions of the characteristic ellipse on the ground can be evaluated either by measurement of the angular correlation of radiation from discrete radio sources by means of two interferometers, the baselines of which form a known angle, or the distance between antennae for which the correlation between two records has been reduced to a certain value (usually e^{-1} or 0.5) can be determined for two different directions. A third way to obtain the dimensions is from the delay between the times of recording characteristic features of the pattern at three different points.

From satellite transmissions it is possible to determine the axes of a so-called correlation ellipse from records taken at only two points by employing the experimental evidence that the ionospheric irregularities have the shape of elongated two-axial ellipsoids, oriented along the geomagnetic field lines.

LITTLE *et al.* (1962) have presented evidence that in the auroral zone the irregularities causing radiostar scintillations at 223 Mc/sec at times have lateral dimensions as small as 100 m or less and that the deviation of the electron density from its mean value, ΔN, may be 10^4 electrons per cm^3 or more. This corresponds to a horizontal gradient in the electron density of the order of 100 electrons per cm^3 per m.

LISZKA (1963), employing satellite scintillation near the auroral zone, found values of the long axis varying between 5.6 and 12.5 km, of the short axis between 0.6 and 1.4 km and for the ratio, r, of long and short axes: 8–21. The r-values are in good agreement with those measured from radiostar scintillation at moderate latitudes.

The Drift Velocity of the Diffracting Layer

For determination of the drift velocity, satellite scintillations cannot be used, as the drift velocity is negligible in comparison with the satellite velocity.

A fixed radio source, like a transmitter on the ground or a discrete cosmic radio source (the apparent motion of which may be neglected), has been employed by many researchers.

A detailed description of a method for determination of the size and direction of the drift velocity from the radiostar scintillations can be found in HARROWER (1963).

In middle latitudes east-west drift velocities of the order of 100 m/sec with a wide

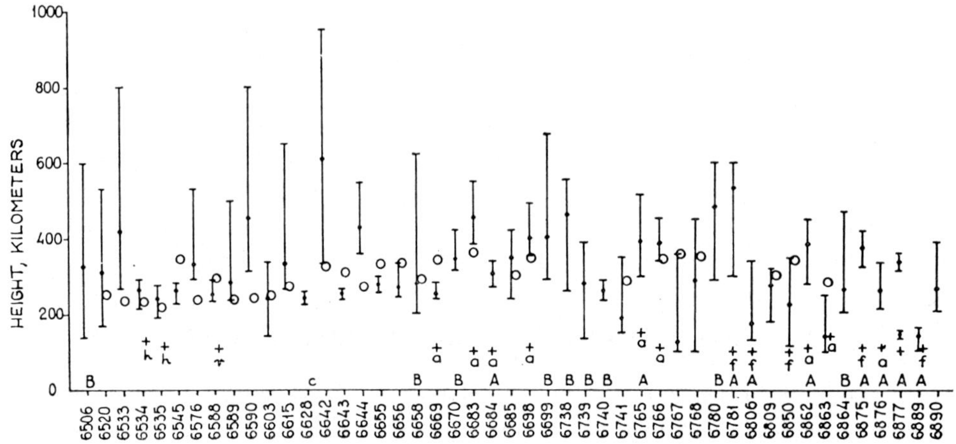

Fig. 5. Heights of the scintillation producing region for 48 passages of Transit IV from period October 10–November 7, 1962. Simultaneous values of *h'F* are indicated by circles and values of *h'Es* by crosses. Symbols used as in *IGY Annals*, Vol. III, Part I, p. 155 (after LISZKA, 1963).

dispersion has been observed during quiet ionospheric conditions. In disturbed periods drift velocities as high as 1000 m/sec or even more have been measured. The drift velocity in north-south direction is one order of magnitude, or more, less than in the east-west direction. The drift direction is mostly towards west before midnight and towards east after midnight – especially during disturbed periods.

The Height of the Scintillation-Producing Irregularities

The height at which the scintillation is produced may be calculated from measurements of the pattern velocity of a satellite of known orbit. Near the auroral zone

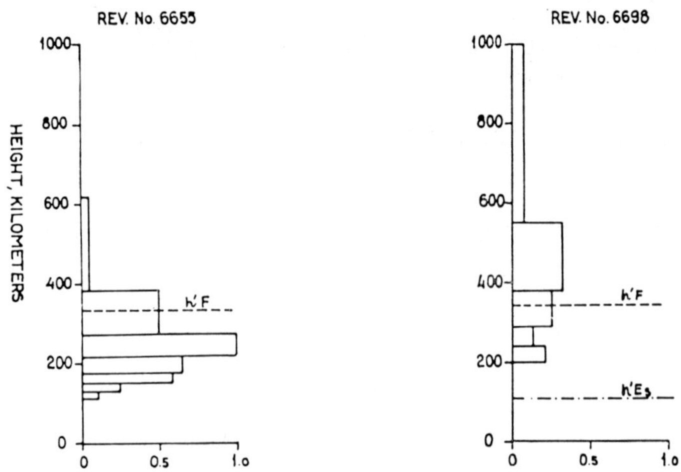

Fig. 6. Examples of height distributions of ionospheric irregularities obtained from distribution of time delays, (after LISZKA, 1963).

irregularities as low down as at 100 km may produce scintillation in 54 Mc/sec satellite transmissions at night.

About 50 fast amplitude records of Transit IV at 54 Mc/sec were reduced for height determination by LISZKA (1963) and the results are shown in Figure 5. Transits to the left of Figure 5 correspond to the forenoon hours and those to the right to night-time. It is seen that daytime scintillation is usually produced at heights above F-maximum, but during the night the height of the scintillation-producing region may be as low as 100 km. The height distribution of the number of scintillations observed per kilometer of height, are given in Figure 6 for two transits.

Mechanisms for Production of the Irregularities

Hitherto nothing has been said about how the irregularities are produced in the ionosphere. This is a matter of much speculation and very little agreement. It has been proposed that the irregularities are caused by atmospheric turbulence. The degree of constriction of the irregularities of electron density perpendicular to the earth's magnetic field lines suggest that the height at which scintillations are produced may not differ very much from the level, where the collisional frequency for ions is equal to the gyromagnetic frequency for ions. This level is about 140 km. However, there is one important difficulty for this hypothesis. Turbulence cannot explain the night-time maximum of scintillation. The intensity of the irregularities should be proportional to the gradient of electron density. On the lower side of the ionospheric layers this gradient maximizes in the daytime. The only level in the ionosphere, where electron density gradients could vary in such a way as to produce a maximum at night is between the E- and F-regions, but the effect is not large enough to explain the observed night time maximum of scintillation.

"Extra terrestrial particles" have been suggested to be responsible for the irregularities.

Another mechanism is that an irregular electric field in the F-region produces appreciable irregularities in electron density there, but it is far from clear whether this effect is efficient enough.

Magnetohydromagnetic waves and leaking from the radiation belts have also been mentioned as possible external sources.

In one of the existing models for the complete magnetic storm process all turbulence in the magnetosphere is considered as having been produced in the outer surface layer of the magnetosphere, from where it is brought into the magnetosphere by convection and by projection along the magnetic field lines.

A two-stream plasma instability has been proposed as a source of irregularities. In this mechanism attracting forces, similar to those between parallel currents in conductors, produce the instability. The qualitative and quantitative predictions of the theory agree with observations of irregularities in the equatorial *Es* layer. This type of instability may be important also for other types of irregularities, e.g. production of the electron density irregularities in aurora.

A type of very longperiodic mechanical waves, so called gravity waves, has also been mentioned as a possible cause.

Finally, it has been pointed out that very low frequency radio waves propagating in ionospheric ducts may accelerate the release of energetic electrons, some of whose energy goes into ionization in the F-layer. In that way a feed back mechanism may be built up which produces irregularities in the F-layer and below.

Although several proposals for responsible mechanisms thus exist, we are still far from an adequate explanation of the scintillation phenomenon.

References

AARONS, J.: 1963, 'Low-angle scintillations of discrete sources', *Radioastronomical and satellite studies of the atmosphere* (ed. by J. AARONS), North Holland Publishing Co., pp. 65–90.

HARROWER, G. A.: 1963, 'Theoretical considerations in the interpretation of radio star scintillation observations', *Radioastronomical and satellite studies of the atmosphere* (ed. by J. AARONS), North Holland Publishing Co., pp. 38–64.

LISZKA, L.: 1963, *Arkiv för geofysik* **4,** 227–246.

LITTLE, C. G., REID, G. C., STILTNER, E., and MERITT, R. P.: 1962, *J. Geophys. Res.* **67,** 1763–84.

PHILLIPS, G. J. and SPENCER, M.: 1955, *Phys. Soc.* B **68,** 481–492.

RATCLIFFE, J. A.: 1956, *Rep. Progr. Phys.* **19,** 188–267.

CHAPTER 17

THEORIES OF AURORA AND OF MAGNETIC STORMS

B. HULTQVIST
Kiruna Geophysical Observatory, Kiruna C, Sweden

Introduction

It has been known that aurora and world wide magnetic storms are very closely connected since 1741, when Celsius and Hiorter discovered it in Uppsala, Sweden. A theory for the aurora must also explain the magnetic storm and vice versa.

All theories which are under discussion today agree that charged particles from the sun are responsible for the occurrence of magnetic storms and aurora on the earth. This was proposed as early as 1881 by Goldstein. It was based on the observation of auroras on the night side of the earth and on their concentration to the auroral zones.

This proposal was supported by the results of laboratory experiments carried out in the last decade of the 19th century by Birkeland in Norway. With the primitive technique of his time he succeeded in shooting electrons towards a spherical terrella containing a magnetic dipole and to observe and photograph zones of precipitation around the poles of the terrella.

Birkeland's experiment stimulated Störmer to make very extensive theoretical studies of the propagation of charged particles in a magnetic dipole field.

Both Birkeland and Störmer investigated the behaviour of particles of one sign in a dipole field. Schuster pointed out in 1911 that a beam from the sun consisting of charged particles of only one sign (either only electrons or only protons) would be very widely dispersed before reaching the earth due to the repulsive forces between the particles. For the density to be high enough at the earth it would have to be unreasonably high when the particles left the sun. To avoid this difficulty Lindemann proposed in 1919 that the beam of charged particles emitted from the sun is electrically neutral, i.e. contains the same number of electrons and positive ions per unit volume. This idea was then developed in two different ways by Chapman and Ferraro from 1923 on and by Alfvén from 1939. All models of aurora and magnetic storms which are seriously discussed today are based on a neutral beam of charged particles (plasma) from the sun being responsible for the effects observed on the earth. Such plasma have now also been directly observed by means of space vehicles.

The production of the plasma beam on the sun and the propagation of the beam to the earth will not be dealt with there. They have been discussed in the sections on solar physics and interplanetary medium of this course. In this short lecture I have chosen to review three of the large number of proposed models, which have played and play a most important part in the discussion on this subject, which – it should be pointed out – is one of intense speculation and discussion. The theoretical models that will be reviewed are those of Chapman and Ferraro, Alfvén, and Axford and Hines.

The Theory of Chapman and Ferraro

Chapman and Ferraro first investigated a highly idealized case: how a neutral unmagnetized plasma with infinite plane boundary surface interacts with the magnetic

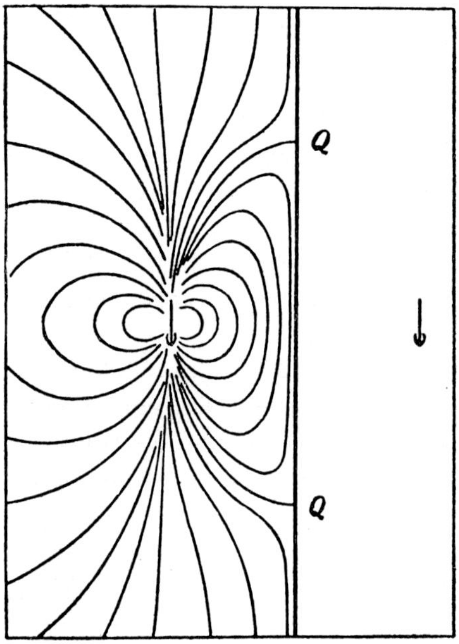

Fig. 1. Diagrammatic sketch of the combined field of the geomagnetic dipole and of the electric currents induced in the plane surface layers of the advancing neutral ionized stream, on the assumption that these currents completely shield the interior of the stream from the geomagnetic field; in this case the magnetic field of the induced current-system is the same as that of the image (shown on the right) of the geomagnetic dipole outside the plasma (after CHAPMAN and BARTELS, 1940).

dipole field when moving towards the earth with a velocity of about 1000 km/sec perpendicular to the plane boundary (cf. Figure 1).

The currents induced in the plasma moving in the geomagnetic field produce a magnetic field equivalent to that of a magnetic dipole identical with the geomagnetic one, but situated at the same distance behind the boundary of the plasma as the geomagnetic dipole is in front of it. Two bar magnets with their magnetic-moment vectors parallel repel each other and equivalent repelling forces are produced also by the currents induced in the plasma approaching the earth. The plasma is braked and it stops when its kinetic energy density ($\frac{1}{2}\varrho v^2$) is equal to the energy density of the magnetic field ($B^2/8\pi$). As can be seen in Figure 1 the geomagnetic field is compressed by the plasma. The field line density (the field strength) is increased at the earth's

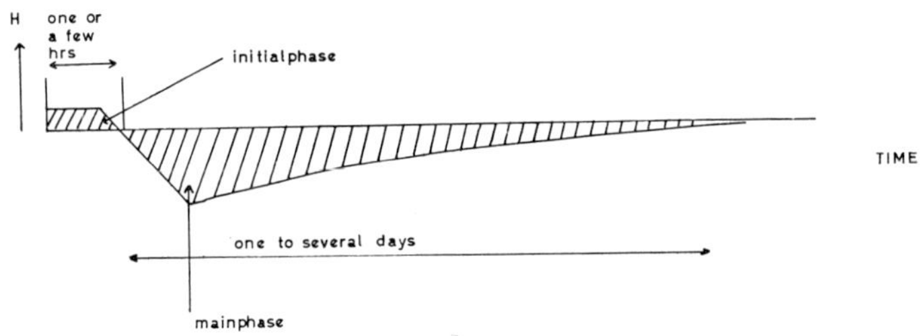

Fig. 2. A smoothed record of the average geomagnetic storm in the horizontal component.

surface. This corresponds to the initial phase of the average magnetic storm, which has the appearance shown in Figure 2 when observed in the horizontal component (H). The picture sketched above is a highly idealized one. In the case of the earth and a solar plasma the plasma will pass the earth at some distance away from it. The earth will cut a hole in the plasma (cf. Figure 3). In the inhomogeneous geomagnetic field the positive ions will drift towards the morning (left hand) side of the hole cut by the geomagnetic field in the plasma seen in Figure 3, and the electrons will drift towards the evening side. This is equivalent to a westward-directed ring current, which decreases the horizontal component of the geomagnetic field at the Earth's surface and is assumed to be responsible for the main phase of the magnetic storm.

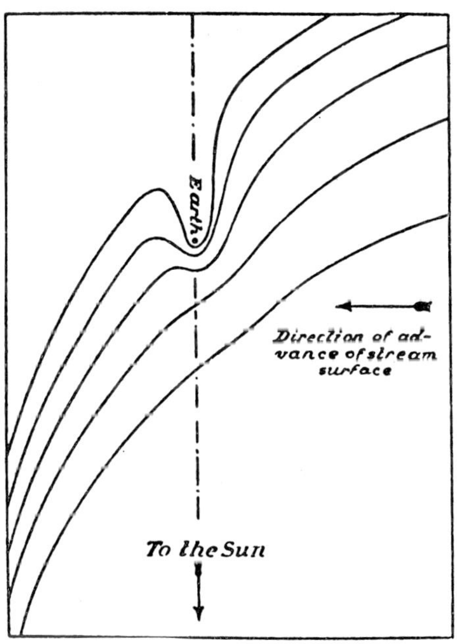

Fig. 3. Sections of the earth's equatorial plane of the advancing front surface of a corpuscular stream, illustrating diagrammatically the formation of a hollow space round the earth by the action of the geomagnetic field on the neutral ionized stream (after CHAPMAN and BARTELS, 1940).

The theory does not include any detailed process for the production of aurora. It was only referred to by the fact that for an ion density in the beam of the order of $100/\text{cm}^3$ and the beam velocity mentioned before (1000 km/sec) the plasma will be stopped in front of the earth at a distance in the equatorial plane of 6 to 7 earth radii. The dipole lines passing through the equatorial plane at such a distance from the dipole cut the earth's surface along two circles at a geomagnetic latitude close to that of the observed auroral zones. This in combination with the fact that the conductivity along the magnetic field lines is much higher than perpendicular to it was used as a support for the hypothesis that particles are precipitated in one way or the other along the field lines from the equatorial region to the auroral zones.

A ring current of the kind Chapman and Ferraro imagined is not stable. For it to be stable an electric field perpendicular to the magnetic field is needed. The existence of such an electric field was postulated by Martyn. He modified the Chapman–Ferraro theory by introducing a radial electric field directed into the plasma, corresponding to an excess of positive charges on the inside and of negative charges on the outside of the plasma surface currents (seen from the earth). This electric field was assumed to be the energy source for the acceleration of the primary auroral particles. If the magnetic field lines are equipotential lines – which they may be thought to be if the conductivity is infinitely high (cf. the section on Axford's and Hines' theory, below) – the electric field will be projected down to the earth's atmosphere. In Martyn's model the electric field in the auroral zone was directed towards the pole. As the magnetic field is nearly vertical there, the combination of the two drives a Hall current along the auroral zone. In this way the theory explains the observed electric jet current along the auroral zone during magnetic storms.

Alfvén's Theory

Both in the original Chapman–Ferraro theory and also in Martyn's modification of it the solar plasma was assumed to be unmagnetized. It is however now known that the plasma is magnetized. Direct observations of magnetic fields in solar beams have been made from space probes in the last few years. The first to introduce a magnetic field of the plasma into the theory was Alfvén.

According to the theory of relativity the electric field depends upon from which coordinate system it is measured. For small velocities the transformation equation is

$$\boldsymbol{E}' = \boldsymbol{E} + \frac{\boldsymbol{v}}{c} \times \boldsymbol{B}$$

(in Gaussian cgs units) where $\boldsymbol{v}$ is the velocity of the coordinate system measured in the coordinate system in which the electric and magnetic fields are given by $\boldsymbol{E}$ and $\boldsymbol{B}$.

In the coordinate system of the plasma beam the electric field is zero. If it were not, charge separation would take place until the polarization field compensated the initial electric field and the resultant field, $\boldsymbol{E}'$, would be zero. As $\boldsymbol{v}$ and $\boldsymbol{B}$ are not parallel in the equatorial plane, the electric field in the plasma, measured in the earth's coordinate system, $\boldsymbol{E}$, is not zero, but $\boldsymbol{E}=-\boldsymbol{v}/c\times\boldsymbol{B}$. Alfvén assumed for simplicity that $\boldsymbol{E}$ is perpendicular to the geomagnetic dipole axis and directed towards west on the day side of the Earth. The last assumption is equivalent to assuming that the magnetic field in the solar plasma is parallel to the geomagnetic field in the equatorial plane. This assumption was made to reach agreement between some results of the theory and observations concerning diurnal variation of some auroral characteristics.

Alfvén considered the motion of one single charged particle in the combined

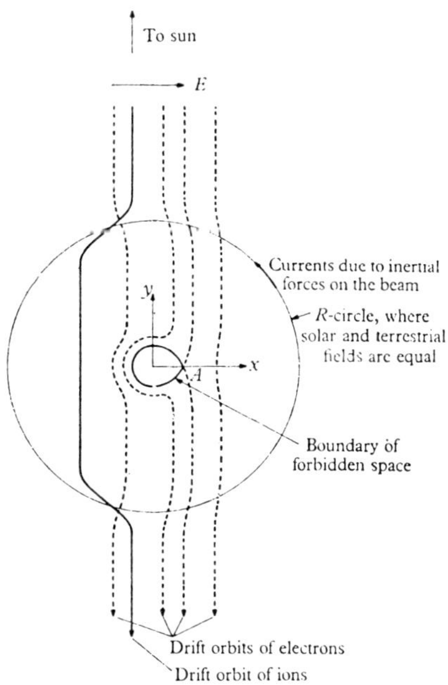

Fig. 4. Currents and particles orbits in the geomagnetic equatorial plane with an electric field perpendicular to the geomagnetic dipole axis (after BLOCK, 1958).

magnetic and electric fields. He thus neglected the interaction between the particles in the plasma, i.e. in the first hand the effect of space charges. The magnetic moment of the motion of the charged particle can be shown to be invariant under conditions fulfilled in this case (the first adiabatic invariant; cf. the lectures of Herlofson in this course). Applying these conditions the trajectories of electrons in the equatorial plane shown in Figure 4 can be derived. There is a forbidden zone which the electrons cannot reach. Its size is determined by the electric field strength, the magnetic moment of the earth and the initial energy of the particle. For reasonable values of E (i.e. of v and B) the forbidden zone projected along the geomagnetic field lines is found to be located at auroral zone latitudes. The projected forbidden zone is not symmetric around the pole but has the shape shown in Figure 5. The protons have a forbidden zone different from that of electrons. Alfvén has shown that the motion of the electrons around the forbidden zone is most important for production of the ring current. As seen in Figure 4 the electrons move eastward around the earth and thus produce a westward current which decreases the horizontal component of the magnetic field on the earth's surface – the main phase of the magnetic storm.

The real trajectories of the particles are trocoids, i.e. the particles rotate around the field lines at the same time as the center of rotation moves along the orbits shown in Figure 4. The radius of gyration is so small that it cannot be seen in the scale of Figure 4. A ring current of this type is stable contrary to the one of Chapman and Ferraro.

In its first version Alfvén's theory was only for the main phase of the storm. It

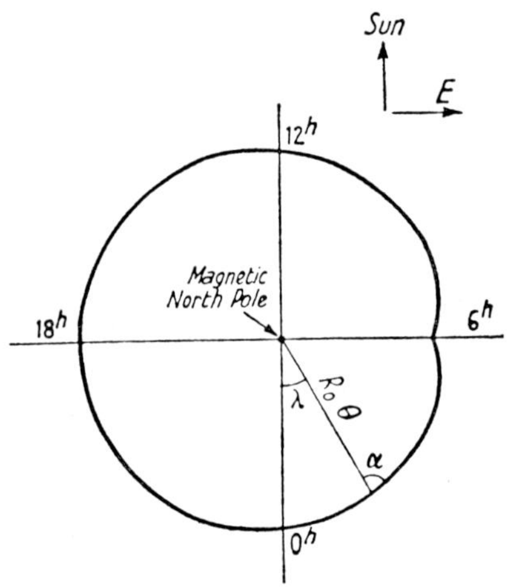

Fig. 5. Auroral curve after ALFVÉN (1950).

was later extended to include also the initial phase. The basis for this was a picture of the interchange from a constant interplanetary magnetic field to the geomagnetic field. For an interplanetary field different from zero there may be a region of higher gradient than elsewhere in the transition region between the two magnetic field regions. Due to their inertia the positive ions will move eastward in this gradient (cf. Figure 4), which will result in an eastward current producing an increase of the geomagnetic horizontal component – the initial phases of the storm. The projection of that region of the equatorial plane where this happens may be an inner auroral zone. If it exists or not is not yet clear. It should be some 5–10 degrees from the pole. There is, however, a difficulty. For reasonable values of parameters the duration of the initial phase will be much too short.

In Alfvén's mechanism an electric field around the earth is the energy source for the acceleration of the primary auroral particles. Alfvén has given reasons for the main part of the voltage drop to be located in the upper atmosphere.

As mentioned earlier, Alfvén neglected possible space charge accumulation during the interaction of the solar plasma and the geomagnetic field. That is to say he assumed that space charges discharge immediately along the magnetic field lines before they have reached such values that they influence the motion of the particles. It has been argued that the space charges may change the trajectories of the particles profoundly. KARLSSON (1963) has therefore considered in some detail the case of maximum space charge accumulation, i.e. no discharge takes place at all but the space charges stay in the equatorial plane. The calculations for this free boundary problem have been made for the stationary case and cylindrical symmetry.

The results found are that the shapes of the particle trajectories as well as of the forbidden zone are different from those in Alfvén's case, but the general character is the same: a forbidden zone is developed. It is, however, circular and centric. The electric field inside the forbidden zone is larger than in the undisturbed beam. The current direction in the auroral zone changes sign at about 06 and 18 local time instead of at 00 and 12 as in Alfvén's case. Observations show that the change of direction takes place at 21 or 22 hours, indicating that nature is somewhere in between the two cases, as may be expected. There are some other observations which are in better agreement with Alfvén's approximation than with Karlsson's.

Alfvén assumed the magnetic field vector of the plasma to be exactly parallel to the equatorial geomagnetic field lines. Of course, one may assume the vector to have an arbitrary direction. Hoyle proposed in 1958 that the auroral particles may be accelerated in neutral points produced where the geomagnetic field and the interplanetary one are equally large but of opposite direction. This case has been considered in some detail by DUNGEY (1961). There is a fundamental difference between the interaction of the plasma with the geomagnetic field when the plasma field is parallel to and antiparallel to the geomagnetic field in the equatorial plane. While in the first case the plasma penetrates into the geomagnetic field up to the forbidden zone, the plasma is stopped already at the neutral line, i.e. where interplanetary and geomagnetic fields are equal, in the antiparallel case. It can probably penetrate to the atmosphere at high latitudes also in this latter case.

Explorer XII was first to measure a sudden change of direction of the magnetic field at about 10 earth radii from the earth in the direction to the sun. It has been proposed that this corresponds to the antiparallel case, mentioned. In principle the magnetic field in the plasma can be thought to have any direction, as this is determined in the first hand by the magnetic field configuration in the disturbed region on the sun, from where the plasma originates.

Before finishing the review of Alfvén's theory it should be mentioned that it has been argued that the electric field of the plasma as seen from the earth does not penetrate into the geomagnetic field any further than to the Chapman–Ferraro boundary between plasma and magnetic field. It is said to be shorted in the boundary layer defining the so-called magnetosphere. This would mean that the plasma would not penetrate the magnetosphere to the forbidden zone. We will discuss this a little more in the next section.

A general difficulty with all mechanisms involving large scale electric fields and large scale charge separation – as Alfvén's mechanism does – is that the charged particles of different sign will be precipitated on different parts of the earth, while observations show that electrons and protons sometimes enter the atmosphere in the same regions.

The Theory of Axford and Hines

The development of magnetohydrodynamic theory has given rise to a number of theories of magnetic storms and aurora, being more or less modern versions of the Chapman–Ferraro theory. While e.g. Alfvén studied the motion of the single particles, the mentioned theories apply macroscopic fluid theory. The two methods should in principle be equivalent. However, due to the complexity of the problem only idealized cases can be studied, and the different approaches often differ considerably. For reviews of these fluid theories the reader is referred to e.g. GOLD (1962) and PARKER (1962). Here will only be described the most recent and extensive fluid theory, that of AXFORD and HINES (1961).

The theory of Axford and Hines is not a mathematically developed model. They make a number of assumptions, and apply a few physical relations in a discussion of a qualitative kind, adjusting the model to observed facts. The most important assumptions made are the following:

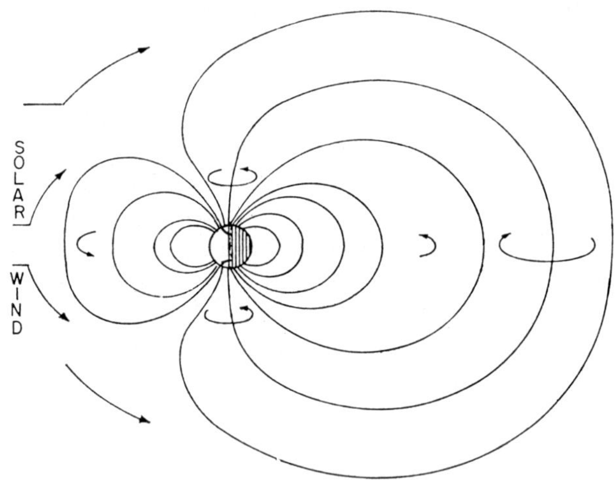

Fig. 6. A section of the magnetosphere, taken on the noon geomagnetic meridian. The solar wind (which is directed away from the sun) moves from left to right in the diagram, causing high-latitude field lines to be swept into the lee of the earth to form a geomagnetic tail. The direction of rotation of the magnetosphere is indicated (after AXFORD and HINES, 1961).

The geomagnetic field is assumed to be enclosed in a limited volume – the magnetosphere – with no field lines going to infinity (see Figure 6). The magnetosphere is assumed to rotate with the earth.

Axford and Hines postulate the existence of a viscous-like interaction between solar wind and the magnetosphere by means of which a small part of the momentum of the solar wind is transferred to the magnetosphere. How this momentum exchange occurs does not matter for their model. They consider turbulence in the boundary layer of the magnetosphere as a probable cause, which view is supported by observations in space vehicles (as by the magnetometers in Explorer XII) of irregular fluctuations of the interplanetary magnetic field outside some 8 to 14 earth radii from the earth.

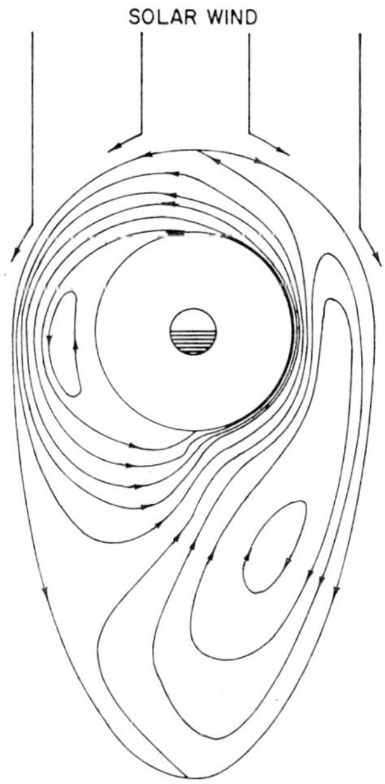

Fig. 7. The type of motion within the magnetosphere to be expected in the presence of rotation of the earth and a viscous-like interaction between the solar wind and the surface of the magnetosphere (after AXFORD and HINES, 1961).

It is also assumed that there are no serious stability problems for convective motion inside the magnetosphere.

Due to the postulated viscous interaction the plasma in the boundary layer of the magnetosphere convects to the rear part of it. All plasma cannot stay there, and it is assumed that in the stationary case a return flow takes place through the interior of the magnetosphere.

The magnetic field lines are assumed to be frozen into the matter and will convect together with the matter. This is the way it is expressed in the magnetohydrodynamic language. Using ordinary electromagnetic terms the convection of the field lines can be qualitatively understood in the following way. The convection in the surface layer is perpendicular to the magnetic field lines in the equatorial plane and has a component perpendicular to the field lines over most of the surface. Therefore an electric field is induced, $\boldsymbol{E} = -\boldsymbol{v}/c \times \boldsymbol{B}$. Thus both $\boldsymbol{v}$ and $\boldsymbol{B}$ are perpendicular to the electric field, i.e. convection takes place so that the magnetic field lines are equipotential lines. Since B-lines are equipotential lines, voltage between two lines is the same all

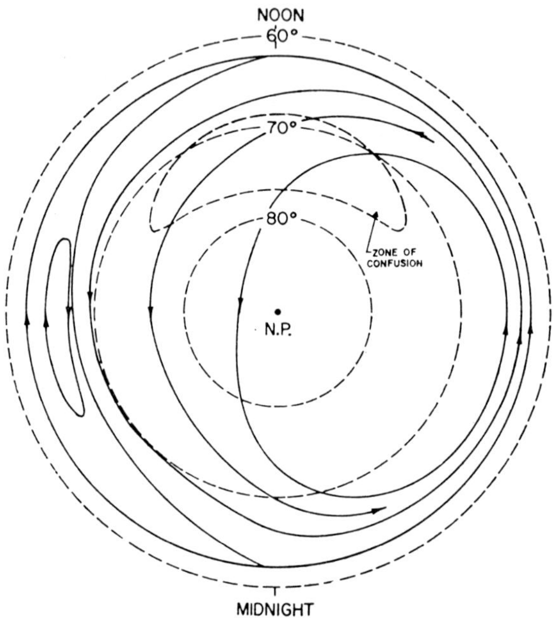

Fig. 8. The pattern of motion at ionospheric levels, obtained by mapping the streamlines of Figure 7 down onto the northern hemisphere along the lines of force of the geomagnetic field (after Axford and Hines, 1961).

along the lines. Thus the electric field induced in the boundary surface of the magnetosphere is mapped along the field lines down to the ionosphere, where the combined electric and magnetic fields produce a Hall drift of the charges, i.e. the plasma convects also in the ionosphere.

The insulating lowest atmosphere plays an important part in this process. Only because of its existence the field lines and the matter bound to them can convect. If the plasma reached down to the solid conducting earth the field lines would have been anchored in the earth. The existence of the insulating atmosphere makes it possible for the field lines to slip around freely. The insulator "cuts" the field lines.

Equipotential lines in an equatorial cross-section of the magnetosphere are closed loops. Therefore the flow lines also form closed loops.

In the stationary case the energy density of the convective motion is assumed to be much less than that of the geomagnetic field except at the boundary. Therefore, the convection cannot deform the magnetic field very much in the interior of the magnetosphere. The flux tubes have to change cross-section when they convect in order to adjust to the local field strength. This produces changes of pressure, density and temperature of the matter when it moves. In the lower ionosphere viscosity and ohmic losses will slow down the convection. The resistance to ionic drift is stronger than that to electrons. Therefore, the ions will drift slower than the electrons and

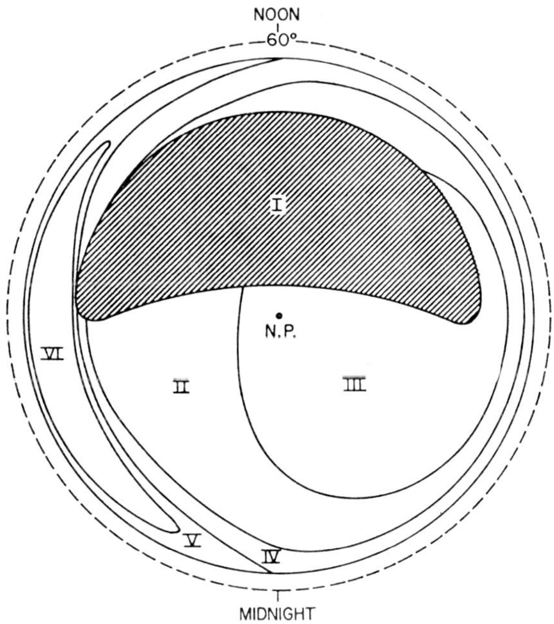

Fig. 9. The distribution of turbulence in the magnetosphere as observed at ionospheric levels (after AXFORD and HINES, 1961).

currents along the flow lines are produced. In this way the S_D current system of magnetic disturbances is assumed to be set up. On the basis of a large number of observational results Axford and Hines propose that the convection in the magnetosphere follows the flow pattern of which the equatorial cross section is shown in Figure 7. The inner boundary of the convection cells corresponds to approximately 62° geomagnetic latitude ($L=4.5\ R_e$). It is assumed to be stabilized by the energetic particles of the outer Van Allen belt. The rotation of the earth is thought to be responsible for the concentration of the stram lines on the morning side, which is shown in Figure 7.

To the flow pattern in the equatorial plane shown in Figure 7 corresponds the flow pattern in the ionosphere of the polar caps shown in Figure 8. There is a region in Figure 8 called the zone of confusion by Axford and Hines. It is the zone in which the boundary between the magnetic field lines closing on the day side and on the night side of the earth is located. All field lines of the surface of the magnetosphere leave the ionosphere within these zones. The boundary between dayside and nightside field lines is thought to change position depending on the strength of the flow of solar plasma etc. and on the whole the situation in these zones is rather variable and unspecified. The "zones of confusion" are assumed to be located inside the auroral zones.

The turbulent matter in the boundary layer is supposed to be convected into the interior. The decay of the turbulent energy in the course of the convection is thought to be more than offset by the compression. The compression is assumed to be strongest in regions II, IV and V of Figure 9. They are the regions where most intense disturbance effects are observed.

A few aspects of the theory of Axford and Hines will be commented on.

It is assumed in the theory that the plasma from the sun is unmagnetized. This gives the condition for the boundary of the magnetosphere: $\frac{1}{2}\varrho v^2 = B^2/8\pi$. If the plasma is magnetized that condition is not relevant. The particles can pass that region by gaining energy from the electric field associated with the magnetic field of the plasma.

According to Alfvén and Karlsson the magnetic-field free solution is a singular solution, different from the asymptotic solution obtained by first including a magnetic field and then letting it go to zero.

In this connection it is sometimes said that the surface of the magnetosphere is an equipotential surface in which the possibly existing electric field of the solar plasma is short-circuited so that it does not penetrate into the magnetosphere. This reasoning seems, however, to be far from clear both in point of principle and from the point of view of the time needed for short-circuiting the field as the plasma passes by.

An important part of the model is that the geomagnetic field lines are equipotential lines. This is necessary for the convection to take place in the whole magnetosphere. It has been shown by Persson that an electric field along the magnetic field lines does exist, unless special conditions are fulfilled in the plasma. This argument is valid unless the earth is surrounded by a closed equipotential surface.

The model of Axford and Hines is of the fluid type. It is not clear if such a model is relevant in the case earth-solar plasma, as the mean free path in the interplanetary medium 1 AU from the sun is probably of the order of 1 or 10 AU.

Only this short review of some of the main lines of thought in three of the most discussed theories of aurora and magnetic storms has been given here due to lack of space.

References

The Chapman-Ferraro theory

CHAPMAN, S. and BARTELS, J.: 1940, *Geomagnetism*, Clarendon Press, Oxford, Chapter XXV.
CHAPMAN, S.: 1963, *Geophysics* (ed. by DE WITT, HIEBLOT and LEBEAU), Gordon and Breach, p. 373.

Alfvén's theory

ALFVÉN, H.: 1950, *Cosmical Electrodynamics*, Clarendon Press, Oxford.
ALFVÉN, H.: 1955, *Tellus* **7**, 50–64.
BLOCK, L.: 1956, *Transactions IAU Symposium* 6, Cambridge University Press, 1958, pp. 312–324.
KARLSSON, S.: 1963, *Phys. Fluids* **6**, 708–722.

Axford's and Hines' Theory

AXFORD, W. I. and HINES, C. O.: 1961, *Can. J. Phys.* **39**, 1433–1464.

Magnetohydrodynamic theories

GOLD, T.: 1962, *Space Sci. Rev.* **1,** 100–114.
PARKER, E. N.: 1962, *Space Sci. Rev.* **1,** 62–99.

Other models

DUNGEY, J. W.: 1961, *Phys. Rev. Letters* **6,** 47–48.
SINGER, S. F.: 1957, *Trans. Am. Geophys. Union* **38,** 175–190.

CHAPTER 18

MAGNETIC DISTURBANCES AND VISUAL AURORAE

A. OMHOLT
Institute of Physics, University of Oslo

1. Introduction

The study of magnetic disturbances, and in particular visual aurorae, has two aspects which in many ways differ appreciably. The first one is to consider the aurora as an indication of some phenomenon going on in the magnetosphere around the earth. One may look upon the earth's atmosphere as a fluorescent screen which indicates where a particle from space hits the earth. Just as a nuclear physicist looks at a scintillation counter, and knows that it has been struck by a fast particle when it lights up, the auroral physicist may look upon the sky. Whenever he sees the aurora he knows that the atmosphere of the earth is being struck by energetic particles, which come from outer space and are moving towards the dipole situated inside the earth, governing the movement of these particles. Thus the study of aurorae and magnetic storms may be regarded as the study of one end of a phenomenon which takes place between the sun and the earth.

The other aspect is that of the ionosphere. One may be quite disinterested in where these particles come from, but concerned about the fact that they disturb the quiet ionosphere and thus disturb radio communications and the quiet, normal state of our upper atmosphere. This aspect of the study includes the investigation of the ionization mechanisms, of the currents set up by the aurora and of the detailed excitation mechanisms of the aurora itself.

In these two first lectures we shall give an elementary review of the most important phenomena, with both aspects in mind. It is intended that these lectures shall serve as background information for the coming lectures in this course.

We shall first describe in very general terms the magnetic storms and disturbances, particularly those related to the appearance of aurorae. Then we shall describe the appearance of aurorae as observed visually or by photographic methods. Finally we shall take a brief look at the auroral spectrum.

2. Magnetic Disturbances*

The magnetic field of the earth is to the first approximation that of a dipole with a magnetic moment $M = 8.1 \times 10^{25}$ gauss cm^3, and whose axis is inclined by an angele of about 11.5° to the geographical axis. It is often convenient to refer magnetic and

* For more comprehensive reviews see Vestine (1960); Chapman and Bartels (1940), and Chapman (1951).

auroral phenomena to a coordinate system which is centered at the geomagnetic axis pole, that is the point where the geomagnetic dipole axis intersects the earth's surface. It should be noted that this geomagnetic pole does not coincide with the dip-pole, which is the point where the magnetic field is vertical. This is because the real earth magnetic field departs considerably from the dipole approximation, particularly close to the earth's surface. But the higher order terms in the real earth magnetic field fall off very rapidly with distance from the center of the earth, thus at a distance of some earth radii the dipole approximation is very good.

A particular magnetic field line can be described by the well-known relation

$$r = r_e \cos^2 \varphi ,$$

when r is the radius vector from the dipole, r_e the radius vector to the point where the

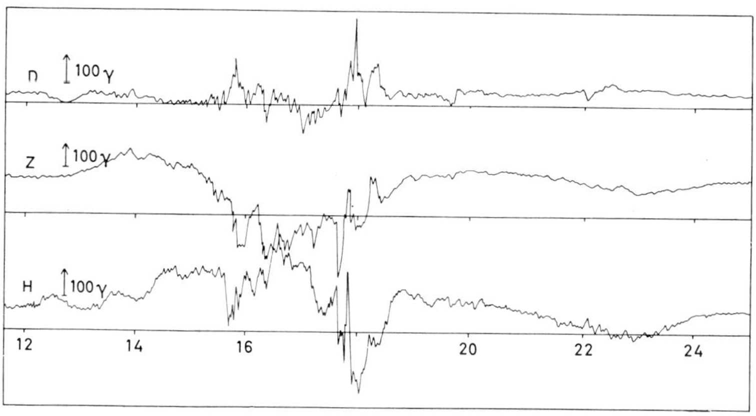

Fig. 1. Magnetic records from the auroral zone during a moderately disturbed night. *D*, *H*, and *Z* are the declination, the horizontal and vertical components.

field line intersects the geomagnetic equatorial plane (vertical to the magnetic dipole axis) and φ the angle between r and r_e.

For the study of the cause of magnetic disturbances, particularly those appearing in the polar regions and which are connected to the appearance of aurorae and ionospheric storms, it is important to know the conductivity in the ionosphere. This is because it is generally believed that most of these disturbances are due to a current system flowing at an altitude of about 100 to 150 km.

The conductivity per electron-ion pair is given on page 289. Considering the fact that aurora produces enhanced ionization mainly at heights between 100 and 150 km, we should expect most of the currents to flow here.

Even on quiet days the earth magnetic field at any particular spot varies slowly during the day. It also varies with seasons, with sun-spot cycles and with the phase of the moon. These effects are generally believed to be due to currents generated in the ionosphere at a height of 100–150 km. Such currents vary because the degree of ionization will vary with the solar activity. Variations due to the moon are supposed to be caused by the so-called dynamo effect. If the moon sets up tides in the ionosphere, a particular volume of air which rises or falls down will be subject to a changing magnetic flux. Thus electric fields are induced. We shall not concern ourselves with these variations since it is outside the scope of this lecture.

Visual aurorae and associated perturbations of the ionosphere are mostly associated with fairly strong magnetic disturbances. While the lunar and solar daily variations are of the orders of some tens of $\gamma(1\gamma=10^{-5}$ gauss), the magnetic disturbances associated with aurorae are of the order of 100 to 1000 γ. A typical, but moderately disturbed day is shown in Figure 1. The magnetic storminess is defined as the deviation from the expected smooth curve.

Conventionally the magnetic storminess is split up in two parts D_{st} and S_D, which

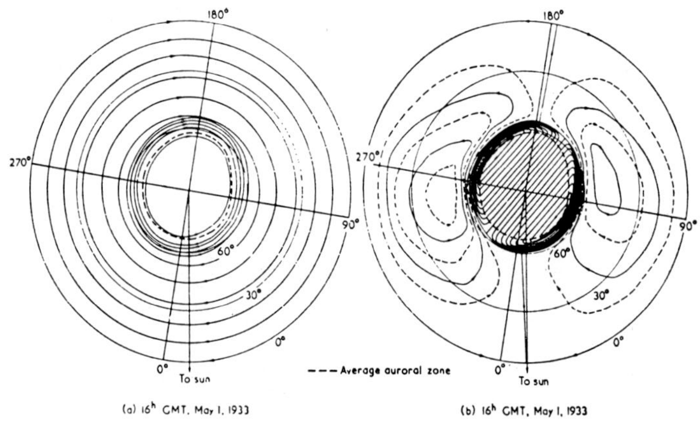

Fig. 2. Partial current systems D_{st} and S_D, respectively, main phase of storm (100000-A flow between successive, full-drawn current lines) (Reproduced from J. A. Ratcliffe (ed.), *Physics of the Upper Atmosphere*, Academic Press, p. 502).

added give the total magnetic disturbance. D_{st} is simply the average of the storminess for a particular geomagnetic latitude, averaged over all longitudes. This is the part of the magnetic storm which is independent of the observer's longitude, i.e. independent of local time. The other part, S_D, is the remainder of the storminess and is thus that part which does depend on local time. The latter part is much greater than the former.

We shall not go into detail how this analysis is being made, the important thing for our purpose is to note that it seems possible to explain all these variations by a closed current system in the ionosphere. Examples of such analyses are shown in Figures 2 and 3. On the first figure it is shown an analysis of a storm into the two components D_{st} and S_D. In Figure 3 the total current system is shown without splitting it up in two parts. Each current line represents 100000 A, thus there flow millions of ampères in the ionosphere during a particular magnetic storm. The important thing in the auroral theory is that it seems possible to explain all these disturbances by closed current systems in the ionosphere. Thus it seems unnecessary to involve

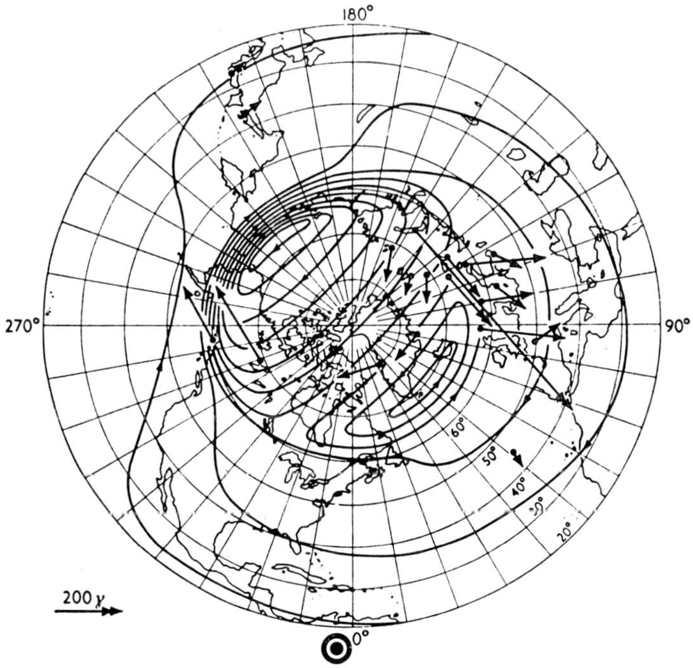

Fig. 3. Disturbance field at 16 hours 50 minutes on May 1, 1963, represented by a combination of current system and current lines (reproduced from J. A. Ratcliffe (ed.), *Physics of the Upper Atmosphere*).

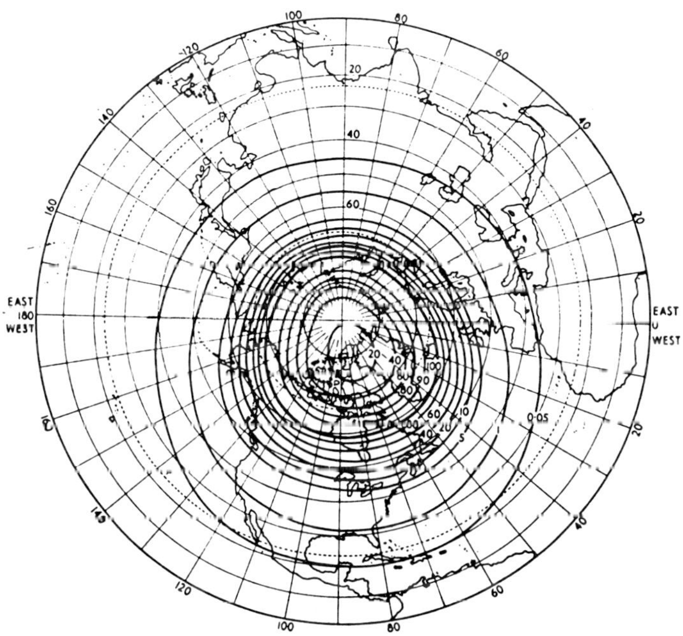

Fig. 4. Estimated percentage frequency of days with occurrence of aurora, clear nights, northern hemisphere (after VESTINE, 1944).

currents which come from outer space along the magnetic field line into the ionosphere, then run horizontally along the conducting layer at 100 to 150 km height, and then return to space again. However, it should be borne in mind that the analysis does not exclude such a possibility.

We should note that, roughly, the current density is highest along the zones of maximum occurrence of aurorae. Thus the lines on Figure 2(a) have a suggesting similarity with the auroral zones, i.e. the zones where aurorae are most frequent, shown in Figure 4. This is in agreement with the fact that, at high latitude, aurora is always accompanied by magnetic storms. However, exceptions seem to exist when looking at it the other way. It happens that great magnetic storms occur, with appreciable local magnetic disturbances, but with no aurora visible from the point of observation. In these cases there are very strong magnetic storms over large parts of the polar area and the aurora occurs very far south. There are also, however, reported cases of aurorae far south which are not accompanied by local magnetic disturbances.

When considering the conductivity in the ionosphere during aurorae, we must bear in mind that there are several effects which make the situation rather complex. For example an aurora enhances the conductivity over certain limited regions. If an electric field sets up currents across this region polarization will immediately occur, because at the limits of the enhanced ionization, electrons and ions will accumulate. The polarization field will then modify the current again. How big this polarization field can be is very difficult to know. This is because whenever space charges occur these may discharge along the magnetic lines of force. Because of the fairly high density of thermal electrons and ions, an excess of electrons may be cancelled very easily by a slow movement of the thermal electrons up along the magnetic lines of force. Similarly an excess of ions may be first cancelled by a net current of electrons downwards. Whether there then will be an excess of ions on the top of the ionosphere or these will readily move out in space is then the next problem. As a whole the problem is very complicated and would need rather elaborate and careful computations which do not seem to have been carried out yet. These problems will be further discussed in the following chapter.

3. Visual Aurorae

The appearance of visual aurorae is so aptly described in the literature that it seems scarcely necessary to repeat it in detail here. We shall only point out some essential features which may be important for our further discussions. For further and more complete information the reader is referred to Störmer (1955), Chamberlain (1961), Bates (1960a) and Harang (1951). The aurora is classified in many forms, according to its shape and stability. An auroral display usually starts with a fairly homogeneous arc, which later goes over into a rayed arc where the rays may move, often violently, in the east-west direction. Later, as the whole display brightens, many forms may occur, and vivid horizontal motions occur together with a great intensification of the whole aurora. The most convenient method of recording aurora is to photograph

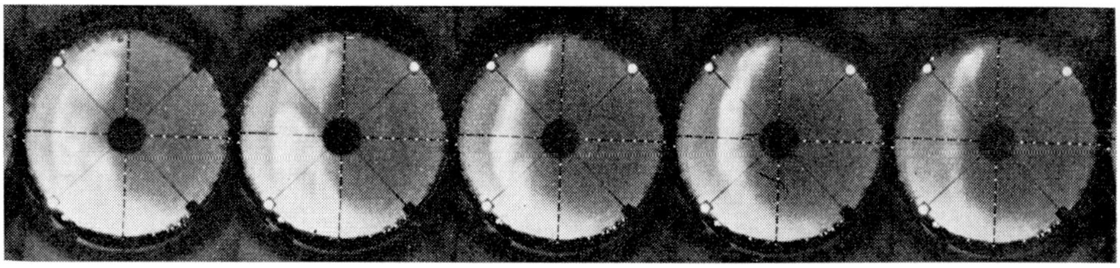

Fig. 5. A series of all-sky pictures of an active, rayed aurora over Tromsö, 1 min, between start of each exposure.

it though an all-sky camera. An all-sky camera photographs the whole sky through a concave or convex mirror. Thus the whole hemisphere is projected on one picture. A series of all-sky pictures illustrating an active phase of a typical auroral display, is shown in Figure 5.

The heights of the aurora have been extensively measured by triangulation, notably by Störmer (1955). The aurora appears from a height of about 90 km. up to several hundred kilometres. The lower boundary of distinct aurorae is fairly easy to measure by photographic triangulation from the ground, and a great number of such measurements have been made. Some of the most typical results are displayed in Figure 6. Auroral rays usually stretch to much greater heights than do arcs and bands. In addition the rays have a greater lower border.

Some typical luminosity distributions with height are illustrated in Figure 7, and in Figure 8 are given the distance to the lower and upper half-intensity points as a function of the height of the maximum intensity point. As is seen the half-width of the auroral luminosity increases systematically with the height of the maximum luminosity up to a certain altitude. From there on it seems to be reasonably constant. We shall discuss these luminosity curves and their interpretation in the next chapter.

In the auroral zone, particularly early in the night, the auroral forms are generally aligned in the east-west direction. The long rays stretching far upwards follow the earth magnetic field. Both these things were the first strong indications that the appearance and shape of aurora were governed by the earth magnetic field.

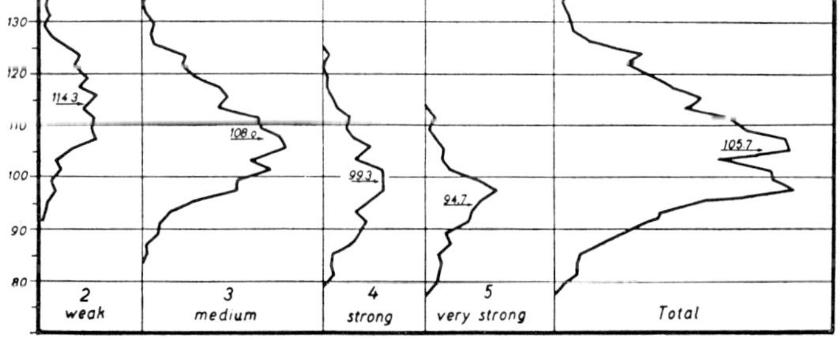

Fig. 6. Lower limits of auroral arcs, grouped according to intensitiy (after Harang, 1951).

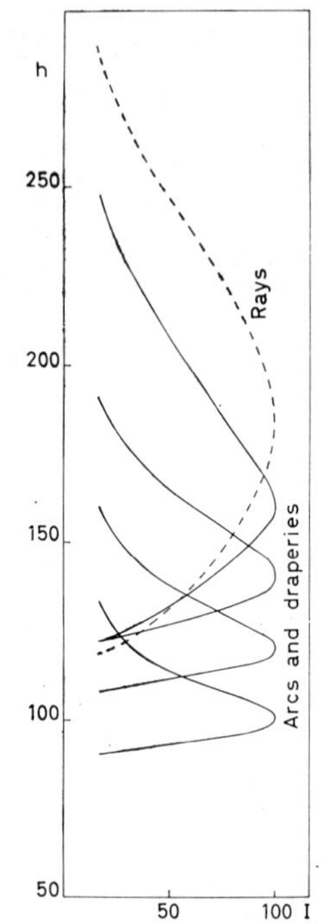

Fig. 7. Average luminosity distributions for aurorae with different lower borders.

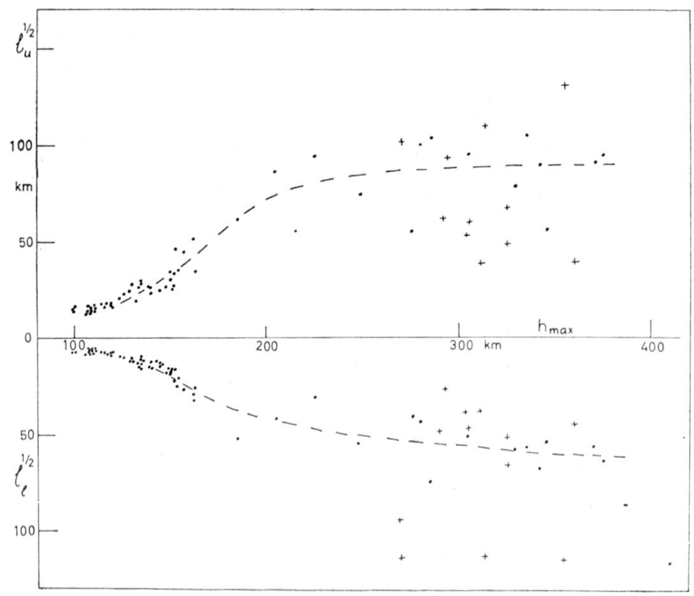

Fig. 8. The distances of the lower ($l_l^{\frac{1}{2}}$) and upper ($l_u^{\frac{1}{2}}$) half intensity points in aurora as functions of the maximum intensity point.

Later at night the aurora may obtain many strange forms, and there is certainly no rule that an arc or a band then should be aligned in the geomagnetic east-west direction. On the contrary, it may often go north-south or curve up in strange shapes. Closer to the magnetic poles it seems that the direction of the auroral arcs and bands very often may differ appreciably from the geomagnetic east-west direction.

The intensity of the aurora may vary several orders of magnitude. The brightest aurora observed is of the order of 1000 times the weakest one which can be detected by the eye, and which has an intensity just detectable above the night-glow emission level. The commonly adopted unit for classification of auroral intensities is the kR (kilo Rayleigh). 1 kR is equivalent to a total emission of 10^9 photons over 4π steradians from a column of light which is 1 cm^2 in cross-section and extends through the aurora. When the emission of the green oxygen line at $\lambda = 5577$ Å, which is one of the strongest spectral features in the visible region, is 1 kR the aurora is said to have a brightness coefficient I. Brightness coefficients II, III and IV correspond to emissions in the green oxygen line of 10, 100 and 1000 kR respectively.

The frequency of the appearance of aurora has been studied for many years and the so-called isochasms, i.e. lines through points on the earth where the aurora is observed with the same frequency, are established with a fairly high degree of accuracy. The isochasms for the northern polar cap were given in Figure 4. HULTQVIST (1961) has shown that these isochasms have almost exactly the same form as projections of circles in the geomagnetic equatorial plane on to the earth's surface along the geomagnetic field lines. When we think of how these isochasms have been derived, this makes the isochasms almost trivial. The isochasms show the frequency of the occurrence of the aurora, but averaged over days, seasons and years. This means that the effect of the direction to the sun, the inclination of the geomagnetic axis relative to the axis of rotation, and the effect of the eleven year solar cycle, have been all cancelled out. When we project the auroral zone out to the equatorial plane, along the geomagnetic lines of force, we arrive at a distance from the earth center of about 7 earth radii. Here the geomagnetic field is very nearly a dipole field, hence it is symmetric around the geomagnetic axis. It seems evident that whatever it is that happens out here and produces the aurora, it must be symmetric around the geomagnetic axis, when the phenomenon is averaged in that way which is used to obtain the isochasms. If it is out here things happen which make particles move with high velocities along the field lines towards the atmosphere, and the average of these phenomena is symmetric, then the isochasms necessarily must be projections of circles in the equatorial plane on to the surface of the earth along the magnetic field lines. If this picture is correct the only significance of the isochasms is that the important things which accelerate the particles towards the poles happen so far out in space that the dipole approximation to the earth magnetic field is valid.

Thus there seems to be much more significance in studying particular storms or averaging in such a sense that important relations are maintained. A significant step in this direction was made by DAVIS (1960, 1962). He analysed the aurora over the northern hemisphere during one particular night and found that in a coordinate system with origo at the north geomagnetic axis pole and with a fixed direction relative to the sun this pattern remained almost constant. Analysis of this kind seems to be much more relevant to the problem, and should be pursued further.

The occurrence of the aurora shows, in particular at low latitudes, a strong variation in phase with the eleven year sunspot cycle. Thus the number of aurorae is much higher during sunspot maximum than during sunspot minimum. When only the frequency of aurorae is considered there seems to be little or no effect in the auroral zones themselves, but when the auroral intensity is considered, there is undoubtedly also such an effect there, although it is not yet proved by statistics. Strong aurorae and magnetic disturbances have also a 27 days recurrence tendency. This is in accordance with the time of revolution of the sun, and shows that the phenomena on the sun, which govern the appearance of aurorae and magnetic disturbances, are as longlived as 27 days or more.

4. Connection Between Visual Aurorae and Magnetic Disturbances

The connection between a typical auroral display and the associated local magnetic disturbances is roughly sketched in Figure 9. Single auroral arcs are usually associated

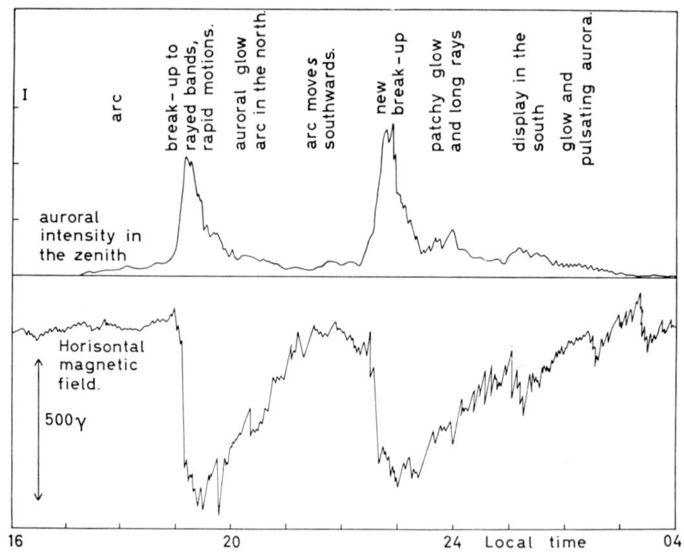

Fig. 9. Magnetic disturbances and associated auroral phenomena.

with only very moderate disturbances, only a few tens of γ. Since the main disturbance is in the geomagnetic north-south direction and the auroral arc is lying in the east-west direction, it is tempting to conclude that the disturbance may be due to a line current running along the arc. This seems to be quite possible, and if one computes the field from an infinite line current and compares this with the magnetic disturbances observed, one finds that typical line currents should be of the order of 10^4 A. If we assume a cross-section of the conductive region of about 10×30 km^2, we obtain a typical current density of about 3×10^{-9} A/cm^2.

Strong magnetic disturbances may be of the order of 1000 γ or so, but these are mostly associated with situations when stronger or weaker aurorae occur over most of the sky. Thus there may be an enhanced conductivity over a very large region. Also in these cases the disturbances are greatest in the geomagnetic north-south direction. If one assumes that a sheet current of infinite length and infinite width flows in the east-west direction and produces these disturbances, and again assumes a thickness of the conducting layer of about 30 km, one arrives at a typical current density of about 5×10^{-9} A/cm^2, which is essentially the same as for the arc. Of course these are very rough estimates and give only the order of magnitude of the currents which may flow in the aurora. As pointed out earlier, these currents must flow in the height between the lower limit of the aurora and about 140–150 km. In the next Chapter we shall look more closely into the magnitude of the conductivity which may be expected and the electric field necessary to produce these currents. The analysis of these phenomena is also complicated by the fact that the magnetic disturbances induce earth currents, which again contribute considerably to the total disturbance field measured (VESTINE, 1960; CHAPMAN and BARTELS, 1940; CHAPMAN, 1951).

That a great magnetic disturbance is correlated with sheet currents is indicated by the fact that during such conditions the correlation distances on the earth are very large. Thus observations from stations which are separated 100 km or so in the north-south direction may often be almost identical.

There seem also to be exceptions to the rule that visual aurorae and magnetic disturbances are associated. Thus it happens in the auroral zone that one may observe very strong magnetic disturbances without any visual aurora. In these cases the magnetic storms are generally very strong, and the aurora occurs very far south. It seems also in these cases to be a general correlation between the appearance of aurorae and magnetic storms, but little or no local correlations at high altitudes.

5. Pulsations

There seems to be a number of pulsations associated with auroral and geomagnetic phenomena. Thus the aurora may show very rapid pulsations. (See OMHOLT, 1963 and BERGER, 1963 and references therein). A single pulse of light from a very narrow region may last much less than a second. On the other hand there often occur pulsating aurorae with periods up to a few minutes. The most rapidly pulsating aurorae seem to be more frequent at low latitudes than at high latitudes. Thus in the auroral zone at Tromsö rapidly pulsating aurorae are very rare, whereas aurorae which pulsate with a period of a few minutes are not uncommon, in particular late at night or early in the morning after a big auroral display. There are also observed pulsations in the geomagnetic field within the same range of frequencies. These pulsations in the geomagnetic field are sometimes associated with pulsations in the auroral luminosity, but not always. Similarly there often occur pulsations in the auroral luminosity without any signs of pulsations in the geomagnetic field. With high altitude balloons it has been observed that X-ray emission is associated with aurorae. This is due to bremsstrahlung generated when the auroral primary electrons are stopped in the upper atmosphere. Also in this X-ray emission there have been observed pulsations (TREFALL, personal communication). These pulsations seem sometimes to be associated with pulsations in the geomagnetic field, but in most cases not.

It seems very puzzling that these kinds of pulsations are sometimes correlated, but most often not. Perhaps this may be explained by assuming different energies of the incident primary particles, since the relative intensities of the X-ray emission, the visual auroral emission, and the geomagnetic disturbances all to a certain extent depend on the specific energy (giving the height where the electrons are stopped and the ionization performed).

6. The Auroral Spectrum*

The dominant emissions in the auroral spectrum are from nitrogen and oxygen. In Table I is given a list of the most important emissions, the transitions they represent and their approximate intensities. Both the relative and absolute intensities should be regarded with some caution since these are not too accurately measured. The intensities given in Table I are those of an aurora of brightness coefficient III. This means that the emission in the green oxygen line 5577 Å is 100 kR.

TABLE I

TYPICAL ABSOLUTE INTENSITIES OF SOME IMPORTANT AURORAL EMISSIONS. AURORA OF BRIGHTNESS COEFFICIENT III.

Atom or molecule	*Spectral feature and transition*	*Wavelength*	*Emission in kR***
N_2	1 Positive bands $B^3\Pi \to A^3\Sigma$	red-infrared	2000
	2 Positive bands $C^3\Pi \to B^3\Pi$	violet-u.v.	100
	Vegard-Kaplan bands $A^3\Sigma \to X^1\Sigma$	ultraviolet	150
N_2^+	1 Negative bands $B^2\Sigma \to X^2\Sigma$	blue-u.v.	200
	Meinel bands $A^2\Pi \to X^2\Sigma$	infrared	2500
O_2	Atmospheric bands $b\Sigma \to X^3\Sigma$	infrared	400
	Infrared atmospheric bands $a^1\Delta \to X^3\Sigma$	infrared	2500
$N(2p^3)$	$^2P \to {}^2D$	10400	100
	$^2P \to {}^4S$	3466	7
	$^2D \to {}^4S$	5199	1†
$O(2p^4)$	$^1S \to {}^1D$	5577	100
	$^1S \to {}^3P$	2972	3
	$^1D \to {}^3P$	6300/64	50†

* For a more complete description the reader is referred to CHAMBERLAIN (1961) and BATES (1960b), in which further references are given.

** Mainly from data collected by CHAMBERLAIN (1961).

† Strongly varying.

The spectrum is generally consistent with the fact that the main atmospheric components at auroral levels are N_2, N, O_2 and O. Exceptions to this are the hydrogen lines Hα(6563 Å) and Hβ(4861 Å). These are occasionally observed in the auroral spectrum. However, they are very much broader than all other spectral emissions, and the emission of these lines is undoubtedly due to protons sometimes coming into the atmosphere during aurora. This is conclusively shown by the fact that when pointing a spectrograph towards magnetic zenith, one obtains a Hα-line which is displaced towards shorter wavelength and which also is considerably broadened. The half width of the line is approximately 20 Å. When pointing the spectrograph at right angles to the magnetic lines of force, the line is symmetric and about 10 Å wide. The emission occurs from neutralized primary protons which have come down along the magnetic lines of force, spiralling along these with a certain pitch angle. We shall not elaborate on this here, since this will be more thoroughly treated in Chapter 20.

Apart from some forbidden lines, to which we shall return shortly, there are no really drastic changes in the auroral spectrum with the height of the emission. This is borne out but by measurements scanning in vertical direction across auroral arcs and along auroral rays. Spectra from widely different heights do not seem to show any really pronounced differences. There are however, certain minor variations which may be due to differences in the atmospheric composition. It is generally believed that the First positive bands of N_2 is enhanced in very low aurorae. Very low aurorae very often show a very strong red lower border, and a few spectra which have been obtained really do seem to indicate that the red colour is due to an enhancement of the First positive bands relative to the other emissions.

The behaviour of the forbidden atomic lines is particularly noteworthy. If one looks at the relative intensities of the two forbidden NI lines, the doublet at 5.199 Å and the line at 10400 Å (cf. Table I), it is seen that the former is much weaker than the latter. However, the latter corresponds to a transition down to the upper term of the former. This means that the green doublet should be at least as strong as the infrared one, probably much stronger since there undoubtedly also is a direct excitation to the ^2D term in addition to the excitation through the ^2P term. We note that ^2P term has a lifetime against radiation of about 4.2 sec, whereas that of term ^2D is about 26 hours. This means that the nitrogen atoms in the ^2D term on the average experience a great number of collisions before they radiate their energy as emission. During these hours they may suffer collisions which deactivate them down to the ^4S term without radiation. This quenching effect is apparently very strong in this case. A similar thing happens with the oxygen atoms in the ^1D term of OI. In ordinary auroral arcs the red lines are usually weaker than the green oxygen lines, whereas the contrary should be the case if there were no deactivation by collisions. The lifetime for atoms in the ^1D term is about 110 sec, and it is generally believed that there occur deactivating collisions in the lower part of the auroral region. Looking at height profiles of various emissions one notes that the main emissions increase to a maximum and above this level decrease again more slowly. However, the red oxygen line s usually increase steadily up to very great heights.

The finite lifetime of the metastable terms which give rise to the forbidden transitions is to a large extent responsible for the colours in the rapidly changing aurora. When the primary excitation mechanisms vary very rapidly, as it may do in aurora (pulses of light lasting much less than 1 sec have been observed), the forbidden lines lag behind the other emissions because the atoms in a metastable state on the average remain there for a considerable time before they emit their energy as photons. This effect also enables one to actually study the effective lifetime of metastable states in the atmosphere, and draw conclusions about the quenching effect of the atmospheric atoms and molecules.

7. Excitation Processes*

It is now generally accepted that almost all aurorae are caused by primary particles coming from outer space into the ionosphere. It is almost certain that most of these are electrons, although protons do contribute to a certain extent. The primary electrons (or protons) collide with the atmospheric atoms and molecules and produce ionization and excitation. When secondary electrons are produced, their energies are of the order of magnitude 100 eV or so. These secondaries again collide with the atoms and molecules and produce higher order secondaries. Thus, as is shown by laboratory experiments, there is produced one ion pair in the ionosphere per 35 eV of the energy carried into the atmosphere by the primaries. A 10 keV electron thus produces about 300 ion pairs or so.

That the secondary electrons have fairly high energies, makes low energy electron excitation very important in the aurora. Moreover, it makes it very likely that the auroral spectrum will not differ very much if the aurora were produced by primary protons rather than by primary electrons, apart from the emission of hydrogen lines. It seems difficult from the main auroral spectrum to distinguish between electrons or ions as primary particles. The dominating role played by primary electrons is proved by the absence of doppler-shifted lines from hydrogen or heavier elements in most aurorae.

Undoubtedly more complicated secondary processes also play a role in the excitation of the auroral spectrum, but it seems likely that their importance is not very great. There is nothing in the auroral spectrum which is against the concept of many primary electrons and a few protons, and that secondary electrons give a major contribution to the excitation. This is also indicated by a few rocket experiments (cf. Chapter 20).

8. Temperatures from Auroral Spectra**

As is well known the rotational structure of band spectrum offers a very good opportunity of measuring the temperature of the mitting gas. This is particularly so if the excitation is due to electrons, because the electrons cannot change the angular momentum of the molecules very much. Thus, if they are in thermal equilibrium before

* For a more complete discussion the reader is referred to CHAMBERLAIN (1961) and BATES (1960b).
** For a more complete description the reader is referred to CHAMBERLAIN (1961) and HUNTEN (1961).

excitation, the state of rotation is not significantly changed by the excitation. The First Negative nitrogen bands are excellently suited for this purpose, and they have been widely used to measure temperatures in the aurora. Although there are certain minor difficulties in interpreting the profiles of the bands, because the spectrographs and the spectrometers used have a very high light power and a fairly low resolution, the major difficulties lie in ascertaining at which height the emission originates. This confused the situation for a very long time. With the fairly poor observational technique available, auroral spectra showed for many years the same temperature almost without exception. Since aurorae definitely occur at various heights, this indicated that the temperature of the atmosphere did not change very much above 100 km. Recent work by the Canadian group (HUNTEN, 1961), has shown that this is not the case, and when proper observation technique is being used and the results are properly treated, a definite temperature increase with height is revealed. Indeed, it has been possible to offer a temperature curve for the polar ionosphere.

The doppler width of atomic lines gives the gas temperature. Some successful measurements of the doppler widths of the red and green forbidden oxygen lines in the auroral spectrum have been made. The forbidden lines are chosen because they are emitted from atoms which have suffered many collisions before they emit their energy. This ensures that thermal equilibrium exists between the excited atoms and the atmosphere. The doppler width of the atomic lines has to be measured by interferometers, and the technique is very difficult and has not yet been very extensively applied.

References

BATES, D. R.: 1960a, 'General character of auroras', *Physics of the Upper Atmosphere* (ed. by J. A. RATCLIFFE), Academic Press, New York.

BATES, D. R.: 1960b, 'The auroral spectrum and its interpretation', in *Physics of the Aurora and Airglow* (ed. by J. A. RATCLIFFE), Academic Press, New York.

BERGER, S.: 1963, *Plan. Space Sci.* **11**, 867.

CHAMBERLAIN, J. W.: 1961, in *Physics of the Aurora and Airglow* (ed. by J. A. RATCLIFFE), Academic Press, New York.

CHAPMAN, S.: 1951, *The Earth's Magnetism*, Methuen and Co., London.

CHAPMAN, S. and BARTELS, J.: 1940, *Geomagnetism*, Clarendon Press, Oxford.

DAVIS, T. N.: 1960, *J. Geophys. Res.* **65**, 3497.

DAVIS, T. N.: 1962, *J. Geophys. Res.* **67**.

HARANG, L.: 1951, *The Aurorae*, Chapman and Hall Ltd., London.

HULTQVIST, B.: 1961, *Plan. Space Sci.* **8**, 142.

HUNTEN, D. M.: 1961, *Ann. Géophys.* **17**, 56.

OMHOLT, A.: 1963, *Plan. Space Sci.* **10**, 247.

STÖRMER, C.: 1955, *The Polar Aurora*, Clarendon Press, Oxford.

TREFALL, H.: Private communication.

VESTINE, E. H.: 1944, *Terr. Mag.* **49**, 77.

VESTINE, E. H.: 1960, 'The upper atmosphere and geomagnetism', in *Physics of the upper atmosphere* (ed. by J. A. RATCLIFFE), Academic Press, New York.

CHAPTER 19

AURORAL EFFECTS ON THE IONOSPHERIC E-LAYER

A. OMHOLT
Institute of Physics, University of Oslo

1. Introduction

In this chapter we shall discuss in more detail the disturbances of the ionospheric E-layer caused by the particles impinging upon the atmosphere during aurora. We know that particles produce the luminosity we observe, and it is also a well-known fact that there are produced a number of electrons and ions, i.e. that the ionization of the E-layer is enhanced during aurora. X-rays produced as bremsstrahlung from the primary electrons may contribute to the ionization at lower heights, in the D-layer. It is also evident that the enhancement of the ionization in the E-layer must increase the conductivity, which in turn makes it more easy for Nature to produce the strong currents whose magnetic effects we observe as magnetic disturbances. We shall also very briefly discuss the possible heating of the upper atmosphere by the aurora and associated phenomena.

2. Ionization Caused by the Aurora

Theoretical and experimental evidence leave little doubt that there is not only emission of light from the upper atmosphere during aurora, but also that the atmosphere itself becomes strongly ionized. A study of the relevant cross-sections for electrons in gases and a reasonable estimate of the processes which go on in the upper atmosphere during aurora, strongly supports the view that the rate of ionization in a given volume of air is approximately proportional to the luminosity, i.e. the number of photons emitted from this volume of air. The First Negative bands of N_2^+ are a direct result of ionization of N_2, and it has been found that the total rate of ionization is about 25 times the rate of emission in photons in this band (CHAMBERLAIN, 1961; OMHOLT, 1959). This means that the study of the luminosity distribution in aurora gives us a means of studying the rate of ionization. If we also can determine the actual electron density, information on the recombination coefficients can be obtained. Some relevant data for aurorae of various brightnesses are collected in Table I. In this table it is assumed that the equivalent thickness of the emitting layer of aurora is about 25 km, in agreement with the luminosity distribution function in aurorae (cf. Figure 7 and Table I on page 276 and 285).

Measurements of the luminosity distribution with height in aurora have been made of distinct auroral forms which can be triangulated with sufficient accuracy from the ground. The vertical half-width (distance between the lower and upper point

TABLE I

SOME CHARACTERISTICS OF AURORAE OF VARIOUS BRIGHTNESSES

Auroral brightness coefficient	*I*	*II*	*III*	*IV*
Emission of First Negative $N_2{}^+$ band in kR	1.5	15	150	1500
Total rate of ionization, in 10^9 ion pairs/cm² (column) sec	40	400	4000	40000
Rate of ionization at height of max. luminosity, in 10^3 ion pairs/cm³ sec	15	150	1500	15000

where the intensity is half of the maximum intensity) is not less than 20 km in any aurora. This lower limit is applicable to very low aurorae. With increasing height of the maximum intensity point this distance increases, and for very high aurorae with the maximum luminosity at about 200 km the half-width has increased to more than 100 km (cf. Figures 7 and 8 on page 276). This therefore leaves little doubt that at least distinct and limited auroral forms have a vertical extension with a half-width not less than 20 km. Considering the frequency of aurorae with various heights the most likely half-width is about 25 km.

As is evident from this, and from the data given in Table I, one should expect a fairly thick ionized layer to be produced during aurora with a rate of ionization in the maximum intensity point ranging from 15 to 15000 ion pairs/cm³ sec.

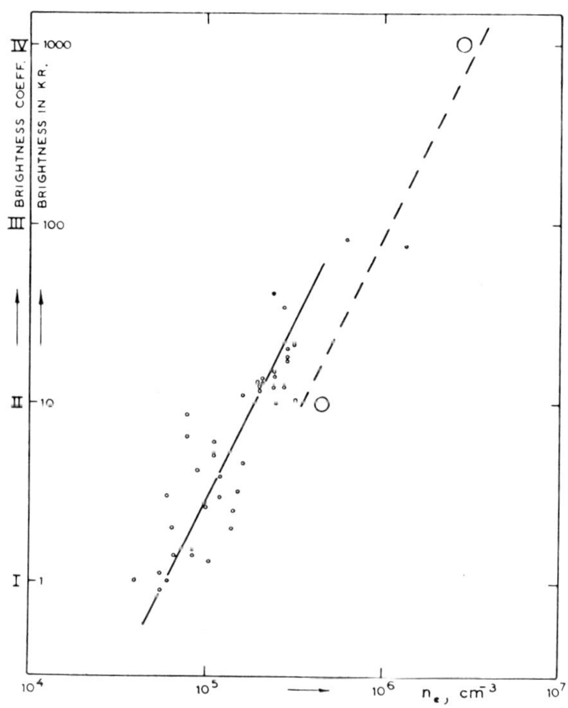

Fig. 1. Brightness of aurora in the zenith versus maximum electron density of the associated E_s-layer (after OMHOLT, 1960).

If the aurora extends over a very wide horizontal region, both in east-west and north-south, we should expect a fairly thick and dense E_s-layer to be formed. This is also the case in many instances. Simultaneous observations of the luminosity and the critical frequency show that there also is a reasonably good correlation between the two quantities, in particular if we bear in mind that the height distribution of auroral forms vary a good deal, which means that the ratio between the rate of ionization at the maximum point and the total luminosity we measure with a photometer from the ground changes. Also the height of this maximum point changes, and if the recombination coefficient changes with height the same rate of ionization may result in different maximum electron densities. These points borne in mind, the correlation between the maximum electron density measured by radio methods and the luminosity measured by photometers must be regarded as quite good. Results of such measurements are reproduced in Figure 1.

Systematic observations of aurorally associated E_s-layers show, however, that the E_s-layer is not always thick. On the contrary, some observers claim that these E_s-layers usually are very thin. Judged from the ionograms one sometimes finds a thickness of only a few kilometres (PIGGOTT, private communication). From our knowledge about the basic processes in the ionosphere during aurora, it is rather difficult to understand how such thin layers can be formed. It must be admitted that the luminosity distribution function with height is derived from aurorae which have a very limited north-south extent; otherwise it cannot be measured from the ground. It could be that extensive auroral layers have a markedly different altitude distribution, but this seems not very likely. Such a thin E_s-layer is even hardly compatible with the assumption of monoenergetic electrons coming almost vertically in along the magnetic lines of force. They could perhaps be produced by protons, which give a much more narrow luminosity distribution with height. The nature of the thin E_s-layer is a problem which probably has to be solved by means of sounding rockets, making simultaneous measurements of the electron density, the auroral luminosity and, if possible, the energy and angular spectrum of the impinging particles.

There is also another puzzle in this picture. If we take the average measured electron density for an aurora of a given intensity, using the values in Table I for the rate of ionization in the maximum point, and use the ordinary recombination law, we arrive at a recombination coefficient of about 2×10^{-6} cm^3/sec. This is two orders of magnitude higher than the recombination coefficient usually assumed for the E-layer. And now it must be borne in mind that the recombination coefficient of 3×10^{-8}, which is deduced from solar eclipses, sunset and sunrise phenomena, is a dynamic coefficient, actually giving the recombination coefficient of the slowest recombining ions (NICOLET, 1962). 80% of the ions produced are probably N_2^+, and 20% O_2^+ or O^+. If we adopt a recombination coefficient of 10^{-7} cm^3/sec for N_2^+ and 3.10^{-8} cm^3/sec for the O^+/O_2^+ content, we are still more than an order of magnitude off the observed value (NICOLET, 1962; OMHOLT, 1960).

It is in this situation that it is tempting to invoke formation of negative ions during the night. Negative ions formed at daytime are destroyed by the solar radiation, which readily detach the electrons from the negative ions as soon as they are formed. This does not occur during the night. However, theoretical results indicate that the formation of negative ions at these heights is too slow compared with the rate of collisional detachment; thus negative ions should be unimportant also at night at heights above 100 km (NICOLET, 1962). There also remains the possibility that some dynamic or diffusion processes decrease the electron density at the maximum point. It must be remembered that it is only the electron density in the maximum point which is measured by the radio methods. Thus a diffusion upwards and downwards of the ion-electron content would reduce this peak.

3. X-rays Produced by Aurora

It is now well known that X-rays are produced in aurora as bremsstrahlung from the primary electrons. This has been observed by many observers (cf. CHAMBERLAIN, 1961). These X-rays penetrate much lower down into the ionosphere than the primary electrons themselves, and therefore it was first thought that these X-rays might be responsible for the ionization in the D-layer during aurora. For electrons penetrating into air, the average fraction f_0 of their energy which is converted to X-rays is given by (*Handb. Phys.* **30**):

$$f_0 = 7 \times 10^{-9} E_0\,, \tag{1}$$

where E_0 is the specific energy of the electrons in keV. This energy is distributed on X-ray energies E according to the equation

$$\frac{\mathrm{d}f_0}{\mathrm{d}E} = 13 \times 10^{-9}\left(1 - \frac{E}{E_0}\right); \qquad (E < E_0)\,.$$

This X-ray emission is adequately described in *Handb. Phys.* **30**. The directional distribution of these X-rays is given by a certain polar diagram relative to the direction of the electron. This polar diagram is fairly wide when the electrons have an energy about 10 keV, the most probable energy for the majority of auroral electrons. As we shall see in the next chapter, it is also reasonable to assume that the electrons are fairly evenly distributed in all pitch angles (i.e. the angle between the path of the electrons and the magnetic lines of force). This means that altogether we may expect that the X-rays are reasonably isotropically distributed in all directions downwards.

An order of magnitude estimate gives the result that for an aurora of brightness coefficient II, produced by electrons of specific energy 10 keV, the X-rays should produce about 4×10^4 ion pairs/cm^2 (column) sec below the aurora. Approximate computations further show that this ionization is fairly evenly distributed in heights down to about 65 km (OMHOLT, 1960). It is rather doubtful whether this rate of ionization is sufficient to produce the electron density in the D-layer necessary to explain the radio absorption observed during aurora. It seems necessary to also invoke other sources for ionization, and the only one which then lends itself readily is a high energy tail of the primary electron spectrum. It is observed that there is a rather wide distribution of the electrons in energies (cf. next chapter), and it is not at all unlikely that there is a sufficient number of high energy electrons which can penetrate down to 90–60 km, which is the height where the absorption of radio waves during aurora occurs. Since the required rate of ionization is not very high the ionization in the D-layer (whether by X-rays or electrons) would give rise to an extremely weak luminosity only, hardly observable from the ground with the stronger aurora higher up as background.

4. Electric Conductivities in the Ionosphere During Aurora

To the first approximation the electric current is given by the equation (Ratcliffe and Weekes, 1960)

$$\mathbf{j} = \mathbf{T} \cdot \mathbf{E}. \tag{1}$$

If the magnetic field is assumed to be aligned along the Z-axis, T is given by

$$\mathbf{T} = \begin{vmatrix} \sigma_1 & -\sigma_2 & 0 \\ \sigma_2 & \sigma_1 & 0 \\ 0 & 0 & \sigma_0 \end{vmatrix} \tag{2}$$

where

$$\sigma_0 = \left(\frac{n_e}{m_e \nu_e} + \frac{n_i}{m_i \nu_e}\right) e^2, \tag{3}$$

$$\sigma_1 = \left[\frac{n_e}{m_e \nu_e}\left(\frac{\nu_e^2}{\nu_e^2 + \omega_e^2}\right) + \frac{n_i}{m_i \nu_i}\left(\frac{\nu_i^2}{\nu_i^2 + \omega_i^2}\right)\right] e^2, \tag{4}$$

and

$$\sigma_2 = \left[-\frac{n_e}{m_e \nu_e}\left(\frac{\omega_e \nu_e}{\nu_e^2 + \omega_e^2}\right) + \frac{n_i}{m_i \nu_i}\left(\frac{\omega_i \nu_i}{\nu_i^2 + \omega_i^2}\right)\right] e^2. \tag{5}$$

Here n_e and n_i are the electron and ion densities (which are equal in the ionosphere) m_e and m_i are the electronic and ionic mass, ν_e and ν_i are the collision frequencies for electrons and ions, ω_e and ω_i are the gyro frequencies of electrons and ions (given by $\omega = |e| B/m$), and e the electronic charge (in emu).

In deriving these equations considerable simplifications have been made. No account has been taken of the energy distribution of the electrons and ions, and the collision frequency of the electrons and ions has been regarded to be independent of the velocity of the particle in question. In the actual ionosphere there is a velocity distribution of the electrons and ions, consequently the collision frequency for a particular particle depends on its energy. At great heights also electron-electron, electron-ion, and ion-ion collisions become important in determining the collision frequency, because here the ratio between electrons and neutrals is higher.

For our consideration, however, the given equations are sufficient. This is because important parameters, such as the electron and ion densities and their distribution with height, as well as the boundary conditions are rather uncertain. As long as these quantities are not very well known, Equations (1) to (5) suffice for our purpose.

Apart from the electron and ion density, the only important parameters which vary greatly with height are the collision frequencies. Going downwards from great heights $\sigma_0' = \sigma_0/n_e$, which is the conductivity per ion pair along the magnetic lines of force, decreases inversely with the collision frequency (roughly inversely with the air density). $\sigma_1' = \sigma_1/n_e$, which is the conductivity along the electric field if this is directed at right angle to the magnetic field, is zero at great heights. Going downwards the ion collision frequency increases, hence the conductivity increases (because of the second term in Equation (4)), and goes through a maximum where the gyro frequency

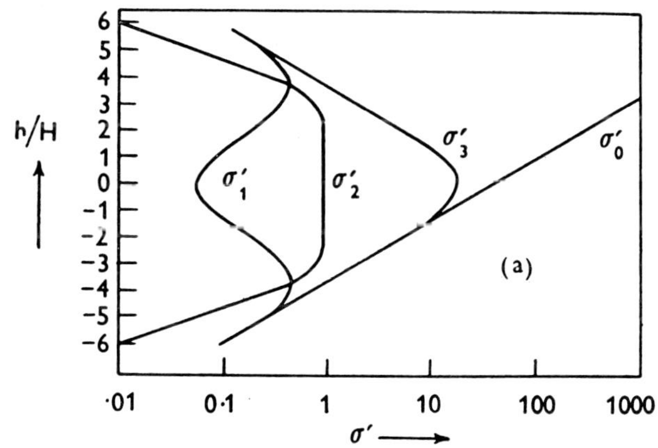

Fig. 2. The conductivities σ_0', σ_1', σ_2', and σ_3' per ion pair. The conductivities in emu are given by multiplying the magnitudes shown here by 1.6×10^{-20} F_{gauss}. The atmosphere is supposed to have a constant scale height, H, and heights are expressed in terms of h/H measured from the level where $\nu_e\nu_i = \omega_e\omega_i$. It is further supposed that $\nu_e\nu_i$ is independent of height, and $\nu_i\omega_i = 1500\, \nu_e\omega_e$ (after CHAPMAN, 1956).

and collision frequency for ions are equal. It then decreases again and rises to a new maximum where the gyro frequency and the collision frequency for electrons are equal. These two maxima occur at heights of approximately 180 and 90 km. $\sigma_2' = \sigma_2/n_e$ which is called the Hall-conductivity, gives the current in the direction of $\mathbf{E} \times \mathbf{B}$. This current is caused by the drift of electrons and ions transverse to both the electric and magnetic field (SPITZER, 1956), with the velocity $u - E/B$. At great heights this conductivity is zero, because both electrons and ions drift in the same direction with the same velocity. Going downwards, the collision frequency of the ions increases and the numeric value of the second term in Equation (5) decreases. Consequently the two terms become unequal and the numerical value of the conductivity increases until the second term in Equation (5) is vanishingly small. This occurs at the height of approximately 150 km. From then on the conductivity is approximately constant down to a height of about 100 km, where the numerical value of the first term begins to decrease.

The maximum values of σ_1' and σ_2' are approximately 1.6×10^{-11} A cm^2/V and 3.2×10^{-11} A cm^2/V respectively. The variation of the conductivities is also shown qualitatively in Figure 2.

From Figure 2 we note that we should expect an enhanced ionization over fairly wide altitude range, with a half-width of about 20–50 km or so. The total conductivity of the ionized layer has to be computed with the electron density as a function of height taken into account. This is still an unknown function, but from the luminosity curves we may guess that there exists a region of high conductivity between about 100 and 150 km. The two conductivities σ_1' and σ_2' change appreciably over this region. If an electric field is applied over an aurora, we may expect that also the direction of the current will change with height. This makes the problem extremely complicated. At present, without detailed knowledge of the electron density profiles, it is only possible to make rough estimates of the electric field responsible for these currents. If we take an aurora of brightness coefficient II we may, according to Figure 1, expect electron densities of the order of $10^5/\text{cm}^3$ over a reasonably large height interval. This means that we get total conductivities of the order of magnitude 3×10^{-6} A/V cm, and bearing in mind (cf. the above chapter) that the current densities in aurorae probably are of the order of a few times 10^{-9} A/cm^2, we arrive at the conclusion that an electric field of about 10^{-3} V/cm, is operating to drive these currents in the ionosphere, regardless of the direction in which these run compared with the direction of the electric field.

If we have an aurora which is very limited in one direction, such as an auroral arc, we may also expect that polarization effects may be of importance, although this question is still far from settled. Suppose an auroral arc is running in the east-west direction, and with an electric field running from east to west. This electric field will drive currents in its own direction, given by the conductivity σ_1, but it will also give rise to Hall-currents given by the conductivity σ_2. This Hall-current will give rise to a polarization, since positive charges will accumulate to the north, and negative charges to the south (if the aurora is in the northern hemisphere with **B** downwards). If such polarization charges are allowed to be built up, they will build up an electric field which exactly cancels the Hall-effect of the original field. Since currents can flow freely in the vertical direction within the ionosphere the polarization field E' will be given by

$$\int \sigma_1 \, \mathrm{d}hE' = \int \sigma_2 \, \mathrm{d}hE. \tag{6}$$

But this new electric field will in turn give rise to a Hall-current running westwards. The total current integrated over height is then given by

$$\int \sigma_1 \, \mathrm{d}hE + \int \sigma_2 \, \mathrm{d}hE' = \left(\int \sigma_1 \, \mathrm{d}h + \left[\int \sigma_2 \, \mathrm{d}h\right]^2 / \int \sigma_1 \, \mathrm{d}h\right) E = \Sigma_3 E. \tag{7}$$

The new integrated conductivity Σ_3 is similar to the Cowling conductivity

$$\sigma_3 = (\sigma_1 + \sigma_2^2/\sigma_1),$$

which is given in Figure 2 ($\sigma_3' = \sigma_3/n_e$), and may be larger than both $\int \sigma_1 \mathrm{d}h$ and $\int \sigma_2 \mathrm{d}h$ (Boström, 1964; Omholt and Lorentsen, 1964).

It is an open question to what extent such polarization charges can be built up in the ionosphere. It must be remembered that the conductivity at these heights along the magnetic field is quite high (cf. Figure 2). Thus it must be expected that the charges are spread out in a vertical direction (along the field lines) very rapidly, and there is established a nearly vertical sheet of space charge running parallel to the original electric field and to the magnetic field. These space charges may discharge through the bottom of the ionosphere, where conditions may be different from those at auroral heights or through connection to the magnetosphere.

If the electric field is directed across the arc, for example northwards, the current along **E** will accumulate positive charges to the north and negative charges to the south. If the electric field does not vanish, these charges must be transported away, thus there must be vertical currents upwards and downwards, which probably are closed somewhere out in space. In this case the Hall-currents will run along the (infinitely long) arc, and space charge accumulation will probably not be as important as in the first case, since the positive and negative space charges will be very far apart.

The origin of the electric field which drives the currents in the ionosphere is still an open question. We may bear in mind that the conductivities along the magnetic field lines are quite high, thus if a sufficient electron and ion density is preserved all along the magnetic field lines out to the magnetosphere, an electric field out here will be transposed along the field lines to the ionosphere. But so little is known about this that we shall abstain from further speculations.

5. Electric Discharges and the Auroral Luminosity

From time to time it has been suggested that the electric currents which are manifested through the magnetic storms may contribute to the excitation of the auroral luminosity (CHAMBERLAIN, 1961). This would mean that the electric discharge raises the energy of the electrons to such levels that optical excitation may be performed (which means that the electrons have to acquire at least an energy of 2 eV). However, detailed discussions of this effect show conclusively that this cannot be the case (CHAMBERLAIN, 1961; OMHOLT, 1959). The main reason is that the conductivity during aurora is too high, and the field required to drive the current is too low. Thus the potential over one mean free path (which is the important parameter in this question), becomes too low, and consequently the electron temperature remains low. There is also another, rather qualitative argument: the electron energy distribution depends strongly on the energy acquired by the electrons per mean free path. As the mean free path changes substantially over one scale height, it will be considerably different in the lower and upper parts of the aurora. This would again mean that the electron energy distribution would be very different in the lower and the upper part of the aurora, with a corresponding great difference in the optical spectrum. Such a difference has not been found; on the contrary, the spectra are very similar in the top and bottom layers of the aurora.

6. Heating Effects

The heating effect on the atmosphere by the particles producing the aurora is probably negligible. If we consider an aurora of brightness coefficient II, we may assume that the total input in the atmosphere is about 1.4×10^{13} eV/cm^2 sec or 20 erg/cm^2 sec. (From Table I, assuming that about 35 eV is required per produced ion pair (cf. *Handb. Phys.* **30**)). Of these we may take about 20% to be converted into heat (this gives at least the order of magnitude). Divided over a height interval of 30 km this gives less than 10^6 eV/cm^3 sec. The heat capacities of the atmosphere at the heights of 100, 120, 140 and 160 km are 2×10^9, 1×10^8, 2×10^7, and 9×10^6 eV/cm^3 respectively. This shows that the time required for heating the atmosphere one °K is of the order of magnitude 100 sec at auroral altitudes. A study of the detailed heat balance shows that the temperature may actually rise to 1000 °K or more if such a heat source could work on the same air volume for a sufficiently long time. However, we must bear in mind that the aurora has a very limited duration, and that there are considerable winds at these heights. Thus a particular volume of air would hardly be exposed to such a heating for any length of time. Consequently we may conclude that the air will be heated a few degrees only.

Joule-heating may be of some more importance (COLE, 1962, 1963). If we, in accordance with the data given in the proceding section, take the current density to be 3×10^{-9} A/cm^2 and the electric field strength to be 10^{-3} V/cm, we get a maximum heat dissipation of 3×10^{-12} W/cm^3 or 2×10^7 eV/cm^3. This is substantially more than the direct input by primary particles, and the heating effects may be appreciable, particularly at heights around 150 km, at the upper boundary of the conducting layer, where the density is low. There is no spectroscopic evidence of a temperature rise at auroral levels (100–140 km) during a disturbed night.

References

BOSTRÖM, R. Y.: 1964, 'A Model of the Auroral Electrojets'. Preprint from the Royal Institute of Technology, Stockholm, Sweden.

CHAMBERLAIN, J. W.: 1961, *Physics of the Aurora and Airglow*, Academic Press, New York.

CHAPMAN, S.: 1956, *Nuovo cimento* 4 (Suppl.), 1385.

COLE, K. D.: 1962, *Australian J. Phys.* **15,** 273.

COLE, K. D.: 1963, *Plan. Space Sci.* **10,** 129.

Handbuch der Physik **30.**

NICOLET, M.: 1962, 'Theories on electron production and recombination in the D- and E-layers' in *Electron Density Profiles in the Ionosphere and Exosphere* (ed. by B. MAEHLUM), Pergamon Press, Oxford, 1962.

OMHOLT, A.: 1959, *Geofys. Publ.* **21,** No. 1.

OMHOLT, A.: 1960, 'Ionization by auroral particles' in *Electromagnetic Wave Propagation* (ed. by M. DESIRANT and J. L. MICHIELS), Academic Press, New York.

OMHOLT, A. and LORENTSEN, H.: 1964, *Current Densities in Visual Aurora.* No. **7,** U.S. Air Force Contract AF 61 (052)–680. The Norwegian Institute of Cosmic Physics, Blindern, Norway.

PIGGOTT, W. R.: private communication.

RATCLIFFE, J. A. and WEEKES, K.: 1960, 'The ionosphere' in *Physics of the Upper Atmosphere* (ed. by J. A. RATCLIFFE), Academic Press, New York.

SPITZER, L.: 1956, *Physics of Fully Ionized Gases*, Interscience Publ., New York.

CHAPTER 20

INTERPRETATION OF VISUAL AURORA

A. OMHOLT
Institute of Physics, University of Oslo

1. Introduction

In this chapter we shall look more thoroughly into various problems mentioned in the preceding chapters. We shall first consider problems related to the morphology and dynamics of the aurora. Next we shall discuss the luminosity curves of the aurora and what conclusions can be drawn about the angular and energy distribution of the electrons. Finally the interpretation of the Doppler broadened hydrogen lines will be discussed, since these yield considerable information about primary protons in aurora.

2. Morphology of the Aurora

As was mentioned in Chapter 18, the conclusions which can be drawn from the study of auroral isochasms are rather meagre. This is because the isochasms are derived by averaging over years, over seasons, and over the time of day. Thus all effects of the direction to the sun, the tilting of the geomagnetic axis relative to the ecliptic, and of the solar cycle, are averaged out. During the last few years some work has been done to discuss aurorae and magnetic storms in a frame of reference which does not give these disadvantages. We shall here limit ourselves to discussing some recent papers which may be representative of this trend, thus the treatment given here is far from comprehensive.

Davis (1960) has recently studied the alignment of auroral forms over the entire northern polar cap. In a frame of reference with origo at the geomagnetic north pole and one axis fixed in the direction to the sun, he drew lines which at any place were parallel to the dominating auroral forms. The result of this work for a particular night is shown in Figure 1. Davis found that during one particular night this pattern, seen in this particular frame of reference, was approximately constant. This means that the phenomena in the magnetosphere which may give rise to ejections of fast particles towards the northern and southern hemispheres, are essentially independent of the position of the earth, a result which actually ought to be anticipated. If the phenomena in the magnetosphere are happening in a region where the dipole approximation to the geomagnetic field is reasonably valid one should expect that a particular point in Davis' pattern should follow an isochasm rather than a circle as the earth rotates. This is because the particles follow the magnetic field lines towards the earth, and the isochasms are, as we have seen before, nearly projections of circles. in the equatorial plane on to the earth's surface along the geomagnetic field lines And a fixed

Ortner & Maseland (eds.), Solar Terrestrial Relations.

point in Davis' coordinate system would follow a circle seen from the solid earth.

Davis' pattern has a very great similarity to the current systems which can explain the geomagnetic disturbances during aurorae (cf. Figure 2, on page 272). This suggests that currents are running along the auroral forms in the ionosphere, which is not unreasonable, since the aurora itself gives rise to an enhanced ionization, and consequently to an enhanced conductivity.

In a recent paper by AKASOFU (1963) the dynamics of the aurora over a large

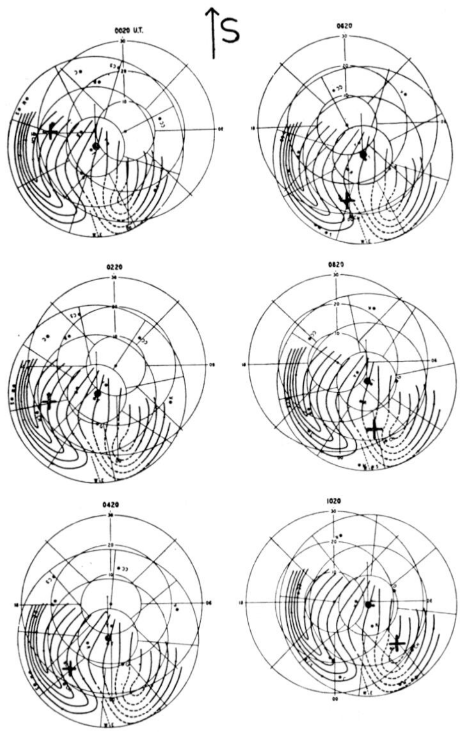

Fig. 1. Alignment and character of auroral forms over the polar cap for one night. ——— homogeneous forms; - - - - - rayed forms. The cross shows how a particular fixed station at the earth moves under this auroral pattern during the night. The dot gives the geomagnetic axis pole and *S* gives the direction of the sun (after DAVIS, 1960).

area of the polar cap was studied. During the early part of the night he found that long auroral arcs could stretch continuously in the east-west direction over a length of several thousands of km. This is in agreement with results obtained by Russian workers. FELDSTEIN *et al.* (1962) found that a homogeneous arc over northern Siberia could stretch over a length of at least 3000 km. They also found that the intensity variations of this arc occurred synchronously at all places. Akasofu's result suggests that not only the intensity variations are synchronous, but also the break-up of the homogeneous forms and the change to rapidly moving aurora occur at the same time all along the auroral zone. However, the break-up may have a different character at various places, regarded in the frame of reference with constant direction relative to the sun. This may explain the fact that there is a certain diurnal trend in the auroral forms seen from a particular station. When the break-up comes, its intensity and character are different for different local times. Thus it is less pronounced early at night than in the morning, when a break-up results in a wide spread diffuse aurora, showing a rather confusing picture. For a further study of these problems the reader is referred to the original papers (DAVIS, 1960, 1962; AKASOFU, 1963) and to the references therein.

3. Morphology of the Hydrogen Emission

The spectrum from distinct and bright auroral forms shows no or very little trace of the hydrogen lines. The general interpretation of this is that protons play an insignificant role as suppliers of energy to these auroral forms (CHAMBERLAIN, 1961; OMHOLT, 1962b). Consequently it is reasonable to assume that these forms are mainly produced by incident fast electrons. Occasionally, however, there occur in the auroral spectrum hydrogen lines which are intense compared to the other spectral emissions. These are the Balmer lines Hα(6563 Å) and Hβ(4861 Å). It seems certain, at least in the auroral zone, that this enhancement of the hydrogen lines is associated with the appearance of a very weak auroral glow covering a large part of the sky (OMHOLT, 1962b; GAL'PERIN, 1962). This glow has usually a very large extension in the north-south direction, but a fairly sharp northern and southern boundary aligned in geomagnetic east-west direction. The available data on the basic processes occurring when protons bombard air are not precise enough to permit definite conclusions, but it seems very likely that this weak glow may be produced by incident protons alone, without any significant contribution from fast electrons (OMHOLT *et al.*, 1962). If this conclusion is correct, it seems that there occur two types of aurora, one caused by electrons, and one caused by protons. The electrons probably produce the ordinary auroral forms showing a distinct pattern in the sky, homogeneous or rayed. The protons seem, however, mainly associated with weak glow, barely visible to the eye. If this interpretation is correct, there is also the question of where and when protons impinge, relative to the zones of electron precipitation. There is at present some weak indications that the proton glow occurs mainly south of the electron aurora, particularly early at night. Around midnight the two types of aurora seem more mixed together, and there is some very weak indication that on the morning side the proton aurora occurs north of the electron aurora (OMHOLT *et al.*, 1962; STOFFREGEN and DERBLOM, 1962).

If improved laboratory data or rocket investigations could help to settle the question of the role of protons in aurora showing hydrogen lines, very useful morphological studies could be made from the ground with photometers and spectrometers (cf. Section 7).

4. Conjugate Aurora

It is very important for auroral theories to ascertain whether aurorae occur simultaneously at magnetically conjugated points, i.e. points on the earth's surface which are lying on the same geomagnetic field line. These studies have been hampered by the lack of conjugated observing stations, and the relatively few results obtained are somewhat contradictory (OMHOLT, 1962b). More recently DE WITT (1962) has studied the occurrence of aurora at the Campbell Island and Farewell, which are conjugated to within 90 km. This means that the all-sky cameras at these two stations cover fairly large conjugated areas. The most important results obtained by De Witt can be summarized as follows:

(1) The brightness of the aurorae occurring in the conjugated areas vary synchronously.

(2) The auroral forms and their motions are similar in the two conjugated areas.

(3) The break-up of homogeneous forms and the transition to rayed and moving forms occur simultaneously in the two areas.

If these results are representative there is indeed a great similarity between the aurorae occurring at conjugated points. This means that the geomagnetic field line in question is not disrupted during the magnetic storm, or it is disrupted in such a way that both its northern and southern parts are connected to the same source of particles in the magnetosphere. It is obvious, however, that much more investigation of this kind is necessary before firm conclusions can be drawn.

5. Auroral Motions

It should be stressed that the motion of visual aurora has nothing to do with motions of the air masses in the atmosphere. The primary particles come in along the magnetic field lines and their drift in horizontal direction due to electric field or magnetic gradients is negligible compared to their vertical velocity. The movement of, for example, a thin auroral ray is thus due to the fact that the primary particles do not follow each other exactly along the same field line. Those particles entering a little later than the first ones come in along a slightly different field line and give rise to a visual impression of a movement, in a similar way as on a movie film. Any particular particle gives up its energy in less than 1/100 of a second, and any drift of the air masses would have negligible effect on the luminosity. Thus the motion of visual aurora reflects changes in the conditions in the magnetosphere or outer space. On the other hand the motions associated with radio aurora reflect motions or changes in the electron densities (irregularities) produced by the primary particles, but which have a certain lifetime, depending on recombination and diffusion. Thirdly, the magnetic disturbances are a result of the integrated effects of both electron and ion movements over large areas. Therefore these three kinds of movements should be treated separately and only a very moderate correlation between them may be expected.

Measurements of the drift of whole auroral forms show that east-west velocities from 100–200 m/sec may occur. The drift in the north-south direction is considerably slower. The result obtained by Canadian and American workers (CHAMBERLAIN,

1961) did not give a systematic and consistent picture. As shown by STOFFREGEN (1961), this is probably because the latitude effect was not properly taken into account. Stoffregen showed that at low latitudes the auroral forms drift mainly westwards early at night, with a reversal around midnight. In the auroral zone the east-west drift is not as pronounced, the direction does not show any particular trend. North of the auroral zone, however, the trend was reversed, the auroral forms drifting mainly eastwards early at night and westwards late at night.

It is premature to associate any of these drifts with any particular feature in auroral theories, but it is evident that these things ultimately must be incorporated in a final auroral theory. They still belong to the unsolved mysteries of the aurora.

If we look at particular features within an auroral form, such as small rays often occurring in bands, one finds that these may move rapidly within the main aurora, preferably in the east-west direction. Velocities up to 40 km/sec have been measured by photoelectric techniques. (OMHOLT, 1962a) The structure of such features may be very narrow. AKASOFU (1961) has measured horizontal cross-sections of rays as narrow as about 200 m only. To the same class of problems belong the curling up and shape of auroral forms in rapid motion. This phenomenon has a suggestive similarity to some laboratory measurements conducted by Kühl and Webster (WEBSTER, 1957; GARTLEIN, 1959; OMHOLT, 1962b). Again it should be stressed that it is premature to include these details in any auroral theory, but that they ultimately must find their place in the theory.

6. Energy Spectrum of Primary Electrons and Luminosity Curves of Aurora

In Chapter 18 it was mentioned that the aurora had its lower limit at about 100 km and that the luminosity stretched upwards from this height. The height may differ appreciably from one aurora to another, this is also the case with the vertical extension of the luminosity; both indicate variations in the energy spectrum and possibly the

TABLE I

PENETRATION DEPTHS OF ELECTRONS IN THE EARTH'S ATMOSPHERE

E (keV)	v (10^9 cm/sec)	*range** (*cm air STP*)	h** (km)
0.3	1	0.01	135
2.6	3	0.04	115
10	6	0.22	97
25	9	1.13	88
50	12	3.84	83
100	17	12.7	75
200	22	39.7	70

* From CHAMBERLAIN (1961).

** h is the approximate height where the electrons are stopped if their path is vertical (JOHNSON, 1961).

angular spectrum of the incident electrons. Since the density of the atmosphere up to auroral heights is now fairly well-known, it is possible from the luminosity distribution curves and the absolute height of the aurora to make some estimates about the energy of the primary particles. As a rough guide we may use the numbers given in Table I.

Table I shows the approximate penetration depths of electrons of various energies, provided they are incident vertically on the atmosphere. An electron which is injected with a certain pitch angle is stopped at a greater height, because the air mass penetrated per unit height interval is greater. Since most of the auroral luminosity occurs at heights above 100 km it is evident that if all the electrons came in vertically the aurorae would mostly be produced by electrons with energies less than 10 keV. The fact that most of them do not penetrate vertically, but have a certain pitch angle, does not alter this conclusion drastically. We may assume that the overwhelming part of the electrons has energies of the order of 10 keV or less. This is also borne out by computations made by REES (1963), and which we shall now look into in more detail.

Already by considering the luminosity curves of aurorae, it became quite clear that the aurora could not be produced by monoenergetic particles penetrating vertically into the atmosphere (CHAMBERLAIN, 1961). From laboratory data it was quite clear that such a beam of particle would produce a much narrower luminosity distribution with height than observed. Even very crude considerations therefore made it clear that one must at least have dispersion in either the energy or in the angular distribution of the incident particles.

This is confirmed by a few rocket measurements. Measurements on the energy distribution and angular distribution of the primary electrons in aurora have been made by various groups, and the results are reviewed by REES (1963). Although none of them is very accurate, these measurements as a whole indicate that the angular distribution of the primary particles is isotropic, and that the energy distribution follows approximately an exponential law. The various observations fit roughly the energy distribution function $\exp(-\varepsilon/\varepsilon_0)$, where ε is the specific energy of the electrons, with ε_0 varying from 5 keV to 22 keV. The upper limit concerns electrons with energies higher than 30 keV only. There is a great variability in the numerical results, as must be expected from the variability in the auroral luminosity curves and the auroral heights.

Very recently REES (1963) has calculated the luminosity distribution of aurorae caused by electrons with different energy and angular distributions. Adopting an isotropic angular distribution and an exponential energy distribution, he gets a luminosity curve whose shape does not fit badly with the observed ones. However, to get the lower limit of the aurora at about 100 km, in agreement with observations, he needs a value of ε_0 as low as 1.4 keV. This represents a very steep energy distribution curve, with very few electrons of high energy, and is not in agreement with the rocket observations.

The main difficulty in the computations is to take the scattering of electrons properly into account (CHAMBERLAIN, 1961). Rees bases his work on theoretical computations and laboratory work, and it seems rather unlikely that his basic data are so seriously in error that it can explain the discrepancy between the theoretical results and the rocket experiments. On the other hand, the rocket experiments are very few and rather tentative, and it should also be borne in mind that the techniques are very difficult. It seems therefore reasonable to view these results with some caution until many more observations have been made. The great variability in the optical aurora also shows that many observations are needed to get a representative picture of the energy and angular distribution and their variations.

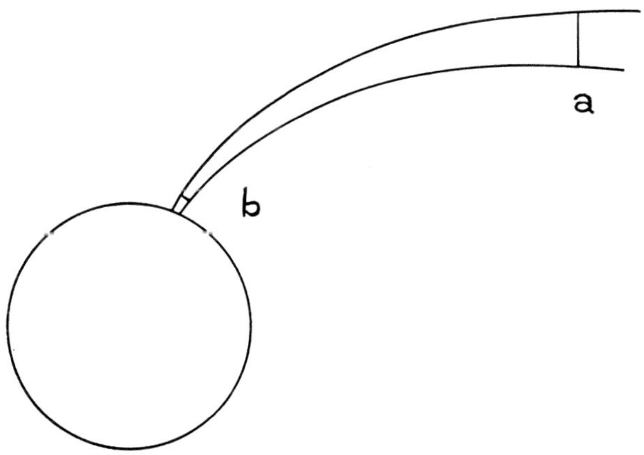

Fig. 2. Shape of magnetic flux tube.

Another difficulty in these computations is that from a theoretical point of view it is difficult to distinguish between various angular and energy distributions. This. is because electrons of a specific energy give a contribution to the luminosity which is concentrated over a much narrower height interval than the total luminosity of the aurora. In consequence, as an example, a certain number of electrons penetrating vertically into the atmosphere may give almost the same contribution to the total luminosity as another number of more energetic particles penetrating with a certain pitch angle, which is such that they are stopped at the same height in the atmosphere as those with lower energy penetrating vertically. Therefore any electron included in the distribution function may be replaced by more energetic celetrons penetrating at higher angles or less energetic electrons penetrating at lower pitch angles.

As we shall see, however, the assumption of an isotropic angular distribution is justified not only from experiments, but also from theoretical considerations. Let us consider a magnetic flux tube limited by magnetic field lines and by two surfaces *a* and *b* vertical to the field lines, as indicated on Figure 2. Any particle passing through *a* towards the earth must pass *b*, provided it is not magnetically reflected between these two surfaces. The condition for this is given by the adiabatic equation

$$\frac{\varepsilon_a \sin^2 \vartheta_a}{B_a} = \frac{\varepsilon_b \sin^2 \vartheta_b}{B_b}, \tag{1}$$

where ε, ϑ and B are respectively the kinetic energy, the pitch angle and the magnetic field.

At b the pitch angle must be less than 90°, consequently the pitch angle ϑ at a must fulfil

$$\sin^2 \vartheta_a < \frac{\varepsilon_b B_a}{\varepsilon_a B_b}. \tag{2}$$

If a is at a distance of a few earth radii, the ratio B_a/B_b is less than 10^{-2}, and the angle ϑ_a must be less than about five degrees, provided there is no acceleration of the particles between a and b. This means that all the particles which pass through the surface b must be fairly closely aligned along the lines of force at a, to within a few degrees.

If we denote the particles penetrating through a per sec and sterad, as a function of the pitch angle ϑ_a, by $N_a(\vartheta)$, the influx into the tube in question at a is given by

$$\Phi_a = 2\pi \sin \vartheta_a N_a(\vartheta)\, \delta\vartheta_a, \tag{3}$$

within the interval of ϑ_a given by $\delta\vartheta_a$. Similarly the total outflux at b is given by

$$\Phi_b = 2\pi \sin \vartheta_b N_b(\vartheta)\, \delta\vartheta_b. \tag{4}$$

Provided the particles fulfil the conditions mentioned above, these two quantities must be equal, and differentiating Equation (1) with respect to ϑ_a and ϑ_b and substituting for $\delta\vartheta_a/\delta\vartheta_b$ one obtains:

$$N(\vartheta_b) = \frac{\varepsilon_b B_a}{\varepsilon_a B_b} \frac{N_a(\vartheta)}{\cos \vartheta_a} \cos \vartheta_b. \tag{5}$$

Since it is reasonable to assume that there will not be very much change in the angular distribution over the small angular interval at a, given by Equation (2), it seems reasonable to assume that the distribution function $N_a(\vartheta)$ is nearly isotropic. Further, since for this small allowed interval of ϑ_a $\cos \vartheta_a \approx 1$, one gets:

$$N_b(\vartheta) = N_0 \cos \vartheta_b. \tag{6}$$

The intensity $\eta(\vartheta)$ of particles, i.e. the flux through a surface at right angles to the direction of the particles, is given by

$$N(\vartheta) = \eta(\vartheta) \cos \vartheta. \tag{7}$$

From Equation (6) it then follows that $\eta(\vartheta)$, which is the measured distribution function used by Rees, should be isotropic. This result may be derived rather directly from Liouville's theorem.

If accelerations of the particles along the field occur on their way from a to b, this result may be no longer valid. For example, if there is an important acceleration over a very narrow range close to b, all the particles will obtain a much lesser pitch angle after the acceleration, and will be confined to a more narrow cone around the magnetic field lines. If, on the other hand, all the accelerations of the particles take place far out in space, close to the equatorial plane, it seems very likely that we should expect an isotropic distribution of the electrons which actually reach the ionosphere, i.e. those which are not reflected on their way through the magnetic field. For a more comprehensive discussion of these problems the reader is referred to the book by CHAMBERLAIN (1961).

7. Interpretation of the Hydrogen Lines

The emission of the hydrogen lines Hα(6563 Å) and Hβ(4861 Å) from aurora is due to fast protons entering the upper atmosphere and being neutralized through collisions with the atmospheric atoms and molecules. This is evidenced by the great

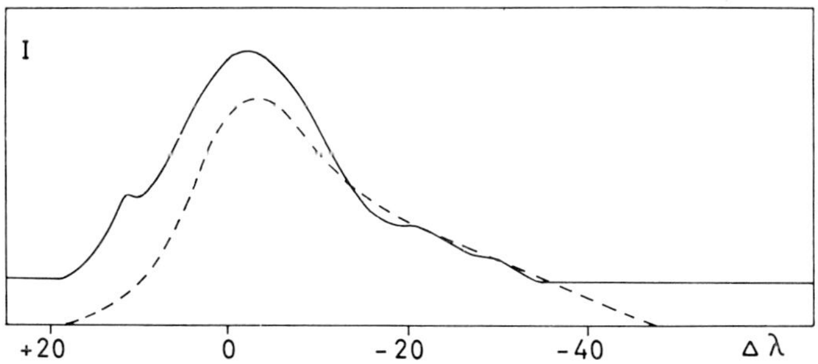

Fig. 3. Hα profiles in the direction of the magnetic zenith. The profile drawn with solid line is probably slightly contaminated by N_2-bands.

Doppler broadening and shift of the hydrogen lines observed in the spectrum (CHAMBERLAIN, 1961; OMHOLT, 1962b). We shall here first describe the appearance of the hydrogen lines, then briefly discuss the elementary processes occurring when fast protons enter the atmosphere, and finally discuss the information one can obtain from a detailed study of the observed profiles compared to the theoretical results.

A. OBSERVED HYDROGEN PROFILES

When the hydrogen lines are observed in aurora they are almost always greatly broadened and shifted towards shorter wavelengths. The shape of the line depends very strongly on the direction of observation. Observations in the direction of the magnetic horizon, i.e. at right angles to the magnetic field lines, yield a line which is symmetrically broadened around the zero-line. The half-width of the line is then about 15 Å, and there are apparently very little variations in the shape and width of the line.

Observations in the direction of magnetic zenith, i.e. upwards along the magnetic field lines, yield a very different profile. One then obtains a profile which is strongly shifted to violet and is rather asymmetric. The half-width is usually of the order of 20 Å, and the shape is very close to that shown in Figure 3. Observations in directions in between these two give lines with shapes somewhere in between the pure horizon and zenith profiles. It is much more difficult to interpret observations at arbitrary angles than observations in the magnetic horizon or magnetic zenith.

Up to very recently no significant variations in the shape and width of the hydrogen lines were observed. Observations made at different places and at different times gave, within the limits of accuracy, the same profiles. Very recently it has been possible to measure significant variations in the profiles. These variations are, however, not very frequent, and of moderate size. The horizon profile has been observed to vary between 15 and 23 Å in width (JOHANSEN and OMHOLT, 1963). The variations in the zenith profile observed so far are very small. The observations were obtained with a rapid scanning spectrometer, which gives the profile in less than 4 minutes. The variations observed occurred over very short periods of time, and this may explain why earlier observations have failed to detect such variations. The earlier observations have mainly been made with spectrographs, requiring quite a long exposure time. Thus the profile has usually been averaged over an hour or more.

B. EXCITATION OF THE HYDROGEN LINES IN AURORA

The processes occurring when a beam of protons enters the atmosphere are believed to be the following (CHAMBERLAIN, 1961):

$$\begin{aligned} H^+ + X &\rightarrow H' + X \\ H' &\rightarrow H + h\nu \end{aligned} \tag{8}$$

followed by

$$H + X \rightarrow H^+ + X + e \tag{9}$$

followed by (8), or (8) is followed by

$$\begin{aligned} H + X &\rightarrow H' + X \\ H' &\rightarrow H + h\nu \end{aligned} \tag{10}$$

which again may be followed by (9) or (10). In these equations X signifies any atmospheric atom or molecule.

If N denotes the total number of primary particles coming into the atmosphere with a given energy, of which N_p are protons and N_H hydrogen atoms, then the ratio between N_p and N_H is governed by the equations:

$$\frac{\mathrm{d}N_p}{\mathrm{d}l} = N_H n \sigma_l v - N_p n \sigma_c v \tag{11}$$

where l is the distance along their path, n is the atmospheric number density, v is the velocity of the protons, and σ_l and σ_c are the cross-section for the processes (9) and (8) respectively. Crude computations on the actual numbers involved show that the left hand side of Equation (11) is much less than each of the quantities at the right hand side, so that

$$\frac{N_p}{N_H} = \frac{\sigma_l}{\sigma_c}. \tag{12}$$

The electron loss and capture cross-sections for protons in air are fairly well known, except for the very lowest energies. Therefore the ratio between protons and hydrogen atoms in such a beam can be established with a fairly high accuracy. The processes (8) and (10) can be split up in a number of different ones according to the excited state of the hydrogen atom at the right hand side of the equation. In a similar fashion as above, taking all possible excited states into account, one can in principle compute the number of photons being emitted in a particular spectral line from such a beam of protons and hydrogen atoms. One then, of course, has to take into account

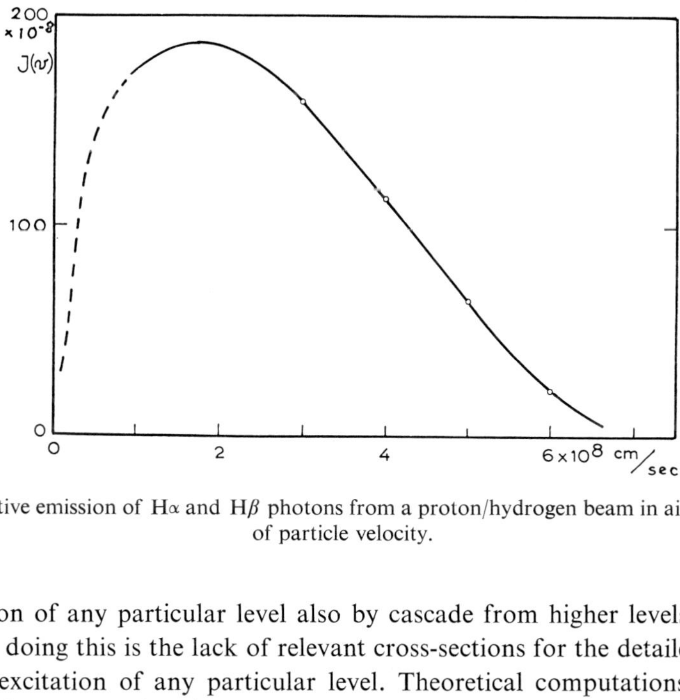

Fig. 4. Relative emission of Hα and Hβ photons from a proton/hydrogen beam in air as a function of particle velocity.

the excitation of any particular level also by cascade from higher levels. The main difficulty in doing this is the lack of relevant cross-sections for the detailed processes leading to excitation of any particular level. Theoretical computations have been carried out for protons bombarding hydrogen atoms, but no data exist for protons bombarding air. This illustrates aptly the difficulties one runs into in atmospheric physics because of lack of basic laboratory data.

Two methods of approach have been used to tackle this problem. In the first method one makes estimates of the detailed cross-sections for air from those for hydrogen by comparing the total loss and capture cross-sections for protons in air and in hydrogen (CHAMBERLAIN, 1961). Using these estimated cross-sections for protons in air one can compute the emission of, for example, Hα and Hβ by a proton/hydrogen beam in air. The other method is to compute for protons in hydrogen the fraction of processes (8) and (10) which leads to emission of Hα and Hβ, and then adopt this fraction as valid for protons in air also (OMHOLT, 1959). Both methods lead to approximately the same result. Since the cross-sections are functions of the proton velocity, these computations have to be carried out for all velocities in question.

In this manner one has computed the emission of Hα and Hβ quanta from a beam of protons and hydrogen atoms in air as a function of their velocity. The resulting emission curve is shown in Figure 4. At low velocities the curve is believed to be rather uncertain, whereas its accuracy at higher velocities probably is reasonable. The total emission of Hα quanta per incident proton is probably around 50, provided the initial energy is above 100 keV. This emission curve, together with the ionization curve for protons in air, enables one to estimate whether or not a particular aurora which shows an Hα line may be produced entirely by protons or not, since one can estimate the expected intensity ratio between the hydrogen lines and the main spectral components for various proton energies. The result is that most bright aurorae are not produced by protons, whereas the glow mentioned in the first lecture may be produced by protons alone. The uncertainties in the basic cross-sections and in the energy spectrum of the protons prevent, however, definite conclusions regarding this point. Its solution awaits improved laboratory data or direct rocket investigations.

C. INTERPRETATION OF THE Hα PROFILES

By assuming a certain energy and pitch angle distribution of the protons it is possible to compute the resulting profile observed in directions of magnetic zenith or magnetic horizon (CHAMBERLAIN, 1961). We shall not go into details here, but it should be pointed out that the relation between the horizon and zenith profiles is dominated by the pitch angle distribution, and less influenced by the energy distribution. To see qualitatively how the pitch angle distribution affects the line profiles, let us first consider a distribution function $N(\vartheta)$ which is isotropic up to $\vartheta=\pi/2$ and zero for greater values of ϑ (cf. Section 6). This would yield a zenith profile which in shape is exactly the one half of the horizon profile, because the total influx then is isotropic, except that protons with negative velocity components along the lines of force are lacking. Thus an observer looking along the lines of force will see exactly the same angular distribution as one looking at right angles to the lines of force, except that the negative velocity components towards the first observer are lacking. If the protons are more narrowly concentrated around the lines of force, the high velocity tails in the horizon profiles will be diminished, and the profile will become narrower. But on the zenith profile this has the effect of removing protons with very low Doppler shift, and the profile becomes broader and gets a dip at zero velocity. Following this line of argumentation, it can be seen that the narrower the distribution is around the magnetic field line, the narrower is the horizon profile, whereas the wider is the zenith profile.

Detailed computations on model distributions show that the observed profiles can be reconciled with an energy distribution approximately proportional to ε^{-2}. If one adopts an angular distribution $N(\vartheta)$ proportional to $\cos^n \vartheta$ one finds that n varies and usually lies in the range 1–6 (CHAMBERLAIN (1961), JOHANSEN and OMHOLT, 1963). These results must still be regarded as rather tentative, because of the lack of good basic data. But it seems reasonably safe to conclude that the intensity distribution $\eta(\vartheta)=N(\vartheta)/\cos\vartheta$ is usually not isotropic, as it should be according to the theoretical picture outlined in Section 6, and as seems to be the case for electrons. One may then speculate whether the protons are subject to acceleration very close to the ionosphere, but it seems rather premature to draw any conclusions on this yet.

It should be stressed that observations of the hydrogen lines may be very useful if more knowledge about the relation between these lines and the spectrum of the incident protons were available. Such information can either be obtained by laboratory measurements of the relevant cross-sections for the basic processes, or it can be obtained through direct rocket and satellite measurements. If a sufficient number of good rocket and satellite sample measurements of the energy distribution, angular distribution and total intensity of the proton streams were available together with observations of the resulting Hα lines, one could by trial and error methods check the basic emission functions and obtain information on this. Later observations of the Hα line from the ground could then be used for synoptic studies of the influx of protons on the upper atmosphere, with considerably less work than involved in rocket and satellite observations.

References

Akasofu, S.-I.: 1961, *J. Atm. Terr. Phys.* **21,** 287.
Akasofu, S.-I.: 1963, *J. Geophys. Res.* **68,** 1667.
Chamberlain, J. W.: 1961, *Physics of the Aurora and Airglow*, Academic Press, New York.
Davis, T. N.: 1960, *J. Geophys. Res.* **65,** 3497.
Davis, T. N.: 1962, *J. Geophys. Res.* **67,** 75.
Feldstein, J. I., Khorosheva, O. V., and Lebedinsky, A. I.: 1962, *J. Phys. Soc. Japan* **17,** Suppl. A, 249.
Gal'perin, Y. I.: 1962, *Plan. Space Sci.* **10,** 187.
Gartlein, C. W.: 1959, *Ann. Géophys.* **15,** 31.
Johansen, O. E. and Omholt, A.: 1963, *Plan. Space Sci.* **11,** 1223.
Johnson, F. S.: 1961, *Satellite Environment Handbook*, Stanford University Press.
Omholt, A.: 1959, *Geofys. Publ.* **20,** No. 11.
Omholt, A.: 1962a, *Plan. Space Sci.* **9,** 285.
Omholt, A.: 1962b, *Plan. Space Sci.* **10,** 247.
Omholt, A., Stoffregen, W., and Derblom, H.: 1962, *J. Atm. Terr. Phys.* **24,** 203.
Rees, M. H.: 1963, *Plan. Space Sci.* **11,** 1209.
Stoffregen, W.: 1961, *J. Atm. Terr. Phys.* **21,** 257.
Stoffregen, W. and Derblom, H.: 1962, *Plan. Space Sci.* **9,** 711.
Webster, R. N.: 1962, *J. Geophys. Res.* **67,** 1347
Webster, H. F.: 1957, *J. Appl. Phys.* **28,** 1388.
De Witt, R. N.: 1962, *J. Geophys. Res.* **67,** 1347.

PART IV

SPACE INSTRUMENTATION

CHAPTER 21

THE SPACE ENVIRONMENT

A. P. WILLMORE
University College, London

During its launching and afterwards, during its life in orbit, an artificial satellite is subjected to a series of mechanical and thermal rigours which together comprise what is rather loosely called the space environment. The launch phase environment will be experienced by a sounding rocket also, but the orbital phase with a duration of only a few minutes can generally be largely ignored. In the course of this chapter we shall discuss the space environment and the problems which it raises, mainly from the view-point of the space scientist with an experiment to conduct. It will be convenient to distinguish the launch phase, with severe mechanical and possibly thermal stresses and the orbital phase, with its extended exposure to extreme temperatures, micrometeorites, low pressures and a variety of damaging electromagnetic and corpuscular radiations. Before a spacecraft is launched it will have to have demonstrated its ability to withstand all or most of these ambient conditions for at least a substantial fraction of its intended useful operating life. This will require a sequence of elaborate tests over a period of many months.

1. The Launch Phase – Mechanical Stresses

The major mechanical problems in the design of experiments to be conducted in spacecraft are concerned with the launch phase. The mechanical stresses are of course due to the characteristics of the rocket motor itself. The achievement of orbital velocities of about 8 km/sec in a time of approximately 100 seconds must involve a mean acceleration of 8 g and peak accelerations are likely to be several times higher. Small sounding rockets of the type used for meteorological purposes have very short burning times of the order of 1 sec and these often have peak accelerations up to 100 g. A second source of mechanical stress is the ignition of the motor itself, which is done by setting off a small explosive charge called the igniter. This generates an impulsive shock which typically has a peak value of 50 g and a duration of 30 msec. Shocks of similar amplitude may also be encountered during handling and transport. Finally, the motor can be considered as a kind of organ pipe resonator which is excited by the turbulent motion of the exhaust gases. In consequence vibration is generated over a very wide frequency range, extending with significant amplitude from about 10 c/sec to 3000 c/sec. The actual levels, and the nature of the spectrum, are very variable with time, and also from motor to motor even within the same type. However, for test purposes, the spectrum will be approximated in some way as shown in Table I for a typical case.

TABLE I

VIBRATION SPECTRUM OF X-248 MOTOR,
APPROXIMATED FOR ENVIRONMENTAL TEST PURPOSES

Frequency range, c/sec	*Acceleration amplitude*, g.
5–50	1.5
50–500	7.1
500–550	14.0
550–650	57.0
650–2000	14.0
2000–3000	36.0
3000–5000	14.0

The figures apply to thrust axis vibration at flight levels, the duration of vibration being maintained for 2.5 minutes. The levels quoted here are applicable to solid fuel motors, those for liquid motors are not so great. There is no doubt that of the three sources of mechanical stress, by far the most important is vibration. It is generally safe to say that equipment capable of withstanding normal vibration levels will be quite unaffected by shock or steady acceleration.

The design of experiments to endure vibration follows well-known mechanical principles. A major factor involves spacing the resonant frequencies of the various components. The main spacecraft structure will itself exhibit a number of resonances, the principal ones being in the frequency range 50 c/sec to 200 c/sec. The amplification of the vibration level at the resonant frequencies will typically be about 2–3 times the input level given in Table I. Below the first of these the structure behaves as a rigid body, so that the input acceleration is transmitted to the experiments without significant change of amplitude. As the frequency rises above them, the experiments are decoupled to an increasing extent from the source of vibration by elastic deformation of the structure, so that the acceleration amplitude begins to fall. In consequence, between 200 c/sec and 800 c/sec the average amplitude is less than the input (although in limited frequency ranges near structure resonances there may be some amplification) and then above 800 c/sec a general attenuation occurs. The latter is particularly marked with fibre-glass structures owing to the large internal damping of this material.

Vibration levels of this order, when maintained for several minutes, are very destructive. Unfortunately it is impossible to design mounts with sufficient internal damping to reduce significantly the amplification at the main resonances near 100 c/sec, and isolating mountings at these frequencies are out of the question on account of the large displacement amplitude of around 1 cm. Isolating mountings can be used only in the case of occasional resonances at high frequencies that are particularly troublesome for some reason. Thus the equipment must simply be designed to endure the environment.

2. The Launch Phase – Thermal Stresses

The payload is protected from the major effects of aerodynamic heating by the nose fairing, though this may itself rise sufficiently in temperature to transfer significant quantities of heat to the interior and, indeed, may itself contaminate the payload surface by evaporation of resinous materials if made of fibre-glass. However, it is sometimes necessary to mount small components protruding from the rocket surface and these may be simultaneously subjected to large aerodynamic stresses and high temperatures.

3. The Orbital Phase – Evacuation

Figure 1 shows, as a function of altitude, the time required to form a mono-molecular layer on an initially gas-free surface. This time is short at sub-orbital altitudes, but above this, surfaces which are periodically scraped by mechanical motions may be expected to become essentially gas-free. Mechanical bearings are in consequence liable to seize up as a result of chemical forces between the metals unless carefully lubricated with solid lubricants such as graphite or molybdenum disulphide. For lightly stressed bearings at low speeds of rotation, polytetrafluoride or related plastics may be used. Alternatively mechanical assemblies can be enclosed in pressurized, sealed containers. These may be filled with helium so that leaks can be detected by

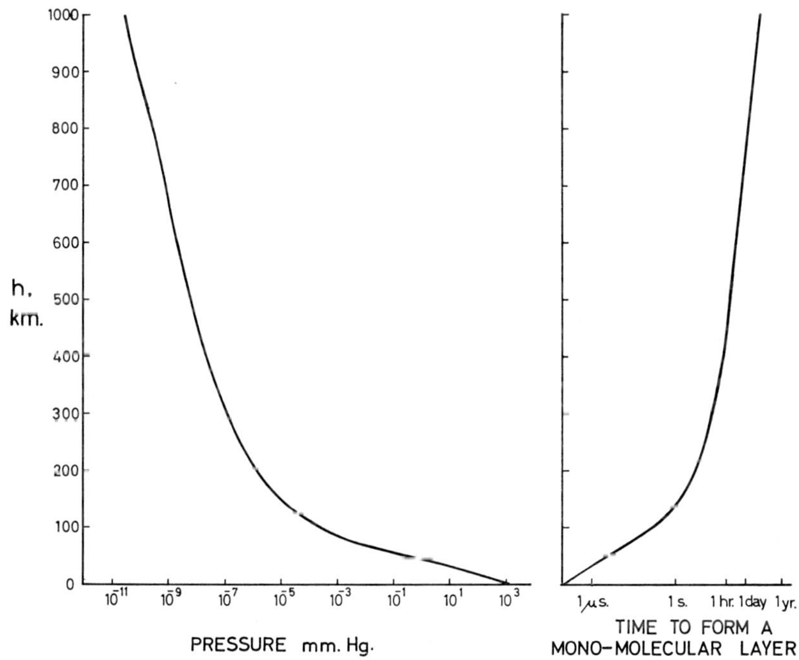

Fig. 1. The time required to form a mono-molecular layer on an initially gas-free surface, as a function of altitude.

placing the container in a vacuum chamber and using an ordinary helium leak detector.

Plastic materials themselves may suffer degradation of their physical properties under prolonged evacuation, generally as a result of the loss of a relatively volatile plasticizer. Materials such as polyvinyl chloride should be avoided for this reason. Others such as nylon may lose some weight without significant modification of their mechanical properties, whilst others again such as polytetrafluoride and some related co-polymers, lose little weight at all.

4. The Orbital Phase – Temperature

The heat capacity of a sounding rocket is great enough that no temperature control

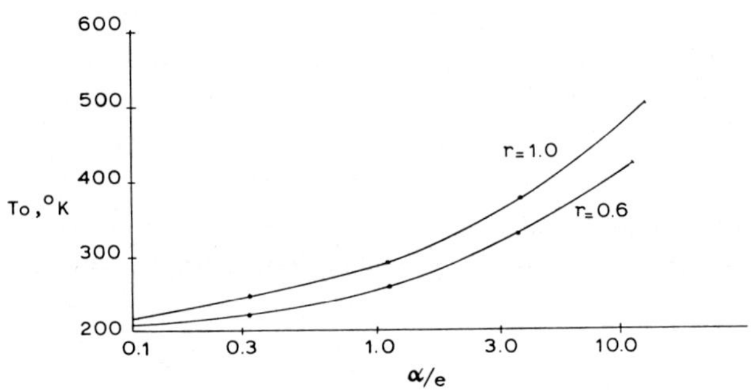

Fig. 2. Variation of mean orbital temperature T_0 with α/e.

is needed during its short flight. The mean orbital temperature of a satellite can be determined by means either of an active or a passive control system. Only the latter is generally practicable with a small, simple satellite and since the two methods are rather similar in principle only passive control will be described in detail. This depends on the fact that the satellite receives energy primarily from sunlight and loses it by radiation in the infra-red at wavelengths of around 10μ. Thus the temperature can be adjusted by using various materials to vary the ratio of the mean absorptivity α of the surface in the visible to the mean infrared emissivity, e.

It is readily shown that the mean temperature, T_0, of a spherical satellite in a low altitude orbit is given by:

$$T_0^4 = \frac{r \cdot S}{4\sigma} \cdot \frac{\alpha}{e} + \frac{T_e^4}{2}$$

where r is the fraction of the orbit when the satellite is sunlit; S is the solar constant; σ is Stefan's constant. T_e is the effective radiation temperature of the earth, about 230 °K. T_0 is plotted, as a function of α/e, in Figure 2 for limiting values of r. It is clear that for T_0 near 300 °K, typical variations in r from 0.6 to 1.0 produce a variation of 26 °C in T_0. Moreover we require $\alpha/e \sim 1.5$. In obtaining this result we have neglected earth's albedo and the effect of internal power dissipation if the satellite has solar paddles. An allowance for the effect of these may reduce α/e to 1.0 to 1.3. The actual values of α and e alone do not effect the mean satellite temperature so long as α/e is constant. However, the magnitude of the temperature fluctuations near the skin of the satellite over a single orbit is directly proportional to α or e and for this reason it is generally desirable to minimize the actual values. In addition to the temperature range of 26 °C due to the variations of time in sunlight, the mean temperature can vary from a variety of other causes, of which the most important are: (a) variations in earth's albedo (temperature change 5 °C); (b) turning internal power on or off (temperature change 10 °C); (c) non-spherical spacecraft (temperature change depends on variation in cross-section presented to the sun); (d) uncertainties and changes in α/e (temper-

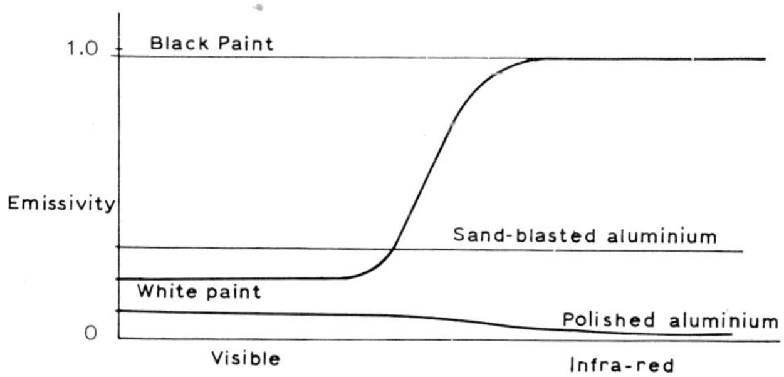

Fig. 3. Typical variation of emissivity with wavelength for different substances.

ature uncertainty ± 10 °C); (e) in interplanetary probes, variation of solar constant.

In all, the mean satellite temperature may vary from its predicted value by ± 30 °C or more. In addition, temperature gradients and fluctuations in the satellite may be quite large, particularly for components mounted on or near the skin – components on the distant side of the spacecraft from the sun may be 20 °C lower than the mean, whilst on the sunward side temperatures may be the same amount higher than the mean. If this results in an intolerably large temperature range for the equipment, active control of the temperature, usually by variation of α/e, must be employed.

The variation of absorptivity with wavelength is shown diagramatically as a function of wavelength for several materials in Figure 3. Polished metals have $\alpha/e \sim 4.6$ so that these tend to maintain a high mean temperature, whilst all other materials have $\alpha/e < 1$. Thus, α/e values just over unity are obtained by mixing metal and painted surfaces. A particularly elegant solution was devised by HAAS *et al.* (1963) for the Vanguard satellites, consisting of a layer of silicon monoxide about 10000 Å in thickness on evaporated, polished aluminium. Since the dielectric layer is transparent in the visible, α has the low value characteristic of aluminium, and it is made optically thin in the infra-red so that e can be adjusted to bring α/e down to unity or less by increasing the thickness of the film. Silicon solar cells have α and e

each near unity and are often used for thermal control, although this leads to large temperature fluctuations unless thermal shielding is employed in the interior of the spacecraft.

5. The Orbital Phase – Radiation

A. SOLAR ULTRAVIOLET LIGHT

Prolonged exposure to solar ultraviolet light causes degradation of mechanical properties of many plastics, particularly materials such as nylon. Others such as polytetra-

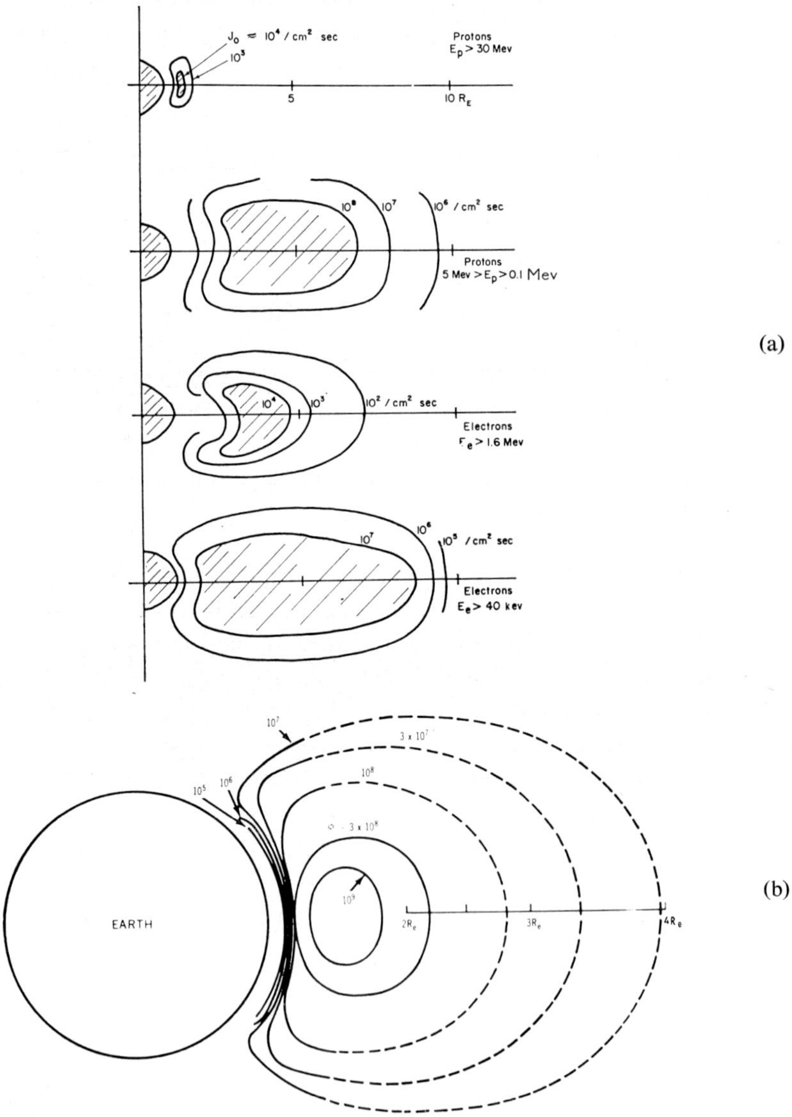

Fig. 4. The flux of charged particles at an earth satellite due to the trapped particles of the Van Allen and artificial radiation belts.

fluoride may satisfactorily withstand prolonged irradiation. Plastics subject to ultraviolet damage can often be protected by a layer of stable paint.

B. CHARGED PARTICLES

The principal flux of charged particles at an earth satellite is, of course, that due to the trapped particles of the Van Allen and artificial radiation belts. Maps of these are shown in Figures 4 (a) and (b). The components most sensitive to radiation damage are solar cells, because of the small mass of shielding material, and transistors. The effects of damage are reduction in the efficiency of conversion of optical to electrical energy in the solar cells due to a reduction in the diffusion length of the carriers, and a reduction of gain and increase of leakage currents in the case of transistors. Resistance to a moderate amount of radiation damage can often be achieved by a careful avoidance of marginal designs. The existence of these will generally be revealed by testing over an extreme range of temperatures and power supply levels.

Solar cell degradation has been extensively studied and typifies well some of the features of transistor damage also. In a silicon junction, carriers produced by red light are formed at a considerable depth in the base material. Whether these contribute to the current generated then depends upon the probability that they will reach the junction before re-combining, i.e. on the diffusion length of the carriers. Cells with a maximum sensitivity for blue light, or those with short diffusion lengths (i.e. those of low efficiency) will thus be most resistant to radiation damage. Moreover, *n-p* cells, because electrons have greater diffusion velocities than positive holes, can be expected to be more radiation resistant than *p-n* cells by a factor of three or four. All these features are observed in practice. Moreover, as the efficiency of the cell is reduced by radiation, its sensitivity to further irradiation diminishes, the relationship between efficiency and dosage being logarithmic. After an initial rapid reduction in efficiency the cell then shows a relatively constant value thereafter. It is possible to use a sufficiently large solar cell array to give satisfactory operation for extended periods with partially degraded cells.

Table II shows the lifetime until significant damage has occurred in the various belts for *n-p* and *p-n* silicon cells with various amounts of shielding. It will be seen

TABLE II

LIFETIMES OF SILICON SOLAR CELLS AGAINST DAMAGE IN THE EARTH'S RADIATION BELTS*

Type of cell	*Amount of Shielding*, gm/cm^2	*Principal damage due to*	*Approximate time in days for significant degradation*
n–p	none	low energy protons	0.1
	1.0	fission electrons in artificial belt	100
p–n	none	low energy protons	0.02
	1.0	fission electrons in artificial belt	20

* Assuming the radiation flux is always that corresponding to the maximum; in practice the lifetime will be greater by an amount which depends on the orbit.

that lifetimes of the order of 1 year can be achieved with shielded *n-p* cells without much damage.

The effects on transistors are more complicated and have been relatively little studied. The most comprehensive investigation was that undertaken for the Telstar satellites (Peck and Schmidt, 1963). It is known that in addition to volume effects, surface effects associated with ionization occur in transistors in gas-filled cases. Some of the effects are irreversible, whilst with others some recovery occurs if the radiation and the operating voltages are removed. The reduction in diffusion length occurs as for a solar cell and results in a reduction in current gain and an increase in leakage current. These effects are least for transistors in which the diffusion length is initially shortest, i.e. those with a good high-frequency response. It would be expected that *n-p-n* transistors would show a similar superiority to *n-p* solar cells, but this does not seem to have been observed in practice. The importance of transistor damage in satellites already launched is not very well known, although several such as Ariel I have shown failure modes in orbit which are strongly suggestive of transistor damage.

6. The Orbital Phase – Meteoritic Particles

Meteors and micrometeorites cover a continuous spectrum in size, the flux of particles of a given mass falling roughly as the mass. The impacts of particles therefore vary from occasional impacts of large particles to relatively frequent impacts of small particles. The first of these may produce penetration of the satellite structure, battery packs, etc., whilst the small ones produce surface erosion. The rate of these processes is extremely variable between low values characteristic of sporadic meteors, to much higher values in regular meteor showers. The effects are rarely important in periods of the order of a year, experiments to detect surface erosion having been unsuccessful. The only known case of satellite damage is that of micrometeorites which punctured the ECHO I balloon, which had an unusually thin skin of large surface area.

References

Hass, G., Drummeter, L. F., and Schach, M.: 1959, *J. Opt. Soc. Amer.* **49**, 918.
Peck, D. S. and Schmid, E. R.: 1963, *Nature* **199**, 741.

Additional References

Clauss, F. J. (ed.): 1960, *Surface Effects on Spacecraft Materials*, Wiley, New York (for properties of materials in the space environment).
Conn, J. H. and Sutton, J. F.: 1961, 'Report on Environment Vibration Test, etc.', NASA Report 321.2 (JC) S-51-11 (for a detailed account of the vibration environment in a typical small satellite).
Cooley, W. C. and Janda, R. J.: 1963, 'Handbook of Space-Radiation Effects on Solar-Cell Power Systems', NASA Report SP-3003 (for an account of solar-cell damage).
Parker, E. N. (ed.): 1963, *Materials for Missiles and Spacecraft*, McGraw-Hill, New York (for properties of materials in the space environment).
Schach, M. and Kidwell, R. E.: 1963, 'Thermodynamics of Spaceflight', NASA, GSFC Report No. X-633-63-50 (for a more detailed account of satellite thermal design).

CHAPTER 22

ROCKET AND SATELLITE SYSTEMS

A. P. WILLMORE

University College, London

The equipment carried in a rocket or satellite to perform a set of specific experiments is generally only a small fraction of the total, the remainder providing ancillary services to the scientific payload, such as data storage and transmission, power generation and distribution, command reception, tracking facilities, stabilization and so on. The experiments must of course operate within the limitations of what it is technically feasible at any time for these services to provide. Evidently an appreciation of the current state of development of the various components of the spacecraft system is essential right from the start of the design of an experiment. The purpose of this chapter is to discuss these limitations as they exist at the time of writing, and to project the probable developments of the next few years. Figure 1 shows a block diagram of the entire system for a small, relatively simple satellite, Ariel I (*NASA report,* 1963). It will be useful to bear this in mind as an illustration for the remainder of the chapter, although techniques are widely divergent between different satellites, and none can be considered typical.

1. Power Generation and Distribution

For a sounding rocket, where the duration of flight and so the total consumption of electrical energy are relatively small, the power source consists invariably of chemical energy stored in primary or secondary cells. High power levels, of at least a hundred watts, can readily be accommodated in this way. For the longer life typical of a satellite, this arrangement is far less satisfactory either from the standpoint of total weight or of operating convenience, and some kind of electrical generator is used. Normally at mean power levels up to a few hundred watts, this consists of a solar cell supply.

Two types of cell are currently in use: *n*–on–*p* cells, which consist of a thin layer of *n*-type silicon semi-conductor on a *p*-type base, and *p* on *n* cells in which the *p*- and *n*-type semi-conductors are reversed. In either case, the thickness of the semi-conductor layer is only a few microns, so that sunlight can penetrate through the junction into the base material. To reduce corpuscular radiation damage, to which the cells are very susceptible (see Chapter 21, Section 5B) they are covered with an absorbing layer of glass, fused silica or sapphire about 1 mm or less in thickness.

The cell converts solar energy into electricity with an efficiency of about 11% – it may be as high as 13% but high efficiency cells are particularly sensitive to radiation damage. The output from 1 square metre of cells arranged normal to the sun is thus about 160 watts. If the solar cells can be maintained in this orientation, and the orbit

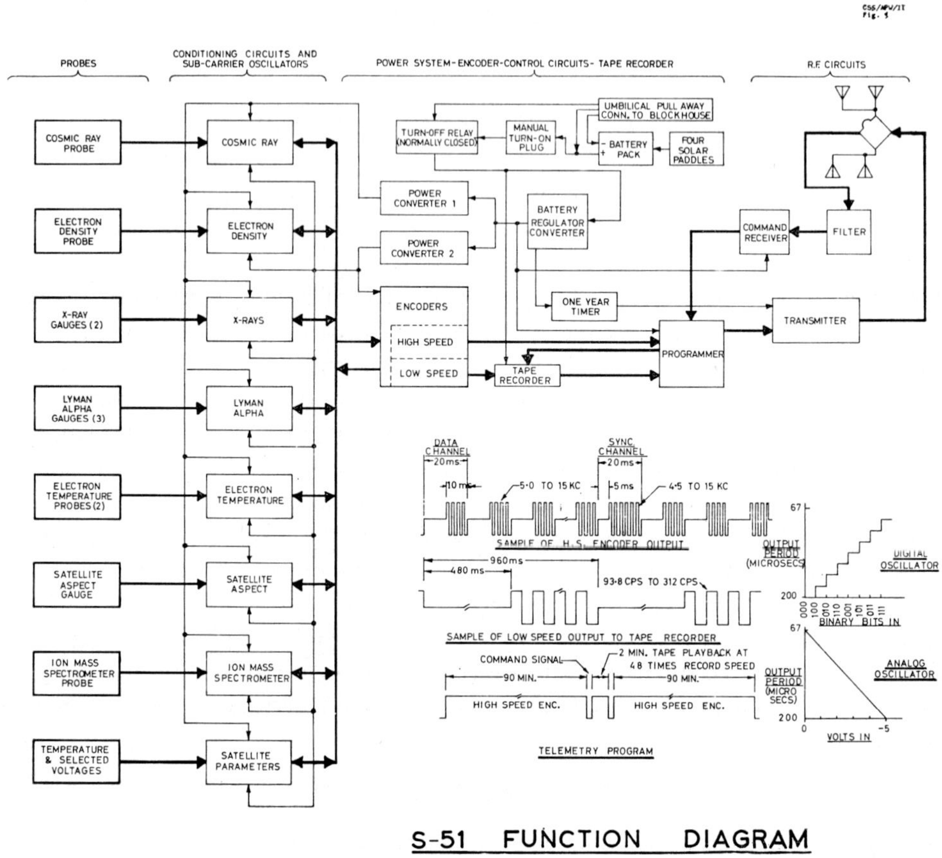

Fig. 1. Function Diagram of the entire system of Ariel I.

is such that the satellite is never in the earth's shadow, this is the available power output. On the other hand, a near-earth orbit may be shadowed for up to 40% of the time and in a typical small unstabilized satellite the effective area of the cells presented towards the sun may in some orientations fall to as little as one-sixth of the total area. In this case the power output may be as low as 11 watts/m^2. Since 1 m^2 is about the right order for the area of a typical solar cell array, we see that the power available may vary from a few hundred watts for a large, stabilized satellite, to only twenty watts or less for a small, unstabilized satellite. The cells are voltaic, and such an array may be connected up to form a battery with an emf of about 20 volts and an internal impedance of one ohm or less.

This power is of course available only when the satellite is sunlit. If operation in the earth's shadow is required, then temporary power storage must be provided in secondary chemical cells, generally in nickel-cadmium cells with sealed steel cases, on account of their relatively high efficiency at low temperatures. These cells present a problem when the orbit becomes entirely sunlit for a period of several days, since they soon become fully charged and the energy subsequently put into them is then converted into heat. As a result the cells tend to overheat to a damaging extent. To obviate this, a charging controller which reduces the charging current as the temperature either of the satellite or of the battery packs rises, is generally used. A more elegant and satisfactory system would be to use a current integrator to determine the total charge drawn from the batteries, and re-charge until this was zero, but such a system has not been used.

Another precaution must also be taken. In the event of the mean load power exceeding the mean output of the solar cells, due to some fault developing either in the array or in the load, the batteries would steadily discharge and might eventually reverse polarity, when further charging would destroy them. The battery voltage is therefore sensed and the load removed if it falls too low, either automatically by means of an under-voltage controller or by ground command, for a sufficient period to allow for re-charging. Also, because the reliability of a battery pack is often less than that of other components of the satellite system, redundant batteries with a change-over mechanism operated either automatically or on ground command, may be fitted.

The potential difference across the battery terminals may vary by several volts from the condition where almost fully-charged batteries are still under discharge, to that where partially-discharged batteries are discharging to supply the load, and this variation will probably necessitate a voltage-regulating converter. The output from this will supply *dc*-to-*dc* converters which provide the spacecraft system with the correct operating voltages. By this time, at least 25% of the mean power generated by the solar cell array is likely to have been dissipated. The remainder must be shared by the spacecraft systems – particularly the data storage and transmission system which is necessarily power-consuming – and the experiments. The portion available to an experiment may vary from 30 watts for an exceptionally large satellite, to a much more typical value of 0.5 to 1 watt.

It may be asked whether other types of generator would not provide larger amounts of useful power. In fact, small nuclear or radio-isotope power generators have almost developed to the stage where they are competitive with solar cell arrays even in the power range near 100 watts and such power supplies may prove of value in the next few years.

2. Data Storage and Transmission Systems

The simplest, and in some respects, one of the best media for spacecraft data storage is the photographic emulsion. On the other hand, unless elaborate on-board processing is undertaken (for example, the development and re-transmission of photographs of the moon's surface in Lunik III), recovery of the film after exposure is required which limits its usefulness in general to the sounding rocket. Most sounding rockets and many satellites do not, in fact, contain any other data storage than that which may happen to be provided, in the form of binary registers, within an experiment, but rely rather on the direct transmission of data as it is gathered. This practice is very wasteful for a continuously-operating satellite, since a considerable fraction, perhaps 75%, of the time when the data are being gathered the satellite is not within receiving range of a telemetry ground station. If data storage is provided, it generally takes the form of a tape recorder with a large replay/record speed ratio, so that data can be stored over the duration of one or more orbits, but re-played whilst within range of a single receiving station. In some cases, core storage of the kind used in digital computers shows some advantage and it is for example planned to use dual 4096 word core stores (each word containing 25 binary digits) on the Orbiting Astronomical Observatory satellites.

From the standpoint of telemetry power, it makes no difference whether the data is continuously transmitted, or stored and transmitted at a high rate, provided the average power is computed over a long enough period. On the other hand, if the stored data is backed up by a continuous transmission in real-time, as occurred on Ariel I, either two telemeters must be used (of wide and narrow bandwidths) with a consequent doubling of the mean power requirement, or the data storage system must be restricted to a very low data rate. On Ariel I, where the latter course was adopted, the tape recorder could be utilised satisfactorily only by including in each experiment a certain amount of in-board data processing (STILTZ, 1961).

Here we shall attempt no more than a brief survey of a diverse and complex subject.

The rate at which information is transferred by a radio telemetering system (e.g. the number of measurements of given accuracy which can be recorded at the ground with essentially complete reliability) depends on the following factors:

(a) the transmitter power;
(b) the distance between transmitter and receiver;
(c) the directivity of the transmitting and receiving aerials;
(d) the intensity of background noise;
(e) the wavelength of the radio carrier wave.

(a) and (b) are fixed respectively by the design of the satellite and its orbital parameters; (d) is ultimately limited either by the strength of celestial radio sources or by technical considerations discussed below, so that only (c) and (e) remain at our disposal in optimizing the system. In consequence, once the optimum wavelength has been selected and the maximum usable directivity of the aerial system determined, the rate of information transfer is largely determined by the two factors not within the control of the telemetry engineer, transmitter power and transmission range. Of course, both the means by which the information is coded for transmission, and the detailed design of the system have some influence on the extent to which the theoretical possibilities can actually be realized in practice.

Unless the satellite is stabilized in geocentric co-ordinates (so that the transmitting aerial is always directed towards the earth) the satellite aerial is generally required to have a nearly omni-directional pattern. The directivity of the receiving aerial depends mainly on the requirement for following the satellite so that it remains within the aerial beamwidth. Aerial gains of about 20 db are normal for earth satellites, although at greater ranges the low angular velocity of the spacecraft may make much higher gains usable.

Figure 2 shows that the two principal components of the receiver background noise are at low frequencies the emission from celestial radio sources, and at high frequencies noise generated within the receiving system itself. The sky noise diminishes with increasing frequency whilst the receiver noise increases, so that in the frequency range between 100 Mc/sec and 1000 Mc/sec the total noise is a minimum. This is

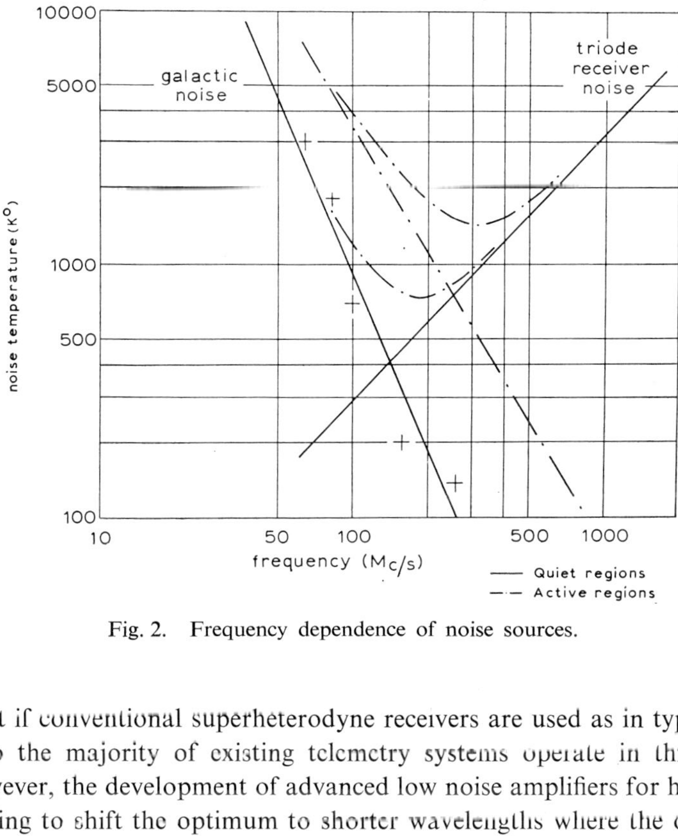

Fig. 2. Frequency dependence of noise sources.

true at least if conventional superheterodyne receivers are used as in typical current practice, so the majority of existing telemetry systems operate in this frequency range. However, the development of advanced low noise amplifiers for high frequencies is tending to shift the optimum to shorter wavelengths where the overall noise level is then lower. This, then, is a region in which significant improvements can be anticipated in the next few years.

It has been remarked that the method of data coding has some influence on the information transmission rate. If amplitude modulation of the carrier is employed, the carrier amplitude being an analogue of the quantity telemetered, in the presence of noise such that the signal-to-noise power ratio is N, the accuracy of measurement becomes $N^{-1/2}$. Thus, for a measuring accuracy of 1% a very good signal-to-noise ratio is necessary. It is however found that if the carrier *frequency* is made to vary as an analogue of the data, the same accuracy can be obtained (with the same time available to make the measurement) with a signal-to-noise power ratio of only 10–30. For this reason all telemetry systems employ frequency, phase or pulse-code modulation either of the carrier wave, or of one or more sub-carriers which then modulate the radio wave. Each of these systems is then capable of giving a roughly similar sensitivity at the threshold of reception.

The telemetry system is invariably required to transmit data from more than one source or channel, so that a form of multiplexing is required. The two common systems are well exemplified by the I.R.I.G. system which is widely used in the U.S., and the U.K. 465 Mc/sec, 24-channel system. The I.R.I.G. system employs frequency multiplexing, the data channels being transmitted on a series of sub-carriers in different frequency bands which simultaneously modulate the radio carrier. The 465 Mc/sec system employs a single sub-carrier frequency which is time-multiplexed by connecting it sequentially to modulation from up to 23 data channels in turn by a motor-driven commutator. The 24th position of the commutator is used to transmit a synchronising signal. Large bandwidths of between a few c/sec and at least several hundred c/sec can be obtained with the I.R.I.G. system, though at the expense of possible cross-talk between channels, whilst the mechanical switch limits the 465 Mc/sec system to about 80 samples/sec on each channel although the accuracy and linearity are on the other hand excellent.

These two systems are essentially typical of all those employed in spacecraft, though at high data rates or in satellites, mechanical may be replaced by electronic commutation. The data rates are best expressed for uniformity in equivalent binary digits, reckoning a sample of 1% accuracy as equivalent to a seven-bit binary number. A typical sounding rocket telemetry, with a transmitter power of 5–10 watts and a range of 400–500 km will transmit 15000–25000 bits/sec, whilst an earth satellite system with a transmitter power from 0.25 watts upwards, and a range of 1000 km to 100000 km, will transmit from 10 to 100000 bits/sec, with a typical value for a small satellite in a near-earth orbit in the range 1000–10000 bits/sec.

We have not so far distinguished between the transmission of analogue (i.e. continuously-varying) or digital (i.e. binary-number) information, although the results of experimental observations may be of either character. The choice of either an analogue or a digital telemetry system is mainly a matter of convenience, particularly as the data from an entire spacecraft will normally be of both kinds, so that whatever type of telemetry system is chosen means for encoding information of the other type must be employed. However, from the point of view of the experimenter, there is no doubt that the combination of a digital telemetry system with a ferrite core data storage system enables the most sophisticated kind of experiment to be carried out, including building-in a relatively complex programme of observations.

3. Data Reduction

A striking feature of a satellite experiment is the very large amount of experimental data which is gathered, amounting in a normal satellite lifetime to hundreds of millions of telemetry samples. To process these by hand is entirely impractical, though few data processing systems which completely eliminate hand reduction have so far been established – a diagram of one system which does so will be found in the seminar entitled 'The Ionospheric Measurements from Ariel I'.* It may seem strange that a section on data reduction should find its way into a chapter on spacecraft systems; it does so to emphasize the point that the entire processing chain, from the beginning of the experimental apparatus to the end of the data reduction system ought to be regarded as the single entity that it actually is. This means to say that the design of an experiment must be carried out with the requirements of the data reduction clearly in mind. In fact it is often the case that little thought is given to the details of the data reduction until after launch, a practice which may add significantly to the already large cost of the computer operations.

In general, telemetry data will be recorded on magnetic tape at a number of stations distributed over the globe, such as the U.S. Minitrack network, supplemented by the two British stations at Singapore and the Falkland Islands. These tapes will then be brought for processing to a central location. The processing required is in general highly specific to an experiment or set of experiments so that it is advantageous that the first stage of processing should prepare the data for entry to a general-purpose computer (generally a digital computer) to avoid a large investment in specialized processing equipment. Accordingly, the first processing operation must convert the data, if analogue, into digital form and then prepare it, on magnetic tape, in a suitable format for a large, fast computer. The subsequent operations are best exemplified by referring to the seminar mentioned above.

References

N.A.S.A.: 1963, Report SP-43, 'Ariel I – The First International Satellite'.
STILTE, H. L. (ed.): 1961, *Aerospace Telemetry*, Prentice Hall, London.

* See Chapter 38, p. 466.

CHAPTER 23

SOME EXPERIMENTAL TECHNIQUES OF SPACE RESEARCH – IONOSPHERIC MEASUREMENTS

A. P. WILLMORE
University College, London

Despite the relatively short period, still less than two decades, since the inception of space research, a wide variety of techniques has been used in experiments conducted from rockets and satellites. The great majority of these have been adapted, often with only trifling modifications, from well-known laboratory experiments. In the course of this chapter and the next some of these techniques are briefly described, with the intention rather of providing typical examples than a comprehensive description. Other descriptions of important experiments will be found elsewhere in this book, particularly in the Seminar by Professor R. L. F. Boyd, (Chapter 38, p. 453) and in Chapter 2 by Professor C. de Jager (p. 26).

1. Effect on the Ionosphere of a Moving Satellite

The orbital velocity of a satellite is roughly one-tenth of the mean thermal velocity of the ionospheric electrons, but ten times the mean positive ion velocity. Thus the satellite motion is hypersonic and a substantial variation of the density of the medium near the satellite from the ambient value might be anticipated. This naturally raises the question of the validity of measurements of density and temperature made from a satellite, particularly in the case where the medium close to the satellite is sampled, as in measurements by Langmuir probes. The problem is by no means simple, but it has been studied to some extent both theoretically and experimentally (in the Explorer VIII and Ariel I satellites). A very illuminating account has been given by GUREVICH (1962) which whilst undoubtedly over-simplifying the problem nonetheless reproduces the main features of the measurements.

To an observer seated on the satellite, the electron flux will appear almost isotropic – only slightly greater on the forward side than elsewhere. The positive ions, on the other hand, will appear as a beam nearly homogeneous in energy. (It will be seen however that the relatively small thermal velocities of the ions are very important in the development of the satellite wake). The Debye length in the ionosphere is always short – near the F-region maximum it will be about 0.2 cm so that the positive ion and electron densities must be nearly the same at all points, except in a thin sheath round the satellite or where the charge density is greatly reduced relative to the ambient value. On the other hand the mean free paths for collisions between the charged, or the charged with neutral particles are very long so that collisions can be neglected.

It is found that the density distribution about the satellite is largely determined by the positive ions, the reason being that these have a relatively large energy (1 eV to 5 eV, compared with 0.1 eV to 0.2 eV for the electrons) in a frame of reference moving with the satellite, so that they are little affected by the electric field round the satellite. The electron density distribution then adjusts itself to the positive ion distribution in a way which is consistent with the requirement of quasi-neutrality of the plasma. Since the satellite rapidly acquires a negative charge from the mobile electrons, it takes up a mean potential equivalent to a few times the electron temperature, negative with respect to the ionosphere. The satellite presents in consequence

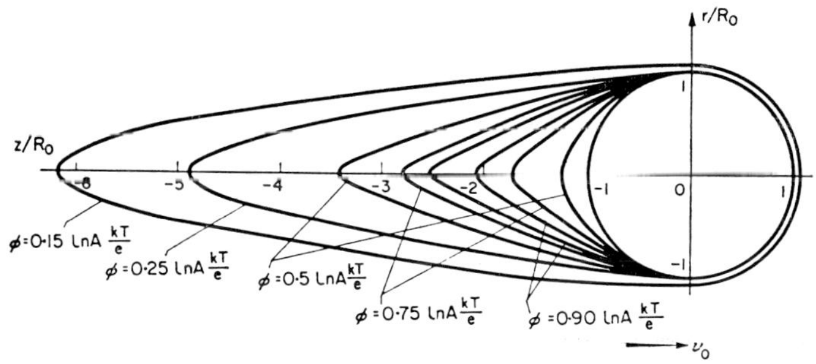

Fig. 1. Equipotential surfaces near a spherical body with a metal surface absorbing all the incident particles.

an absorbing surface to the positive ions and those which are intercepted by the forward motion of the satellite are absorbed, leaving an evacuated wake behind the satellite. Into this depleted region move electrons, charging it negatively, and, at a rate largely governed by their thermal velocities, the positive ions.

The result is that on the forward side of the satellite the charged particle density is very close to the ambient value, whilst in the wake it falls to a minimum value which is two or more orders of magnitude lower. Figure 1 shows the potential distribution calculated by Gurevich. The density can readily be obtained from Boltzmann's equation

$$n_e(a) = n_e(o) \exp - \frac{eV}{kT_e}$$

where $n_e(o)$ is the ambient electron density; T_e is the electron temperature, and V is the potential at a. It will be seen that the wake region is several satellite radii in length (corresponding approximately to the ratio of the satellite velocity to the mean thermal velocity of the positive ions). By mounting probing electrodes on booms one or several metres in length, accurate measurements of ionospheric density can be made, except possibly when the boom lies along the wake. This will also enable accurate measurements of the ambient temperature to be made.

2. Measurements with Langmuir Probes

These are electrodes of arbitrary shape and size whose voltage current characteristic is measured. From this a good deal of ionospheric information can be obtained, in ways which are discussed in the Seminar by R. L. F. Boyd. In particular measurements of the density and temperatures of the electrons and of the positive ions can be made, and some information can even be obtained about the mass distribution of the positive ions. Generally speaking, the analysis of the probe characteristic is carried out on the assumption that the mean free path is much longer than the probe dimensions, which restricts the measurements to the region above 90 or 100 km. However measurements have also been made in the D-region (BOURDEAU *et al.*, 1959), which is difficult to study by any measuring technique, by making use of the fact that when the mean free path is very short, the current to the probe is governed by the mobility of the charged particles in the atmosphere. This is a most promising technique for the future.

3. Mass Spectrometry in the Ionosphere

The simplest form of mass spectrometer that can be used in the ionosphere is a Langmuir probe, operated in such a way that the energy distribution of the positive ions is obtained. Such a device has two important advantages, that the ion temperature can also be measured conveniently and that the efficiency of collection is very unlikely to depend significantly on the ion mass, so that the relative composition of the ionization is obtained accurately. On the other hand the mass peaks are broadened considerably by the ion thermal motions so that the mass resolution is poor, sufficient only to discriminate between the major ionic constituents, which at satellite altitudes are H^+, He^+ and O^+.

For measurements from rockets, or where a better resolution is necessary, some form of mass spectrometer must be used. Generally this has been based on the Bennett radio-frequency mass spectrometer (BENNETT, 1950) which has been extensively used in the laboratory. The principles of the instrument can be gathered from Figure 2. The ions are accelerated by an entrance grid to a mean energy V_0 which is large compared with the thermal energies. Then they pass through a group of three grids which are maintained at the mean potential V_0. Small radio-frequency potentials are applied in the same phase to the first and third grids. An ion passing through in a time short compared with the radio frequency cycle will lose the energy gained in the first space whilst passing through the second, unless it happens to pass the central grid when the electric field is passing through zero, in which case the mean electric field and also the energy gained are very small. Similarly, an ion whose transit time is many r.f. cycles will gain little energy because of the periodic reversals of the electric field. However, an ion whose transit time is comparable with, but shorter than, the r.f. field and which enters the system at a phase near that at which the radio frequency voltage is zero as it passes the centre, will gain considerably in energy. In fact, at the optimum phase, an ion whose transit time is 0.74 r.f. cycles gains energy 1.45v, v being the amplitude of the r.f. potential.

The grids are followed by a drift space and then a second group of three grids identical to the first. The length of the drift space relative to the grid spacing is so adjusted that an ion whose velocity corresponds to the maximum energy gain will cross the drift space in an integral number L of rf cycles, and will thus gain further

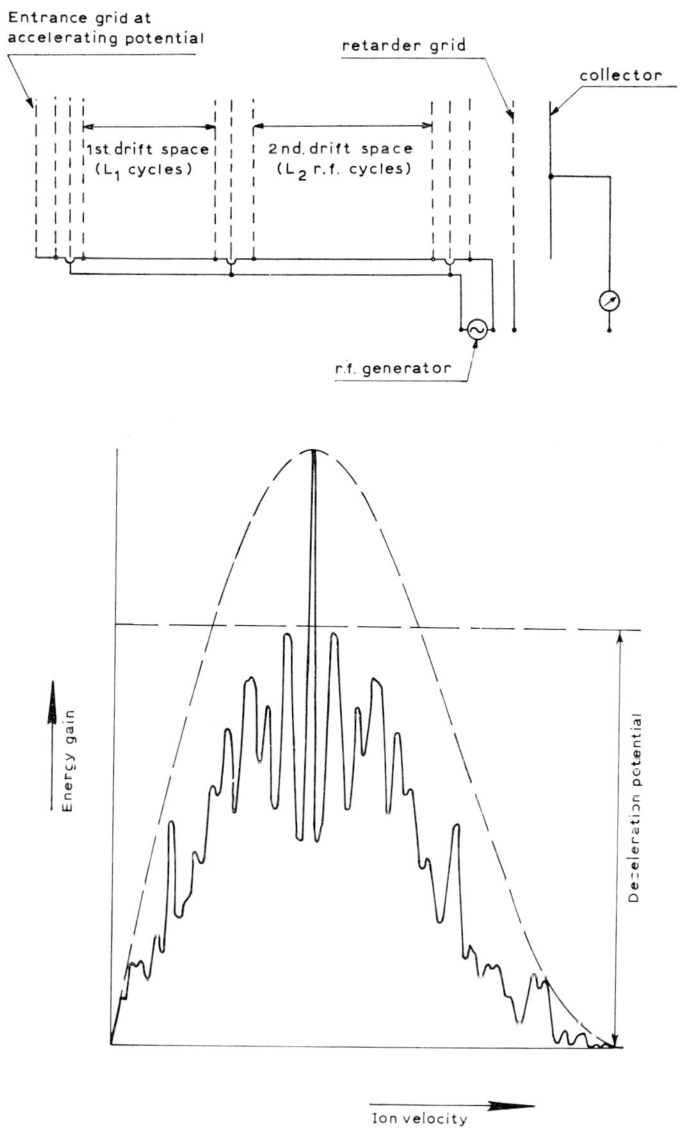

Fig. 2. Scheme and energy gain curve of radio-frequency mass spectrometer with $L_1 = 9$ cycles and $L_2 = 7$ cycles.

energy 1.45v at the second set of grids. Ions whose velocity is $(L_1 \pm 1)/L_1$ times the optimum will also gain energy at each grid, but the energy gain is not quite so great since they do not pass through the radio-frequency fields in the optimum transit time; on the other hand, ions with velocities $(L_1 \pm 0.5)/L_1$ will lose at the second group of grids all the energy gained at the first.

A second drift space now follows the first. The length, corresponding to L_2 rf cycles for the optimum velocity, is so chosen that the subsidiary maxima at $L_1 \pm 1$ cycle in the first drift space approximately coincide with the minima at $L_2 \pm \frac{1}{2}$ cycle in the second. This substantially depresses the secondary maxima with respect to the main peak. The energy gain curves for $L_1=9$, $L_2=7$ are shown in Figure 2b. The minima on either side of the main peak correspond to ions whose drift time along the entire instrument is $((L_1+L_2) \pm 0.5)$ rf cycle. This is thus the order of magnitude of the mass resolution of the instrument. Unless a third drift space is employed, the ions now pass through decelerating grid so biased that only particles whose energy gain is near the maximum can reach the collector C and contribute to the collector current. The potential of this grid is normally adjusted to suppress the highest secondary peaks or "harmonics" (although these have been used to assist in an unambiguous identification of the ionic mass). A wide variety of instruments with two, three or even four drift spaces has been used for different applications, but two drift spaces with $L_1=7$, $L_2=5$ is more typical of rocket instruments. This is sufficient to give a mass resolution of about 1 in 40.

The principal advantages of the Bennett spectrometer for space applications are its high sensitivity, since the large entrance aperture offsets the absorption due to many grids, and the avoidance of magnetic fields which are both difficult to generate in a rocket or satellite and which may lead to a dependence of efficiency on mass due to stray magnetic fields near the entrance. On the other hand, the path length in the instrument is necessarily long, of the order of 20 cm so that the requirement of a long mean free path limits its use to the region above 100 km. Moreover the adjustment of the various voltages on the grids must be done with care, since the energy gain is normally made small compared with the mean energy V_0.

4. Radio-Frequency Probes (Aerial Impedance Measurements)

Last amongst the techniques which depend on measurements of the ionosphere in the immediate proximity of the satellite, we have those in which the dielectric constant of the ionosphere is determined. If we neglect both the geomagnetic field and collisions, the dielectric constant of the plasma is given by

$$\varepsilon = 1 - \omega_p^2/\omega^2 \tag{1}$$

where $\omega_p=(Ne/\pi m)^{1/2}$ is the plasma angular frequency; (N is the electron density, and e/m the charge to mass ratio of the electron, ω is the angular wave frequency). (See, for example, SPITZER, 1956).

Thus, by measuring ε the electron density can be found. It is in principle possible in various ways to deduce other parameters of the plasma than this, but at the present time rf probes share with radio-propagation techniques the limitation (as compared with Langmuir probes) that the electron and ion temperatures cannot be measured.

The experiment can be conducted in either of two ways. Either the probing frequency ω can be varied and the point at which $\varepsilon = 0$ detected, or a fixed value of ω greater than the maximum value of ω_p anticipated can be used, and ε measured at this frequency. Since frequency is a quantity which is readily measured with accuracy, the first method would at first sight seem preferable, but it in practice requires corrections to be made for the part of the circuit not immersed in the ionosphere, which are difficult to make. Moreover there are often strong reasons discussed below, for wishing to operate at a much higher frequency than the plasma frequency.

The most accurate version of this type of experiment used to date is probably that due to Sayers. Two coarse circular grids, 10 cm in diameter and 10 cm apart, form a parallel plate capacitor whose free-space capacitance can be accurately measured or calculated. The capacitor forms one arm of a rf capacitance bridge which is balanced prior to flight. The out-of-balance voltage produced by the presence of ionospheric plasma in the capacitor is amplified and resolved into in-phase and quadrature components. Provided the dielectric constant of the plasma is near unity, a condition which is satisfied at sufficiently high operating frequency, the two components are proportional to the change in real and imaginary components of ε. The change in the real component is directly proportional to N whilst the loss component, negligible except in the D- and lower E-regions, is proportional to collision frequency. It is necessary for the plasma between the grids to have the ambient density. To ensure this a sweep potential is applied to both grids of sufficient amplitude to ensure that space potential is reached; this point can be recognized since it corresponds to maximum electron density between the plates. The frequency used in this experiment is 10–30 Mc/sec.

When the wave frequency is high compared with the plasma, gyration and collision frequencies, and when the geometric form of the capacitor is simple, the analysis of this experiment presents no difficulties. However, if the frequency used is not high, the simple form (1) of the expression for ε cannot be used and instead an allowance must be made for the effect of collisions and the geomagnetic field. The result is a complicated expression for the capacitance of the system which cannot usually be evaluated in closed form and which depends on the orientation of the capacitor with respect to the geomagnetic field. If the entire capacitor is not at the same potential, the case with some configurations that have been used, the variation of electron density with position is an additional complicating factor. All of these problems can be overcome by the use of a sufficiently high radio-frequency and a sweep potential, whilst with careful design retaining the ability to measure charge densities as low as 10^3 cc^{-1}. The condition that the probing frequency be larger than the collision frequency may then be satisfied into the D-region.

5. Radio Frequency Propagation Measurements

Closely related to the technique described in the last paragraph is a wide variety of methods of studying the propagation of radiowaves in order to deduce the value, at various locations, of the ionospheric refractive index at a known wave frequency. One of the most powerful forms of the technique is the top-side sounder experiment carried by Alouette I. The satellite carries a pulse transmitter and, so long as the wave frequency exceeds the local value of the plasma frequency, but not that of the F2 maximum, reflections are received from the upper side of the ionosphere. The satellite transmits to the ground ionograms which are precisely analogous to those produced by a ground-based vertical incidence sounder, but which refer to the top-side ionosphere. However, the majority of propagation experiments employ a path from rocket or satellite to ground, though propagation in either direction may be used.

In the past, propagation techniques have provided the most accurate measurements of electron density over the wide range encountered in the ionosphere. They are rapidly being overtaken in these respects by probe techniques which are capable of giving much additional information. However an extremely powerful feature is the ability to make measurements when the mean free path of the electrons is of the order of 10 cm or less, i.e. in the D-region. Especially in the study of high latitude phenomena, where under disturbed conditions strong ionization may be produced at low altitudes, radio propagation measurements will retain great importance. Fortunately the most serious sources of error, which are often important in the F-region, can often be made negligibly small for D-region observations. An additional feature is that the measurements are relatively insensitive to local disturbances of the plasma by the spacecraft, so long as these are only slowly changing.

There are four parameters of a radiowave that can be observed – frequency or phase, amplitude, direction of propagation and plane of polarization. Each of these may be affected by propagation in the ionosphere, so that observations of each may provide ionospheric information and in fact each forms the basis of at least one experimental technique which has actually been used. The problem is essentially that of determining, at a large number of points between the ground observing station and the spacecraft, the value of the ionospheric refractive index. Since this is proportional to $\varepsilon^{1/2}$, we see from Equation 1 above that it is a function of the electron density. If, however, we allow for the effect of collisions between the electrons and other atmospheric constituents, an imaginary component of ε, which is proportional to the collision frequency, results. If the wave is to be propagated over significant distances, it is necessary for the wave frequency ω to be somewhat greater than the collision frequency, a condition which can be satisfied for all altitudes in the ionosphere by using a sufficiently high frequency. The effect of collisions is that the electron gas attenuates the wave, an effect which is important in passing through the D- and E-regions. The presence of the geomagnetic field in the plasma causes it to become doubly-refracting. Generally the two rays which are propagated can each be used to obtain a measure of electron density, thus providing a valuable cross-check, though in the case of observations of the rotation of the plane of polarization (the Faraday effect) the measurement consists in comparing the phase path of the two rays, and this check is absent.

Of the many variants of the different techniques which are available, we shall limit ourselves to describing only two which illustrate the underlying similarities of them all.

A. THE SEDDON EXPERIMENT

This technique, originally devised in the U.S.A. by J. C. Seddon, is one of the most widely used. It can be regarded as a comparison of the Doppler shifts of two harmonically-related carrier waves due to the spacecraft velocity.

Let us consider an ionosphere which is horizontally stratified, containing no magnetic field, and through which a spacecraft is moving vertically. In this case the transmission paths to a ground station vertically beneath the spacecraft are straight lines parallel to the rocket velocity. The spacecraft carries two transmitters, the first operating on as low a frequency as possible but high enough to penetrate the ionosphere from the greatest height where measurements are to be made – say 12 Mc/sec. The second operates on a high harmonic – e.g. the sixth – of this, 72 Mc/sec. For the second frequency f_2, the refractive index of the ionosphere is always near unity, so the Doppler shift is:

$$\Delta f_2 = f_2 \cdot \frac{v_r}{c_0}$$

$$= 6 f_1 \cdot \frac{v_r}{c_0}$$

v_r is the rocket velocity, (c_0 is velocity of light in vacuo.)
However, for the low frequency wave f_1 we have

$$\Delta f_1 = f_1 \frac{v_r}{c_0 \{1 - (\omega_p^2/(2\pi f_1))^2\}^{1/2}}.$$

On the ground, both both transmissions are received, and the difference in frequency, f_B, between the higher and the sixth harmonic of the lower frequency waves is measured.

$$f_B = f_2 - \Delta f_2 - 6 f_1 + 6 \Delta f_1$$
$$= -\frac{6 f_1 \cdot v_r}{c_0} [1 - \{1 - (\omega_p^2/(2\pi f_1)^2\}^{1/2}] \sim \frac{3 v_r}{4\pi 2 c_0 f_1} \cdot \omega_p^2 .$$

Since v_r is known from measurements of the rocket trajectory, the plasma frequency and so the electron density can be obtained.

In practice, the restrictions on the spacecraft path and the geomagnetic field imposed in the last paragraph are not satisfied. In consequence measurements are made on both the ordinary and extra-ordinary rays and an allowance must then be made for the differing refraction of all four waves which causes them all to follow different paths from transmitter to receiver. This is because we really measure the difference in the rate of change of phase path for the low frequency and the reference waves. This depends on the integrated electron density along the wave path; and from the changes in this quantity as the spacecraft moves we deduce the vertical distribution of N, assuming the ionization to be disposed in horizontal layers. Provided these calculations are accurately performed, it is in the last assumption that the principal errors arise.

Obviously, the errors will be least when the spacecraft moves nearly vertically and does not rise to great altitudes, conditions which are satisfied in the case of rocket observations on the D- and lower E-regions. However, in the case of transmissions from a satellite, where the path may move horizontally by hundreds of kilometres, far greater errors can be expected. Efforts have been made to reduce these, for example, by estimating the horizontal density gradients by means of spaced ionosondes, but it appears likely that the method is best used to estimate the average value of some F-region parameter, such as the thickness of an equivalent parabolic layer.

B. MEASUREMENTS OF FARADAY ROTATIONS

In this experiment, instead of comparing the changes of phase path for waves of two different frequencies, the comparison is made for the ordinary and extra-ordinary rays. This simplifies the experimental arrangement considerably. The principle of the method can most easily be understood by considering the propagation of a plane wave through an infinite slab of homogeneous plasma, in which there is a magnetic field B directed perpendicularly to the sides of the slab. The incident plane wave can be resolved into two oppositely-rotating circularly polarised waves. In the medium, these become the ordinary and extra-ordinary waves. Because the electrons have a preferred direction of rotation about the magnetic field, their interactions with the two waves differ and so the waves have different phase velocities. Thus in passing through the slab, a phase difference accumulates and when the two waves re-combine the plane of polarisation has been rotated when compared with the incident wave. In fact, for the conditions postulated here, we have for the refractive indices of the two waves

$$\mu^2 = 1 - \frac{\omega_p^2}{\omega(\omega \pm \omega_g)}$$

where $\omega_g = 2\pi eB/mc$ is the gyrofrequency of the electrons. If $\omega \gg \omega_g$, $\omega \gg 2\omega_p$, the difference in the two indices is

$$\Delta\mu \sim \frac{\omega_p^2}{2^2}\left(1 + \frac{\omega_g}{\omega} - 1 - \frac{\omega_g}{\omega}\right) = \frac{\omega_p^2 \cdot \omega_g}{\omega^3}.$$

The travel time from transmitter to receiver is, for either wave,

$$T = \int_{\text{path}} \frac{\mu}{c_0} \mathrm{d}s$$

where ds is an element of the ray path and the integral is taken over the entire path. If the difference in travel times is ΔT for the ordinary and extra-ordinary rays, then the phase difference is $\omega\Delta T$ and the rotation of the plane of polarization is

$$\phi = \tfrac{1}{2}\omega\Delta T$$

If we assume that the two rays travel along the same path,

$$\phi = \frac{\omega}{2}\int\limits_{\text{path}} \frac{1}{c_0}\frac{\omega_p^2\omega_g}{\omega^3}\,\mathrm{d}s$$

$$= \frac{1}{2c_0\omega^2}\int\limits_{\text{path}} \omega_p^2\omega_g\,\mathrm{d}s\,.$$

By measuring the rate of change of ϕ with a linearly polarised aerial, the variation of $\omega_p^2\cdot\omega_g$ along the path is obtained. (In general, an allowance must also be made for the angle between the ray path and the geomagnetic field.) From this the density variation along the ray path is found and so, by assuming horizontal stratification of the ionosphere, the vertical density distribution. As with the Seddon experiment, the assumption that the two rays travel the same path is not essential, having been made here only to simplify the calculation, but the asumptions of horizontal stratification is inherent in the method, and the same observations regarding the resultant errors are applicable in both cases.

The rotation of the plane of polarization is observed by means of a linear aerial, since each rotation by 2π causes the current in the aerial to go through a maximum and a minimum. The time intervals between minima then give the rotation rates. If, as is commonly the case, the transmitting aerial is rotating, then the spin rate must of course be added to, or subtracted from, the observed rotation rate. In the case of a sounding rocket experiment, where the height intervals corresponding to rotations of 2π may be rather large, the use of a high spin rate of perhaps 5 rps enables better height resolution to be obtained. The spin rate of the rocket can easily be determined with high accuracy, using for example, detectors for the sun or the geomagnetic field.

From observations on a satellite it is possible, in addition to determining the density variations over the height range covered by the satellite, to obtain some information on the total electron content of the ionosphere below the satellite. In general, the number of rotations of the plane of polarisation is ambiguous by multiples of 2π. To find the total number of rotations, a point on the trajectory can sometimes be found where the transmission path is essentially normal to the magnetic field. At such a point, the rotation of the plane of polarisation is zero. A method of more general application is to represent the density distribution by means of adjustable parameters, and vary these to reproduce the observed variation of the plane of polarisation during the satellite passage.

For a more detailed account of the propagation of radio waves in a plasma, the reader should consult SPITZER (1956), whilst a discussion of the different radio propagation studies which can be made will be found in AARONS (1963).

References

AARONS, S. (ed.): 1963, *Radio Astronomical and Satellite Studies of the Ionosphere*, North Holland Publishing Co., Amsterdam.
BENNETT, W. H.: 1950, *J. Appl. Phys.* **21**, 143.
BOURDEAU, R. E., WHIPPLE, E. C., and CLARK, J. F.: 1959, *J. Geophys. Res.* **64**, 1363.
GUREVICH, A. V.: 1962, *Plan. Space Sci.* **9**, 321.
SPITZER, L.: 1963, *Physics of Fully Ionized Gases*, Interscience, New York.

General References

MOLOZZI, A. R. (ed.): 1963, *Instrumentation of the Ionospheric Sounder*, etc., Proceedings of the COSPAR Symposium, Warsaw (for a description of the Alouette topside sounder).
The Proceedings of the Royal Society Conference, May 1963, in course of publication (for an account of the Ariel I experiments).

CHAPTER 24

SOME EXPERIMENTAL TECHNIQUES OF SPACE RESEARCH – MEASUREMENTS OF SOLAR ELECTROMAGNETIC RADIATION

A. P. WILLMORE

University College, London

Whilst the corpuscular emission of the sun is undoubtedly of great importance in the ionosphere, principally under disturbed conditions but also under quiet conditions (see for example Chapter 39 by A. P. Willmore), the major influence of the sun occurs as a result of its emission of electromagnetic radiation. Also for the study of the sun itself this is still the most powerful tool available. For wavelengths below 3000 Å, atmospheric absorption is significant throughout the ultraviolet and X-ray spectrum down to the normal shortwave limit of the solar spectrum near 5 Å and far beyond. The radiation between 2000 Å and 3000 Å is mostly absorbed in dissociating ozone, and can be observed above the ozone layer at about 70 km. The radiation between 1000 Å and 2000 Å is absorbed in dissociating molecular oxygen near 100 km, and so can be observed at heights above 130 km. Radiation of shorter wavelengths is absorbed in ionizing various atomic constituents at altitudes which increase with decreasing wavelength to the He II resonance line at 304 Å, which must be observed from altitudes above 230 km. At still shorter wavelengths the radiation is more penetrating, until at 2–3 Å an altitude of 80 km is sufficient.

The problem of observing the solar spectrum from a spacecraft can be divided into three parts: first the sun has to be acquired and the observing instrument pointed towards the sun with an accuracy which must often be 1 arc minute or better, then means must be provided for selecting the required portion of the spectrum and finally the selected photons must be registered in some way. These three topics – pointing, wavelength selection, and photon detection – will each be discussed separately.

1. Solar Pointing

Rockets and satellites do of course approach very closely the ideal case of a rigid body falling freely in inertial space, and so possess a considerable degree of inherent stability, particularly when spinning. They are, however, subjected to a number of perturbing torques whose compensation is the function of the stabilizing system, once the sun is acquired. The principal sources of disturbance are the following:

(a) *Aerodynamic forces*. These are zero for a spherical spacecraft, and are negligible for an orbit whose perigee height is greater than 500 km. They are often important for rockets, or near perigee for low altitude satellites.

(b) *Gravitational forces.* These arise from the interaction of a satellite with an aspherical mass distribution and the gradient of the geogravitational field. Their origin can easily be seen for a satellite in the form of a dumb-bell whose axis is at an angle (not zero or $\pi/2$) to the vertical. The centre of mass moves with the orbital velocity appropriate to its distance from the geocentre, so that the outer mass of the dumb-bell has too great, and the inner one too low, a velocity for its distance. Thus there is a net outward force on the outer mass and a net inward force on the inner mass, resulting in a torque on the dumb-bell. This torque can be made zero by adjusting the moments of inertia about the principal axes to be equal; otherwise it may be the dominant one for a large satellite.

Fig. 1. The Orbiting Solar Observatory of the U.S. National Aeronautics and Space Administration; the first one was launched in March, 1962.

(c) *Magnetic torques.* These arise from the interaction of the geomagnetic field with the permanent and induced magnetic moments of the satellite. By careful adjustment of compensating magnets, the moments can be made negligibly small, otherwise for a small satellite the magnetic torque is likely to be the dominant one.

(d) *Solar radiation pressure.* This is zero for a spherical satellite, otherwise whilst small it may not be negligible if the satellite has large solar paddles on long booms.

As a result of the above torques, a spacecraft possessing gyroscopic stability corresponding to a spin rate of, say, 30 rpm will typically exhibit precession of the spin axis at a rate up to a few degrees per day if no exceptional pains are taken in its construction. This means that an active, closed-loop type of control system must be employed to stabilize the optical instruments.

Any such system must of course include both means for detecting the pointing error of the optical axis of the spacecraft e.g. an aligned assembly of photo-electric cells designed to produce error signals, and for applying correcting torques to the spacecraft (or to the optical instruments) to rotate or precess the axis to the desired position. If the mass of the spacecraft is large compared with that of the portion of the instrumentation which must be aligned with the sun, it can be used as a counter weight, the instrument section being moved with respect to it to take up the required orientation. If the entire spacecraft is to be moved, torques can be applied by means of gas jets, by introducing a magnetic moment with current loops or magnetized rods, thus producing magnetic forces by the interaction with the geomagnetic field, or finally by means of inertia wheels, motor-driven flywheels which can spin up, thus imparting an equal and opposite angular momentum to the spacecraft.

At the present time, two very successful solar stabilization systems have been employed, the bi-axial "sunseeker" developed for Aerobee rockets by the University of Colorado and the Orbiting Solar Observatories of the U.S. National Aeronautics and Space Administration, the first of which was launched in March, 1962. A brief description of this satellite, shown in Figure 1, will serve to illustrate the techniques which have been employed.

The main body of the satellite, comprising 65% of the weight, is a rotating octagonal prism called the wheel. This carries photoelectric detectors and gas jets (the spherical fibre-glass reservoirs for these can be seen at the end of short booms) and a primary stabilization system which after launch orientates the spin axis so that the sun lies within $\pm 3^{\circ}$ of the plane of rotation of the wheel and maintains it there subsequently, whilst also maintaining the spin rate near 30 rpm. The gyroscopic stability of the wheel is then sufficient to ensure that corrections by the gas jet system are required only at intervals of a few days. Mounted perpendicularly to the wheel plane is a semi-circular solar paddle or "sail" which supports the main optical instruments, the optical axis being approximately perpendicular to the plane of the sail. The sail is free to rotate on the wheel and is coupled to it by an electric motor, which is arranged to drive the sail at just the speed required to maintain its orientation fixed in space with the optical axis in the direction of the sun. The pointed instrument section contains a secondary set of photo-electric detectors and a fine control system. The instruments can be rotated about an axis lying in the plane of the sail and perpendicular to the wheel axis by means of a second electric motor, and so can be maintained with the optical axis on the centre of the sun's disc with an accuracy of ± 1 arc minute, or executing a programmed scan across the disc after the fashion of a television raster.

There are, of course, a number of important observations which do not require any kind of stabilized spacecraft, principally observations of the total emission from the sun integrated over rather wide wavebands. In this case, relatively large photon fluxes are obtained, and the rather crude methods of wavelength selection which are employed (see Section 2) are generally insensitive to the direction of arrival of the photons. However, if high dispersion instruments are used, or if it is desired to measure the emission from a localized region such as a flare, then a stabilized vehicle of the kind described here becomes essential.

2. Wavelength Selection

Up to the present time, grating spectrographs have invariably been used in high resolution studies of the solar spectrum from spacecraft. For wavelengths down to that of the Lyman-α line at 1216 Å, aluminized diffraction gratings can be used at small incidence angles, but at shorter wavelengths the reflectivity rapidly diminishes. Since the intensity of the spectrum is also generally falling, this loss is normally partially offset by operating the grating at grazing incidence. Even so the reflectivity at the 304 Å He II resonance line is unlikely to exceed a few per cent. Since the

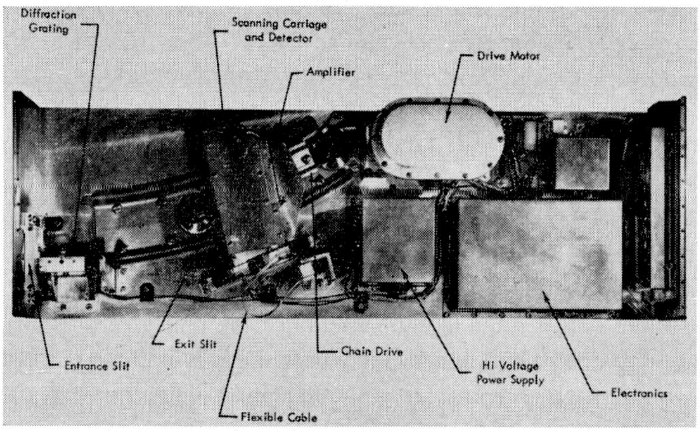

Fig. 2. A photo-electric spectrograph employed by Neupert to obtain solar spectra from 100 Å to 1200 Å with a resolution of 1 Å.

visible portion of the spectrum is much brighter than the wanted portion below 1200 Å; and since the instrumental efficiency is also higher in the visible, scattered light becomes a serious difficulty. If a photo-electric detector is used, it is often possible to arrange that its response is small or zero at wavelengths considerably longer than those it is desired to observe. When the detector is a photographic emulsion, it may be necessary to precede the main spectrograph by a grating whose function is to predisperse the spectrum and so prevent unwanted light from entering the main spectrograph entrance slit. Figure 2 shows a photo-electric spectrograph employed by Neupert and to obtain solar spectra from 100 Å to 1200 Å with a resolution of 1 Å. A description of some other instruments will be found in the chapter by Professor C. de Jager dealing with solar rocket and satellite experiments.

Instruments of this kind have been used to study the spectrum from wavelengths of less than 100 Å up to the visible with a resolution which is generally of the order of a fraction of 1 Å. Grating instruments are probably not well suited at shorter wavelengths, mostly because of the interference of scattered light, but crystal spectrometers using crystals with rather large atomic spacings are being developed. Up to the present, the majority of measurements below 100 Å and some at greater wavelengths which require special mention, have been made with relatively broad-band instruments employing transmission filters, or photon detectors which are inherently wavelength sensitive. Since it is extremely difficult to obtain windows which transmit between the long-wave cut-off of thin organic films, which will transmit to 70 or 80 Å, and the short-wave cut-off of lithium fluoride at about 1150 Å, the method is not applicable between these wavelengths.

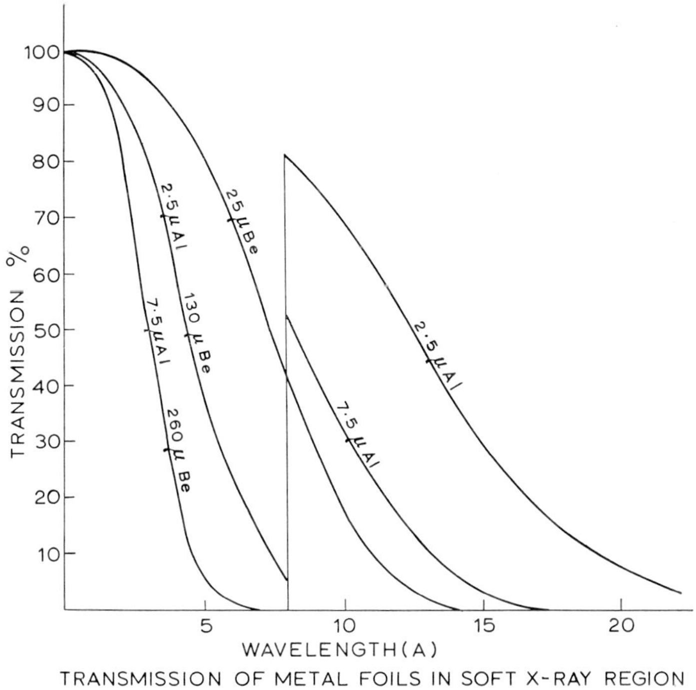

Fig. 3. The transmission of two balanced filters of aluminium and beryllium in the soft X-ray region from 2 to 20 Å.

We shall limit ourselves to a description of one technique which has been used for wavelengths below 16 Å. In this region, thin aluminium foils of differing thicknesses form very convenient filters, as shown in Figure 3. They transmit, however, in two bands on either side of the *K*-edge at 8 Å. By using in addition beryllium foils of such a thickness as to exhibit the same transmission for the shorter wavelength band only, the intensity in each band can be determined by subtraction. This technique resembles somewhat the Nicholl filter, a similar differential device consisting of filters made of adjacent elements in the Periodic Table. The Nicholl filter is, however, used at much shorter wavelengths; in the 2–20 Å region the choice of suitable elements is very restricted and we must be content with considerable bandwidths. The filter system just described forms the basis of a simple device (Pounds and Bowen, 1962) for monitoring the X-ray spectrum of the sun by exposing a small piece of X-ray film through a number of aluminium and beryllium foils, the film being carried to 150 km or more in a Skylark sounding rocket. The film is protected by a very robust mounting and can be recovered intact even when the rocket has no parachute.

3. Photon Detectors

The ideal photon detector for spacecraft studies of the sun must satisfy several criteria. Except for the simplest of experiments, the quantum efficiency should be high or the observing time required for accurate intensity measurement becomes excessive. Unless the detector is intended to be operated in a wavelength-sensitive manner, the sensitivity should vary slowly with wavelength over a wide range – preferably according to a predictable law. The problem of obtaining absolute spectrum intensity measurements in the extreme ultraviolet is by no means satisfactorily solved. It is likely in the next few years that absolute instrument calibrations will be obtained at a limited number of wavelengths – reliable interpolation must then be possible at other wavelengths. At the shorter wavelengths, particularly in the X-ray region, a wide dynamic range and high sensitivity is required, for the flux may vary with solar activity over a range of several orders of magnitude.

If the instruments could be recovered after making the observations, a choice is possible between photographic and photoelectric detectors. In general, the advantages of the photo-electric detector are excellent linearity of response, good reproducibility and, for some experiments, fast response. The photographic emulsion, on the other hand, has an unrivalled storage capacity. Whilst an entire spectrum can be taken at one time photographically, the exit slit and a photo-electric detector must generally be made to scan the waveband at a relatively slow rate. This will often offset to a considerable extent the much higher quantum efficiency of the photo-electric detector. With the development in the future of suitable image-storing photocells this disadvantage may disappear. At the present time, the two types of detector are complementary rather than competitive. For intensity measurements, and for studies of intensity variations with solar activity, photo-electric detectors are superior. On the other hand, for high resolution spectral studies designed to resolve and identify spectral lines or to resolve line profiles, photographic emulsions are best used. Of course, for most satellite observations recovery is not available and only photo-electric detection is possible.

Several suitable photographic films exist for use in the 1–3000 Å region. At the longer wavelengths the sensitivity of the emulsion is enhanced by incorporating a suitable fluorescent dye in the gelatin super-coat. This converts the ultra-violet photons into photons of longer wavelength where the emulsion has a maximum sensitivity. Such an emulsion is Eastman Kodak 103-0.

At wavelengths below 100 Å, absorption in the gelatin supercoat itself becomes of increasing importance, and an X-ray film in which this is omitted becomes preferable. At a wavelength which differs between emulsion types, but is typically in the vicinity of 15 Å, a remarkable change in the properties of the emulsion occurs. The familiar, approximately logarithmic response to varying exposure is replaced by a linear response, and the sensitivity becomes almost independent of wavelength. This corresponds to a region where the absorption of a single photon in a grain is sufficient to render it developable (as compared with the situation at visible wavelengths, where a number of photons must be absorbed by each grain). This much improves the usefulness of the emulsion for X-ray observations.

Two types of photo-electric detector are available – gas filled ionization chambers or counters, and photo-multipliers. (Scintillators are not suitable for use above 1 Å and have therefore been used only to observe the short-wavelength tail of the flare spectrum.) Ionization chambers and counters which can be used over a wide spectral range, except between 80 Å and 1150 Å where no window materials are available, have the important property that they can be given a band-pass response and used for wavelength discrimination. One way in which this can be done is by a suitable choice of window material and gas filling. Many detectors of this kind have been devised and used by H. Friedman of the U.S. Naval Research Laboratory. A particularly important example is the nitric oxide chamber which is used to measure Lyman-α line intensity. This is an ionization chamber with a lithium fluoride window which transmits at wavelengths longer than about 1150 Å, whilst the photo-ionization threshold of the nitric oxide filling is 1350 Å so that the chamber responds only to radiation between these wavelengths. In this band, 90% of the energy of the solar spectrum is concentrated in the Lyman-α line which therefore effectively determines the response of the counter. The quantum efficiency of the counter, which may be as high as 50%, is determined by the absorption in the window, which is easily measured, and the proportion of the photons absorbed in the gas which produce ionisation rather than dissociation which is known to be 80%. The chambers are therefore easily standardised. A somewhat similar principle has also been used by Friedmann and his co-workers at X-ray wavelengths. It is of course possible to use these detectors either with a low or zero gas gain as ionization chambers, or with a high gas gain as Geiger counters, depending on the sensitivity which is required.

The principal disadvantage of this kind of detector is encountered when the spectral intensity is changing rapidly with wavelength or the form of the spectrum is imprecisely known, conditions which both occur in the X-ray region below 50 Å. In this case it may be difficult to distinguish changes of spectral slope from changes of spectral intensity, since each produces an alteration in the output of the detector. Moreover, to deduce the spectral intensity the spectral slope must be assumed, though it is generally not known with much accuracy. Below 20 Å a great improvement results from using the detector as a proportional counter.

In this case, a high proportion (10–50%) of the incident photons produce photo-electrons with nearly the same energy. These then produce secondary electron-ion pairs at the rate of one for each 25 or 30 electron volts of quantum energy possessed by the photon (the exact energy required per ion pair depending somewhat on the nature of the gas filling). Thus, the charge collected on the counter wire after gas multiplication is on an average accurately proportional to the quantum energy of the photon, and the X-ray spectrum can then be determined by means of a pulse height analyser. The obtainable wavelength resolution is limited by statistical fluctuations in the number of ion pairs produced by monochromatic photons. At 12 Å, the mean number of ion pairs per photon is about 30, and the spread in pulse heights is about 20%. Above 20 Å, the wavelength resolution becomes too poor to be of much use. It is difficult to find combinations of window material and gas filling which enable wavelength ranges greater than 4 : 1 to be covered by a single counter;

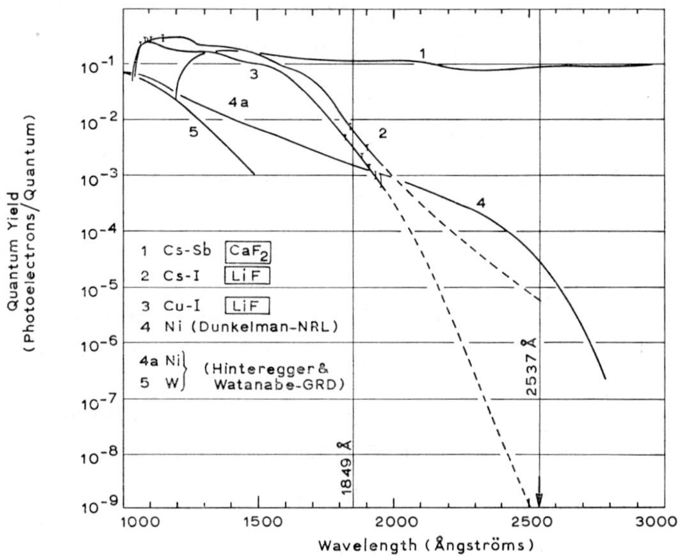

Fig. 4.

this range can then usefully be split into a number of parts giving a resolution of about 10% by a multi-channel analyser. An arrangement of this kind was successfully used in the Ariel I satellite.

Of all the available photon detectors, the one with the widest operating wavelength range is the windowless photomultiplier. Its characteristics are largely determined by the material used for the photocathode. Many different materials have been evaluated for the purpose and photon efficiencies between a few per cent and as high as 20% or 30% are obtainable over the entire wavelength range from 1 Å to visible wavelengths (though not with a single cathode). Some typical examples are shown in Figure 4. An important feature is that by selecting a cathode whose work function corresponds to the longwave limit of observation, the response to longer wavelength scattered light can be eliminated. For this reason tungsten photocathodes, which cut off at 1400 Å, have been widely used. By the addition of a window of, for example, lithium fluoride, a short wave cut-off can also be introduced, giving the photomultiplier a band pass response. Such devices are not likely to be of much importance in solar measurements on account of their wide bandwidths, though they have been used to determine ultraviolet colour indices of stars.

The principal experimental problems with windowless photomultipliers are the selection of electron multipliers which do not degrade when exposed to atmospheric conditions prior to launch, and which also will withstand the vibration environment during launch. A system of exceptional stability and rigidity which has been widely used is the Bendix photo-multiplier shown in Figure 5. The electrons emitted from

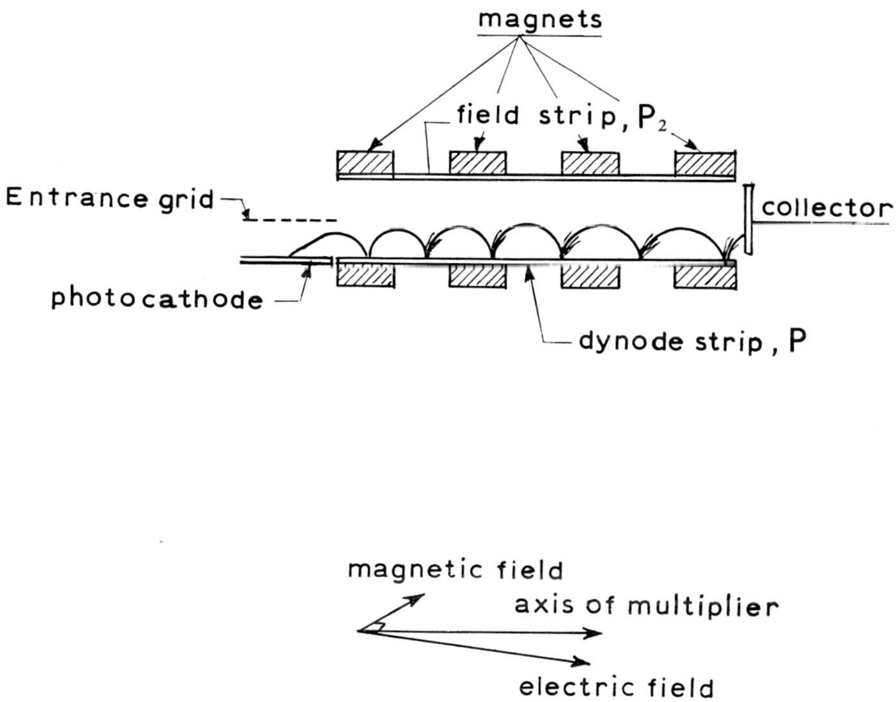

Fig. 5. (a) Diagram of Bendix photomultiplier; (b) Arrangement of electric and magnetic fields in multiplier.

the photocathode start to move in a cycloidal path in crossed electric and magnetic fields. The secondary emitting multiplier surfaces are coated with tin oxide which, being also a semi-conductor, serves to establish the electric field along the length of the multiplier.

By operating the side plate P, at a slightly higher potential than P_2, the electric field is tilted with respect to the multiplier axis so that the electrons strike the secondary emitting surface with sufficient energy to produce a net multiplication. The stage gain is low, but many impacts can be employed resulting in an overall gain of the order of 10^6. The price paid for the simplicity of the system is that the spread in pulse heights at the collector is extremely large, requiring careful design in the following pulse amplifier.

The greatest single difficulty in using photo-multiplier detectors at the present time is that of obtaining accurate absolute calibrations for the wavelengths less than 1000 Å, a difficult problem which can undoubtedly be solved but only at the expense of much careful and detailed laboratory work.

General References

1. For descriptions of the O.S.O. and O.A.O., see the handbooks: *Experimenter's Manual for Orbiting Solar Observatory*, N.A.S.A. Report X-623-63-187, August 1963, *O.A.O. Spacecraft Handbook*, Published by Grumman Aircraft Engineering Corporation.
2. For a review of observations of solar ultra-violet and X-ray emission, see H. Friedman, *Rep. Prog. Phys.* **28** (1962) 163
3. Pounds, K. A. and Bowen, P. J.: 1962, *Mon. Not. Roy. Astron. Soc.* **123**, 348.

PART V

RETROSPECT

CHAPTER 25

THE ALPBACH DISCUSSIONS IN RETROSPECT

NICOLAI HERLOFSON

The Royal Institute of Technology, Stockholm

Much of the discussion centered round the interaction between a flow of solar particles and the geomagnetic field.

The geomagnetic field lines through the auroral zones on the earth's surface reach to distances of 6 or 7 earth radii in the equator plane. Out to this distance the permanent field of the earth predominates – beyond twice this distance, at 10 to 15 earth radii, the interplanetary magnetic field becomes decisive.

The high energy particles in the Van Allen radiation belts are largely located inside the torus-shaped surface formed by the geomagnetic field lines through the auroral zones. A very readable account is given by VAN ALLEN (1963). The last fourteen lines of the review are now out of date, and revision of this part is found in articles by O'BRIEN, VAN ALLEN, LAUGHLIN and FRANK (1962). Diagrams of particle flux in the radiation belts are found in a publication of the *National Academy of Sciences*, Washington.

From time to time considerable fluxes of softer particles flow along the geomagnetic field lines into the auroral zones. These particles move in from distant parts of the geomagnetic field, while the inner region occupied by the radiation belts is normally inaccessible ("forbidden") to them.

The stream of auroral particles is related to the influx of solar particles. But whether the auroral particles are actual solar particles reaching the atmosphere, or local particles accelerated by secondary processes, is a matter of considerable dispute. Direct evidence is lacking, and the theoretical difficulties are largely unresolved.

AXFORD and HINES (1961) emphasize that the earth's dipole field combined with a homogeneous interplanetary field directed along the geomagnetic axis contains a special set of circular field lines, all meeting in two singular points S_1 and S_2 in the figure above. If no electric field can be maintained along magnetic field lines, the sphere formed by the singular field lines will act as a Faraday cage and short-circuit the electric induction field of any magnetized plasma attempting to cross this boundary. In a stationary state, large scale diffuse electric currents in the region *A* outside the boundary will deflect the solar particles past the magnetosphere, in much the same way as a liquid streams past a solid sphere. According to Axford and Hines there will be no plasma flow across the magnetospheric boundary, and the aurora must be a secondary result of momentum transfer to plasma inside the boundary.

On the other hand ALFVÉN (1963) maintains that differences in the pitch angles of ion

Ortner & Maseland (eds.), Solar Terrestrial Relations.

and electron orbits can permit electric field components along the magnetic field lines. This would prevent the short circuit at the magnetospheric boundary envisaged by Axford and Hines, and allow the solar particles to stream freely into the magnetosphere till they are impeded at B by the growing gradient of the geomagnetic field acting on the circular component of their motion.

It is worth noting that while a Faraday cage can easily eliminate static fields,

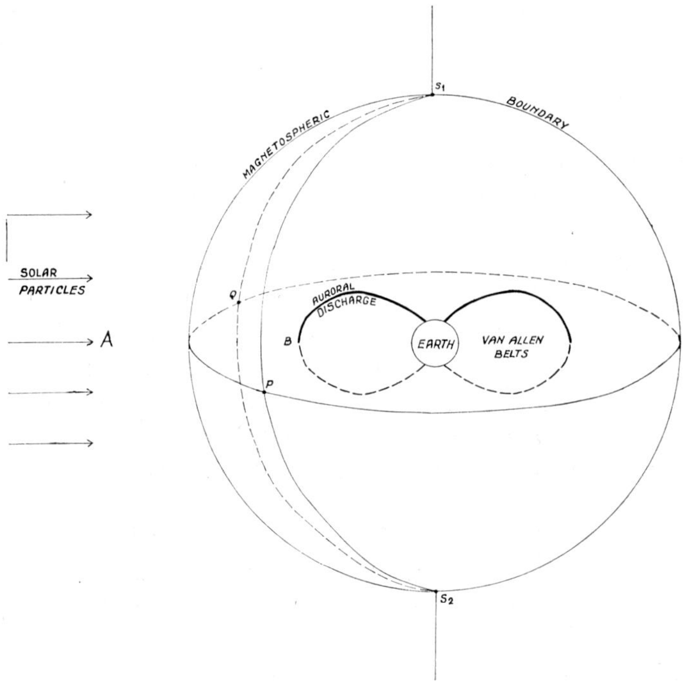

Fig. 1.

it is much more difficult to screen time variable fields. If a solar plasma stream tries to invade the magnetosphere over the arc between P and Q, electric currents must flow between P and Q via the singular points at S_1 and S_2. This circuit has considerable electric inertia, and still more so since the velocity of light is reduced to the local hydromagnetic velocity near the circuit.

An alternative approach would be to investigate whether a hydromagnetic signal and the electric currents from a transient flow at P or Q would reach the key points at S_1 and S_2 before the flow has arrived at B, where the diamagnetic braking sets in anyhow.

A systematic review article by C.-G. Fälthammar on fields and particles in the magnetosphere will be published in the proceedings of the COSPAR symposium on 'High latitude particles and their effects on the polar ionosphere' by Logos Press, London 1965.

References

Alfvén, H.: 1963, *Space Sci. Rev.* **2,** 862.
Axford, W. I. and Hines, C. O.: 1961, *Can. J. Phys.* **39,** 1433.
Fälthammar, C.-G.: 1965, in *High Latitude Particles and Their Effects on the Polar Ionosphere,* Logos Press, London.
O'Brien, B. J., Van Allen, J. A., Laughlin, C. D., and Frank, L. A.: 1962, *J. Geophys. Res.* **67,** 397–403 and 1209–1225.
National Academy of Sciences (Washington): 'A Review of Space Research', *Publication* 1079, pp. 7–17.
Van Allen, J. A.: 1963, *Space Science* (ed. by D. P. Legalley), John Wiley, New York, Ch. 7.

SEMINARS

PART VI

SOLAR PHYSICS

CHAPTER 26

A ROCKET EXPERIMENT FOR MEASURING SOFT X-RAYS FROM THE SUN

H. MASELAND

Space Research Section of the Utrecht Observatory, Utrecht, The Netherlands

Introduction

The purpose of the experiment is to study the soft X-rays of the sun in the 10–60 Å region. According to the theory (*Symposium,* 1955) the soft X-ray emission of the sun originates mainly from the corona, while the calculated spectrum consists of strong lines superimposed on a continuum. Because of the absorption of the greater part of this radiation at heights below 100 km, very little – and that only by rocket and satellite experiments – is known. Although *i.a.* Friedman, Mandelštam and Tousey got results in this field, measurements over many years are still necessary to obtain more information about the quiet and disturbed corona. A comparison of theory and results is in the mean time reported by G. ELWERT (1962).

Instrumentation

The X-ray photons are detected by a proportional counter, filled with a mixture of Argon and Methane at 60 cm Hg. The window consists of a Mylar-film with a thickness of about 6 microns. A tension of 1800 V is applied to the central wire of the counter and is supplied by a unit, consisting of a voltage stabilizer, a converter and a voltagemultiplier. The output pulses, which are of the order of 0.5 mV, are amplified and fed to three discriminators with three different triggerlevels. The two voltage regions between these triggerlevels correspond with the two efficiency regions of the proportional counter (10–20 Å and 40–60 Å). The outputs from these three discriminators are combined in two anti-coincidence circuits, which in this manner give two series of pulses corresponding to two different energy levels of X-ray photons.

The pulses are counted in scalerchains. The scalers for the high level pulses (10–20 Å-region) have a capacity of 2^{14} and for the low level pulses (40–60 Å-region) a capacity of 2^{10}. The 24 outputs of the scalermultivibrators are connected to the commutator of the telemetry.

As the angle of inclination of the photons hitting the counterwindow vary with the altitude of the rocket, the data provided by the counter should be corrected for this angle. Therefore an aspect-sensor has to be mounted near the counter and with its optical axis parallel to the counter. The aspect-sensor is a hollow cylinder (about 6 cm long); a ground glass divides the cylinder in two parts. The cylinderpart (about 1 cm long) at the front end of the cylinder, to be illuminated by the sun, acts as a rim and throws a shadow upon the glass dependent on the angle of inclination of the sun. At the other end of the cylinder a photodiode senses the illuminated part on the glass and gives an output, dependent on the light-shadow ratio of the glass and thus on the angle of inclination of the sun. The output of the aspect-sensor is fanned out and connected to a number of contacts, in a row, in the commutator of the telemetry.

The equipment is fully transistorized with the exception of a voltage regulator tube in the high tension unit. The circuitry is mounted upon epoxyfibre boards and coated, with an epoxyresin. The boards are assembled in two cans, one for the high tension unit, the other for the information circuitry. The scans are filled with a silicone rubber. Special care has to be taken to make the equipment completely airtight. This is especially of importance for the high tension unit and its connectors to avoid the corona effect. (cf. PRICE, 1958; GILESPIE, 1953; HARPER, 1961; and SEIDEL, 1959).

The Experiment

For a higher reliability of the experiment two photon-counters each with their electronic circuitry as described above, are mounted diametrically opposite each other in the nosecone. The Véronique rocket, in which this experiment will be flown is expected to have a spin of about 1 rev/sec so that once during each revolution the number stored in both scalerchains of each counter will be increased; in this way it is not necessary to reset the scalers, because an amount of equal readings or readings of a slowly increasing number dependent on the background will be obtained between two illuminations of each counter.

As mentioned above the photons are discriminated according to their energy into two levels. Another way of discrimination will be possible during the flight when the rocket goes upwards from 90 km to 150 km and down. The measured intensity is dependent on the wavelength as the shorter wavelengths penetrate more easily into the atmosphere. By measuring the intensity as a function of the height of the rocket, together with the known absorption coefficients of the atmosphere it is possible to get information about the energy distribution in this part of the spectrum.

The equipment is completed now and will be flown in the Véronique early 1964. The same experiment but with modified circuitry will be flown in an ESRO-Centaur-rocket, probably in the middle of 1965.*

References

Symposium on Solar Eclipses and Ionosphere, London, 1955.

ELWERT, G.: 1962, 'Comparison between theory and measurements of the X-ray emission of the sun', *Proc. Inst. Conf. on the Ionosphere,* Bartholomeus Press.

PRICE, W. J.: 1958, *Nuclear Radiation Detection,* McGraw Hill.

GILESPIE, A. B.: 1953, *Signal, Noise, and Resolution in Nuclear Counter Amplifiers,* Pergamon Press Ltd., London.

HARPER, C. A.: 1961, *Electronic Packaging with Resins,* McGraw Hill.

SEIDEL, G.: 1959, *Gedruckte Schaltungen,* Verlag Technik Berlin und Berliner Union, Stuttgart.

* The results of the launching from Hammaguir-base on Nov. 4, 1964 are submitted to *J. Geophys. Res.* (added in proof).

CHAPTER 27

HIGH RESOLUTION SOLAR SPECTROSCOPY AT THE JUNGFRAUJOCH SCIENTIFIC STATION

N. GREVESSE

Institut d'Astrophysique, Université de Liège, Cointe-Sclessin, Belgium

In 1958, a solar spectrograph was installed by the *Institut d'Astrophysique* of the University of Liège at the Jungfraujoch International Scientific Station in Switzerland, at an altitude of 3580 m.

It consists of a 30 cm coelostat, an horizontal telescope having a focal length of 12.2 m and a prism-grating vacuum spectrometer of the Ebert-Fastie type (MIGEOTTE, 1960; DELBOUILLE and MIGEOTTE, 1960). This instrument is equipped with an original Babcock grating (dimensions: 20.4 × 13.5 cm) ruled at 600 lines per mm and blazed at 2.7μ (first order). (See Figure 1.)

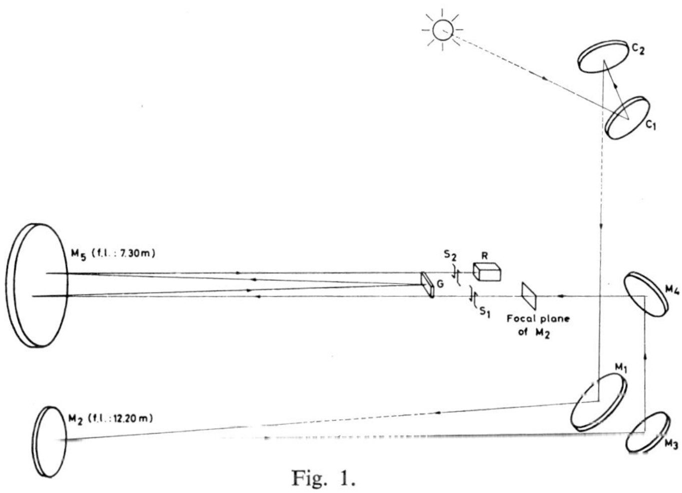

Fig. 1.

The spectrum is recorded by turning the grating on a vertical axis. The solar image (diameter: 115 mm) can be displaced by means of motors acting on the mirror M_2 so as to bring on to the entrance slit the region of the solar disk to be studied. Up to the present, the observations have been done by direct recording, using as detector a 13-stage infrared photomultiplier.*

* Made in the *Laboratoire de Physique Astronomique, Observatoire de Paris,* under the direction of Professor A. Lallemand.

Between 1958 and 1960, L. Delbouille and G. Roland have observed several times the solar spectrum between 7500 Å and 12000 Å. Part of the results of their observations have just been published in the form of a Photometric Atlas (DELBOUILLE and ROLAND, 1963). This Atlas consists of 165 plates. Each of these sections, covering about 30 Å, is a photographic reproduction at a scale $\frac{1}{2}$, of the original records taken at the center of the solar disk. In the region studied 10450 absorption lines (solar and

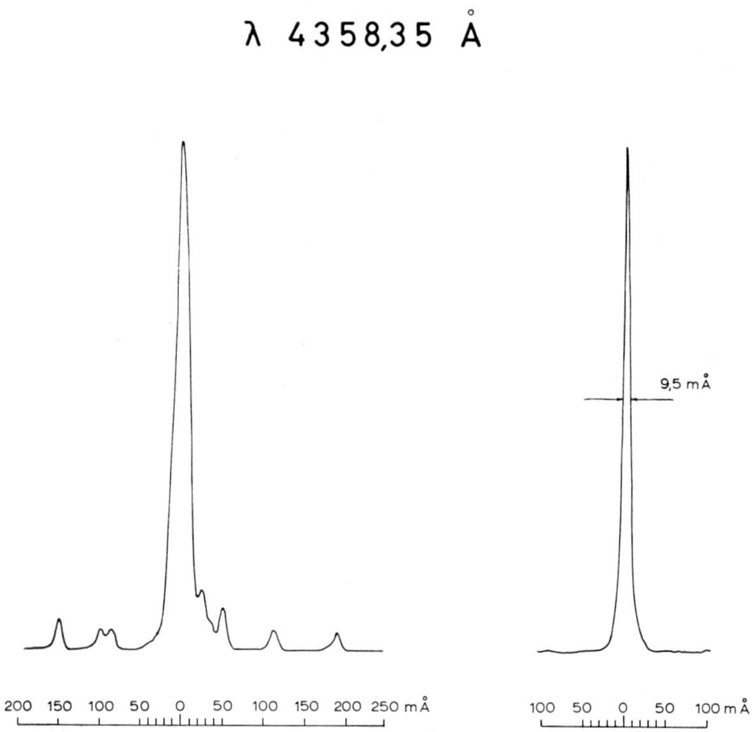

Fig. 2.

telluric) have been found on the records as compared with 4675 for the same region in the table of BABCOCK and MOORE (1947). Actually the identifications are undertaken by W. S. Benedict (Johns Hopkins University, Baltimore, Maryland, U.S.A.), O. C. Mohler (University of Michigan, U.S.A.) and J. W. Swensson (Institut d'Astrophysique, Cointe-Sclessin). W. S. Benedict has already identified 2100 lines of CN and 3000 lines of water.

Solar lines which are almost entirely blended by the water vapor absorption lines of the earth's atmosphere at lower altitude can be observed from the Jungfraujoch and many more details are usually found in the structure of the telluric bands (DELBOUILLE and MIGEOTTE, 1960).

The same instrument has also been used by C. de Jager from the *Sterrewacht 'Sonnenborgh'*, Utrecht (Netherlands) and L. Neven from the *Observatoire Royal de*

Belgique, Uccle (Belgium) to study the profiles and the center-to-limb variations of the profiles of many solar lines (DE JAGER and NEVEN, 1962, 1963), by O. Namba (*Sterrewacht 'Sonnenborgh'*, Utrecht, Netherlands) to study the profile of the infrared He I line in solar faculae (NAMBA, 1963).

Recently, the spectrograph has been modified to double pass the grating in order to increase the dispersion and the resolution (EVANS and WADDEL, 1962). Two small

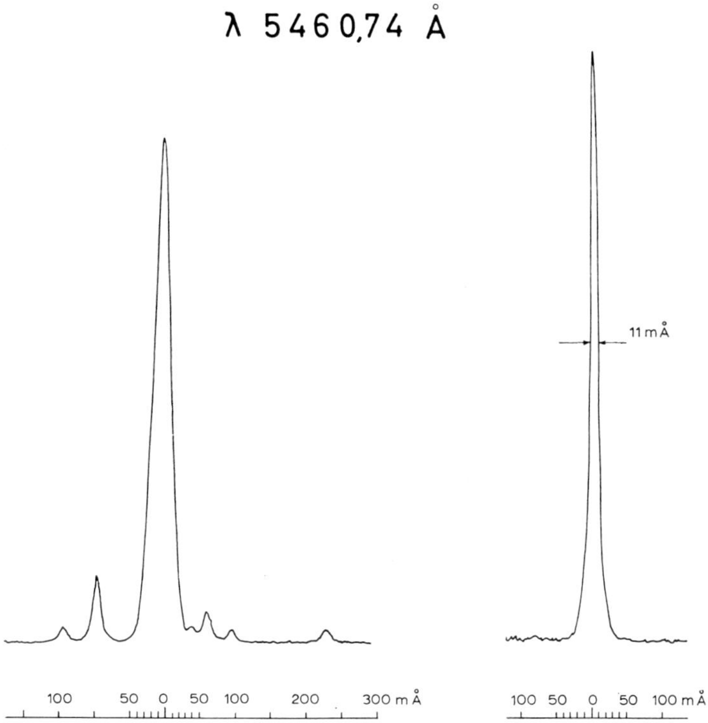

Fig. 3.

auxiliary flat mirrors were added, the first reflecting the light beam, through an intermediate slit placed in the focal plane of the spectrometer mirror, on a second one which sends it back to the grating. Under these conditions, the intensity of the ghosts produced by the grating is much reduced and the grating can be used in high orders.

The two following figures reproduce the profiles of the λ 4358 and λ 5461 emission lines of Hg and Hg_{198} (see Figure 2 and 3 *).

The first solar spectra obtained after such a modification have been published (DELBOUILLE, NEVEN and ROLAND, 1963). Because of the better resolving power, lines which do not appear in the Göttingen Atlas can be seen on the Jungfraujoch tracings and the lines are deeper (BRÜCKNER, 1960). This fact is of great importance

* I am much indebted to Dr. L. Delbouille and Miss G. Roland for giving me these line profiles before publication.

for the further understanding of the emission processes in the outer photospheric layers. Actually, one knows that deviations from local thermodynamical equilibrium certainly occur in the photospheric layers, but so many parameters are still uncertain that we cannot say exactly to what amount. E. A. Müller and J. P. Mutschlecner have computed the "effects of deviations from LTE on solar abundances" (MÜLLER and MUTSCHLECNER, 1964). Their conclusions are in perfect agreement with the conclusions of A. UNSÖLD (1963) discussing the reasons for the validity of LTE assumption in the solar photosphere. On the other hand, J.-C. Pecker (PECKER *et al.*, 1959–62) has found deviations from LTE. His computations are based on measurements of central intensities of lines from the Utrecht Atlas (MINNAERT, MULDERS and HOUTGAST, 1940).

It would then be very interesting to recompute the deviations from LTE in the outer solar photospheric layers with the new central intensities measured at the Jungfraujoch and with the most reliable values of the other parameters (such as *f* values) which might be available in the future.

L. Delbouille, G. Roland and L. Neven intend to use the double pass spectrograph to map a new atlas of the visual and ultraviolet regions, by direct photoelectric record (with a Lallemand 20-stage photomultiplier). They are planning to record simultaneously analog tracings and digital data on punched paper tape.

References

BABCOCK, H. D. and MOORE, C. E.: 1947, 'The Solar Spectrum λ 6600 to λ 13495', Carnegie Institution Publication 579, Washington.

BRÜCKNER, G.: 1960, *Photometric Atlas of the near Ultraviolet Solar Spectrum* 2988–3629 Å, Universitätssternwarte Göttingen.

DELBOUILLE, L. and MIGEOTTE, M.: 1960, *J. Opt. Soc. Amer.* **50,** 1305.

DELBOUILLE, L., NEVEN, L., and ROLAND, G.: 1963, *J. Quant. Spectr. Rad. Trans.* **3,** 189.

DELBOUILLE, L. and ROLAND, G.: 1963, 'Photometric Atlas of the Solar Spectrum from λ 7498 to λ 12016', *Mém. Soc. Roy. Sci. Liège,* Special Volume **4**.

EVANS, J. and WADDEL, J.: 1962, *Appl. Opt.* **1,** 111.

DE JAGER, C. and NEVEN, L.: 1963, *Ann. Obs. Roy. de Belgique,* 3ième série, **9,** 2.

DE JAGER, C. and NEVEN, L.: 1962, *Bull. Astron. The Netherlands* **16,** 307.

MIGEOTTE, M.: 1960, 'La station d'observation solaire de l'université de Liège au Jungfraujoch', Fédération des Sociétés Scientifiques, No. 3.

MINNAERT, M., MULDERS, G., and HOUTGAST, J.: 1940, *Photometric Atlas of the Solar Spectrum from λ 3612 to λ 8771*, Sterrewacht 'Sonnenborgh', Utrecht.

MÜLLER, E. A. and MUTSCHLECNER, J. P.: 1964, 'Effects of Deviations from LTE on Solar Abundances', *Astrophys. J., Suppl.*, Vol. IX, No 85.

NAMBA, O.: 1963, *Bull. Astron. The Netherlands* **17,** 93.

PECKER, J.-C. *et al.*: 1959–1962, Series of papers on 'Écarts à l'équilibre et abondances dans les photosphères solaire et stellaires', *Ann. Astrophys.* **21-24.**

UNSÖLD, A.: 1963, *Z. Phys.* **171,** 44.

PART VII

INTERPLANETARY MEDIUM

CHAPTER 28

SCIENTIFIC RESULTS OF THE MARINER II VOYAGE TO VENUS

LEVERETT DAVIS, JR.
California Institute of Technology, Pasadena, California

Mariner II was launched on August 27, 1962, passed at a distance of only 41000 km from the center of Venus on December 14, and transmitted its last scientific and engineering data to earth on January 2, 1963. It was remarkably successful for the first vehicle of its type whose launching was satisfactory. A large part of the emphasis on this first flight was engineering and a great amount of important technological information was obtained from it. The history of this voyage, including a discussion of the engineering and scientific problems and results, is surveyed by JAMES (1963). My discussion this morning will be restricted to the scientific results obtained to date by the seven groups of experimenters. The emphasis given each experiment will be governed by my own interests and what I happen to know.

A very interesting scientific by-product of the engineering technology is information on the scale of distances and some of the masses in the solar system. Radio signals sent from the earth and received by Mariner were rebroadcast back to give a very accurate Doppler measurement of its radial velocity relative to the earth. The 22000 Doppler observations, together with the simultaneous radar observations of Venus made by the Goldstone tracking station, will enable an exceedingly accurate computation to be made of the astronomical unit (abbreviated AU), which determines the scale in cm of all distances in the solar system. From the deflection of the trajectory as it passed Venus, preliminary calculations have shown that the previous value for the mass of Venus, 0.8148 ± 0.0004 times the mass of the earth, can be refined to 0.81485 ± 0.00004. The Doppler radial velocity is affected also by the motion of the earth around the center of mass of the earth-moon system and from this a refined value of the mass of the moon can be obtained. This work was carried out by T. W. Hamilton, J. F. Koukol, N. A. Renzetti, D. W. Trask and J. D. Anderson.

During the encounter period (*Preliminary Reports*, 1963), Venus was scanned by a microwave radiometer at wavelengths of 13.5 and 19 millimeters and by an infrared radiometer at wavelengths of 8.4 and 10.4 microns. There were three scans, one on the dark side, one nearly along the line separating the light and dark sides, and one on the light side. The microwave radiometer showed that the surface temperature was about 700 °K and the infrared radiometer showed that the temperature at the top of the clouds was about 250 °K. Thus it is now definite that the high radio frequency temperatures previously recorded are at the surface rather than in the ionosphere, and the greenhouse model for the Venusian atmosphere gains added support. There were no temperature differences detectable between the light and dark sides. The cloud cover in the regions scanned seems quite continuous with no breaks detected as wide as a few percent of the diameter of the disk. There was no significant amount of CO_2 in the upper clouds. The experimenters for the infrared radiometer were L. D. Kaplan, G. Neugebauer and C. Sagan; the experimenters for the microwave radiometer were A. H. Barrett, D. E. Jones, J. Copeland, and A. E. Lilley.

The experiment of W. M. Alexander for the observation of interplenatary dust grains yielded only a couple of definitely confirmed impacts, many fewer than had been expected (*Preliminary Observations*, 1962). No thorough analysis of the results in terms of particle density is available.

An ion chamber and three Geiger counters were used by H. R. Anderson H. V. Neher, and J. A. Van Allen to observe energetic particles (*Preliminary Reports*, 1963; *Preliminary Observations*, 1962; ANDERSON, 1963). From their observations it may be concluded that there is no great flux of protons in the range from 10 to 100 MeV since the observations in space are about what would be expected from the flux of particles observed above 100 MeV by ion chambers flown on balloons. On several occasions low energy cosmic rays were observed, enabling us to extend our knowledge of the time constants for their appearance and escape to the lower energies observable in space. Several Forbush decreases in the galactic cosmic ray flux were observed both by Mariner II and on earth, confirming and extending the earlier proof by Pioneer V that on such occasions cosmic rays are screened out from a very large volume of space in the solar system. Aside from such transient increases and decreases in the cosmic ray flux associated with solar disturbances, the general level remained constant to within 2 percent between the orbit of earth and the orbit of Venus. You will recall that the flux of low energy galactic cosmic rays, those with energies in the neighborhood of 1 BeV, varies over the 11 year sun spot cycle. Presumably this is due to some screening mechanism outside the orbit of earth, which excludes low energy particles more effectively when the sun is active than when it is quiet. It is tempting to conclude that this absence of change in the cosmic ray flux between the earth and Venus means that the entire screening mechanism is located outside, perhaps far outside, the orbit of the earth. However, it is just possible that there is some further screening between the earth and Venus, but that during the four months of the flight the general level of cosmic rays at any fixed distance from the sun increased a bit, as expected at this phase of the sun spot cycle, and that this increase just compensated for the decrease due to the additional screening. Unfortunately, observations of the low energy cosmic ray flux at the earth adequate to check this point do not seem to be available.

In my opinion, the most important result of the Mariner II experiment was the conclusive demonstration (*Preliminary Observations*, 1962) that, at least at this stage of the solar cycle, the solar wind flows out from the sun all the time essentially as in Parker's well-known model, although with a considerably lower density than was popular a few years ago. The plasma detector of Mariner II measured the current due to charged particles coming along directions lying within about 5° of the line to the sun and having energies in a narrow band. An electrostatic analyzer selected in succession 10 energy bands that correspond to proton velocities distributed in a geometric series between 231 and 8224 km/sec. The main conclusions to be drawn from a preliminary analysis of the data are that measurable solar wind blew essentially radially outward from the sun at all times, that most of the time the velocity lay between 350 and 750 km/sec, and typical densities were a few per cm^3 (perhaps ranging from less than 1 to as much as 30). Particles were present simultaneously in several channels; if this is interpreted as a manifestation of the temperature of the gas, values of the order of 2×10^5 °K are usually obtained. A less orthodox alternative is that this is in part the manifestation of waves with frequencies greater than about 1 sec^{-1}. There is also clear evidence for the presence in the solar wind of alpha particles as well as protons. The density ratio seems to fluctuate and the relative abundance of alpha particles seems to be lower than in the photosphere. A very striking feature of the plasma data is the fact that the velocity showed a series of peaks, each lasting from 2 to 4 or 5 days, that the level of geomagnetic disturbance correlated well with the solar wind velocity, and that the peaks showed a strong tendency to recur at approximately 27-day intervals. The experimenters for the plasma probe were M. M. Neugebauer and C. W. Snyder.

A triaxial fluxgate magnetometer was used to measure the vector interplanetary field at intervals of 37 sec throughout most of the voyage (*Preliminary Observations*, 1962). Unfortunately the space craft magnetic field shifted during launch from the prelaunch value by an amount that is not known accurately and showed substantial later shifts. Thus there is always some uncertainty in the zero from which the interplanetary field is measured and, while I believe that a reasonable interpretation of the data can be presented, anyone is free to modify it by making a different estimate of the space craft field. Much of the time the interplanetary field has a general order of magnitude of 5 gamma and shows relatively little change over periods of many minutes. There are many periods of disturbance when the field varies much more rapidly, time constants ranging from fractions of an hour through a few minutes to perhaps only a few seconds being apparent. On these occasions the field changes may be as much as 5 or 10 gamma and the field itself may sometimes be even larger in magnitude. If one looks at individual hourly averages of the field components, no particular structure of the interplanetary field is apparent. If one averages over periods of 10 to 20 hours, there is an indication that the field may have a basic spiral structure in the plane of the ecliptic, the field making an angle of about 45° with the radius from the sun. Superposed on this regular field are fluctuations or waves that deflect the field in a random way from its mean direction. Typically the deflection may be through 20° to 40°; it is made up of a superposition of waves with periods ranging from several hours down to fractions of an hour in quiet times and down to only a minute or less in disturbed times. I must emphasize that this model is still very tentative. It appears somewhat more certain that the component of the field along the radius from the sun seems to have one typical average value during one-half of the solar rotation and another average value, most likely of equal magnitude but opposite sign, over the other half of the solar rotation. This pattern shows a rough 27-day periodicity. No effects ascribable to the presence of a Venusian magnetic field could be detected in the neighborhood of Venus. This requires that the magnetic moment of Venus be less than 1/10th, or perhaps only 1/20th that of the earth. It is by no means excluded that Venus has essentially no magnetic field. This result is consistent with the recently established low rotation rate of Venus; but considering our uncertain quantitative knowledge of dynamo theory, neither result can be said to require the other. The magnetometer experimenters were P. J. Coleman, Jr., Leverett Davis, Jr., E. J. Smith, and C. P. Sonett.

Next let us consider a tentative model in terms of which some of these observations may be discussed. Some parts of this model may be novel, but some have been borrowed from others. It must be emphasized that the model is provisional and many details are likely to be wrong. But it is the best I can do at the moment.

Consider first the region within 10 or 20 solar radii of the sun. Near the sun the magnetic field dominates the plasma flow; the field must be essentially force-free except for oscillations and waves whose velocity of propagation is much faster than the expansion that becomes the solar wind. The gas can move freely only along the tubes of force. Hence there is likely to be a tendency for the gas to well up from the photosphere into the upper corona in places where the magnetic field extends to large radii in a relatively regular, nearly vertical pattern. The entire upper corona may be filled by gas that spreads out from a few such sources and in quiet times the entire solar wind will come from them. The magnetic field in interplanetary space is drawn from only a few regions and its radial component changes sign much less frequently than would be the case if all the irregular fields of the photosphere were sampled as the sun rotates. There may be some tendency for the alpha particles to settle relative to the protons in this welling-up region, which might account for a low density of alphas in the solar wind.

As one goes out to the region where the plasma momentum begins to become more important than the more rapidly decreasing magnetic field stresses in determining the motion, the tubes of force may form nozzles out of which the gas is ejected. The flow will not be quite radial, as in Parker's familiar model of the solar wind, but will be modified by the field. This may explain the high velocity, relatively narrow jets of low density wind observed near the earth by the Mariner II plasma probe. If there were merely a local hot spot on the sun, the angular width of the jet might be expected to be greater and there might be a stronger tendency for the density to increase more rapidly than the velocity. If the cross-sectional area of the nozzle is not merely proportional to the square of the distance from the center of the sun, it should be possible to vary density and velocity independently. The appearance of the same jet on several successive rotations of the sun indicates that the field patterns in the upper corona are quite stable. Near the sun the field and gas must co-rotate with the sun as they expand, provided the tubes of force remain connected to the sun. At intermediate distances the interaction between the field and the angular momentum of the wind is complicated and it is not clear how co-rotation shifts over to the essentially radial flow that should be present near the earth. But there do seem to be good possibilities for angular motions near the sun that will complicate the calculation of radial velocities based on the time lag between central meridian passage of a photospheric or chromospheric feature and arrival at the earth of a disturbance supposedly associated with the feature. If propagation is not radial, the disturbance may originate from the feature at some time other than central meridian passage.

Near the earth – in fact, starting inside Mercury's orbit – the momentum of the solar wind dominates the magnetic forces, which can no longer influence the average motion significantly. If the field maintains any continuity as it expands, it should be stretched out and sheared into the well-known spiral when conditions are steady. Near the earth it would be expected that if the wind has a velocity in the 400 to 500 km/sec range, the spiral would make an angle of about 45° with the radius from the sun. There may, of course, be superposed waves and oscillations. The magnetometer observations suggest that during quiet times these have characteristic lengths in a frame moving with the wind of from 10^5 to $10^{6.5}$ km, and that during disturbed conditions there are many fluctuations with a scale of only $10^{4.5}$ km or less. There appears to be rough equipartition between the magnetic energy density and the kinetic energy density associated with the fluctuations about the average of the mass motion.

As one goes farther out in the solar system, the spiral should tend to become very flat so that the average direction of the field is nearly normal to the radius vector. Here the energy density of the magnetic field and the kinetic energy density both vary inversely as the square of the distance from the sun and if there is equipartition, the waves in the field lines should continue to make angles of 20° to 40° with the average direction of the field lines. As the solar wind comes to the point where its momentum cannot further push back the galactic magnetic field, the wind must pass through a shock transition in which its velocity decreases and its field strength increases. As layer after layer of flattened spirals pile up, the fact that the lines of force originate at different points on the sun and hence have different senses means that they cannot merge into a simple uniform field. Instead the field is likely to be quite irregular as in the region just outside the earth's magnetosphere where the solar wind piles up and forms a stagnant region as its flow is impeded by the geomagnetic field. This region at the interface between the solar wind and the interstellar region will not continue to increase in thickness indefinitely. The irregular field structure will decay because of its finite conductivity. Its outer layers will be swept away by the relative motion of the solar system and the interstellar gas. As its field lines merge with the interstellar field, the plasma contained in the tubes of force in the transition region may drain out of the tubes into the interstellar region, leaving the field essentially alone to maintain the pressure balance. This will further increase the field strength in the transition region. It is tempting to think that this irregular thatch of field lines forms the barrier that tends to keep low energy galactic cosmic rays out of the solar system and thus produces the eleven-year cycle of cosmic ray intensity.

This lecture was prepared while I was a guest of the *Max Planck Institut für Physik und Astrophysik*. I am very grateful for the hospitality I received. The preparation of the written version was supported in part by NASA Grant NsG-426.

References

Anderson, H.: 1963, 'Mariner II: High-Energy Radiation Experiment', *Science* **139**, 42–45.

James, J. N.: 1963, 'The Voyage of Mariner II', *Scientific American* **209**, 70–84.

Preliminary Observations, 1962: 'The Mission of Mariner II. Preliminary Observations: Neugebauer, M. and Snyder, C. W.: 'Solar Plasma Experiment'; Van Allen, J. A. and Frank, L.: 'The Iowa Radiation Experiment'; Alexander, W. M.: 'Cosmic Dust'; Coleman, Jr., P. J., Davis, Jr., L., Smith, E. J., and Sonett, C. P.: 'Interplanetary Magnetic Fields'; *Science* **138**, 1095–1100.

Preliminary Reports, 1963: Mariner II. Preliminary Reports on Measurements on Venus: Frank, L. A., Van Allen, J. A., and Hills, K. H.: 'Charged Particles'; Chase, S. C., Kaplan, L. D., and Neugebauer, G.: 'Infrared Radiometer'; Barath, F. T., Barrett, A. H., Copeland, J., Jones, D. E., Lilley, A. E.: 'Microwave Radiometers'; Smith, E. J., Davis, Jr., L., Coleman, Jr., P. J., and Sonett, C. P.: 'Magnetic Field'; *Science* **139**, 905–910.

CHAPTER 29

ON THE PENETRATION OF INTERPLANETARY PLASMA INTO THE MAGNETOSPHERE

H. ALFVÉN

The Royal Institute of Technology, Stockholm, Sweden

Theoretical and experimental evidence indicates that the usual magnetospheric models may need essential modification. Theoretically it has been concluded that the low-density magnetospheric plasma, in contrast to an ideal magnetofluid, may support strong electric fields even along the magnetic field lines. This can profoundly influence the penetration of interplanetary plasma. Laboratory experiments show that a plasma can penetrate much closer to a terrella than what would correspond to the magnetosphere boundary in closed magnetosphere models.

References

The lecture was a summary of the following publications:

Alfvén, H., Danielsson, L., Fälthammar, C.-G. and Lindberg, L.: 1964, 'On the Penetration of Interplanetary Plasma into the Magnetosphere', *Natural Electromagnetic Phenomena Below 30 KC/S*, (Proceedings of a NATO Advanced Study Institute, Bad Homburg, Germany, July 22–August 2, 1963), edited by D. F. Bleil, p. 33, Plenum Press, New York; also *Proceedings of the Plasma Space Science Symposium*, Washington, 11–14th June 1963, D. Reidel Publishing Co., Dordrecht, Holland.

Alfvén, H. and Fälthammar, C.-G.: 1963, *Cosmical Electrodynamics*, Clarendon Press, Oxford, pp. 162–168.

Persson, H.: 1963, *Phys. Fluids* **6**, 1756.

CHAPTER 30

GEOMAGNETICALLY TRAPPED PROTONS

A. M. LENCHEK
Observatoire de Paris, Meudon

Abstract. We begin with a discussion of the dynamics of trapped particles. The generation of a proton component of the radiation belt by the action of cosmic ray albedo neutrons is then discussed. We conclude by pointing out a number of investigations, chiefly experimental, that might usefully be made in the future.

Dynamics of Trapped Particles

The exact trajectory of a charged particle in the earth's dipole magnetic field requires a numerical integration of the equation of motion. Consequently, this method is impractical for the study of trapped particles, which make a very large number of north-south oscillations as well as drifting around the earth.

A very great simplification of the problem is made by the use of the *guiding center approximation*, introduced by ALFVÉN (1950). Provided that the magnetic field varies "sufficiently slowly" in space and time, the motion under this approximation is well represented by a circling motion about a guiding center which tends to follow the line of force. By "sufficiently slowly" one means the following: denote the gyroperiod by t_g, given by

$$t_g = 2\pi m\gamma / qB \tag{1}$$

where m = rest mass, $\gamma = (1-\beta^2)^{-1/2}$, $\beta = v/c$, v = velocity, q = charge and B = magnetic field strength. Denote the gyroradius by a, given by

$$a = m\gamma v \sin\alpha / qB \tag{2}$$

where α is the angle between $\mathbf{v}$ and $\mathbf{B}$, called the pitch angle (Figure 1). Then if

$$\frac{dB}{dt} \ll \frac{B}{t_g} \tag{3}$$

and

$$|\operatorname{grad} B| \ll B/a \tag{4}$$

the guiding center approximation is valid. Under these conditions the gyration about the line of force generates a closed current loop. The magnetic moment of the current loop

$$\boldsymbol{\mu} = -\mathbf{B} m v^2 \gamma \sin^2\alpha / 2B^2 \tag{5}$$

is an *adiabatic invariant*. That is, the magnitude of $\boldsymbol{\mu}$ is a constant in the limit of infinitely slow variations of the field (ALFVÉN, 1950; CHANDRASEKHAR, 1960).

Ortner & Maseland (eds.), Solar Terrestrial Relations.

In a field such as the dipole field the constancy of the magnetic moment leads to the phenomenon of *reflection*. From Equation (5)

$$\tfrac{1}{2}mv^2\gamma\frac{\sin^2\alpha}{B} = \text{constant}. \tag{6}$$

Since the static magnetic field does no work on the particle, v and γ are constant, giving

$$\frac{\sin^2\alpha}{B} = \text{constant}. \tag{7}$$

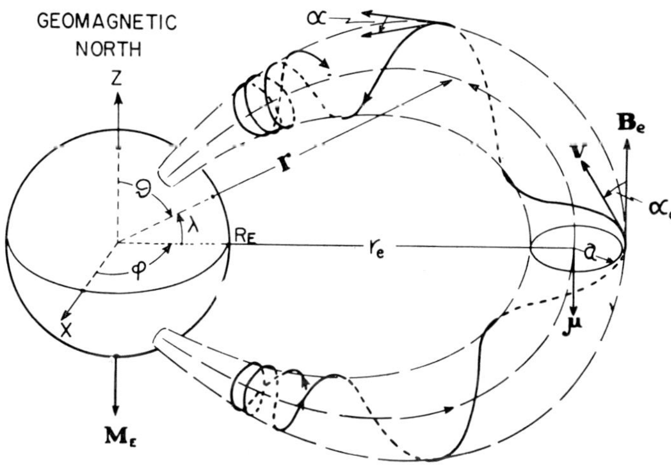

Fig. 1. Motion of a trapped particle in a dipole field, illustrating a tube of force, the gyration about the tube of force and the oscillation from mirror point to mirror point. The azimuthal drift is not shown.

Thus if α_e and B_e denote the values as the particle crosses the equatorial plane (where B has its minimum value) the pitch angle will continuously increase as the particle moves into the region of stronger B and will attain 90° at a point where the field strength is

$$B_m = B_e/\sin^2\alpha_e\,. \tag{8}$$

At this point the Lorentz force

$$\mathbf{F} = q\,\mathbf{v}\times\mathbf{B} \tag{9}$$

has a component directed toward the region of smaller field strength (for particles of either sign) and the guiding center is reflected. This point is therefore called the mirror point (Figure 1).

Notice that the location of the mirror point on a given line of force is independent of the charge, mass or energy of the particle. It depends only on the equatorial pitch angle, α_e.

The period of the north-south oscillation from mirror point to conjugate mirror point and back again is called the bounce period t_b and is given closely (HAMLIN, KARPLUS, VIK and WATSON, 1961) by

$$t_b = \frac{4r_e}{v}(1.30 - 0.56\sin\alpha_e) \tag{10}$$

where r_e is the distance from the dipole to the line of force in the equatorial plane (Figure 1). Thus t_b varies from ~ 0.1 to ~ 1 second for relativistic particles.

Superimposed on the north-south oscillation is a drift about the dipole axis. The velocity of the drift is

$$\mathbf{v}_d = \mathbf{f} \times \mathbf{B}/B^2 \tag{11}$$

where $\mathbf{f}$ is the force per unit charge (other than the $\mathbf{v} \times \mathbf{B}$ Lorentz force) seen in a frame moving with the guiding center (ALFVÉN, 1950). For particles trapped in the static geomagnetic field $\mathbf{f}$ has two components. One is the diamagnetic interaction between the magnetic moment of the spiralling particle and the magnetic moment of the earth. The other is the centrifugal force arising from the motion of the guiding center along the curved line of force (SPITZER, 1956; NORTHRUP and TELLER, 1960; LEW, 1961). The drift velocity varies with latitude and particle energy for a given line of force. Averaging over the bounce period, LEW (1961) finds the drift period to be expressible as

$$t_d = 1.03 \times 10^4 \frac{1 + E'}{E'(2 + E')} \frac{m_e}{m} \frac{R}{r_e} \frac{G}{F} \text{seconds} \tag{12}$$

where E' = kinetic energy/rest energy, m_e = electron mass, m = particle mass, R = earth radius, r_e = equatorial radius of line of force and G/F is a function of equatorial pitch angle varying from 1.0 for $\alpha_e = \pi/2$ to 1.5 for $\alpha_e \sim 0$. In the non-relativistic limit, particles of the same energy have equal drift periods. For the trapped protons in the range ~ 1 MeV to ~ 100 MeV, t_d is in the range ~ 1 minute to ~ 1 hour in the inner belt. The direction of the drift is westward for protons and eastward for electrons.

The motion of a trapped particle is seen to be described by three periodic motions with periods t_g, t_b and t_d. The adiabatic invariant μ is associated with the first of these periodic motions, the gyration about the line of force. Two other adiabatic invariants are known, associated with each of the other two periodic motions.

Associated with the bounce motion is the so-called *second adiabatic invariant*

$$J = \oint m\mathbf{v} \cdot \mathbf{dz} \tag{13}$$

where $\mathbf{dz}$ is an element of arc length along the line of force and the integral is over a complete bounce cycle (CHANDRASEKHAR, 1960). If the first invariant is conserved, J may be replaced by an equivalent quantity I, given by

$$I = \oint \sqrt{1 - B(z)/B_m}\,\mathrm{d}z \tag{13'}$$

which has dimensions of length.

Associated with the drift motion is the *third adiabatic invariant* (NORTHRUP and TELLER, 1960)

$$\Phi = \int \mathbf{B} \cdot \mathbf{dA} \tag{14}$$

a flux integral taken as follows. The guiding center oscillates between the two mirror points. As the guiding center drifts east-west each mirror point traces out a curve about the dipole axis (the "mirror trace") while the guiding center itself tends to sweep out a "magnetic shell". The surface used in Equation (14) is that bounded by a mirror trace.

Each of the three adiabatic invariants is a constant if the magnetic field varies sufficiently slowly over the oscillation period which is associated with the invariant. In Table I we give representative periods over which the magnetic field variation should be small in order to allow conservation of the corresponding invariant.

The orbit of a particle is completely specified by the values of B_m and I: the value of B_m specifies that the mirror trace lies on a certain surface of constant field strength and the value of I selects a unique mirror trace from among all the possible traces. MCILWAIN (1961) has taken advantage of this property of the field to develop a coordinate system that has proved to be of enormous value in mapping particle distributions. Early attempts to map the measured intensities were severely hampered by the very steep spatial gradients. For example, at low altitudes in the inner zone a change in geocentric range of 3 per cent results in an intensity change of over a factor of 10. Since the shape of the true field deviates more than 3 per cent from any possible dipole representation, attempts to use a dipole model produce highly disordered results.

McIlwain's approach is to regard any point in space as the mirror point of a hypothetical (test) charge. The value of B at the point is thus regarded as B_m. The quantity I is then computed from the point to its conjugate *using the true field in the calculation.* The point is then labelled by B and I. One next turns to a perfect dipole field and seeks the point having the same values of B and I. In the dipole field this point is found on a particular magnetic shell having a particular value of r_e. One computes $L \equiv r_e/R$, R being the earth radius. This value of L is now assigned

TABLE I

REPRESENTATIVE PERIODS OF SMALL MAGNETIC FIELD VARIATIONS

Invariant *	*Associated period*	*Typical value of the associated period in the inner belt for* 100-*MeV protons*	*Typical value of the associated period in the inner belt for* 1-*MeV electrons*
μ	gyroperiod	10^{-3} sec	10^{-6} sec
I	bounce period	10^{-1} sec	10^{-1} sec
Φ	drift period	~ minute	~ hour

* To allow conservation of a given invariant the field should vary only slightly during the corresponding time shown at the right.

to the point in question *in the true field*. Thus a point in real space may be labelled by its B–L values. A shell of constant L is eccentric in location relative to the earth's center and is somewhat distorted. However, if a particle conserves its first two invariants, then the guiding center always remains on this shell. When particle intensities are plotted as functions of B–L rather than altitude-latitude-longitude a remarkable symmetry appears, vivid evidence of the magnetic control of the particles.

Since the geomagnetic field fluctuates on time scales of fractions of seconds to days, the constancy of certainly the second and third invariants, and possibly also the first invariant, is a subject of the greatest importance as well as much uncertainty. We shall return to this question a little later. For the moment we continue to study the motion in a time-invariant field.

Consider a number of particles which are released at the equatorial plane with a variety of pitch angles. According to Equation (8) those with the smallest equatorial pitch angles will mirror closest to the earth where the field strength is largest. Imagine that the earth has no atmosphere. If B_s is the value of B at the earth's surface then any particle with an equatorial pitch angle smaller than that given by

$$\sin \alpha_c = (B_e/B_s)^{1/2} \tag{15}$$

will strike the earth and be absorbed. The angle α_c thus defines a cone of directions at the equatorial plane within which no trapped particles will be found. This cone is called the *loss cone*. In the presence of the atmosphere the value of α_c must be modified to correspond to reflection at some critical altitude beneath which the atmosphere is too dense to allow periodic north-south oscillation.

We are now in a position to consider the *loss mechanisms* in a time-invariant magnetic field.

In passing through the outer atmosphere, charged particles make Coulomb interactions with the atoms, ions and free electrons. These Coulomb deflections change the pitch angle and dissipate energy. The fluctuations in pitch angle are of a random nature and produce a *diffusion* of the mirror point up and down along the line of force (WALT and MACDONALD, 1962). If the equatorial pitch angle random-walks into the loss cone the particle is lost. At the same time particles drift monotonically downward in energy and may be lost from the observable belt when their energy drops below detector thresholds.

When pitch-angle diffusion and energy loss are compared one finds that the effect of diffusion is negligible *for protons* (SINGER, 1958). That is, during its lifetime a proton maintains its initial value of α_e. (For electrons diffusion and energy loss are comparable). A lifetime may be defined for protons as, for example, the time over which the energy decreases from E to E/e (LENCHEK and SINGER, 1962). For $E \lesssim 300$ MeV in hydrogen or $E \lesssim 100$ MeV in helium or oxygen this time is

$$T_{1/e}(\text{energy-loss}) = k \cdot 10^{10}\, E^{3/2}/\varrho \,\text{sec} \tag{16}$$

where E is in Mev, ϱ is the number of atoms or ions cm^{-3} averaged over the spiral path and k is a coefficient given in Table II.

TABLE II

COEFFICIENT K FOR SEVERAL ATOMS AND IONS

Atmosphere	*k*
H	13.7
H^+	6.4
He	7.5
He^+	4.5
O	2.0
O^+	1.74

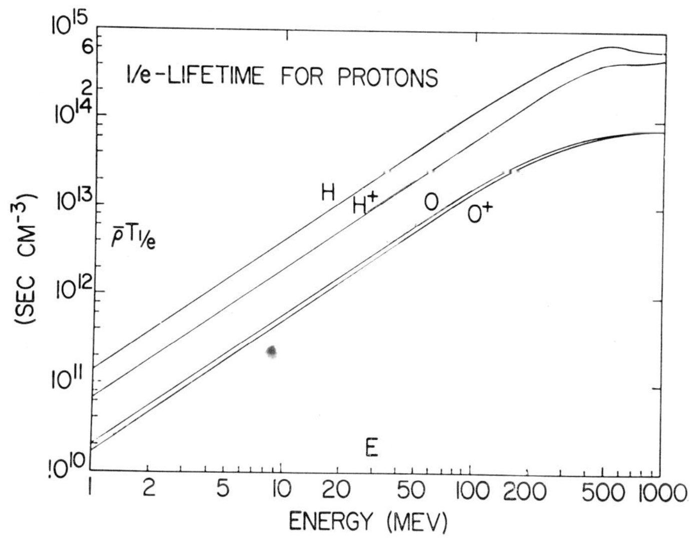

Fig. 2. 1/*e* – lifetime for protons, the time for reducing the energy *E* to *E*/*e* by coulomb interactions or for reducing the number of protons to 1/*e* by absorption in non-elastic nuclear interactions.

Nuclear interactions give a lifetime

$$T_{1/e}(\text{nucl. inter.}) = 1/\varrho \sigma v_n \tag{17}$$

where σ_n is the cross-section for nonelastic interactions. (Elastic scattering is negligible because at low energies, where the scattering is isotropic, energy-loss is a much faster loss process, and at high energies the scattering becomes very strongly peaked in the forward direction and the mean free path for accumulating a significant root-mean-square deflection becomes very long compared to the mean free path for nonelastic interaction). These two lifetimes may be combined to give the 1/*e*-lifetime shown in Figure 2.

At energies below ~1 MeV charge exchange dominates over energy loss. In an interaction such as

$$H + H^+ \rightarrow H^+ + H$$

a fast proton and a thermal atom are replaced by a fast atom and slow proton. The fast atom then flies out of the magnetosphere. These loss processes are summarized in Figure 3.

We now return to the problem of the constancy of the adiabatic "invariants". Two sorts of effects involving variation of the "invariants" may be envisaged. The field may fluctuate rapidly enough to violate the condition given by Equation (3) or the corresponding condition for one of the other "invariants". Secondly, even in a

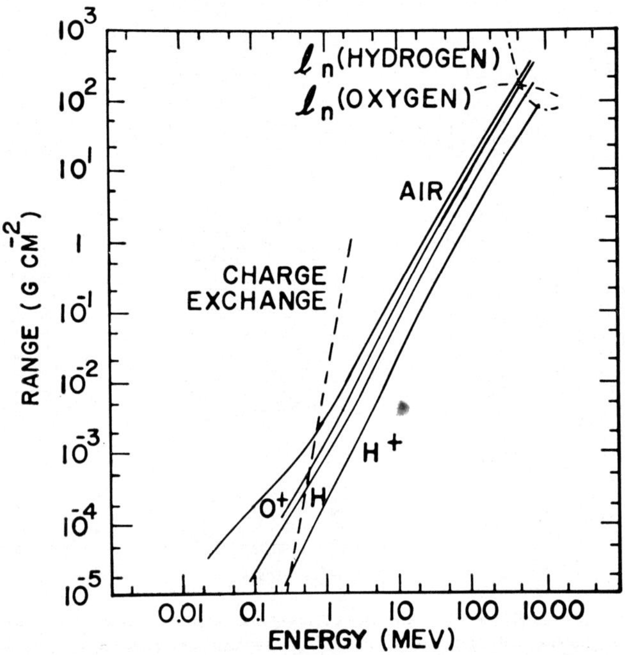

Fig. 3. Range and mean free path of protons. The solid curves are ranges for energy dissipation in coulomb interactions. The curve labelled 'charge exchange' gives the mean free path in atomic hydrogen. The curves labelled l_n give the mean free paths for non-elastic interaction.

perfectly static field the inhomogeneity may introduce a slow variation of, say, the magnetic moment (cf. Equation (4)). Thus a particle's mirror point may not remain fixed long enough for the collision processes to become dominant.

An experimental test of the latter possibility is known. Gibson *et al.* (1960) confined a quantity of Ne^{19} in a magnetic mirror machine. A detector sensitive to the positrons generated by the radioactive decay of the Ne^{19} looked into the loss cone from outside the machine. The trapping time was measured and found to be in good agreement with the time predicted by assuming perfect conservation of the magnetic moment and Coulomb scattering in the gas within the machine. The positrons made, on the average, a number of bounces between mirrors $\gtrsim 10^8$ and a number of gyrations $\gtrsim 2 \times 10^{10}$. A relativistic particle trapped in the earth's field for 10^8 bounce periods will have a lifetime of the order of one year.

The criterion for conservation of the magnetic moment in a static field, following from Equation (4), may be expressed in terms of the so-called "Alfvén discriminant"

$$\chi \equiv \frac{a}{|B/\text{grad}\, B|} \tag{18}$$

as $\chi \ll 1$. In the experiment of GIBSON *et al.* χ was ~ 0.02. We shall return to this question a little later; at this time we turn to the effects of fluctuations of the field.

A classical study of the effect on the magnetic moment of a changing magnetic field (HERTWECK and SCHLUTER, 1957) shows that the adiabatic approximation of constant magnetic moment holds to a very high approximation, even if the gyroperiod changes appreciably over a few gyroperiods. But it also indicates that the relative change of magnetic moment very rapidly becomes large when the time scale of the changing field approaches the gyroperiod. That is, when the rate of change of B is in resonance with the undisturbed gyrofrequency, the magnetic moment may be appreciably changed.

Calculations relating explicitly to the geomagnetically trapped particles have been made by, among others: DRAGT (1961), WENTZEL (1962) and HAYAKAWA and OBAYASHI (1963) on the first invariant, PARKER (1961) on the second invariant, and DAVIS and CHANG (1962) and PARKER (1960) on the third invariant. The calculations relating to the second and third invariants show that under the action of sufficiently strong hydromagnetic waves of the proper frequency or the action of large scale fluctuations of the field during magnetic storms, particles may diffuse across lines of force, i.e., L may change. The calculations on the first invariant indicate that particles whose gyration is in phase with the Doppler-shifted frequency of a hydromagnetic wave propagating along **B** may suffer changes in magnetic moment and may thus find their mirror points lowered into the atmosphere. For example, DRAGT (1961) finds that hydromagnetic waves generated in the outer magnetosphere will lead to the existence of a maximum trappable proton energy E_c of ~ 150 MeV at $L=2$ earth radii and varying as $E_c \propto L^{-11}$. Unfortunately, unambiguous observations of hydromagnetic waves in the inner magnetosphere have not yet been accomplished.

The effects of inhomogeneity and time variations of the field may perhaps be summarized as follows: The magnetic moment will not be conserved when the gyroradius is comparable to some typical scale length of the inhomogeneity. Each "invariant" will begin to be affected when the appropriate oscillation period becomes comparable to the time scale or period of the field variation. The effects of large scale variations of the field must be very important in the outer parts of the magnetosphere where the relative changes in field strength are large. In the inner region where the fast protons are observed, loss by action of hydromagnetic waves will be dominant if the waves are present with sufficient amplitude, with sufficiently high frequency of occurrence and with the proper frequencies. However, this loss process must always compete with loss in a static field due to an excessively large gyroradius and loss due to collision processes.

2. Geomagnetically Trapped Protons and Their Origin

We are now ready to consider the geomagnetically trapped protons and their origin. The radiation belt was discovered in February 1958 by VAN ALLEN and co-workers (1958). A Geiger counter carried on Explorer I detected an intense flux of ionizing particles at ~ 2000 km altitude in the latitude range $\pm 30°$. The early interpretation of the observed high counting rate was that one had detected particles of *solar* origin, perhaps electrons in the keV to MeV range. These early conclusions were necessarily very tentative because, as Prof. Herlofson has pointed out in his first lecture*, the observations were made with a single Geiger tube which is, by itself, unable to identify the particles. The possibility existed that the particles were protons with energies in the 100-MeV range.

SINGER (1958) and VERNOV and co-workers (1959) independently suggested that the radiation was, in fact, fast protons produced by the cosmic ray albedo neutrons. A cosmic ray albedo neutron is a neutron that has leaked out of the atmosphere, after having been produced by a nuclear interaction between a cosmic ray particle and an atmospheric nucleus. Some of these neutrons, with energies extending from thermal energy to at least a GeV, will decay within the magnetosphere, producing protons, electrons and neutrinos. Neutrons in the 100–MeV range will produce electrons which are chiefly below 1 MeV and protons with energies essentially equal to the energy of the parent neutron. Those protons which are injected with equatorial pitch angles outside of the loss cone will be trapped.

An estimate of the intensity of the proton belt may easily be made. From a knowledge of the number of nuclear interactions in the atmosphere and studies of the secondary nucleons emitted in high-energy nuclear interactions, the number of neutrons leaking out of the atmosphere may be estimated. SINGER (1958) obtained a figure ~ 0.1 neutrons $cm^{-2}\ sec^{-1}$ with energies > 50 MeV. Call this J_n. The decay density Q is then

$$Q = J_n/v\tau \sim 10^{-14}\ cm^{-3}\ sec^{-1}$$

taking an average velocity $v = 10^{10}$ cm sec^{-1} and a neutron lifetime $\tau = 10^3$ sec. The density of trapped protons will then be on the order of

$$N \simeq QT$$

where T is the lifetime. Assuming adiabatic trapping, $T = T_{1/e}$, the lifetime of a 100 MeV proton in a helium atmosphere of 10^5 atoms cm^{-3}, the density at 2000 km altitude HANSON (1962), is $\sim 5 \times 10^8$ sec. This gives $N \sim 5 \times 10^{-6}$ protons cm^{-3}. The proton flux $J = Nv$ will thus be $\sim 5 \times 10^4\ cm^{-2}\ sec^{-1}$. This is just about the observed value in the heart of the proton belt. (Note that all these figures refer to particles > 50 MeV).

* See footnote on page V.

The identity of the particles was confirmed to be protons with energies extending up to several – 100 MeV when FREDEN and WHITE (1960) flew a nuclear emulsion through the belt.

A quantitative calculation of the trapped proton intensity as a function of energy, pitch angle and line of force, may be made by solving the appropriate transport equation. Let $n(E, \alpha_e, L, t)$ be the number of protons in a unit volume, with equatorial

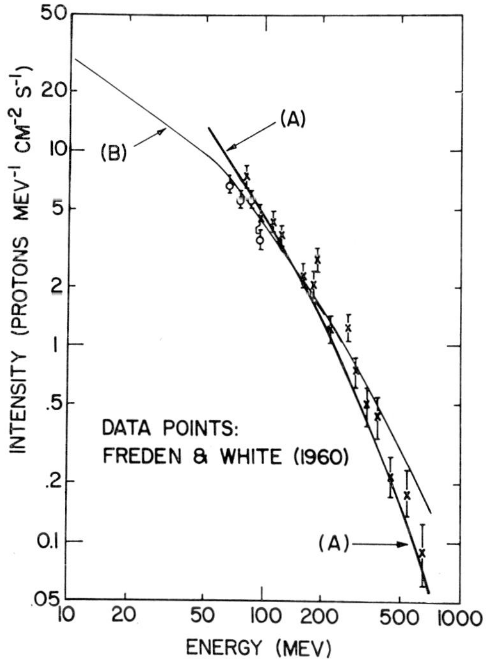

Fig. 4. The data are from two nuclear emulsion flights to ~ 1200 km, L ~ 1.3 (FREDEN and WHITE, 1960). Curve A is calculated by LENCHEK and SINGER (1962) taking into account the anisotropy of the albedo neutrons and is normalized to the data by dividing the computed intensity by 2.1 (E_c = 1000 MeV). Curve B is calculated by Freden and White assuming an isotropic albedo.

pitch angle within a unit solid angle, within a unit energy interval, on a shell described by L and at time t. The function n is governed by the transport equation (RAY, 1960; FREDEN and WHITE, 1960; LENCHEK and SINGER, 1962)

$$\frac{\partial n}{\partial t} = -\frac{\partial}{\partial E}\left(n\frac{\mathrm{d}E}{\mathrm{d}t}\right) - nv\sigma\varrho + q = 0. \tag{19}$$

We have put $\partial n/\partial t = 0$, implying equilibrium. The first term of Equation (19) represents the effect of energy loss, with

$$\frac{\mathrm{d}E}{\mathrm{d}t} = -4\pi(e^2/mc^2)^2 Z\varrho mc^3 \Lambda/\beta\,, \tag{20}$$

where e and m are the charge and mass of an electron, Z is the atomic number of the atmospheric atoms, ϱ is the number density averaged over the spiral path of the trapped particle, Λ is a slowly varying logarithm term whose magnitude may be taken to be constant at 23 for this problem and $\beta = v/c$, v being the speed of the trapped particle. Notice that ϱ is a function of α_e and L for a given atmospheric model.

The second term of Equation (19) represents the effects of nuclear collisions and charge exchange, σ being the appropriate cross-section, which is generally a function of energy.

The third term of Equation (19) is the unidirectional, differential source strength. This is the number of protons injected tangent to the spiral path per MeV per cm^3, per steradian per second, averaged over the spiral path. Thus q is a function of E α_e and L.

The production of secondary trapped protons by the nonelastic nuclear interactions has been analysed and found to be negligible.

Solutions of this equation using various source functions have been given by HESS (1959), RAY (1960), FREDEN and WHITE (1960) and LENCHEK and SINGER (1962). FREDEN and WHITE employed an *isotropic* neutron flux with a differential energy spectrum $E^{-2.0}$, based on the work of HESS *et al.* (1961). The result is curve B of Figure 4. However SINGER (1960b) pointed out that the neutron intensity above the atmosphere should be strongly *anisotropic*, the faster neutrons being confined to a narrow disc-shaped distribution tangent to the horizon. Now the source function q involves the directional neutron intensity averaged over all the directions toward which the spiralling velocity vector points, and evaluated at the point where the vector intersects the earth. When the velocity vector makes a large zenith angle as it contacts the earth, both high energy and low energy neutrons are seen. But when the velocity vector makes a relatively small zenith angle only relatively low energy neutrons are seen. Consequently, the injection of the higher energy protons is less favored and the spectrum is steepened.

Employing the anisotropic neutron intensity

$$j_n(E, \phi) = \begin{cases} 0.32\, E^{-1.8} & \text{for} \quad \phi > \dfrac{\pi}{2} - \vartheta(E) \\ 0 & \text{for} \quad \phi < \dfrac{\pi}{2} - \vartheta(E) \end{cases} \tag{21}$$

neutrons/MeV cm^2 ster sec, where E is in MeV, ϕ is the zenith angle and $\vartheta(E)$ is given by

$$\sin\vartheta = 300/p\,, \tag{22}$$

where p is the neutron momentum in MeV/c, LENCHEK and SINGER (1962) carried out the averaging discussed above. The result was equivalent to replacing the anisotropic neutron beam by an *isotropic* intensity with spectrum

$$j_n^{\text{effective}} = 23.5\, \eta_0(L, \alpha_e)\, E^{-2.9}\,. \tag{23}$$

The coefficient η_0 is independent of energy and is on the order of 0.25 in the inner belt ($L = 1.5$).

To see more clearly the reason for using the steeper effective spectrum rather than the original $E^{-1.8}$, one may compute the average of $j_n(E, \phi)$ over the upward hemisphere. Thus

$$\overline{j_n(E,\phi)} = \frac{1}{2\pi}\int_0^{2\pi}\int_{\varphi_{\min}}^{\pi/2} j_n(E,\phi)\sin\phi\,\mathrm{d}\phi\,\mathrm{d}\xi$$
$$= 0.32\,E^{-1.8}\sin\vartheta(E)$$
$$= \text{constant} \times E^{-1.8}p^{-1}. \tag{24}$$

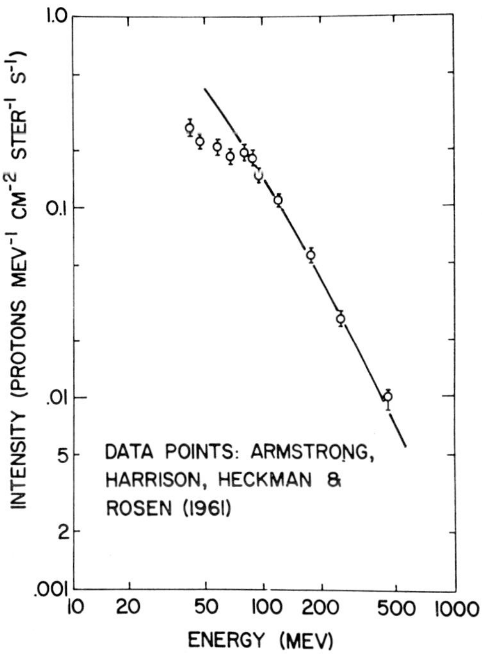

Fig 5 The data are results of a nuclear emulsion flight with apogee at ~ 1125 km, L ~ 1.2 (ARMSTRONG *et al.*, 1961). The solid curve is calculated from the neutron albedo theory (LENCHEK and SINGER, 1962) and is normalized to the data at 100 MeV. For the calculation a cut-off energy $E_c = 1000$ MeV was used.

The lower limit $\phi_{\min}$ is just $\pi/2-\vartheta$. We see that the average directional intensity has an energy spectrum steeper by the factor $1/p$.

The differential spectrum $E^{-1.8}$ is based on earlier work on charged cosmic ray albedo (WENTWORTH and SINGER, 1955).

The resulting trapped proton spectrum is shown by curve A in Figure 4 and also in Figure 5.

While the theory seems to account for the intensity and spectrum of fast protons *near the earth* it must be modified at large L. To see the necessity of this, we may return to the simple $N = QT(1/e)$ picture. Note that $\mathrm{T}(1/e)$ is proportional to $1/\varrho$. Therefore the trapped flux varies as Q/ϱ. To first approximation Q varies as L^{-2}. Thus the intensity of the belt should go on increasing so long as ϱ falls off faster than L^{-2}. An atmospheric model deduced by SINGER (1960a) falls off more rapidly than L^{-2} throughout the entire magnetosphere, which would lead to a radiation belt with intensity continually increasing with increasing L. A more conservative belt would be given by using the co-rotating atmosphere of JOHNSON (1960), which has a relatively constant ϱ beyond $L \sim 4$. But even this atmosphere produces a belt which, while it does exhibit a peak, does not fall off to one-half its peak intensity until $L = 4$. On the other hand the observed 100-MeV proton belt is rather sharply cut off at $L \sim 2$ (FAN, MEYER and SIMPSON, 1961).

The absence of high-energy protons beyond about $L = 2$ has been ascribed to variation of the magnetic moment due to interactions with hydromagnetic waves (DRAGT, 1961; WENTZEL, 1962) and, alternatively, to a "breakdown" of the constancy of the magnetic moment in the *static* dipole field (SINGER, 1959). According to the latter idea, one notes that for a particle of a given energy, the Alfvén discriminant is proportional to L^2. If a critical value χ_c. of the discriminant exists, above which the trapping fails, then protons of a given energy will no larger be trapped beyond the distance at which their discriminant equals χ_c.

That such a critical value does indeed exist is suggested by orbit calculations carried out for mirror geometries (GARREN *et al.*, 1958). SINGER made an empirical determination of χ_c, setting it equal to the discriminant of 75–100 MeV protons at $L = 2$. The result is

$$\chi_c \sim 0.08,$$

which falls within the range $0.02 - 1.0$ set by the positron trapping experiment (GIBSON *et al.*, 1960) and the "limit" of the Alfvén (1950) approximation (Equation (18), $\chi \ll 1$).

A consequence of Singer's theory is the prediction that the maximum trappable momentum varies as L^{-2} or, in the non-relativistic region the maximum trappable energy (cut-off energy) varies as

$$E_c \propto L^{-4}, \tag{25}$$

Observations which tend to confirm this behavior have recently been reported. MCILWAIN and PIZZELLA (1963) measured the energy spectrum of protons in the 30- to 50-MeV range. Expressing the spectrum as

$$j(E) = j_0 e^{-E/E_0} \tag{26}$$

they find

$$E_0 = \text{constant} \times L^{-5.2}, \tag{27}$$

a result which they regard as supporting the hypothesis of the breakdown of the adiabatic conditions. Observations of protons with $E > 350$ MeV (GARMIRE, 1963) also seem to confirm a breakdown of the magnetic moment with increasing L. Orbit calculations by Garmire demonstrate a breakdown of the magnetic moment, with the cutoff energy E_c varying approximately as L^{-4} for particles mirroring at constant altitude and as approximately L^{-2} for particles mirroring near the equatorial plane. The difference between these two results indicates that the pitch angle, neglected in the Singer treatment, plays a role in the breakdown process, the smaller the pitch angle the weaker being the trapping ability of the field.

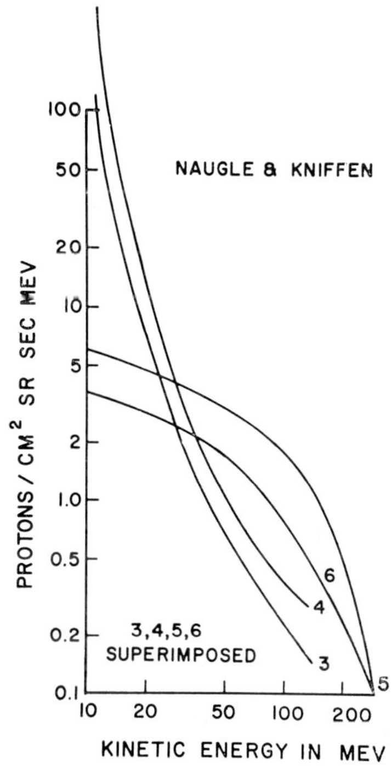

Fig. 6. Energy spectra of trapped protons on four different magnetic shells (NAUGLE and KNIFFEN, 1961). Spectrum 3, $L = 1.79$; Spectrum 4, $L = 1.72$; Spectrum 5, $L = 1.54$; Spectrum 6, $L = 1.47$. The geographic locations of the points can be seen in Figure 7.

Up to this point we have spoken only of neutrons generated by the *galactic* cosmic rays. The energy spectrum of the trapped protons resulting from such cosmic rays agrees fairly well with spectra observed at $L \lesssim 1.5$. However, a measurement at slightly higher L (NAUGLE and KNIFFEN, 1961) shows that the spectrum below about 20 MeV undergoes a rapid change in form at $L = 1.6$ (Figure 6). It was suggested (HECKMAN and ARMSTRONG, 1962; NAUGLE and KNIFFEN, 1961) that *solar* cosmic rays might generate a detectable proton component by producing albedo neutrons in the polar caps. Solar cosmic rays require a treatment separate from that given the galactic cosmic rays since their mean energy is three orders of magnitude smaller and they arrive in intermittent, but intense, bursts. Calculations (LENCHEK, 1962) of this effect show first, the proton energy spectrum produced by the polar albedo is in fair agreement with that measured by Naugle and Kniffen and second, because of the geometry of the problem no "polar" protons should be observed at $L \lesssim 1.6$. That is, a "shadow zone" is formed close to the equator (Figure 7).

Although these two features agree with the observations, the reader should be cautioned on the following point. The energy spectrum predicted by the calculations shows a peak near Θ the "nuclear temperature" of the evaporation reactions which produce the albedo neutrons. On theoretical grounds, this energy was estimated to be

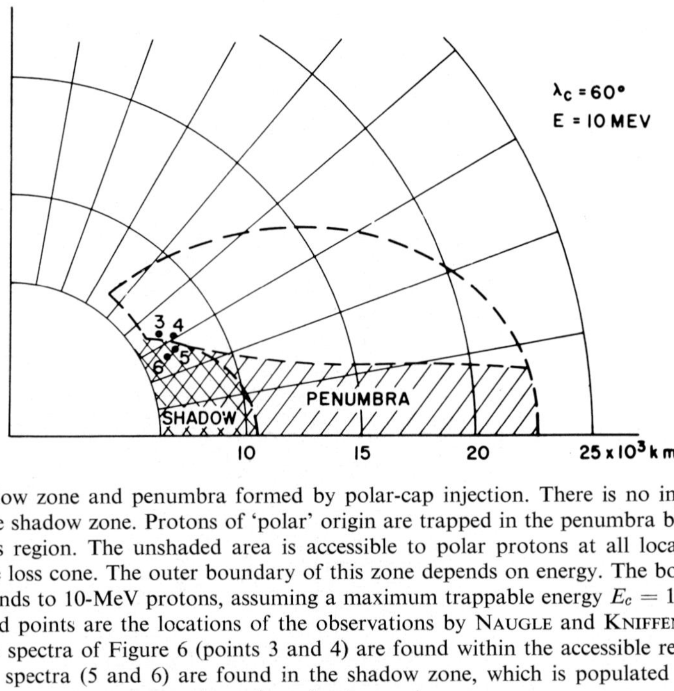

Fig. 7. Shadow zone and penumbra formed by polar-cap injection. There is no injection of fast protons in the shadow zone. Protons of 'polar' origin are trapped in the penumbra but they do not mirror in this region. The unshaded area is accessible to polar protons at all local pirch angles outside of the loss cone. The outer boundary of this zone depends on energy. The boundary drawn here corresponds to 10-MeV protons, assuming a maximum trappable energy $E_c = 1600\ L^{-4}$ MeV. The numbered points are the locations of the observations by NAUGLE and KNIFFEN (1961). Note that the steep spectra of Figure 6 (points 3 and 4) are found within the accessible region while the relatively flat spectra (5 and 6) are found in the shadow zone, which is populated by the albedo from the galactic cosmic rays.

4 MeV (LENCHEK, 1962). Thus the Naugle-Kniffen observations ($E > 12$ MeV) cover only the *tail* of the predicted spectrum. However, BAME *et al.* (1963) have measured a proton spectrum at $L = 2.5$, $B = 0.069$, that extends the Naugle–Kniffen spectrum down to 1 MeV without any sign of a peak. Possibly the correct value of Θ is near 1 MeV, (HESS *et al.*, 1961). Possibly the protons seen by Bame *et al.* are not of albedo origin. The problem deserves more study.

A proton component which is certainly not of albedo origin is that observed by Explorer XII (DAVIS and WILLIAMSON, 1963). In the energy range 100 keV to 5 MeV a flux of $\sim 10^8$ protons $cm^{-2}\ sec^{-1}$ is found at $L = 3.5$. This figure is several orders of magnitude larger than that produced by albedo neutrons, assuming loss only by atmospheric interactions and breakdown of the magnetic moment in the quiescent field. This outer belt component shows fluctuations in intensity associated with magnetic storm events. Beyond $L \sim 4$ the energy density of these protons is a large fraction of $B^2/8\pi$, suggesting that the origin of this component lies in a collective interaction with the magnetic field.

3. Suggestions for Future Investigations

We bring this seminar to a close by mentioning some areas of future work which will undoubtedly shed light on the problems remaining unsolved. Foremost among them is the question of the *conservation of the adiabatic invariants*. This problem can be attacked experimentally as well as theoretically.

If some part of the trapped protons are solar cosmic rays which have somehow been trapped by the field and/or diffused into it, we should expect to find some trapped *alpha-particles*. So far, none has been found in the nuclear emulsion experiments but the total number of particles analysed is small.

Better data on *neutron production* cross-sections are needed, as for example in the reaction N^{14} (p, n), N^{14} (p, pn), etc. In some energy ranges these cross-sections are unknown.

Measurements of the *energy spectrum of fast neutrons* in the atmosphere and above the atmosphere are needed, a particularly difficult task.

Finally, possibly the most challenging and rewarding task of all, the measurement of the *electric fields* in the magnetosphere and the *hydromagnetic waves*, together with the associated *background plasma*. These observations are vital to our understanding not only of the loss of protons but of the acceleration of protons and electrons within the magnetosphere, aurorae, ionospheric and atmospheric dynamics, and the entire magnetosphere. Indeed, when taken together with the fast particles they *are* the magnetosphere.

References

Alfvén, H.: 1950, *Cosmical Electrodynamics*, Clarendon Press, Oxford.

Armstrong, A. H., Harrison, F. B., Heckman, H. H., and Rosen, L.: 1961, *J. Geophys. Res.* **66**, 351.

Bame, S. J., Conner, J. P., Hill, H. H., and Holly, F. E.: 1963, *J. Geophys. Res.* **68**, 55.

Chandrasekhar, S.: 1960, *Plasma Physics*, University of Chicago Press, p. 61.

Davis, Jr., L. and Chang, D. B.: 1962, *J. Geophys. Res.* **67**, 2169.

Davis, L. R. and Williamson, J. M.: 1963, *Space Research* **3**, 365.

Dragt, A.: 1961, *J. Geophys. Res.* **66**, 1641.

Fan, C. Y., Meyer, P., and Simpson, J. A.: 1961, *J. Geophys. Res.* **66**, 2607.

Freden, S. C. and White, R. S.: 1960, *J. Geophys. Res.* **65**, 1377.

Garmire, G.: 1963, *J. Geophys. Res.* **68**, 2627.

Garren, A., Riddell, R. J., Smith, L., Bing, G., Henrich, L. R., Northrup, T. G., and Roberts, J. E.: 1958, *Proc. 2nd. U.N. Int. Conf. on Peaceful Uses of Atomic Energy*, United Nations, Geneve, p. 65.

Gibson, G., Jordan, W. C., and Lauer, E. J.: 1960, *Phys. Rev. Letters* **5**, 141.

Hanson, W. B.: 1962, *J. Geophys. Res.* **67**, 183.

Hayakawa, S. and Obayashi, T.: 1963: *J. Geophys. Res.* **68**, 3311.

Heckman, H. H. and Armstrong, A. H.: 1962, *J. Geophys. Res.* **67**, 1255.

Hertweck, F. and Schlüter, A.: 1957, *Z. Naturforsch.* **12a**, 844.

Hess, W. N.: 1959, *Phys. Rev. Letters* **3**, 11.

Hess, W. N., Canfield, F. H., and Lingenfelter, R. E.: 1961, *J. Geophys. Res.* **66**, 665.

Johnson, F. S.: 1960, *J. Geophys. Res.* **65,** 577.
Lenchek, A. M.: 1962, *J. Geophys. Res.* **67,** 2145.
Lenchek, A. M. and Singer, S. F.: 1962, *J. Geophys. Res.* **67,** 1263.
Lew, J. S.: 1961, *J. Geophys. Res.* **66,** 2681.
McIlwain, C. E.: 1961, *J. Geophys. Res.* **66,** 3681.
McIlwain, C. E. and Pizella, G.: 1963, *J. Geophys. Res.* **68,** 1811.
Naugle, J. E. and Kniffen, D. A.: 1961, *Phys. Rev. Letters* **7,** 3.
Northrup, T. G. and Teller, E.: 1960, *Phys. Rev.* **117,** 215.
Parker, E. N.: 1960, *J. Geophys. Res.* **65,** 3117.
Parker, E. N.: 1961, *J. Geophys. Res.* **66,** 693.
Ray, E. C.: 1960, *J. Geophys. Res.* **65,** 1125.
Singer, S. F.: 1958, *Phys. Rev. Letters* **1,** 181.
Singer, S. F.: 1959, *Phys. Rev. Letters* **3,** 188.
Singer, S. F.: 1960a, *J. Geophys. Res.* **65,** 2577.
Singer, S. F.: 1960b, *Phys. Rev. Letters* **5,** 300.
Spitzer, L.: 1956, *Physics of Fully Ionized Gases,* Interscience, New York, Ch. 1.
Van Allen, J. A., Ludwig, G. H., Ray, E. C., and McIlwain, C. E.: 1958, *Jet Propulsion* **28,** 588,
Vernov, S. N., Grigorov, N. L., Ivanenko, I. P., Lebedinskij, A. I., Murzin, V. S., and Čudakov. A.: 1959, *Doklady AN SSSR* **124,** 1002 (English Translation: *Soviet Physics – Doklady* **4,** 154).
Walt, M. and Mac Donald, W. M.: 1962, *J. Geophys. Res.* **67,** 5013.
Wentworth, R. C. and Singer, S. F.: 1955, *Phys. Rev.* **98,** 1546.
Wentzel, D.: 1962, *J. Geophys. Res.* **67,** 485.

CHAPTER 31

COMETARY AND AURORAL STUDIES AT LIEGE

L. REMY-BATTIAU*

Institut d'Astrophysique de l'Université de Liège, Cointe-Sclessin, Belgium

1. Cometary Studies

As already pointed out by LÜST (1962), the study of comets is one way to obtain information about interplanetary matter (plasma and dust) and more generally, one may say that their random position in space permits their use as test-objects for approaching some aspects of solar activity. Unfortunately, our knowledge of cometary physics is far from being complete and the study of quite a few problems needs to be deepened or even, in some cases, simply undertaken. Of course, a lot of problems could be solved by an adequate comet probe; SWINGS (1962ab) has already reviewed this subject in different papers. At the present time, a group of workers at Liège are trying to contribute to the understanding of several of these problems.

The different molecules or radicals in comets emit their characteristic electronic spectra by fluorescence processes excited by the electromagnetic radiation of the sun, taking into account the absorption of radiation corresponding to the Fraunhofer lines and the radial velocity of the comet respectively to the sun (Swings' effect). However, when studying emission band spectra one has to make assumptions regarding the distribution of the molecules or radicals on the different energy levels; the gas densities, of course, are so low that no thermodynamic equilibrium prevails. Considering the rotational levels of the lower vibrational level of the lower electronic state involved in the transition $B^2\Sigma^+ - X^2\Sigma^+$ of CN, the type of distribution proposed by CARRINGTON (1962), synthetic profiles have been obtained for the (O–O) band in the case of Comet Burnham (1959*k*) and Comet Seki-Lines (1962*c*) and compared to the observed profiles; the agreement is good. However, further investigations have to be pursued for other comets and other emissions before drawing conclusions as to the validity of Carrington's method.

The problem of the excitation processes of the forbidden oxygen red and green lines has not yet received a satisfactory answer (REMY-BATTIAU, 1962). If one supposes that the oxygen atoms are released on the fundamental level ^3P, it is impossible to explain by a fluorescence mechanism, the different intensity ratios observed of the red doublet to the green line. When taking into consideration excitation by electron collisions, the electronic density N_e being: 10^3 elec. cm$^{-3} \leqslant N_e \leqslant 10^5$ elec. cm^{-3}, it is still impossible to explain the fact that occasionally the green line is more intense than the red doublet, even when assuming electronic temperatures as high as 10^5 °K. It seems that

* *Chargé de Recherches F.N.R.S.*

one has to admit that the oxygen atoms are released on the different excited states ^1D and $^1S_0$ and that, in some cases, the ^1D level is de-excited in less than 10^2 sec; in view of this, one should then reconsider the possibility of the fluorescence mechanism.

Spectroscopic investigations of comets have generally dealt with the different discrete emissions usually neglecting the continuum. Nevertheless, it has been shown that this continuum is most probably due to scattering of sunlight by solid particles (HOUZIAUX, 1959), the shape, nature and size distribution of which are still unknown. Until a probe is launched through the head of a dust comet, the only way of obtaining information is in trying to explain theoretically the degree of polarization and the intensity of the continuum at different wavelengths, and to evaluate the total mass of particles required to justify the observed intensities. There is still a definite shortage of precise monochromatic measurements in spectral regions where the continuum alone is present, and till now, just a few observations of this type have been available, concerning mainly the two bright comets Arend-Roland (1957 III) and Mrkos (1957 V). Supposing that the particles are small spherical "bits" of dielectric "ices" of radius a and refractive index n, it has been possible to fit the curve obtained in the case of Comet Arend-Roland (1957 III) by BLACKWELL and WILLSTROP (1957) for the polarization degree as a function of the diffusion angle, with particles whose refractive index $1.25 < n < 1.50$ and sizes distribution of the type a^{-4} with 10^{-6} cm $\leqslant 2a \leqslant 10^{-4}$ cm. On the other hand, VAINU BAPPU and SINVHAL (1960) showed that the polarization degree at λ 4300, 4800 and 5890 Å was of the order of 20 to 25% for Comet Arend-Roland (1957 III) and Comet Mrkos (1957 V). Assuming that the polarization degree P is slowly varying in that wavelength range, for the same size limits and distribution as previously mentioned, one can explain the uniformity of $P(\lambda)$ by taking a refractive index close to 1.50. Unfortunately with these hypotheses, it was impossible to reproduce the curve deduced by HOUZIAUX (1959) from his observation of Comet Arend-Roland (1957 III) for the spectral intensity distribution (REMY-BATTIAU, 1963). Nevertheless, no definite conclusion can be drawn from these considerations until other ranges of sizes and other types of particles (absorbing particles or a mixture of dielectric and absorbing particles) have been considered.

In connection with the relationship between solar activity and cometary phenomena, the anomalous brightness outbursts are certainly of great interest. Because of the continuum type of the few spectra observed during such outbursts, WHITNEY (1955) and PLAVEC (1957) tried to interpret these outbursts, assuming that the increase in brightness was due to reflection of sunlight by small particles ejected from the nucleus. Their papers have been discussed taking into account the fact that for the small dimensions of the particles considered by these authors, the phenomena is not reflection but scattering of light. Assuming respectively a refractive index equal to 1.25 and 1.50 and radii of 10^{-3}, 10^{-4}, 10^{-5} cm, the ejected masses, for densities 0.81, 0.91 and 1.62 g cm^{-3}, lie in the range 10^9 to 10^{13} grams for the outbursts considered (Swift, June 4, 1899; Holmes, Jan. 16, 1893; Pons-Brooks, Sept. 23, 1883 and Jan. 1, 1884). For particles of refractive index $1.27 - i\,1.37$, radius 10^{-4} cm and density 1 g cm^{-3} the computed mass lies in the same range. Assuming velocities of ejection of 0.2,

TABLE I

ABSORPTION ABOVE THE EMISSION LAYERS OF DIFFERENT POSSIBLE ATOMIC EMISSIONS BELOW 3000 Å

Element; number of multiplets; emission altitude	*Spectroscopic notation of the transition*	*Wavelength Å*	*Intensity at the altitude of emission* I_{em}	*Further remarks on the intensities*		
OI	$2p^4\ {}^3P - 3s\ {}^3S^\circ$	1304.8	50	at 120 km, $I = 27.6$;	at 260 km, $I = 26.3$	remains the strongest OI emission at all altitudes.
(6)	$2p^4\ {}^3P - 3d\ {}^3D^\circ$	1027.4	33	at 120 km, $I = 3.6$;	at 260 km, $I = 3$	should probably be enhanced by the solar Lyman-β line.
100 km	$2p^4\ {}^3P_1 - 2p^4\ {}^1S_0$	2972	11	no absorption		
	$2p^4\ {}^3P - 3s'\ {}^3D^\circ$	990.2	11	absorbed by N_2		
OII	$2p^3\ {}^2D^\circ - 2p^4\ {}^2D$	718.5	1.5	at 160 km, $I = 0.2$		
(12)	$2p^3\ {}^4S^\circ - 2p^4\ {}^4P$	833.3	1.5	at 160 km, $I = 0.7$		weak emissions
140 km						
NI	$2p^3\ {}^2P^\circ - 2s\ {}^2P$	1744	0.5	at 120 km, $I = 0.3$		weak emission
(12)						
100 km						
NII	$2p^2\ {}^1D_2 - 2p^3\ {}^1D_2^\circ$	775.9	2.4	at 140 km, $I = 0.3$		
(23)	$2p^2\ {}^3P - 2p^3\ {}^3P^\circ$	916	2.4	at 140 km, $I = 0.6$;	at 240 km, $I = 0.1$	
130 km	$2p^2\ {}^3P - 2p^3\ {}^3D^\circ$	1084.6	2.4	at 140 km, $I = 1.3$;	at 240 km, $I = 0.6$	

0.8 and 1.6 km sec^{-1}, the kinetic energies of the halos observed should be 10^{17} to 10^{23} erg. A cometary nucleus of 1 km radius, of albedo 0.1 and whose effective absorbing surface is one fourth of its total surface, requires a time 10^{-2} to 10^4 hours to absorb an amount of solar energy equal to the computed kinetic energies (REMY-BATTIAU, 1963).

The exact mechanism responsible for the ejection of material from the nucleus and the formation of the tail is not well understood. In that connection problems of ionization of the material ejected and the acceleration of condensations of matter in the tail are being investigated from cometary pictures taken with short time exposure at rather short time intervals with Baker-Nunn camera used for satellite tracking.

Information about the electron density within comet tails could be deduced from occultation by radio-stars. Unfortunately, the ray deviation which the tail should produce would be barely higher than one minute of arc, which is about the value of the resolving power of a good radio telescope.

2. Auroral Studies

At the present time, as far as we know, very little information has been obtained from rocket observations on the far ultraviolet spectrum of the aurora. Prediction of this ultraviolet spectra might help in preparing such experiments and of course would be greatly valuable in disentangling rapidly the essential features of the spectrograms really observed, as soon as such material is secured.

Predictions of the ultraviolet emissions have been published (SWINGS, BOSMAN-CRESPIN and ARPIGNY, 1961) but of course, when observed from the upper part of the atmosphere, let us say by a rocket spectrograph pointing downward (vertically), the intensity of the different possible emissions may be attenuated by atmospheric constituents above the emitting layer. Predictions of this have been prepared for a brightness III aurora, where the intensity of the green [OI] line (λ 5577) in the relative scale adopted is equal to 100. The different relative intensities deduced for the ultraviolet emissions are based on laboratory data (MOORE, 1950; BATES, 1949; NICHOLLS, 1962; BIRGE and HOPFIELD, 1928; TANAKA, 1953; WILKINSON, 1956) and on relative intensities for the visible region given by CHAMBERLAIN (1961).

The results obtained for atomic elements are gathered in Table I. The first column indicates: (1) the element studied, (2) in brackets, the number of multiplets which one should expect to observe below 3000 Å relying on the observations in the visible region, (3) the altitude at which the emission takes place. The second column gives, for the brightest emissions, the spectroscopic notation of the transition, followed in the third one by the corresponding approximate wavelength on which the multiplet is centered, and in the fourth one by the intensity in the relative scale adopted at the height of emission. The fifth column collates indications and remarks concerning the relative intensities of these brightest emissions when observed vertically from above the emission layer. The effect of absorption by atmospheric constituents has been calculated using NICOLET's (1962) model atmosphere and values of the absorption coefficients of O_2, N_2, O and N given by WATANABE, INN and ZELIKOFF (1953),

TABLE II

ABSORPTION ABOVE THE EMISSION LAYERS OF DIFFERENT POSSIBLE MOLECULAR EMISSIONS BELOW 3000 Å

Element; number of multiplets; emission altitude	*Spectroscopic notation of the transition*		*Intensity at the altitude of emission* I_{em}	*Further remarks on the intensities*
$N_2 2P$ (5) 100–120 km	$C^3\Pi_u - B^3\Pi_g$	(2–0)	2.3	no absorption
$N_2 VK$ (22) 140 km	$A^3\Sigma_u^- - X^1\Sigma_g^+$	(1–3)	16	no absorption
		(2–2)	14	no absorption
$N_2 L–B–H$ (52) 100–120 km	$a^1\Pi_g - X^1\Sigma_g^+$	(0–2)	5.8	if emission height = 100 km: at 120 km absorbed by O_2 = 120 km: at 260 km, $I = 2.6$
		(0–3)	5.8	if emission height = 100 km: at 120 km absorbed by O_2 = 120 km: at 260 km, $I = 3.8$
$N_2 B–H$ (17) 100–120 km	$b^1\Pi_u - X^1\Sigma_g^+$	(0–2)	3	if emission height = 100 km: at 160 km, $I = 0.55$ = 120 km: $I = 2.7$
		(0–3)	3	= 100 km: $I = 1$ = 120 km: $I = 2.8$
		(0–4)	3	= 100 km: $I = 0.2$ = 120 km: $I = 2.5$
N_2^+ (29) 110–120 km	$C^2\Sigma - X^2\Sigma$	(3–6)	6	if emission height = 110 km: at 240 km, $I = 4.9$ = 120 km: at 240 km, $I = 5.7$
		(3–8)	6	no absorption
		(3–9)	6	no absorption
		(3–10)	7	no absorption

WILKINSON and MULLIKEN (1952), WATANABE and MARMÖ (1956), LEE (1955), PRESTON (1940), BATES and SEATON (1949, 1950), MATSUNAGA and WATANABE (1961), ITAMOTO Jr and MCALLISTER (1961). Table II contains the same type of informations relative to the main molecular emissions.

One may conclude that the dominating features of the ultraviolet auroral spectrum seen from above the emitting layers should be the OI triplet around 1304.8 Å, the [OI] 2972 Å line, the NII lines the wavelengths of which are greater than 1740 Å and the NII 1084.6 Å emissions, the $N_2 2P$(2–0), (3–1) and (4–2) bands and the N_2VK bands. If N_2LBH is emitted at 120 km, most of the bands of this system will remain visible at higher levels but if emitted at 100 km, all bands should be already greatly absorbed by O_2 at 120 km. If N_2^+ is emitted at 120 km most bands will remain visible at all heights but if emitted around 110 km, all bands whose wavelengths are in between 1350 and 1500 Å will be rapidly absorbed by O_2 at 120 km. All other emissions (of OI, OII, NI, NII, N_2, N_2^+, O_2, O_2^+, NO, NO^+, OH) should be weak or absent.

If hydrogen is present and emitted around 110 km, the Lyman-α line should be intense at all altitudes, Lyman-β should be reduced to half its intensity around 180 km and then remain constant while Lyman-γ and Lyman-δ should be rapidly absorbed by N_2.

References

BATES, D. R.: 1949, *Proc. Roy. Soc. London*, A, **196,** 217.
BATES, D. R. and SEATON, M. J.: 1949, *Mon. Not. Roy. Astron. Soc.* **109,** 698.
BATES, D. R. and SEATON, M. J.: 1950, *Proc. Phys. Soc. London* **63,** 129.
BIRGE, R. T. and HOPFIELD, J. J.: 1928, *Astrophys. J.* **68,** 257.
BLACKWELL, D. E. and WILLSTROP, R. V.: 1957, *Mon. Not. Roy. Astron. Soc.* **117,** 383.
CARRINGTON, T.: 1962, *Astrophys. J.* **135,** 883.
CHAMBERLAIN, J. W.: 1961, *Physics of the Aurora and Airglow,* Academic Press, p. 197.
HOUZIAUX, L.: 1959, *Bull. Acad. Roy. Belg., Cl. Sc.* **45,** 218.
ITAMOTO Jr, F. K. and MCALLISTER, H. C.: 1961, *Hawaii Institute of Geophysics, Contribution* **29.**
LEE, P.: 1955, *J. Opt. Soc. Amer.* **45,** 703.
LÜST, R.: 1962, *Space Sci. Rev.* **1,** 522.
MATSUNAGA, F. M. and WATANABE, K.: 1961, *Hawaii Institute of Geophysics, Contribution* **33.**
MOORE, C.: 1950, *N.B.S. Circular* **488,** section 1.
NICHOLLS, R. W.: 1962, *J. Quant. Spectr. Rad. Transf.* **2,** 433.
NICOLET, M.: 1962, Private communication to Mrs. D. Bosman-Crespin.
PLAVEC, M.: 1957, *Publ. Astron. Inst. Acad. Sc. Czechoslovakia* **30.**
PRESTON, W. M.: 1940, *Phys. Rev.* **57,** 887.
REMY-BATTIAU, L.: 1962, *Ann. Astrophys.* **25,** 171.
REMY-BATTIAU, L.: 1963, *Bull. Acad. Roy. Belg.*, Cl. Sé., **49,** 916.
SWINGS, P., BOSMAN-CRESPIN, D., and ARPIGNY, C.: 1961, *Mém. in 8°, Soc. Roy. Sc. Liège, 5e série,* **5,** 583.
SWINGS, P.: 1962a, *Space Age Astronomy* (ed. by A. J. DEUTSCH and W. B. KLEMPERER), Academic Press, p. 370.
SWINGS, P.: 1962b, *Smithsonian Astrophys. Obs., Special Report* **111.**
TANAKA, Y.: 1953, *J. Chem. Phys.* **21,** 1403.
VAINU BAPPU, M. K. and SINVHAL, S. D.: 1960, *Mon. Not. Roy. Astron. Soc.* **120,** 152.
WATANABE, K., INN, E. C. Y., and ZELIKOFF, M.: 1953, *J. Chem. Phys.* **21,** 1026.
WATANABE, K. and MARMÖ, F. F.: 1956, *J. Chem. Phys.* **25,** 965.
WHITNEY, C.: 1955, *Astrophys. J.* **122,** 190.
WILKINSON, P. G. and MULLIKEN, R. S.: 1952, *Astrophys. J.* **125,** 594.
WILKINSON, P. G.: 1956, *Can. J. Phys.* **34,** 250.

CHAPTER 32

ROCKET EXPERIMENTS, LABORATORY WORK AND EQUIPMENT AT THE ASTROPHYSICAL INSTITUTE OF THE UNIVERSITY OF LIEGE

F. REMY

Institut d'Astrophysique de l'Université de Liège, Cointe-Sclessin, Belgium

1. Rocket Experiments

This first part will be a brief description of two space projects related to auroral and cometary phenomena, which we hope will be performed in the near future within the scope of ESRO.

The first of these projects consists in the launching through an aurora of a high luminosity spectrograph especially designed for the photographic recording of far ultraviolet spectra between 1000 and 3000 Å. In order to diminish its size, this spectrograph is an off-plane modified Wadsworth-type mounting. The concave grating which will be used in this experiment has 1200 lines/mm and the prototype photographic chamber has an aperture ratio $f/4$. It is highly probable that following the same principles, a second version with an aperture ratio $f/2$ will be realizable. After the exposure, the photographic film will reintegrate a shielded box which will be recovered. An experimental arrangement of the optical parts of the spectrograph has now been achieved and tests of the quality of spectra obtained will be started very soon.

The second project consists in the launching at twilight of 40 liters of liquid NH_3 at an altitude of about 240 km. The aim of this experiment is not to form a cometary nucleus of the "icy" type but to study the behaviour of NH_3 in the field of the solar radiation and the production of the fluorescence spectrum of NH_2, one of its photolysis products. A report pointing out the interest of such an experiment has been presented by SWINGS (1962). The main characteristic of this experiment on account of the small available quantity of NH_3, is to release it in the form of small droplets which should sublimate more rapidly than the total amount. Actually, laboratory experiments are being undertaken to find out how liquid NH_3 behaves when suddenly exposed to vacuum, as well as to test the most adequate methods of release of NH_3. On the other hand, theoretical work is being carried out on the mechanism of fluorescence of NH_3 in the field of solar radiation in order to prepare synthetic spectra of the fluorescent emissions.

2. Laboratory Works

As very often pointed out by Swings, and more recently in a report presented at the

COPERS Colloquium on Space Astronomy in July 1962 (SWINGS, 1963), a great deal of experimental work is needed to prepare efficiently space experiments and to fully exploit the results. So, an important part of the activity of the Institute is directed towards spectroscopic works able to furnish basic data required for space science.

The first one is related to the problem of the C_3 radical in comets. The so-called "4050 group" was extensively studied in the emission spectra of comets (HERZBERG (1942ab); SWINGS (1942, 1956)) and finally correctly attributed by DOUGLAS (1951, 1954) to the C_3 molecule. In collaboration with Herzberg and Lagerqvist, a very extended analysis is being done of a high dispersion absorption spectrum of C_3 obtained by flash photolysis of diazomethane. But even for an absorption spectrum, its structure is still very complex. Nevertheless, the analysis of this spectrum has furnished new results (especially concerning the grounds of the complexity of the spectra obtained) which have been presented at a meeting of the Faraday Society in April 1963 (GAUSSET *et al.*, 1963). In order to simplify the spectrum, work has been carried out by BARGER and BROIDA (1962) and WELTNER and WALSH (1962), who succeeded in producing emission spectra of C_3 trapped in rare gases matrix at low temperature. But due to intermolecular effects, the lines are shifted and the comparison with other spectra is rather difficult. As can be easily seen from this short description of the problem, it would be of great interest to produce C_3 absorption spectra starting from different "parent-molecules". For this reason, we are constructing a photoflash equipment of the conventional type; we hope it will be working at the end of this year. The description of typical photoflash equipment may be found in papers published by PORTER (1950) and CLAESSON and LINDQVIST (1957ab).

Also in connection with the study of cometary molecules, we have installed a non-conventional source able to produce carbon-compounds spectra. This source is in fact the combination of a small carbon furnace and a hollow cathode discharge, the carbon furnace standing for the cathode of the discharge.* This source has been constructed to withstand high voltage pulses (of the order of 20 kV), to keep a good vacuum (in order to avoid the *CO* spectrum as impurity) and to allow a rapid adaptation to the study of vacuum ultraviolet spectra. Heated with a low voltage current provided by a variable transformer of 10 kilowatts, the furnace may reach temperatures greater than 2000 °C. A systematic study of spectra produced by different kinds of exciting sources will be undertaken with this installation, namely, with a direct current source, and low and high voltage pulses.

Equipment for "time-resolved" spectroscopic studies is also under development at the Institute. Attempts at a quantitative explanation of the first results obtained are under way and I shall only mention briefly some potentialities of this method. A critical review of methods and results of "time-resolved" spectroscopy has been recently published by LAQUA and HAGENAH (1963). The source used for the excitation of the gas (a radar type RT-31/APS-2F equipped with a 2J22 magnetron) is run at a recurrence frequency of 1 kc/sec and gives sharp pulses of reproducible and known

* A more extensive description of this installation will be published somewhere else by F. Remy.

duration of 2 μsec. What may be studied in this case by the "time-resolved" spectroscopy is the evolution of the intensity of the spectral lines of the afterglow emitted by the excited gas between the exciting pulses. This study may be done electronically with a photomultiplier followed by a "gate" amplifier. By a time-displacement of this gate between the exciting pulses, it is possible to follow the change in intensity of an emission line. In the case of the He I 4471.5 Å line, for example, a strong afterglow appears rapidly after the end of the exciting pulse for a pressure of 10 mm Hg of He, in the discharge tube. For a pressure of 2 mm Hg, this afterglow is very faint and appears after a longer period following the end of the exciting pulse. Another possibility of this "gate" device is, by operating it at a definite and fixed time after the end of the exciting pulse, to scan with the photomultiplier the whole spectrum emitted by the excited gas. It has been found, for example, that the ratio of intensities He I 3888.6/He I 4471.5 is > 1, one μsec after the end of the exciting pulse, but becomes < 1, three μsec after the end of this pulse. By some improvement in the electronics and namely in the width of the "gate" system, it seems possible, with this basic instrumentation, to determine some fundamental physical parameters involved in the excitation phenomena.

Besides the use of electronics, some attempts at photographic "time-resolved" spectroscopy have also been made. One advantage of the photographic emulsion is its ability to record the full "time-resolved" spectrum in one exposure. The source of excitation used (the driver unit of a SCR 584 radar) gives pulses of 0.8 μsec width at a recurrence frequency of 400 c/sec. A slotted disc bearing windows of a suitable shape is run by a motor synchronized with the exciting source. Placed between the discharge tube where the gas is excited and the spectrograph, this device makes it possible to isolate and to photograph parts of afterglows at a given time after the end of the exciting pulse. By changing the phase between the current feeding the synchronous motor and the exciting pulses, it is possible to explore the interval between these pulses. With this kind of source, it has been found that the He II and He_2 spectra are formed during the first 30 μsec after the end of the exciting pulse. In order to change easily the width of the exploring window, a second synchronous motor and another disc of the same shape will be added in front of the existing equipment. The width of the window will be modified by changing the phase between the currents feeding the two motors. If the following experiments are quite satisfactory, this system will be adapted for the study of vacuum ultraviolet afterglows.

In a paper, MICHEL (1957) pointed out the presence of a new spectrum in the Schumann region appearing in a hollow cathode discharge through a rapid stream of water. He obtained this spectrum after an exposure time of four hours with a 1-m focal-length vacuum spectrograph and a resolving power of about 10000. Michel suggested that this spectrum might be due to O_2H or O_2H^+. But as far as we could find out, no further successful work on this emission has been carried out. For this reason, we have tried to determine the best conditions of production of this spectrum to allow a high dispersion study of it. With the 3-m focal-length vacuum spectrograph of the Institute, we succeeded after two hours' exposure to photograph fairly strong spectra. The resolving power obtained was about 50000 and the reciprocal dispersion of about 5 Å/mm at 1850 Å. Owing to the high complexity of these spectra, still larger dispersion and resolving power are needed; for this reason, we are now adapting a new grating on the 3-m spectrograph.

Before going into a short description of laboratory equipment, I would mention that a very extended study of the vacuum ultraviolet absorption spectrum of H_2 is carried on at the Institute. Numerous results of this study have already been published by MONFILS (1960, 1961abc and 1962).

3. Laboratory Equipment

Actually, the main apparatus for spectroscopic research in the vacuum ultraviolet at the Institute, is a 3-m focal-length spectrograph in the Paschen-Runge mounting working at normal incidence with a 15000 lines/inch grating. With this grating, spectra can be obtained between 700 and 4500 Å in the first order with a resolving power of more than 50000 at 3000 Å. We shall soon be able to use a 30000 lines/inch grating which will increase the resolving power in the first order and will also allow the use of a second slit situated on the opposite side of the used portion of the Rowland circle. This slit will make available in the second order, the region between 1000 and 2000 Å. Furthermore, a photoelectric recording device has been placed along the Rowland circle, which will permit the scanning of the spectrum with speeds varying in 12 steps from about 1mm/min to 130mm/min.

The construction of a 21-foot vacuum spectrograph is presently under way. It will consist of an Eagle mounting with the optics completely coplanar. To avoid vibrations, the optical parts of this spectrograph (slit, grating and support of the Rowland circle) will be fixed in the ground independently of the vacuum tank. It will be possible to replace very quickly the photographic recording by a photoelectric scanning of the spectrum. With a 30000 lines/inch grating, it will be possible to reach with this spectrograph the 6th order at 2000 Å.

A high luminosity vacuum ultraviolet monochromator (aperture ratio $f/5$) will be working very soon. One of its first uses will be the calibration of UV photomultipliers and photographic films to be launched in space experiments.

Two new apparatuses for quantitative reduction of laboratory and astronomical spectrograms will complete the existing equipment. The first one, already working, consists in the combination of a microdensitometer (Zeiss G II) which gives the optical density of the photographic emulsion as a function of wavelength, and a device consisting in an X-Y recorder and a curve follower unit, which transforms this density into the intensity of the incident light, according to the calibration curve of the corresponding wavelength region. The second apparatus, the construction of which is under way, is a photoelectric comparator for accurate determination of wavelengths. The basic comparator is a McPherson 100; the optical and electronic arrangements are inspired by those described by TOMKINS and FRED (1951) and STEINHAUS (1959) (but with cylindrical optics), the measurements being made by coincidence of two traces on an oscilloscope screen. An encoder will be attached on the comparator screw to allow punching out measurements of line positions on cards which may be directly handled by electronic computers (PHILLIPS, 1959).

References

BARGER, R. L. and BROIDA, H. P.: 1962, *J. Chem. Phys.* **37,** 1152.
CLAESSON, S. and LINDQVIST, L.: 1957a, *Arkiv för Kemi* **11,** 535.
CLAESSON, S. and LINDQVIST, L.: 1957b, *Arkiv för Kemi* **12,** 1.
DOUGLAS, A. E.: 1951, *Astrophys. J.* **114,** 466.
DOUGLAS, A. E. and CLUSIUS, K.: 1954, *Can. J. Phys.* **32,** 319.
GAUSSET, L., HERZBERG, G., LAGERQVIST, A., and ROSEN, B.: 1963, *Discussions of the Faraday Society* **35,** 113.
HERZBERG, G.: 1942a, *Astrophys. J.* **96,** 314.
HERZBERG, G.: 1942b, *Rev. Mod. Phys.* **14,** 195.
LAQUA, K. and HAGENAH, W. D.: 1963, *Proceedings of the Xth Colloquium Spectroscopicum Internationale*, Spartan Books, Washington D.C., p. 91.
MICHEL, A., 1957, *Z. Naturforsch.*, **12**a, 887.
MONFILS, A. and HERZBERG, G.: 1960, *J. Mol. Spectroscopy* **5,** 482.
MONFILS, A.: 1961a, *Bull. Acad. Roy. Belg. Cl. Sc.* **47,** 585.
MONFILS, A.: 1961b, *Bull. Acad. Roy. Belg. Cl. Sc.* **47,** 599.
MONFILS, A.: 1961c, *Bull. Acad. Roy. Belg. Cl. Sc.* **47,** 816.
MONFILS, A. and ROSEN, B.: 1962, *J. Quant. Spectr. Radiat. Transfer* **2,** 321.
PHILLIPS, J. G.: 1959, *Astrophys. J.* **130,** 308.
PORTER, G.: 1950, *Proc. Roy. Soc. London* **A 200,** 284.
STEINHAUS, D. W.: 1959, *J. Opt. Soc. Amer.* **49,** 468.
SWINGS, P.: 1942, *Rev. Mod. Phys.* **14,** 140.
SWINGS, P.: 1956, *Vistas in Astronomy* **2,** (Ed. A. BEER), Pergamon Press, London, see references mentioned p. 972–973.
SWINGS, P.: 1962, 'Research in Space Science', *Smithsonian Astrophysical Observatory, Special Report* **111.**
SWINGS, P.: 1963, *Space Sci. Rev.* **1,** 762.
TOMKINS, F. S. and FRED, M.: 1951, *J. Opt. Soc. Amer.* **41,** 641.
WELTNER Jr, W. and WALSH, P. N.: 1962, *J. Chem. Phys.* **37,** 1153.

Note added in proof. The first launching, at twilight, of liquid NH_3 has been carried out with success in July 1964 in Sardinia. A description of this experience, of the observation equipment and of the first results obtained are published in a set of papers listed below.

BREDHOL, H., DUBOIS, I., et GAUSSET, L.: *Bull. Soc. Roy. Sc. Liege* (in press).
LÜST, R., ORTNER, J., et ROSEN, B.: *Revue d'Aéronautique* (in press).
REMY, F. and BREDHOL, H.: *Bull. Soc. Roy. Sc. Liège* (in press).
REMY, F.: *Revue Universelle des Mines* (to be published).

PART VIII

SOLAR COSMIC RAYS

CHAPTER 33

SOME EXAMPLES OF ENERGY AND NATURE MEASUREMENTS OF SOLAR PARTICLES

LYDIE KOCH

Service d'Électronique Physique, Centre d'Études Nucléaires de Saclay, France

1. Solar Particles

A. NATURE

Direct cosmic ray study is a probe into the understanding of source and propagation mechanisms of particles in the solar system. Solar cosmic rays include protons between 1 MeV and 10 to 20 GeV, α particles and heavier nuclei, perhaps also electrons and γ rays.

In this range of energy below 10 GeV, the cosmic ray beam is approximately a sample of the nuclear abundances present in the solar system. The quantities given in Table I, column 2, are measured with nuclear emulsions in balloons. The main anomaly is for the light group Li, Be, B, 10^6 times more abundant in cosmic rays; this is due to fragmentation of nuclei of higher Z crossing interstellar matter. The presence of such amounts of these light nuclei indicates that the maximum thickness of matter, mainly H, crossed between source and us is approximately 3.5 g/cm^2. Measurements of solar heavy nuclei show great variation from flare to flare as we shall see later.

TABLE I

NUCLEAR ABUNDANCES (after P. MORRISON)

	Cosmic rays $E \leqslant 10$ GeV	*Mean cosmic composition*
H	100	100
He	15	15
Li		$4 \cdot 10^{-7}$
Be	0 to 0.4	$1 \cdot 10^{-7}$
B		$1 \cdot 10^{-7}$
C		
N	1.2 ± 0.4	0.037
O		0.010
F		0.010
Ne	0.2	0.003
Mg	0.09	0.003
Si	0.07	0.004
Fe		$6 \cdot 10^{-4}$
Co	0.06	to
Ni		$4 \cdot 10^{-3}$
from Ni onwards	$< 10^{-5}$	10^{-6}

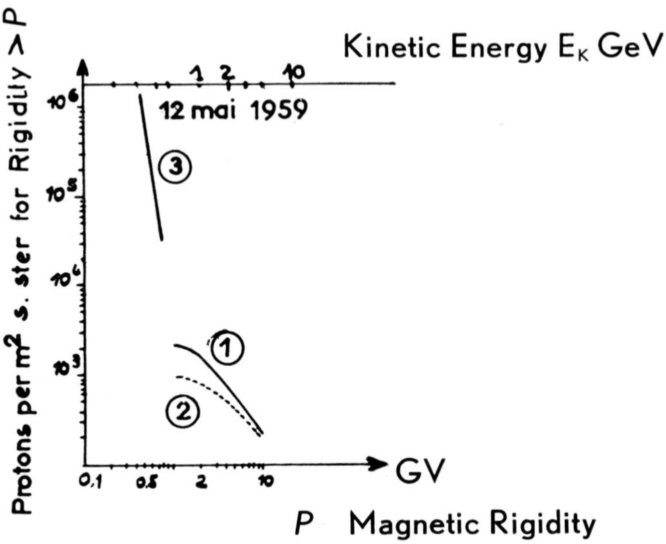

Fig. 1. Integral spectral distribution of cosmic protons during solar minimum (curve 1), solar maximum (curve 2) and solar flare of May 12, 1959 (curve 3). *P* is the magnetic rigidity defined as momentum/unit charge and expressed in units of GV.

B. TIME VARIATIONS

Until now the main result from the study of solar particles is the correlation of the time variations of their intensity and sunspot activity. There are two major solar modulations of the cosmic ray intensity for energies below 10 GeV: one of short duration called the Forbush decrease, at times appertaining to flares, and one of a period of 11 years called a long-time variation as we shall see for protons and heavier

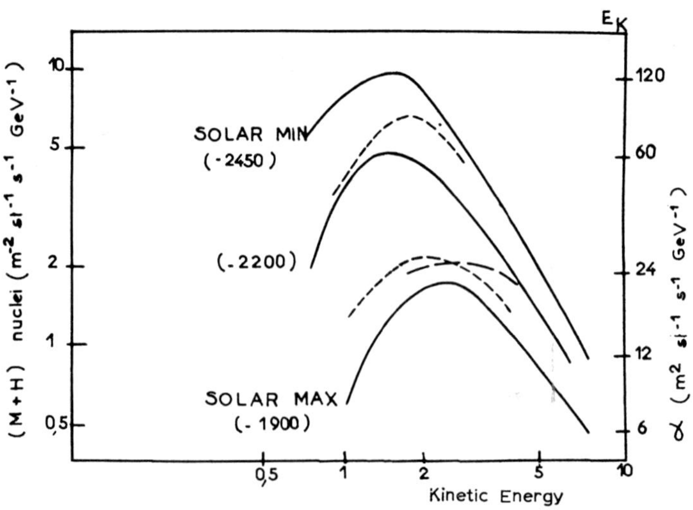

Fig. 2. Differential spectral distribution of α rays and Medium Nuclei.

nuclei in Figure 1 and 2. On the contrary the galactic flux of energy E > 10 GeV has been constant within a narrow margin of 10^8 years.

The double origin of the low energy part of the cosmic beam was demonstrated by Meyer and Vogt in balloon experiments during the last great flares in July 1961.

In the energy above 200 MeV this solar modulation is the dominating process: the flux increases with decreasing solar activity but below 200 MeV the proton intensity decreases with declining solar activity. This leads us to conclude that most of these

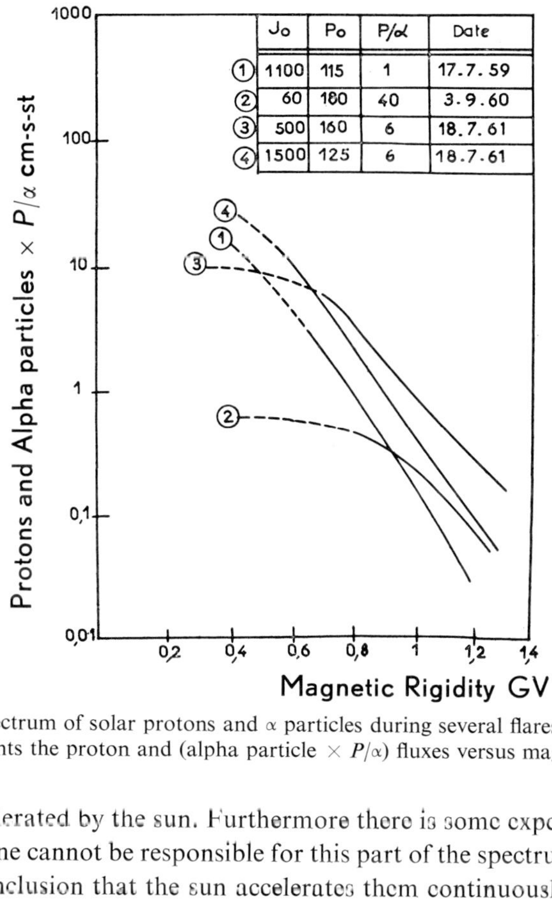

Fig. 3. Integral spectrum of solar protons and α particles during several flares (from Freier). Each curve represents the proton and (alpha particle × P/α) fluxes versus magnetic rigidity.

particles are accelerated by the sun. Furthermore there is some experimental evidence that the flares alone cannot be responsible for this part of the spectrum, which leads us to the further conclusion that the sun accelerates them continuously.

C. SPECTRAL DISTRIBUTION

From the best correlations between measurements obtained so far, the rigidity seems the fundamental parameter for understanding the behaviour of the solar particles. From the work of FREIER and WEBBER (1963) an exponential rigidity spectral representation fits all the measurements of solar cosmic rays of energy $E \geqslant 1$ MeV.

$$\frac{\mathrm{d}J}{\mathrm{d}P} = \left(\frac{\mathrm{d}J}{\mathrm{d}P}\right)_0 e^{-P/P_0}$$

or in integral form $J = J_0\, e^{-P/P_0}$. With $J_0 = P_0\, (\mathrm{d}J/\mathrm{d}P)_0$ and these terms are only functions of time. $\mathrm{d}J/\mathrm{d}P$ is the differential number of particles by rigidity interval.

Both solar protons and α particles are described by the same equation with the same P, and not the same J, at a given time. Propagation in the interplanetary medium changes P and J but not the exponential nature of the spectrum: one obtains similar spectrum for all components at all times (Figure 3). (See note added in proof, p. 410.)

This suggests that the spectrum in the source region may also be an exponential rigidity spectrum with a characteristic P, related to that observed on earth by the propagation conditions existing for the particles.

Now it is possible to deduce the main parameters to be studied in relation to solar particles.

(1) Intensity of radiation versus time until the lowest possible energies.

As G. Wibberenz proposes it will be very interesting to study the maximum intensity for different rigidity levels and to take these maximum intensities for constructing the spectrum instead of putting the intensity at a fixed time.

Calculations show that for three dimensional isotropic processes the maximum intensities will depend only on the distance R from the source as R^{-3}.

Direct observations in space by Explorer XII (Bryant and Clyne) suggest such a mechanism. Simultaneous measurements at various distances from the sun in and outside the magnetosphere are needed.

(2) Rigidity spectrum of various components and correlation with other characteristics of flares (energy released, amplitude of magnetic field, radio emission spectra and related electron spectra).

(3) Ratio of intensity of various nuclei above a given rigidity.

(4) Flux anisotropy at the beginning of the flares.

The very low energy part of the spectrum is of great interest for testing theories of injection and modulation of charged particles in the inner solar system.

2. Use of Solid State Detectors for Cosmic Particles Spectrometry

The experimental study of solar particles is now possible with a new technique, the solid state detectors (SSD) which permit precise measurements of low energy proton

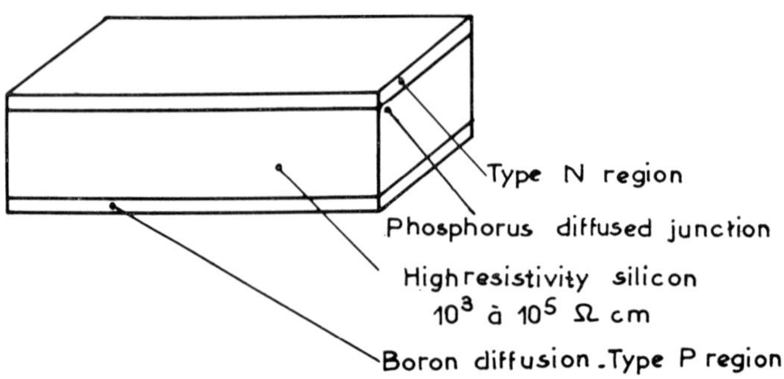

Fig. 4. Solid state detector: silicon PN junction.

spectrum and precise determination of the nature of the particles. This technique was already used for the study of protons in radiation belts and of solar protons between 1 and 15 MeV by Pieper in the Injun Satellite during the July 1961 flares.

A. GENERAL PROPERTIES OF SSD

Developed four years ago, this new type of nuclear particle detector looks like a flat

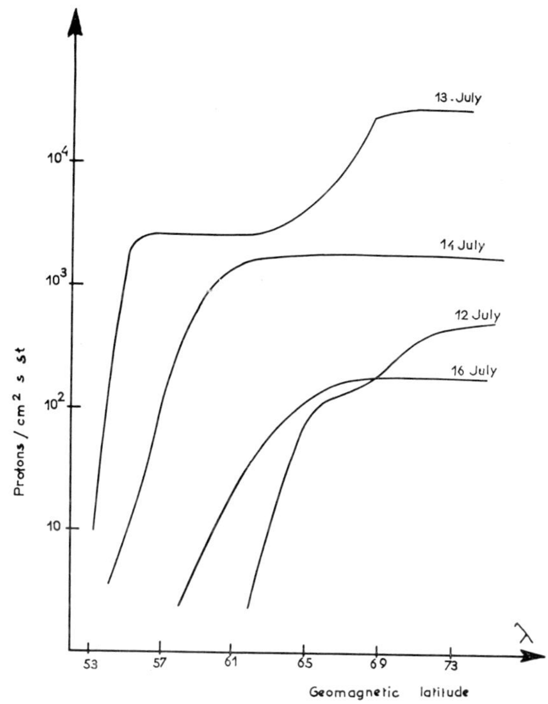

Fig. 5. Low-energy protons measured by the Injun satellite during the period July 12–16, 1962.

cylinder with a surface of a few cm^2 and a height between 0.5 and 5 mm. The material is a semiconductor of high resistivity, silicon in most cases (cf. Figure 4).

On each flat side an electrode is deposited and a difference of potential of about 100 V is applied between these electrodes. When one charged particle crosses the silicon piece an electric pulse appears between the electrodes, due to the collection of all the charges produced. This device is a solid ionisation chamber, and may be used as a detector but also to measure the energy lost by the particle in it, simply by measuring the amplitude of the pulse produced. There are many papers describing the properties of this new kind of detector, including descriptions of how to construct them.

There are several advantages over the ionization chamber or counters:

(1) The mean energy loss per ion pair produced is 10 times smaller (3.5 eV in silicon instead of 35 eV in air) also the precision will be 3 times higher. For example the energy loss in 1 mm of silicon by a 4 GeV proton (minimum ionization energy)

is 281 keV, the number of electronhole pairs produced is approximately 80000. The precision of energy loss measurement will not be limited by statistical fluctuations on the number of pairs but only by the electrical noise and the Landau fluctuations; incidentally the thickness of air equivalent for creation of the same number of ion pairs will be 10 m.

(2) There is no "wall" effect: the solid state detector can operate in vacuum and at very low temperature.

(3) The electric pulse given by a nuclear particle is very short, less than 10^{-8} sec.

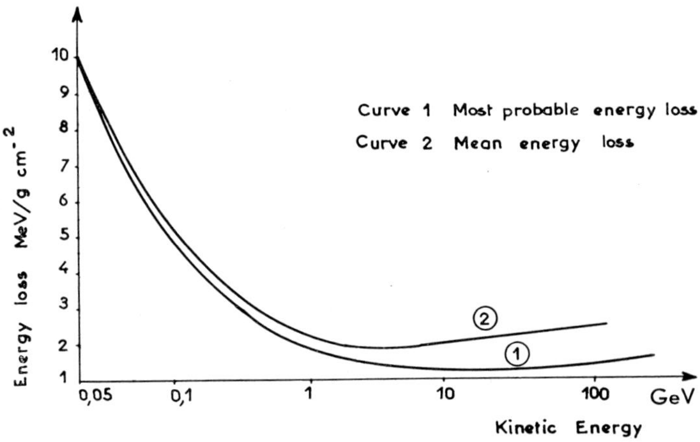

Fig. 6. Specific energy loss for protons in silicon (with density correction).

This permits discrimination between several particles and also counting at the very high rate necessary during flares (cf. Figure 5).

B. PRINCIPLE OF ENERGY MEASUREMENT

Solar particles are constituted mainly of protons, the measure of the specific energy loss dE/dx gives us directly the energy of the incoming proton; dE/dx is given in Figure 6. The two curves correspond to the fact that there is a fluctuation in the energy lost by several mono-energetic particles in the same amount of material, this fluctuation called "Landau effect" is due to the rare collisions where the energy transferred to the knock-on electron is very great. This gives an asymetrical distribution of dE/dx where the most probable energy loss is lower than the mean energy loss.

The thickness of the distribution is: 7% at 20 MeV, 14% at 50 MeV, 23% at 100 MeV, 28% at 500 MeV and above for only one detector.

This fluctuation can be reduced by use of two identical detectors, where only the smaller pulse is taken into account.

Figure 7 shows dE/dx distribution obtained with π-mesons at the minimum ionization energy and protons near the minimum with the synchrocyclotron Saturn.

We have constructed a proton spectrometer for energies between 1 and 500 MeV with two solid state detectors in coincidence which define a small solid angle for

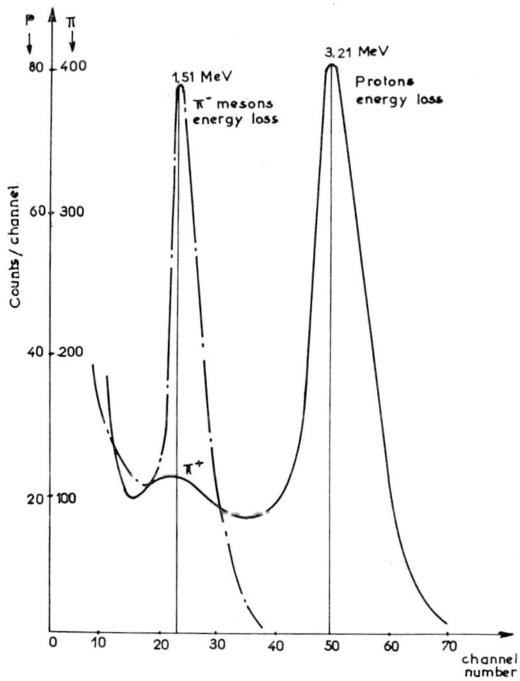

Fig. 7. Line spectrum of pions and protons of momentum 0.787 GeV/c measured with NIP silicon diode, thickness 4.93 mm.

admission of particles, 10^{-2} st, in order to avoid geometrical fluctuations in energy loss (cf. Figure 8).

C. PRINCIPLE OF *Z* MEASUREMENT

If we now consider the problem of how to distinguish the *Z* number of 2 relativistic

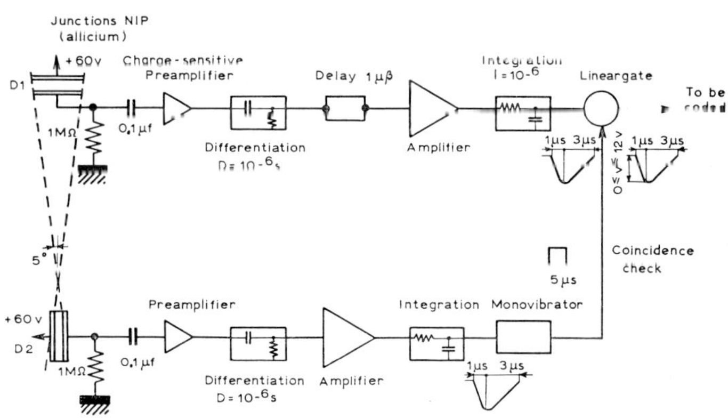

Fig. 8. Proton spectrometer with a telescope of solid state detectors. Block diagram.

particles having energy greater than the minimum of ionisation E_{min}, it seems that solid state detectors present some advantages over other devices.

As we know from previous measurements the sun accelerates He, H_3 (Discoverer XVII) and much heavier nuclei. The main results are published in the Proceedings of the Washington COSPAR Meeting and Kyoto Conference on Cosmic Rays (Sept. 1962). For example Lunik II and Sputnik VII measurements on heavy

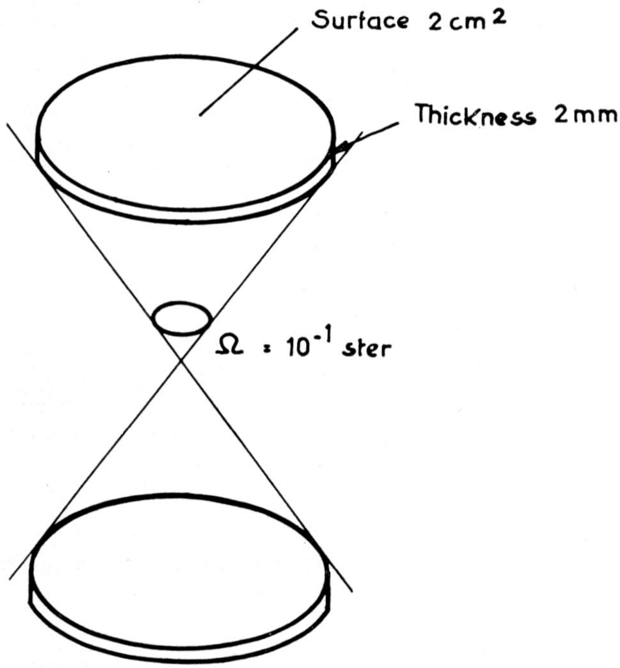

Fig. 9. Telescope of silicon junction detectors for atomic number selection of relativistic cosmic ions.

nuclei up to $Z = 14$ during flares give the ratio of intensity during flare and the mean intensity in quiet periods which increases with Z, a very surprising result which demands confirmation.

Z measurement of particles of energy $E > E_{min}$ can be made very simply by means of the telescope of solid state detectors described previously, but with a greater solid angle for admission of particles (cf. Figure 9). This part of the incoming spectrum can be easily selected by the geomagnetic cutoff $\lambda < 38°$.

Since dE/dx is proportional to Z^2 this provides a ready means of discriminating between the various particles.

Earlier studies of the amplitude of electric pulses in solid state detectors due to U-235 fission particles stopped in silicon, have shown that the proportionality with energy loss is conserved until at least an energy loss of 30 $MeV/mg \cdot cm^{-2}$. As a singly-charged relativistic particle loses 1.3 keV/mg·cm^{-2}, that means that the solid state detectors may be crossed by particles losing 23000 times more energy than singly charged particles, without showing any saturation effects. This means that we could observe particles having $Z=150$, if they can be created.

From the theoretical analysis of the Landau effect in one silicon detector (2 mm thick), discrepancy in the Z selection is given by Figure 10, $\Delta Z=1$ until $Z=20$; $\Delta Z=3$ until $Z=50$. We cannot use this precision in a satellite experiment, of say, six months' duration, because of the very low incoming flux.

For solar heavy particles studies it is necessary to know the nature and spectrum of

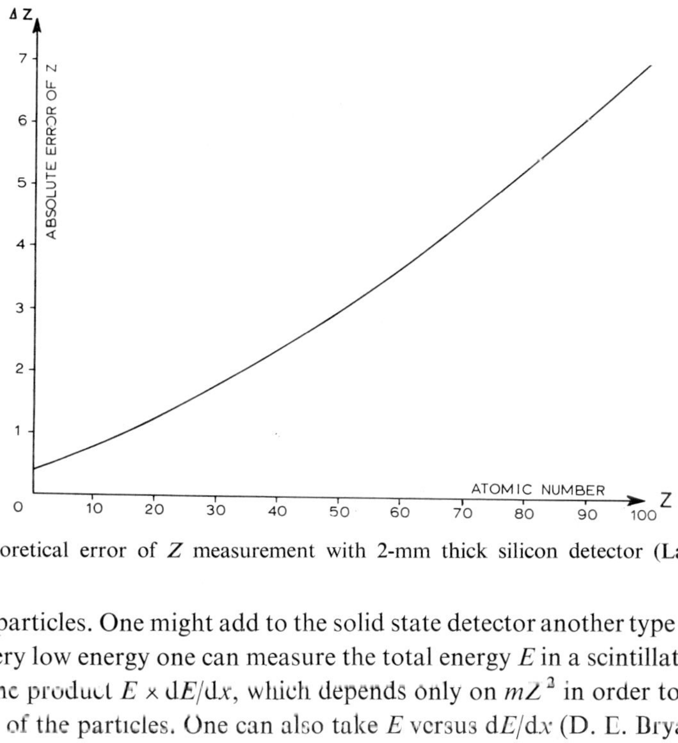

Fig. 10. Theoretical error of Z measurement with 2-mm thick silicon detector (Landau effect).

low energy particles. One might add to the solid state detector another type of detector:

(a) at very low energy one can measure the total energy E in a scintillation counter and make the product $E \times \mathrm{d}E/\mathrm{d}x$, which depends only on mZ^2 in order to obtain the mass and Z of the particles. One can also take E versus $\mathrm{d}E/\mathrm{d}x$ (D. E. Bryant).

(b) at higher energy one can measure the Čerenkov light produced by the particle in a transparent medium.

$$I \propto Z^2\left(1-\frac{1}{\beta^2 n^2}\right), \qquad \beta = v/C$$

where n is the refractive index, I increases with energy.

Ariel measurements of the energy spectrum of heavy cosmic rays were made with this type a counter, which gave the integral spectrum of all nuclei of $Z>6$. A similar method was utilized in Lunik II and Sputnik VII. The method of $\mathrm{d}E/\mathrm{d}x$ measurement by solid state counter associated with Čerenkov counter measurement permits selection of different Z values.

References

BRYANT, D. E. and CLYNE, T. L.: 1963, *Space Research* **3,** 376.
BRYANT, D. E., LUDWIG, G. H., and MCDONALD, F. B.: 1962, *IRE Trans., Nuclear Science* **9,** 376.
FREIER, P. and WEBBER, W. R.: 1963, *Space Research* **3,** 692.
FREIER, P. and WEBBER, W. R.: 1963, *J. Geophys. Res.* **68,** 1605.
MEYER, P. and VOGT, R.: 1963, *Phys. Rev.* **129,** 2275.
MORRISON, P.: *Handbuch der Physik* **24**.

Related Literature

A. COSMIC RAYS

FICHTEL, C. E. and GUSS, D. E.: 1961, 'Heavy Nuclei in Solar Cosmic Rays', *Phys. Rev. Letters* **6,** 495
FREIER, P. S.: 1962, 'The Measurement of the Cutoff Rigidity at Minneapolis Using Solar Protons and Alpha Particles', *J. Geophys. Res.* **67,** 7.
FREIER, P. S.: 1963, 'Measurements of Solar Particles and Protons', *J. Geophys. Res.* **68,** 1805.
GUSS, D. E. and WADDINGTON, C. J.: 1963, 'Observations on the Solar Particle Events of July, 1961', *J. Geophys. Res.* **68,** 2619.
MACDONALD, F. B. and WEBBER, W. R.: 1962, 'Čerenkov Scintillation Counter Measurements of the Light Medium and Heavy Nuclei in the Primary Cosmic Radiation from Sunspot Minimum to Maximum', *J. Geophys. Res.* **67,** 2119.
PIEPER, G. F., SMUDA, A. J., and BOSTRÖM, C. O.: 'Solar Protons and the Magnetic Storm of July 13, 1961', *Space Research* **3**.
RAY, E. C.: 1962, 'Re-entrant Cosmic Ray Albedo', *J. Geophys. Res.* **67,** 3289.
SCHATZMAN, E.: 1963, 'Energy and Mass Spectra of Solar Cosmic Rays', *Space Research* **3**; *Ann. Astrophys.*
SINGER, S. F.: 1958, 'The Primary Cosmic Radiation and Its Time Variation', *Progress in Elementary Particles and Cosmic Ray Physics,* North Holland Publishing Co., Amsterdam.
WEBBER, W. R.: 1962, *Time Variations of Low Rigidity Cosmic Rays During the Recent Sunspot Cycle,* North Holland Publishing Co., Amsterdam. (With 250 references.)

B. SEMICONDUCTOR RADIATION DETECTORS

Proceedings of the VII Annual Meeting, 1961, *IRE Trans.* **8,** 1.
Proceedings of the II International Symposium on Nuclear Electronics, Belgrade, May 1961; AIEA, Vienna, 1962.
HANSEN and GOULDING: 1961, 'Leakage Current in Semiconductor Function Detector', *Nuclear Instruments and Methods* **12,** 249.
MANN, H. M. and MELKONIAN: 1962, *IRE Trans.* **9,** 1.
MESSIER, J.: 1962, 'Silicium de haute résistivité compensé par transmutation nucléaire', *Compt. Rend.* **17,** 2083.
STEINHEIMER, R. M.: 1959, *Phys. Rev.* **115,** 137.

Note added in proof (January 1965). Recent works on low energy proton and alpha particles spectra give very different and more precise results. If the alpha particle data are multiplied by a factor of 6.7 to normalize them adequately to the protons in the rigidity range 1.5–2 GV, an important splitting of the two spectra is noted at lower rigidities and both spectra are falling sharply at the lowest rigidities measured (0.5 GV).

ORNES, I. and WEBBER, W. R.: 1964, *Phys. Rev. Letters* **13,** 106.
FICHTEL, C. E. and GUSS, D. E.: 1964, *Phys. Rev.* **133,** 818.

CHAPTER 34

SOME COMMENTS ABOUT DIFFUSION OF SOLAR PARTICLES IN INTERPLANETARY SPACE

GERD WIBBERENZ

Institut für Reine und Angewandte Kernphysik, Universität Kiel, Germany

The acceleration of charged particles to relativistic energies is one of the most interesting features of cosmic physics. Calculations by SCHATZMAN (1963) have shown that, under conditions valid in the chromosphere of the sun, a hydromagnetic shock wave (that may be connected somehow with the energy release mechanism during a solar flare) may serve as a source of particle acceleration, leading to the right order of magnitude for the total intensity and a power law spectrum.

A description of experimental methods for measuring the energy spectra of charged particles of different nature and a summary of some results has been given by L. Koch during this course; it has been pointed out that the spectra can be changed between sun and earth, because the interplanetary (ipl) magnetic fields influence the motion of a charged particle. Several review articles exist on this subject (ROEDERER, 1963; SIMPSON, 1963; DORMAN, 1961); various authors discuss the conditions of ipl space during special solar events (e.g. MEYER *et al.*, 1956; BISWAS *et al.*, 1962; NEY and STEIN, 1962), to be compared with several models for the magnetic fields (MORRISON, 1956; GOLD, 1959; ELLIOT, 1960, PARKER, 1958, 1961). An extensive summary of experimental results and their interpretation has been given by FREIER and WEBBER (12).

During the last years, satellite experiments have been shown to be very useful for two reasons: (1) The vehicles can stay outside the earth's magnetosphere, so that it is not necessary to consider secondary effects inside the cavity. (2) In contrast to balloon and rocket observations from high latitudes a continuous registration of particle fluxes is possible down to very low energies. Also, the interesting initial phase of an outburst (the first arrival of particles of different energies) can be studied.

Careful analysis of the time-dependence of the energy spectra for different particles should furnish information on acceleration near the sun as well as on diffusion in ipl space. This can be demonstrated by a simple example.

Explorer XII observations of solar particles have been described by BRYANT *et al.* (1962); part of this work will be presented further on, and some conclusions will be drawn for our problem. Explorer XII spent more than half of its orbit outside the magnetosphere; a class 3 flare occurred on September 28, 1961, after a relatively long quiet period of the sun, so that the ipl conditions may be referred to as "normal", i.e. not disturbed by motions of plasma or magnetic fields connected with strong preceding flares.

The time variations of the proton intensities were analysed for energy intervals from 2 to 600 MeV and integral above 600 MeV. All typical time lags from the optical flare increased with the decreasing energy: the beginning of the intensity variation, the onset of isotropy of the radiation, and the moment of maximum intensity (t_m).

For energy groups between 200 and 300 MeV and above 600 MeV a quantitative comparison with a three-dimensional diffusion model has been carried out by BRYANT *et al.* (1962). It is assumed that, from a point source ($R=0$), a total number of particles $N_0(E)$ as a function of proton energy E is released for $t=0$. If the particles diffuse isotropically with diffusion length $D=\lambda v/3$ (λ=mean free path, v=particle velocity) into an infinitely extensive medium, the following result for the density $N(E, t, R)$ of protons with energy E, after a time t, at distance R from the source is obtained (for derivation see e.g. DORMAN (1963), page 288):

$$N(E,t,R) = \frac{N_0(E)}{32\pi\left(\pi(\lambda v/3)\,t\right)^{3/2}} \exp\left(-\frac{3}{4}\frac{R^2}{\lambda v t}\right). \tag{1}$$

Magnetic irregularities are assumed to be the source of the scattering mechanism, with a mean free path λ generally depending on the energy of the particle, $\lambda=\lambda(E)$.

The analysis of BRYANT *et al.* (1962, see for example Figure 17) leads to the following result: for $1.4<t<7$ hours after this solar event the time dependence of the intensity can be described by Equation (1), leading to $\lambda\approx 0.04$ AU for $200<E<300$ MeV as well as for $E>600$ MeV. For $t<1.4$ hours the particle flux is anisotropic, and the interpretation of the time and angular dependence is complicated, probably connected with a general solar magnetic field. For $t>7$ hours deviations from simple diffusion are explained according to Parker by leakage of particles out of the inner solar system, beginning at a distance of 2–3 AU. Due to a change in character of magnetic fields at that distance, the probability of being scattered back in the direction of the sun becomes very small or zero.

For the medium time interval this very simple diffusion model is, of course, not the only one to describe the observed time-dependence of the energy spectra. However, assuming that this model is valid under certain conditions (this has to be proved by further experiments), some quantitative conclusions can be drawn from Equation (1):

Differentiation with respect to t gives the time of maximum intensity as

$$t_m = \frac{1}{2}\frac{R^2}{\lambda v} \qquad \left(\approx 12.5\frac{R}{v} \text{ in the case discussed above}\right), \tag{2}$$

and, omitting some numerical factors

$$N(E,t,R) \sim \frac{N_0(E)}{R^3} f(t_m/t) \tag{3}$$

with $f(x)=x^{3/2}\exp\left(-\frac{3}{2}x\right)$.

It follows from Equations (1) and (3) that:

(a) The "true" spectrum of flare particles $N_0(E)$ as emitted near the sun, can be obtained by taking the maximum intensity for every energy group, $N(E, t_m, R)\sim$

$N_0(E)/R^3$. This procedure is independent of the special energy dependence of $\lambda(E)$. A test of the validity of this assumption will be possible if the absolute intensities are measured with accuracy simultaneously by a satellite near the earth and by a deep space probe, because $N(E, t_m, R)$ should depend only on E and R.

(b) For a fixed time there is a difference between the measured spectra $N(E, t, R)$ and the original spectrum $N_0(E)$,but it is possible to correlate them. Let be $N_0(E) \sim E^{-\gamma}$ for a certain energy interval*; the relation between v and E depends on the mass of the particle, $v \sim E^{1/2}$ for the non-relativistic case; the behaviour of $\lambda(E)$ can be derived from the measured t_m-values, using relation (2). Let us further assume for simplicity that $\lambda \sim E^{\alpha}$ for the energy interval considered (for the example above, $\alpha \approx 0$). Then, introducing the various functions into Equation (1) gives for times sufficiently longer than the t_m-values involved (so that the exponential term can be neglected)

$$N(E, t, R) \sim E^{-\gamma-\gamma'}$$

with $\gamma' = 0.75 + 1.5\alpha$ non-relativistic, $\gamma' = 1.5\alpha$ for $v \approx c$. Thus, the spectral change depends on the particle velocity as well as on the depence of λ on E. It can be seen that not only the steepness, but also the general nature of the spectrum can be changed by the diffusion process, if a sufficiently large portion of the spectrum is analysed thorougly. From an experimental determination of $\lambda(E)$, using Formula (2), further information on the magnetic conditions of ipl space will be available, especially if particles with different mass-to-charge ratios, i.e. with different relations between velocity and rigidity are analysed, as, for example, protons and alpha-particles (see BISWAS *et al.*, 1962; NEY and STEIN, 1962).

Additional Remarks

(1) The discussion above refers to relatively quiet conditions; it is well known that if several flares occur in brief succession, the magnetic conditions of ipl space may be strongly disturbed, leading to two effects: (a) A so-called guiding magnetic field, highly ordered, may favour the propagation of charged particles from sun to earth considerably. (b) A magnetic plasma cloud or bottle from the flare reaching the earth about $1\frac{1}{2}$ days later may contain many trapped particles, especially of lower energies, giving rise to a much steeper energy spectrum. A good example of this effect can be seen in a later phase of the September 28, 1961, event discussed above (BRYANT *et al.*, 1962).

(2) It is not certain that diffusion into space begins immediately near the sun, i.e. for $R \approx 0$. As E. Schatzman has shown during his lectures, particles may be trapped after acceleration up to distances where the magnetic fields become sufficiently weak. A similar result is deduced from Explorer VII measurements by YOSHIDA *et al.* (1963). They conclude from the time variations of 18 and 30 MeV protons for about 30 flares,

* It is not of importance here that the spectra may be described by an exponential rigidity law rather than by a power law in energy, as has been pointed out by FREIER and WEBBER (1963); if the energy interval is not too large, a power law in energy is a sufficiently good approximation.

that there is a finite injection period for these particles, and that three-dimensional diffusion begins at a distance $R \approx 0.2 - 0.3$ AU from the sun. This idea is supported by the fact that, for the events not correlated with a visible flare, the intensities are on the average a factor of 10 smaller. This factor can be explained by the R^{-3}-dependence of Equation (3). This would also explain the observation mentioned by HULTQVIST (1963), that polar cap absorption events (influx of solar protons of medium energy on the polar caps) are strongly correlated with visible flares. For flares on the reverse side of the sun the residual intensity near the earth is too small to produce large effects, if the particles are carried out to a rather large fraction of one AU before they begin diffusing.

(3) MEYER and VOGT (1963) detected protons of solar origin on occasions long after a solar flare; if these protons are not produced continuously by the sun, the diffusion and storage mechanisms indicated above will have to be changed considerably.

References

BISWAS, S., FICHTEL, C. E., and GUSS, D. E.: 1962, *Phys. Rev.* **128**, 2756.
BRYANT, D. A., CLINE, T. L., DESAI, U. D., and MCDONALD, F. B.: 1962, *J. Geophys. Res.* **67**, 4983.
DORMAN, L. I.: 1963, *Progr. Elem. Part. Cosm. Ray Phys.* **7**, 236.
ELLIOT, H.: 1960, *Phil. Mag.* **5**, 601.
FREIER, P. S. and WEBBER, W. R.: 1963, *J. Geophys. Res.* **68**, 1605.
GOLD, T.: 1959, *J. Geophys. Res.* **64**, 1665.
HULTQVIST, B.: 1965, 'Disturbance Effects in the Lowest Ionosphere', see page 228 in this book.
MEYER, P., PARKER, E. N., and SIMPSON, J. A.: 1956, *Phys. Rev.* **104**, 768.
MEYER, P. and VOGT, R.: 1963, *Phys. Rev.* **129**, 2275.
MORRISON, P.: 1956, *Phys. Rev.* **101**, 1397.
NEY, E. P. and STEIN, W. A.: 1962, *J. Geophys. Res.* **67**, 2087.
PARKER, E. N.: 1958, *Phys. Rev.* **110**, 1445.
PARKER, E. N.: 1961, *Astrophys. J.* **133**, 1014.
ROEDERER, J. G.: 1963, *Space Research* **3**, 518.
SCHATZMAN, E.: 1963, *Space Research* **3**, 709; *Ann. Astrophys.* **26**, 234.
SIMPSON, J. A.: 1961, *Science in Space,* McGraw Hill, p. 239ff.
YOSHIDA, S., NAGASHIMA, K., KAWABATA, K., and MORIMOTO, M.: 1963, *Space Research* **3**, 608.

CHAPTER 35

ON THE RELATIONSHIP BETWEEN AURORAL ZONE X-RAY BURSTS AND POLAR MAGNETIC SUBSTORMS

G. KREMSER

Institut für Stratosphärenphysik am Max-Planck-Institut für Aeronomie, Lindau/Harz, Germany

Abstract. It is shown that auroral zone X-ray bursts of not too short duration are associated with the enhancement of the world-wide ionospheric current-system which after SILSBEE and VESTINE (1942) describes the latitude and longitude dependency of the geomagnetic bay, also called polar magnetic substorm (AKASOFU and CHAPMAN, 1963). This explains that the geomagnetic disturbances accompanying the X-ray bursts of local nature can always be recognized in the magnetograms of many stations, and also that they are sometimes very weak and even not recognizable at the very location where the X-ray bursts are measured overhead. The geomagnetic disturbance is, however, always the most pronounced at the stations with local time near midnight irrespective of the location of the associated X-ray burst.

1. Introduction

Auroral zone X-ray bursts which can be measured at balloon altitudes of about 30 km must be interpreted as bremsstrahlung produced by electrons with energies between approximately 10 keV and some 100 keV (VAN ALLEN, 1957; WINCKLER, BHAVSAR and ANDERSON, 1962; O'BRIEN, 1962). The electrons themselves are generally stopped at atmospheric levels around 100 km altitude. There is evidence that these electrons are neither of extra-magnetospheric origin nor simply dumped from the radiation belt. Today it is assumed therefore that they are accelerated by some as yet unknown mechanism, which is initiated and fed by the interaction of the solar plasma with the magnetic field of the earth.

The magnetograms taken near the launching station of the balloon however do not show satisfactory correlations with the radiation measurements. There are X-ray bursts associated with polar magnetic substorms (for the definition of the polar magnetic substorm see e.g. AKASOFU and CHAPMAN, 1963), but also events with nearly no correlation. (Examples in PFOTZER, EHMERT and KEPPLER, 1962)

This – as will be seen – seemingly contradictory fact gave rise to an extended study of the problem with the aid of magnetograms of more than 40 stations around the world covering the periods of 9 balloon-flights with 21 X-ray events. Seven of these flights were made at Kiruna, (Sweden) by a group of the Max-Planck-Institut für Aeronomie, two by BROWN at College (Alaska) and by WINCKLER at Ft. Churchill (Canada) respectively. The latter overlapped with corresponding flights at Kiruna.

In this seminar only a brief survey of the results obtained can be given. The entire discussion will be published elsewhere (KREMSER, 1963).

2. The Polar Magnetic Substorm

In what follows, we shall see that the greater X-ray bursts are always associated with polar magnetic substorms, even if there is no trace of them on the magnetograms registered near the launching station. This results from the fact that these disturbances are a world-wide phenomenon with a distinct longitudinal and latitudinal dependency of the disturbance field.

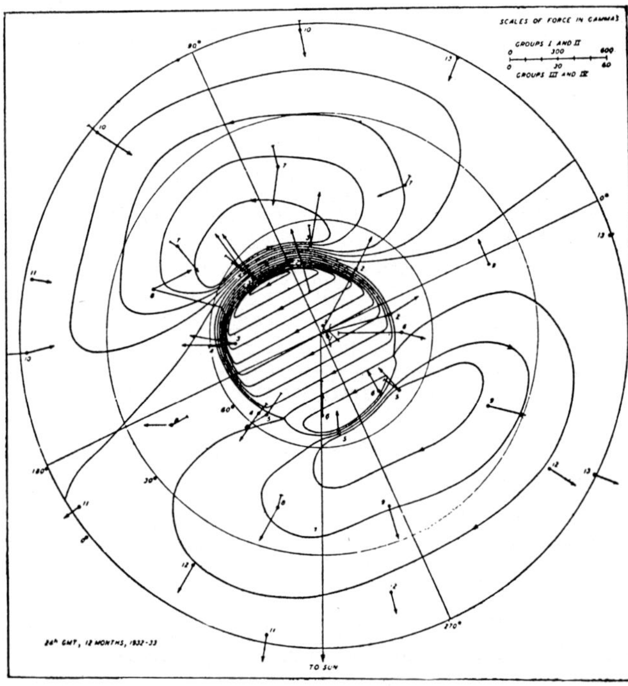

Fig. 1. Current-system of an average polar magnetic substorm (SILSBEE and VESTINE, 1942).

To describe this dependency SILSBEE and VESTINE (1942) have calculated a current-system in the ionosphere, which can just explain the observed magnetic disturbance field of an average bay, as they called this type of disturbances. Their current-system (Figure 1) consists principally of strong night- and daytime-currents in the auroral zone, now called auroral electro-jets, and their widespread return currents which flow partly over the polar cap, partly over the region of the middle latitudes.

Characteristic are the following points: (1) The current-system is fixed with respect to the direction to the sun. (2) The nighttime electrojet is much greater than the daytime one. The maximal currents flow near local midnight and local noon respectively. (3) The currents in the auroral zone and in middle latitudes flow in opposite direction. (4) This fact entails the existence of neutral regions, where opposite currents cancel each other and means that the polar magnetic substorm can only be recognized in regions proximate to the currents but not in the neutral ones.

But regarding this figure we have to keep in mind that there exists no unique relation between magnetic field and current-system, so that the same magnetic field can be realized by different current-systems. Today nobody doubts that the auroral electrojets flow at the time of a polar magnetic substorm, but the existence of the return

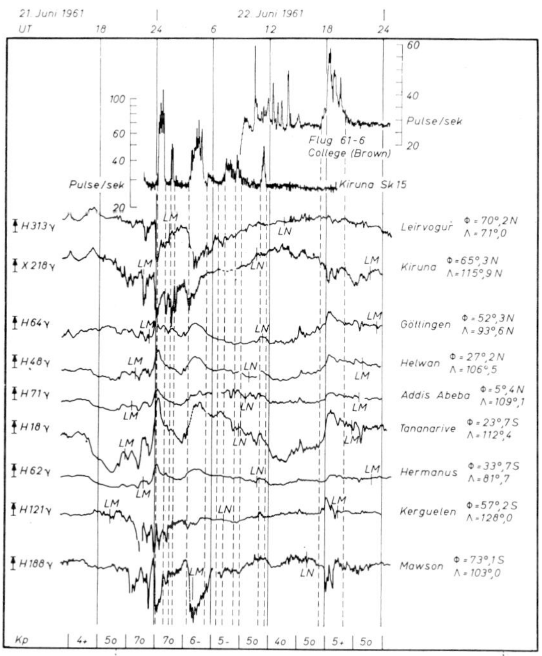

Fig. 2. Horizontal components *H* or north components *X* of the stations of the 'Kiruna Group'. The arrows on the left side are in the positive direction and correspond to the marked field intensity. Co-ordinates are geomagnetic ones (LM = local midnight; LN = local noon).

currents has never been proved experimentally. Therefore this figure cannot be taken to explain the physical background of the geomagnetic disturbance; but it can well be regarded as a simple scheme which shows the longitudinal and latitudinal variation of amplitude and sign of the different components of the magnetic disturbance field.

In connection with our problem we have to state that it is not possible to decide only from the inspection of magnetograms recorded near the launching station if a polar magnetic substorm has accompanied the X-ray burst or not. If we find no traces of a disturbance at a certain station it can have been situated in the neutral region or near the sometimes very weak daytime-currents. On the other hand deflections in the magnetograms of this station can be due to local disturbances only, which need not have anything to do with the world-wide polar magnetic substorm that can only be determined from analysing magnetograms of many widespread stations.

3. Some Examples of the Association of the X-Ray Bursts and Polar Magnetic Substorms

In Figures 2 and 3 the most characteristic features of a polar magnetic substorm

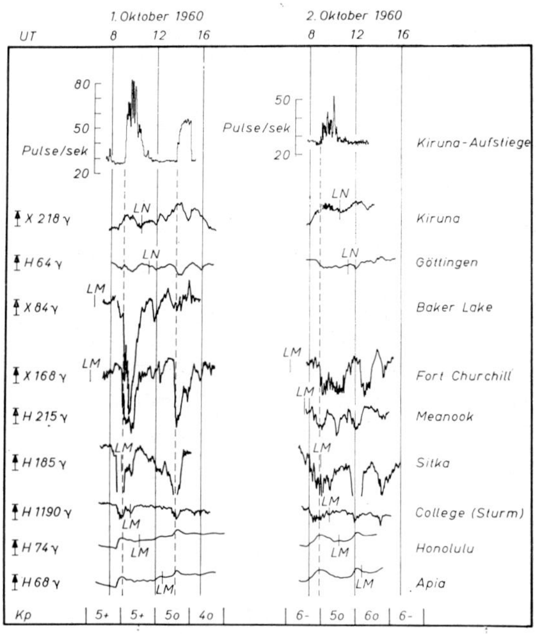

Fig. 3. Parts of the recordings of balloon flights at Kiruna, made on October 1 and 2, 1960, and the corresponding changes in the geomagnetic horizontal component *H* or north component *X* at Kiruna, Göttingen and some stations with local times near midnight. The arrows on the left side are in the positive direction and correspond to the marked field intensity. Co-ordinates are geomagnetic ones (LM = local midnight; LN = local noon).

mentioned above can be seen in the recordings of the horizontal- or northcomponent of the geomagnetic field at the time of X-ray bursts. In Figure 2 the counting rates of Al-walled Geiger-Müller counters during balloon flights at Kiruna and College and the registrations of the magnetic field of stations with geomagnetic longitudes differing not too much from that of Kiruna are compiled. It is easy to see that the greater bursts, lasting for an hour and more, are always associated with polar magnetic substorms. These bursts have drastically taken the shape of a socle with several peaks superimposed. The initial increase is often very abrupt. Sometimes they are preceded by a small precursor.

During the balloon flight shown in Figure 2 the first of these socle type bursts occurred at Kiruna shortly after 00.00 UT. One finds negative deflections in Leirvogur, Kiruna, Kerguelen, and Mawson, positive ones at the other stations. In the current-system of SILSBEE and VESTINE (1942) the negative excursions are caused by westward currents in high latitudes (auroral electrojets), the positive ones by eastward currents in middle and low latitudes (return currents). The disturbances in middle latitudes begin a little before the onset of the disturbance in high latitudes, but the maxima occur approximately together. At that time the initial increase of the socle begins. The abrupt end does not fall in such a pronounced phase of the associated geomagnetic disturbance. The second minor burst is only accompanied by small deflections at Kiruna and Göttingen. But one can imagine that this disturbance has caused the flat decaying part of the great preceding one.

The third burst at Kiruna between approximately 03.00 UT and 05.00 UT is again clearly related to a polar magnetic substorm. The latitudinal dependency of the sign of the excursions is the same as for the great disturbance at 00.00 UT. Therefore it is caused by an analogous current-system. The following minor bursts are not accompanied by so distinct magnetic disturbances.

At the time of the burst at about 11.30 UT we find positive deflections in all latitudes, which can be explained by a very weak daytime current-system so that the current lines of the nighttime current-system are not closed on the nightside, but go around both sides of the earth. This burst was the last measured at Kiruna until the end of the flight at about 19.30 UT.

At College there was registered a group of short "isolated" bursts between 10.00 UT and 14.00 UT. (I call them "isolated" bursts in contrast to the "socle-type" bursts.) During this time no trace of a polar magnetic substorm can be seen on the registrations of this group. From the registrations at the other stations (see KREMSER, 1963) can be seen, however, that they are also associated with magnetic disturbances, but only in a rather limited longitude region with local time near midnight. That means: when the "isolated" bursts occur, only a weak current-system is present in the ionosphere which causes distinct magnetic disturbances only very near its current maximum.

At about 18.00 UT we then find a greater burst at College but no sign of a polar magnetic substorm there (see KREMSER, 1963).

At that time College was situated approximately in a neutral region. According to the current-system of SILSBEE and VESTINE we should find the magnetic disturbance at stations with local times near midnight, that is in the Kiruna-group. And there it is indeed very well pronounced. We even find different peaks in the magnetograms near the auroral zones which correspond exactly with the peaks of the counting rates. It is remarkable that no burst was registered at this time in Kiruna. This fact touches the problem of the extension and local distribution of the primary electron bombardment which cannot yet be explained.

In Figure 3 there are compiled parts of the counting rates during two balloon flights and the corresponding registration of the horizontal components of the magnetic field at Kiruna, where the balloon was launched, at Göttingen, a station of the Kiruna-group in middle latitudes and at stations with local times near midnight. In the current-system of Silsbee and Vestine these stations are situated near the maximum of the nighttime auroral electrojet and its return-current and should show great magnetic disturbances. On October 1, Kiruna was located in the region of the beginning of the daytime electrojet. Therefore, we find positive deflections in the auroral zones and negative ones in middle latitudes. On October 2, the neutral region was nearly overhead Kiruna. In both cases we find only weak disturbances in Kiruna, but pronounced ones at stations in the maximum region of the nighttime electrojet and its return-currents, even near the equator.

Among 21 bursts analysed until now all those which resemble the "socle-type" ones appear together with the polar magnetic substorms.

4. The Onset of the X-ray Bursts

Apart from the mere association of bursts and disturbances the detailed analyses of the magnetograms has shown another characteristic feature concerning the onset of the X-ray bursts which can distinctly be exemplified by the burst at College at about 18.00 UT (Figure 2). This burst has a relatively great precursor, which begins with the onset of the magnetic disturbance at Göttingen, Helwan and Tananarive. The abrupt onset of the socle occurs at the time of a steep decrease in H in the auroral zone and shortly before the maximum of H in middle latitudes. All examined bursts show the tendency of beginning in this phase of the corresponding polar magnetic substorm.

5. Schematical Diagram of the Relationship

The above described relationship between the X-ray bursts and the polar magnetic substorms has led to the schematical diagram shown in Figure 4. It shows the idealized counting rate of a "socle-type" burst and idealized magnetogram traces of a station near the maximum of the polar magnetic substorm. The longitude of these stations is not necessarily the same as the longitude of the balloon location. It depends on the time of day of the occurrence of the burst.

The polar magnetic substorm begins at least in middle latitudes before the X-ray burst with a bay-like excursion in the horizontal component. At high latitudes the steep initial phase of the polar magnetic substorm and the initial increase of the counting rates occur often at nearly the same time. During this phase the disturbance in middle latitudes reaches its maximum. For the decaying phase it was not possible to find a similarly obvious relationship in time between burst and geomagnetic disturbances.

6. Conclusions

We can summarize the following findings:

(1) There is a close relationship between X-ray bursts in the auroral zone of not too

short duration and the enhancement of a world-wide ionospheric current-system which causes the polar magnetic substorm. This system resembles the one derived by SILSBEE and VESTINE (1942) for an average bay.

(2) This has the consequence that magnetic disturbances correlated with X-ray bursts of relatively local character can be recognized at many stations all over the world.

(3) Since the current-system develops also neutral transition regions approximately

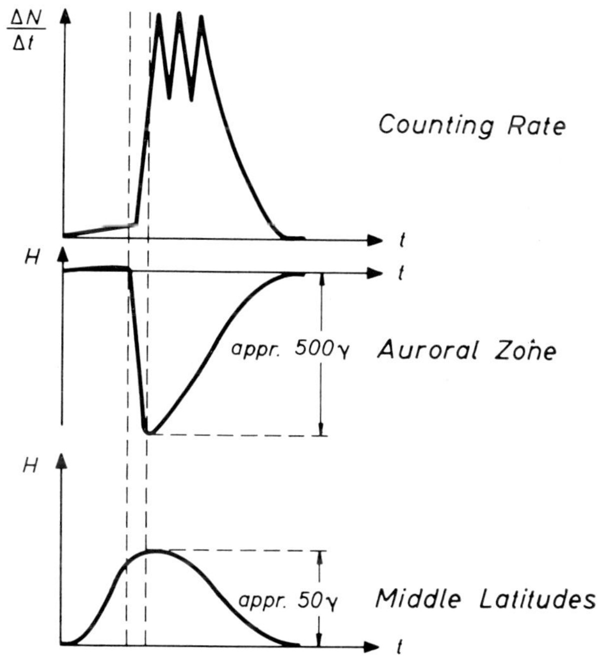

Fig. 4

rigid with respect to the sun earth line, it can be understood that occasionally the associated magnetic disturbances cannot be recognized at the very location where the X-ray burst is observed.

(4) The associated magnetic disturbance is always well pronounced at stations with local time near midnight.

(5) It is very probable that the bombardment of the ionosphere by intense low energy particle-fluxes causing the enhancement of the world-wide current-system coincides also with local bursts of more energetic electrons producing the X-rays. But the results show also that there is no direct dependency between ionospheric currents and X-ray bursts, as the latter occur also in the transition regions. It seems that both events are only connected by the same initiating mechanism, but are then widely independent of each other.

Acknowledgement

The author is deeply indebted to the directors of the geomagnetic observatories who provided him directly with the copies of the magnetograms covering the periods of the mentioned balloon flights in the auroral zone.

He is also grateful to Prof. Dr. A. Ehmert and Dr. G. Pfotzer for initiating this study and for helpful discussions, and expresses also thanks to the balloon group of the institute for providing him with the basic balloon recordings.

References

Akasofu, S. and Chapman, S.: 1963, *J. Geophys. Res.* **68,** 2375.

Kremser, G.: 1963, *Über den Zusammenhang zwischen Röntgenstrahlungs-Ausbrüchen in der Polarlichtzone und bay-artigen erdmagnetischen Störungen.* D7, Dissertation, Göttingen, *Mitt. Max-Planck-Institut für Aeronomie* No. 14 (1964).

O'Brien, J.: 1962, *J. Geophys. Res.* **67,** 3687.

Pfotzer, G., Ehmert, A., and Keppler, E.: 1962, *Mitt. Max-Planck-Institut für Aeronomie,* No. 9.

Silsbee, H. C. and Vestine, E. H.: 1942, *Terr. Magn.* **47,** 195.

Van Allen, J. A.: 1957, *Proc. Nat. Acad. Sci. U.S.* **43,** 57.

Winckler, J. R., Bhavsar, P. D., and Anderson, K. A.: 1962, *J. Geophys. Res.* **67,** 3717.

PART IX

IONOSPHERE AND PLANETARY ATMOSPHERE

CHAPTER 36

POSITIVE ION REACTIONS AND PRODUCTIONS IN THE IONOSPHERE*

MARCEL NICOLET
Centre National de Recherches de l'Espace, Bruxelles

1. Ionic Reactions

Starting from the observational knowledge of the ionosphere, the following positive ions must be considered: H^+, He^+, N^+, N_2^+, O^+, O_2^+, NO^+, Mg^+, Ca^+, ... etc. Further, it should be noted that metallic ions are not present in the whole ionosphere, H^+ and He^+ occur in the upper part of the ionosphere above the F_2 peak, O_2^+, NO^+ and O^+ are the essential ionic constituents of the D, E and F regions and N_2^+ and N^+ are minor ions with maximum concentrations in the F region. Thus, O^+, N^+, NO^+, O_2^+ and N_2^+ must be considered as the basic ions which are involved in the various reactions.

The predominant ion in the F region, O^+, is subject to the following ion-atom interchange reactions:

$$(\gamma_1); O^+(^4S) + N_2(^1\Sigma) \rightarrow NO^+(^1\Sigma) + N(^4S) + 1.12\,\text{eV} \quad (1)$$
$$(\gamma_2); O^+(^4S) + O_2(^3\Sigma) \rightarrow O_2^+(^2\Pi) + O(^3P) + 1.54\,\text{eV} \quad (2)$$
$$(\gamma_3); O^+(^4S) + NO(^2\Pi) \rightarrow NO^+(^1\Sigma) + O(^3P) + 4.37\,\text{eV} \quad (3)$$
$$(\gamma_4); O^+(^4S) + NO(^2\Pi) \rightarrow O_2^+(^2\Pi) + N(^4S) + 0.15\,\text{eV}. \quad (4)$$

The following charge transfer process must be added as a production process.

$$(\gamma_{16}); N^+(^3P) + O(^3P) \rightarrow N(^4S) + O^+(^4S) + 0.93\,\text{eV}.$$

Similarly, a charge transfer process between O^+ and NO is equivalent to reaction (3).

Several reactions involving O_2^+ lead to NO^+, and particularly,

$$(\gamma_5); O_2^+(^2\pi) + NO(^2\pi) \rightarrow NO^+(^1\Sigma) + O_2(^3\Sigma) + 2.81\,\text{eV} \quad (5)$$
$$(\gamma_6); O_2^+(^2\Pi) + N(^4S) \rightarrow NO^+(^1\Sigma) + O(^3P) + 4.22\,\text{eV} \quad (6)$$
$$(\gamma_7); O_2^+(^2\Pi) + N_2(^1\Sigma) \rightarrow NO^+(^1\Sigma) + NO(^2\Pi) + 0.87\,\text{eV}. \quad (7)$$

* For a complete analysis with references, see NICOLET, M. and SWIDER Jr., W.: 1963, *Plan. Space Sci.* **11**, 12.

The processes affecting N_2^+ involve charge transfer processes and ion-atom interchange reactions:

$$(\gamma_8);\ N_2^+(^2\Sigma) + O(^3P) \rightarrow O^+(^4S) + N_2(^1\Sigma) + 1.96\,\text{eV} \tag{8}$$
$$(\gamma_9);\ N_2^+(^2\Sigma) + O_2(^3\Sigma) \rightarrow O_2^+(^2\Pi) + N_2(^1\Sigma) + 3.50\,\text{eV} \tag{9}$$
$$(\gamma_{10});\ N_2^+(^2\Sigma) + N(^4S) \rightarrow N^+(^3P) + N_2(^1\Sigma) + 1.03\,\text{eV} \tag{10}$$
$$(\gamma_{11});\ N^+(^3P) + NO(^2\Pi) \rightarrow N_2^+(^2\Sigma) + O(^3P) + 2.22\,\text{eV}. \tag{11}$$

Finally, ion-atom interchange reactions in which N^+ and ON^+ are involved must be considered by taking into account that the lower state of ON^+ is the excited state $^3\Pi$ of the normal ion NO^+.

$$(\gamma_{12});\ N^+(^3P) + O_2(^3\Pi) \rightarrow ON^+(^3\Pi, NO^+) + O(^3P) + [6.69 - E(^3\Pi)]\,\text{eV} \tag{12}$$
$$(\gamma_{13});\ N^+(^3P) + NO(^2\Pi) \rightarrow ON^+(^3\Pi, NO^+) + O(^3P) + [5.30 - E(^3\Pi)]\,\text{eV} \tag{13}$$
$$(\gamma_{14}^{**});\ N_2^+(^2\Sigma) + O(^3P) \rightarrow ON^+(^3\Pi, NO^+) + N(^4S) + [3.08 - E(^3\Pi)]\,\text{eV} \tag{14}$$
$$(\gamma_{15}^{*});\ N_2^+(^2\Sigma) + O_2(^3\Sigma) \rightarrow ON^+(^3\Pi, NO^+) + NO(^2\Pi) + [4.47 - E(^3\Pi)]\,\text{eV}. \tag{15}$$

Reactions which may occur in the ionosphere require that the excitation energy $E(^3\Pi)$ is less than the energy balance. Thus reactions (12) to (15) must be rejected since $E(^3\ \Pi)$ is more than 7 eV.

The reaction energies derived in (1) to (15) correspond to the following dissociation and ionization potentials.

Dissociation of ionized molecules:

O_2^+, 6.66 eV; N_2^+, 8.73 eV; NO^+, 10.88 eV.

Dissociation of neutral molecules:

O_2, 5.12 eV; N_2, 9.76 eV; NO, 6.51 eV.

Ionization of neutral molecules:

O_2, 12.08 eV; N_2, 15.58 eV; NO, 9.25 eV.

Ionization of neutral atoms:

O, 13.62 eV; N, 14.55 eV; H, 13.60 eV;

2. Photo-ionization Processes

The reactions (1) to (13) must be used with the photo-ionization and recombination rates to write equations for electrons and ion concentrations. The following ionization processes are considered:

$$I(N_2);\ N_2 + h\nu(\lambda < 796\,\text{Å}) \rightarrow N_2^+ + e \tag{16a}$$
$$I_2(N);\ N_2 + h\nu(\lambda < 510\,\text{Å}) \rightarrow N + N^+ + e \tag{16b}$$
$$I(N);\ N + h\nu(\lambda < 852\,\text{Å}) \rightarrow N^+ + e \tag{17}$$
$$I(O);\ O + h\nu(\lambda < 910\,\text{Å}) \rightarrow O^+ + e \tag{18}$$
$$I(O_2);\ O_2 + h\nu(\lambda < 1026\,\text{Å}) \rightarrow O_2^+ + e\,. \tag{19}$$

The photo-ionization of NO is essentially due to the Lyman-α radiation at 1216 Å:

$$I(\mathrm{NO}); \mathrm{NO} + h\nu(\lambda = 1216\ \text{Å}) \rightarrow \mathrm{NO}^+ + e\,. \tag{20}$$

The absorption cross-sections and atmospheric densities are known with sufficient precision as far as the order of magnitude is concerned. Consequently it is possible to state that the following ionization processes occur:

(I) *D region:*

(a) Ionization of nitric oxide by Lyman-α, since O_2, N_2 and O cannot be photo-ionized by λ 1216.7 Å. Other constituents such as Na ($\lambda < 2412$ Å), Ca($\lambda < 2071$ Å), CH_3($\lambda < 1260$ Å), C($\lambda < 1100$ Å) can be photo-ionized by radiations which are unable to ionize O_2.

(b) Ionization by X-rays of $\lambda < 10$ Å, i.e. with cross-sections less than 10^{-19} cm^2. It is known from rocket and satellite measurements that the Lyman-α flux is between 3 and 6 ergs cm^{-2} sec^{-1}, while the X-ray flux is extremely variable with solar conditions. From 2–8 Å the minimum-to-maximum variation is a factor of several hundreds.

(c) Ionization by cosmic rays which is effective in the lower D region, where Lyman-α cannot penetrate and where X-rays of $\lambda < 2$ Å are not important; particularly during solar quiet conditions.

An exact analysis of the ionization problem in the D region requires the determination of the separate effects of the three ionizing sources. In fact, it is necessary to compare the ionizing effect of the stable sources of Lyman-α and cosmic rays as compared with the extremely varying source represented by X-rays.

The ionization rate coefficient of cosmic rays is between (Φ being the geomagnetic latitude)

$$I(\Phi \geqslant 0°) \geqslant 10^{-18}\ \text{sec}^{-1} \quad \text{and} \quad I(\Phi \leqslant 50°) \leqslant 10^{-17}\ \text{sec}^{-1},$$

i.e. with the following values: $\Phi = 0°$, 1.25×10^{-18} sec^{-1}; $\Phi = 30°$, 2×10^{-18} sec^{-1}; $\Phi = 40°$, 4.5×10^{-18} sec^{-1}; $\Phi = 50°$, 1×10^{-17} sec^{-1} and $\Phi = 60°$, 1.25×10^{-17} sec^{-1}.

It is clear that X-rays of $\lambda < 2$ Å emitted during completely quiet conditions cannot lead to an ionization rate coefficient of the order of 10^{-18} sec^{-1}. On the other hand, since the ionization rate coefficient of nitric oxide by Lyman-α reaches values of the order of:

$$I(\mathrm{NO}) = 5 \times 10^{-7}\ \text{sec}^{-1};$$

a small ratio $n(\mathrm{NO})/n(O_2) = 10^{-10}$ leads to an electron production of the same order as the X-ray production under quiet solar conditions. Thus the disturbed solar conditions, and particularly solar flare conditions, produce a D region due to X ray radiation.

(II) *E Region:*

(a) Ionization of molecular oxygen by ultraviolet radiations of $\lambda > 910$ Å, particularly by monochromatic radiations such as Lyman-β at 1025 Å, i.e. with absorption cross-sections between 10^{-19} cm^{-2} and 10^{-18} cm^{-2}. In addition, ionization of atomic oxygen by radiation of $\lambda < 910$ Å and, particularly, by the chromospheric Lyman continuum which can penetrate into the E layer via the windows of molecular nitrogen absorption. The variation with the solar cycle of these ultraviolet radiations should be less than a factor of two.

(b) Ionization by X-rays of $\lambda > 10$ Å and particularly by the radiation in the spectral range $30 < \lambda < 100$ Å.

(c) Ionization of meteor atoms with low ionization potentials.

An exact analysis of the ionization production problem in the E region still requires the determination of the separate effects of the three ionizing sources. The stable source represented by the ultraviolet radiation will have its maximum ionization effect under quiet solar conditions and, particularly, during the minimum of the solar cycle. The X-ray action will vary with solar activity and will become predominant during disturbed solar conditions.

(III) *F Region:*

(a) Ionization of molecular nitrogen for $\lambda < 796$ Å with absorption cross-section greater than 10^{-17} cm^2.

(b) Ionization of atomic oxygen for $\lambda < 796$ Å subject to the absorption of molecular nitrogen.

(c) Ionization of atomic oxygen for $\lambda > 800$ Å with different absorption cross-sections for its different ionization potentials at 910 Å, 732 Å and 665 Å.

An exact analysis of the ionization problem in the F region requires a simultaneous knowledge of the energy of solar emissions and of the absorption cross-sections of O, N_2 and O_2. In particular, the penetration of monochromatic solar radiations between the E and F_1 peaks must be known in order to determine the exact behavior of the electron production.

3. Electron Recombination

The various processes (1) to (15) represent the most important ionospheric reactions in which nitrogen and oxygen positive ions are involved. The processes (16) to (21) correspond to photo-ionization of these constituents. It should be kept in mind that similar ionization processes produce metallic ions and also light ions such as He^+ and H^+. Consequently, the first recombination process to be considered is radiative recombination, such as $X^+ (\text{or } XY^+) + e \rightarrow X (\text{or } XY) + h\nu$. Since the recombination coefficient is always between 1×10^{-12} and 5×10^{-12} cm^3 sec^{-1} in the ionosphere, the radiative recombination is negligible compared with other processes affecting ions. Three-body electronic or ionic recombination processes are also negligible since they cannot play a role in the normal ionosphere.

The essential recombination process in the ionosphere is Bates' process of neutralization due to the dissociation of a molecular ion as follows:

$$XY^+ + e \rightleftharpoons XY^* \rightarrow X^* + Y^*$$

(the * indicating possible excitation). It is clear from the theoretical analysis that it is practically impossible to predict the temperature variation of the dissociative recombination coefficient, α. In other words, if values of α between 10^{-6} and 10^{-9} cm^3

sec $^{-1}$and variations with temperatures such as $T^{-1\pm0.5}$ are assumed, they do not contradict theoretical considerations. Consequently, the absolute value of α and its dependence on T must be found experimentally.

The principal dissociative recombination processes which occur in the ionosphere are:

$$\alpha(O_2)\;;\quad O_2^+ + e \rightarrow O + O \tag{21a}$$

$$\alpha(NO);\, NO^+ + e \rightarrow N + O \tag{21b}$$

$$\alpha(N_2)\;;\quad N_2^+ + e \rightarrow N + N. \tag{21c}$$

Various experimental determinations of the recombination coefficients of these three ions exist, but it is difficult to adopt a correct value for ionospheric purposes. Obviously, a temperature dependence has been introduced in order to follow ionospheric observations, but a coefficient varying with $T^{-1\pm0.5}$ leads to very different values in the F layer.

The night-time conditions should give a value of the recombination coefficient in the E and F_1 regions. However, it is not yet clear what the exact vertical distribution of electrons is in the night-time E and F_1 layers. If there is a peak in the E layer due to meteor ions, the analysis of the normal recombination is perturbed.

Electron concentrations of the order of 10^3 cm^{-3} below the F_2 region require an electron recombination coefficient of the order of 2.5×10^{-8} cm^3 sec^{-1} for a predominant ion such as NO^+, or the presence of meteor-ions with radiative recombination coefficients. A value of α reaching 10^{-7} cm^3 sec^{-1} cannot result in more than 2.5×10^2 electrons at the end of the night. Nevertheless, types of E_S ionization could modify the analysis since the ion behavior is different according to the nature of the ions. For example, meteor ions and nitric oxide are not subject to resonance charge transfer as is the case for O_2^+, O^+ and N_2^+, and their redistributions in the presence of the earth's magnetic field are different. In any event, it must be remembered that the electron-ion collision frequency decreases with temperature ($\propto T^{-3/2}$) and that the normal tendency of the temperature dependence of α_D, therefore, should be, in a complicated way, to decrease with increasing temperatures.

4. Analysis of Ionospheric Reactions

A. GENERAL CONDITIONS

All processes (1) to (21) (after neglecting the effect of negative ions) must be considered in order to determine the essential reactions in the ionosphere. From this, it is clear that only NO^+ can be considered as disappearing by dissociative recombination alone: O_2^+ and N_2^+ concentrations are affected by ion-atom interchange reactions (γ_6 and γ_7) and by charge transfer processes (γ_8 and γ_9); respectively. O^+ is transformed into molecular ions by ion-atom interchange reactions (γ_1 to γ_4). Its transformation into NO^+ is due to reactions with N_2 and NO (γ_1 and γ_3) and into O_2^+ by O_2 and NO (γ_2 and γ_4). However, O^+ production is increased by the charge transfer process between atomic oxygen and molecular nitrogen ions (γ_8). The ratio $n^+(O_2)/n^+(NO)$ is not independent of reactions leading to O^+ or NO^+, since O_2^+ may transform into NO^+ by ion interchange reactions (γ_7 and γ_6).

B. EQUILIBRIUM CONDITIONS

In order to determine the relative importance of reactions (1) to (21), ionization equations are first considered for steady state conditions. They may be written as follows:

$$n^+(N_2) = \frac{n(N_2)I(N_2) + n^+(N)\gamma_{11}n(NO)}{\alpha(N_2)n_e + \gamma_8 n(O) + \gamma_9 n(O_2) + \gamma_{10}n(N)} \tag{22}$$

$$n^+(N) = \frac{n(N_2)I_2(N) + n(N)[I(N) + \gamma_{10}n^+(N_2)]}{\gamma_{12}n(O_2) + [\gamma_{11} + \gamma_{13}]n(NO)} \tag{23}$$

$$n^+(O) = \frac{n(O)[I(O) + \gamma_8 n^+(N_2)]}{\gamma_1 n(N_2) + \gamma_2 n(O_2) + [\gamma_3 + \gamma_4]n(NO)} \tag{24}$$

$$n^+(O_2) = \frac{n(O_2)[I(O_2) + \gamma_2 n^+(O) + \gamma_9 n^+(N_2)] + n(NO)\gamma_4 n^+(O)}{\alpha(O_2)n_e + \gamma_6 n(N) + \gamma_7 n(N_2)} \tag{25}$$

$$n^+(NO) = \frac{1}{\alpha(NO)n_e}\{n^+(O)[\gamma_1 n(N_2) + \gamma_3 n(NO)] + n^+(O_2)$$
$$[\gamma_6 n(N) + \gamma_7 n(N_2)] + n^+(N)[\gamma_{12}n(O_2) + \gamma_{13}n(NO)] + n(NO)I(NO)\}. \tag{26}$$

An adequate approximation for (22) to (26) is introduced by neglecting reactions (3), (4), (11) and (13) in that $\gamma n(NO) \ll \gamma n(O_2)$ or $\gamma n(N_2)$.

In all the ionization equations (22) to (26) there is competition between reactions depending on the neutral constituent concentrations. A simple way to investigate the various effects is to introduce numerical factors having all possible values between 0 and 1. They are:

(a) for loss of atomic oxygen ions:

$$A = \frac{\gamma_2 n(O_2)}{\gamma_1 n(N_2) + \gamma_2 n(O_2)}; \qquad 1 - A = \frac{\gamma_1 n(N_2)}{\gamma_1 n(N_2) + \gamma_2 n(O_2)} \tag{27}$$

(b) for production of atomic oxygen ions,

$$B = \frac{\gamma_8 n(O)}{\gamma_8 n(O) + \gamma_9 n(O_2)}; \qquad 1 - B = \frac{\gamma_9 n(O_2)}{\gamma_8 n(O) + \gamma_9 n(O_2)} \tag{28}$$

(c) for loss of molecular nitrogen ions,

$$C = \frac{\alpha(N_2)n_e}{\alpha(N_2)n_e + \gamma_8 n(O) + \gamma_9 n(O_2)}; \qquad 1 - C = \frac{\gamma_8 n(O_2) + \gamma_9 n(O_2)}{\alpha(N_2)n_e + \gamma_8 n(O_2) + \gamma_9 n(O_2)} \tag{29}$$

(d) for loss of molecular oxygen ions,

$$D = \frac{\alpha(O_2)n_e}{\alpha(O_2)n_e + \gamma_6 n(N) + \gamma_7 n(N_2)}; \qquad 1 - D = \frac{\gamma_6 n(N) + \gamma_7 n(N_2)}{\alpha(O_2)n_e + \gamma_6 n(N) + \gamma_7 n(N_2)}. \tag{30}$$

C. MOLECULAR NITROGEN ION

The ion N_2^+ is a minor constituent in the ionosphere even if its production by ultra-violet radiation $n(N_2)\,I(N_2)=q(N_2)$ is important. Equation (22) becomes, after making the approximations justified before,

$$n^+(N_2) = \frac{q(N_2)}{\alpha(N_2)\,n_e + \gamma_8 n(O) + \gamma_9 n(O_2)} \tag{31}$$

showing that various loss processes dominate, depending on the altitude involved.

Photoequilibrium represents the ionization conditions for N_2^+. Nevertheless, an exact analysis of aeronomic data is difficult since a determination of the factor C in the equation (29) is needed.

D. ATOMIC NITROGEN ION

Atomic nitrogen is a minor constituent and its ion, N^+, exists only above the F_1 layer, being a certain fraction of O^+ ($<10\%$) in the F_2 layer. The N^+ concentration is given essentially by the following approximation of (23):

$$n^+(N) = \frac{n(N_2)\,I_2(N) + n(N)\left[I(N) + \gamma_{10} n^+(N_2)\right]}{\gamma_{12} n(O_2) + \gamma_{16}\,n(O)}\,. \tag{32}$$

The principal loss process of N^+ occurs by ion-atom interchange reactions with O_2 which apparently is very rapid, or by the generally adequate approximation,

$$n^+(N) = \frac{n(N_2)\,I_2(N)}{\gamma_{12} n(O_2) + \gamma_{16}\,n(O)}\,. \tag{33}$$

E. ATOMIC OXYGEN IONS

The principal reactions in which O^+ is involved lead to the following equation for photo-ionization equilibrium, (24)

$$n^+(O)_{eq} = \frac{q(O) + n^+(N_2)\,\gamma_8 n(O)}{\gamma_1 n(N_2) + \gamma_2 n(O_2)} \tag{34}$$

or

$$n^+(O)_{eq} = \frac{q(O) + B(1-C)\,q(N_2)}{\gamma_1 n(N_2) + \gamma_2 n(O_2)} \tag{35}$$

in which $0 \leq B \leq 1$.

In the lower ionosphere B and C decrease to very small values while they increase with height in the F region. The maximum effect for an additional production of O^+ ions should occur for $B=1$ and $C=0$. Such conditions are never reached but are best approached in the F_1 layer. Thus, there is a contribution to the direct photo-ionization of atomic oxygen by the indirect photo-ionization of molecular nitrogen leading to an important production of the O^+ ion. At altitudes corresponding to the E layer, ion-atom interchange reactions (1) and (2) are important enough to transform O^+ ions into NO^+ and O_2^+. If approximate ionization equilibrium can be retained for day-time conditions in the F_1 layer according to ionospheric observations, they must be rejected for the F_2 layer. Daytime and nighttime conditions for F_1 and F_2 layers clearly show that the loss coefficient

$$\gamma_1 n(N_2) + \gamma_2 n(O_2) = \beta(z) \tag{36}$$

becomes small so that

$$n^+(O) = n^+(O)_{t=0}\, e^{-\beta(z)t} + n^+(O)_{eq}\left[1 - e^{-\beta(z)t}\right]. \tag{37}$$

In other words, the low electron concentration in the F_1 layer for night-time conditions means that the principal loss processes (1) and (2) for atomic oxygen ions are large, and that steady state conditions can be used for day-time conditions. In the F_2 layer, the night-time concentration of O^+ ions is such that the steady state is far from being an adequate approximation for real conditions, which depend on diffusion processes. In fact, it is well known that diffusion leads to the F_2 peak and is active above it, so that photoequilibrium cannot be applied in that region.

It is not possible to obtain exact aeronomic conditions without knowing the ratio $\gamma_1 n(N_2)/\gamma_2 n(O_2)$. Again an apparent interpretation of ionospheric behavior may be given with an arbitrary hypothesis since very many parameters are involved for which accurate values are as yet unknown.

F. MOLECULAR IONS O_2^+ AND NO^+

Mass-spectrometric measurements show that O_2^+ and NO^+ are the principal ions in the E layer and that O^+ becomes important only in the F_1 layer. Equilibrium conditions may be considered as a very good approximation to study the general behavior of molecular ions in a sunlit atmosphere, since they disappear via dissociative recombination.

Again neglecting reactions with NO, (25) and (26) are written as follows:

$$\begin{aligned} n^+(NO)\{\alpha(NO)n_e\} &= q(NO) + [1 - A][\gamma_1 n(N_2) + \gamma_2 n(O_2)]\, n^+(O) \\ &\quad + [\gamma_6 n(N) + \gamma_7 n(N_2)]\, n^+(O_2) \end{aligned} \tag{38}$$

$$\begin{aligned} n^+(O_2)\{\alpha(O_2)n_e + \gamma_6 n(N) + \gamma_7 n(N_2)\} &= q(O_2) + A[\gamma_1 n(N_2) + \gamma_2 n(O_2)] \\ &\quad n^+(O) + [1 - B][\gamma_8 n(O) + \gamma_9 n(O_2)]\, n^+(N_2). \end{aligned} \tag{39}$$

Equations (38) and (39) show how $n^+(NO)$ and $n^+(O_2)$ are related to the same loss reactions of atomic oxygen ions. An exact knowledge of the factor A defined by (27) is first required. But the ratio $n^+(NO)/n^+(O_2)$ is subject to the effect of reactions (6) and (7) leading to a transformation of O_2^+ into NO^+.

There is, therefore, an important problem to be resolved before determining the exact behavior of the upper D region and lower E layer.

Analysis of (39) shows that the direct production of O_2^+ ions by photoionization is an important process since all solar radiation of $\lambda < 1026$ Å is involved. But the final concentration $n^+(O_2)$ is related to the effect of its transformation into NO^+ by reactions (6) and (7). Above a certain altitude, (39) leads to

$$\frac{n^+(O_2)}{n^+(O)} \geqslant \frac{\gamma_2 n(N_2) + q(O_2)/n^+(O)}{\alpha(O_2)\, n_e} \tag{40}$$

which becomes, where $n^+(O) \simeq n_e$,

$$n^+(O_2) \geqslant \frac{q(O_2)}{\alpha(O_2)\, n_e} + \frac{\gamma_2 n(O_2)}{\alpha(O_2)}. \tag{41}$$

Equations (40) and (41) should be applied to the F_1 and F_2 layers, respectively; O_2^+ production depends on the ion-atom interchange reaction (2) together with direct photoionization.

The general ionization equations (38) and (39) can be written, using equilibrium conditions (41) and (35),

$$\begin{aligned} n^+(NO)\{\alpha(NO)\, n_e\} = {} & q(NO) + [1 - A]\, q(O) + [1 - C][B - AB]\, q(N_2) \\ & + n^+(O_2)[\gamma_6 n(N) + \gamma_7 n(N_2)] \end{aligned} \tag{42}$$

and

$$\begin{aligned} n^+(O_2)\{\alpha(O_2)\, n_e\} = {} & q(O_2) + Aq(O) + [1 - C][1 - (B - AB)]\, q(N_2) \\ & - n^+(O_2)[\gamma_6 n(N) + \gamma_7 n(N_2)]\,. \end{aligned} \tag{43}$$

This analysis shows that the last term of the second member of equation (42) and (43) cannot be neglected when reactions (6) and (7) are involved. When the transformation of O_2^+ and NO^+ can be ignored, it is clear from (42) and (43) that atmospheric conditions are simplified, since then one can write:

$$n^+(NO)\{\alpha(NO)\, n_e\} = [1 - A]\, q(O) + [1 - C][B - AB]\, q(N_2) + q(NO) \tag{44}$$

and

$$n^+(O_2)\{\alpha(O_2)\, n_e\} = Aq(O) + [1 - C][1 - (B - AB)]\, q(N_2) + q(O_2)\,. \tag{45}$$

Application of (44) and (45) in the F_1 layer requires the knowledge of a large number of parameters. Numerous experimental and aeronomic data are needed before the whole ionosphere will be susceptible to explanation even under photo-equilibrium conditions.

In conclusion, the five equations (31), (33), (34), (44) and (45) show how each ion is related to the various productions $q(N_2)$, $q(O)$, $q(O_2)$, $q(NO)$ and $q(N)$. Any aeronomic analysis must deal with the vertical distribution of the various factors A, B, C and D. An exact knowledge of these factors requires a precise determination of the rate coefficients of reactions (1) and (2), (8) and (9), (6) and (7) and (12) with that of the recombination coefficients $\alpha(N_2)$, $\alpha(O_2)$ and $\alpha(NO)$, i.e. the knowledge of at least 10 parameters when the vertical distribution of neutral constituents (N_2, O_2, O and NO) with their ionization and absorption rate coefficient is known with accuracy, along with the solar flux. Since temperature effects are also important, it is clear that only approximate solutions can be obtained. Some simplification may be introduced when a specific ionospheric region is considered, as shown in the analysis of the theoretical behavior of each ion. Nevertheless, the general complexity of equations demonstrates that contradictory conclusions can be reached when a parameter is not well chosen. More in situ observations, particularly of ion densities, are needed before sufficient aeronomic data will be available to determine rate coefficients of reactions yet unknown from laboratory measurements.

CHAPTER 37

THE ATMOSPHERES OF MARS AND VENUS

S. I. RASOOL
Institute for Space Studies, N.A.S.A. Goddard Space Flight Center, Greenbelt, Mld.
and
Department of Meteorology, New York University, New York, U.S.A.

The nine planets of our solar system can be grouped into two distinct classes: (1) the *terrestrial planets*, which include Mercury, Venus, Earth and Mars; and (2) the *major planets* – Jupiter, Saturn, Uranus, Neptune and Pluto.

The terrestrial planets, of which the Earth is an example, are relatively small in size, nearer to the sun and therefore relatively hot, have higher density, and rotate relatively slower; while the major planets are much larger in size, are farther away from the sun and therefore colder, have very low density, and usually rotate much faster.

Because of these common properties, the study of the *terrestrial planets* becomes simpler as the knowledge of the structure of the Earth's atmosphere can be applied to the study of these planets. Before going on to a detailed discussion of the observed properties of the atmospheres of Mars and Venus, we shall outline the general charateristics of the planetary atmospheres which essentially determine the structure of the atmosphere of any other planet.

1. General Considerations

Temperature, pressure, and composition are the three essential parameters which determine the structure of a planetary atmosphere. Of these, temperature is the most significant because it directly reflects the processes of energy absorption in the atmosphere and at the ground. For an atmosphere in hydrostatic equilibrium the three parameters are interlinked by the equation of state and the equations of hydrostatic equilibrium. If we know two of these parameters, we can deduce the third and thereby define the atmosphere almost completely.

A. COMPOSITION

In order to discuss the composition of a planetary atmosphere, we shall go quite far back into the history of our subject and deal first with the processes by which we believe the planets themselves were formed. By this approach the student of the subject can hope to gain an understanding of the conditions which limit the structure and composition of planetary atmospheres, and he can place his detailed investigation of the atmosphere of the earth and his comparison of the similarities and differences among the planets within the broad context of the evolving solar system.

The birth of the stars depends on the interplay between gravitational and nuclear forces – the weakest and the strongest forces in nature. According to the present ideas stars are formed out of accidental condensations arising in the swirling and turbulent motion of the gas dust that pervade outer space. When a condensed region develops by chance in some part of outer space, the particles in the region are drawn still closer together by the attractive force of gravity, and the condensation develops until eventually the entire mass is highly compressed. This takes about 10 million years. The compression causes the center of the cloud of gas and dust to become quite hot, eventually reaching a temperature of about 10 million degrees, at a pressure of 100 million atmospheres.

At such high temperatures nuclear reactions can ignite spontaneously. In these reactions the hydrogen nuclei at the center of the dust cloud combine to form helium nuclei, releasing at the same time enormous amounts of energy, according to the following sequence:

$$p + p \rightarrow d^2 + e^+ + \nu$$
$$d^2 + p \rightarrow \mathrm{He}^3 + \gamma$$
$$\mathrm{He}^3 + \mathrm{He}^3 \rightarrow \mathrm{He}^4 + 2p$$

yielding a total energy of 26.2 MeV per He^4 nucleus. The energy generated by these nuclear reactions is great enough to keep the star from collapsing further under the force of gravity, and it lives out the rest of its life in a precarious balance between nuclear and gravitational pressures.

Developments of the last decade have led to the conclusion that this reaction that goes on in newly born stars is the first step in a cooking process, or fusion, in which all the elements of the universe are synthesized out of hydrogen as the basic building block. As the fusion continues at the center of the young star, the hydrogen is gradually replaced by helium, until eventually helium becomes the dominant constituent. Since, in this nuclear reaction, hydrogen is the fuel and helium constitutes the ashes, when a large amount of helium has been produced the fusion rate diminishes; and when that occurs, the star begins to collapse again under the force of gravity. The collapse produces a further increase in temperature at the center of the star, until the core reaches 10^8 °K, at which temperature the helium itself starts to burn. With the burning of helium, the energy release is renewed at the center and the gravitational collapse is halted.

The fusion of the helium consists in the combining of three helium nuclei to form the single heavier nucleus of carbon:

$$\alpha + \alpha \rightleftharpoons \mathrm{Be}^8$$
$$\mathrm{Be}^8 + \alpha \rightleftharpoons \mathrm{C}^{12*} \rightarrow \gamma + \mathrm{C}^{12}\,.$$

The fusion of helium nuclei into carbon is again accompanied by the release of a large amount of energy. When a substantial fraction of the helium has been converted into carbon, the next stage of burning begins, through such reactions as $\mathrm{C}^{12} + \alpha \rightarrow \mathrm{O}^{16} + \gamma$, and in this way successively heavier elements are built up from the original hydrogen.

However, when the critical element of iron is reached, the process halts. The iron nucleus is approximately halfway between the lightest and the heaviest elements in nature, and it is also the most stable of all elements. It absorbs energy instead of releasing it, and damps the reaction at the center of the star. When this happens the star implodes under the force of gravity. The implosion produces enormous pressures and temperatures at the center, which lead in turn to an explosion in which the star is blown apart.

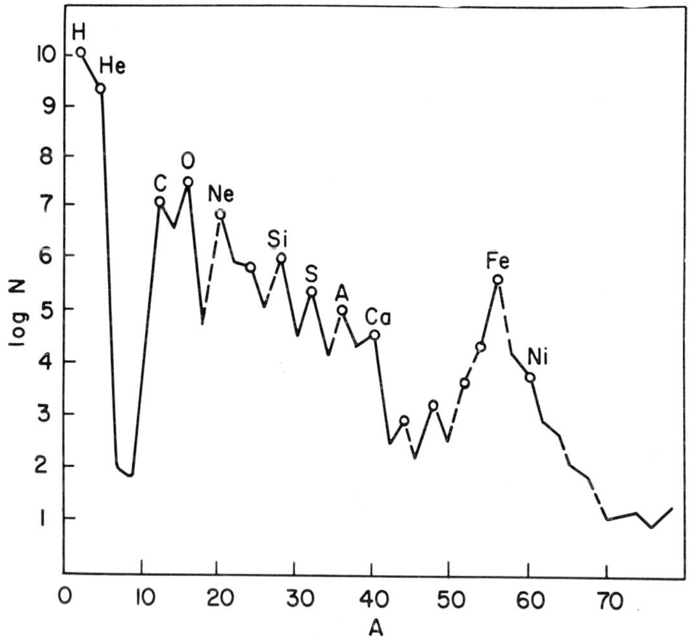

Fig. 1. Abundance of the elements in the sun (after Aller). *A* is the atomic weight and *N* is the relative abundance compared to Si $= 10^6$.

Almost all the matter of the star is ejected into space by the explosion, including the heavy elements the star has been manufacturing during its lifetime. After the explosion these heavy elements mix with the hydrogen gas in the space between the stars, and in the course of time new stars condense from the enriched mixture of gas and dust, as we described previously. Thus, the life history of the star is a cycle of dust to dust, from birth out of the condensation of interstellar matter to final destruction by collapse and explosion.

The exploding star is called a supernova. Such exploding stars are extremely bright; they may have as much as one milliard times the luminosity of the sun.

About 50 supernovae have been photographed with telescopes in the last 75 years. In our own galaxy, only a few occur every thousand years that are bright enough to be seen by the unaided eye. The earliest reported supernova was the brilliant explosion recorded by Chinese astronomers in 1054 AD. At the position of this supernova there is today a great cloud of gas known as the Crab Nebula, which is expanding outward at a speed of 1000 miles per second.

About half of all stars are believed to follow the evolutionary path to the supernova explosion. Those which do not are the stars of modest mass, whose evolution is halted at an intermediate stage of nucleosynthesis, before iron is formed, because their central temperatures never climb to the values required for the synthesis of the heavier elements. The critical mass for supernova development is about 1.4 times the mass of the sun. Stars lighter than these turn downward on a path leading to the white dwarf population.

The preceding account of stellar evolution gives us an indication of what may be expected for the relative abundances of the elements in the universe. Hydrogen is of course the most abundant, helium the next, and then C, N, O, S, Al, Mg, and other elements, decreasing roughly in order of atomic weight, with variations according to the values of the cross sections for the nuclear reactions which determine their rates of synthesis and their equilibrium concentrations. We expect a peak at iron, because of its high nuclear stability; and we expect low concentrations of the elements heavier than iron, which are produced entirely in the final flash of the supernova explosions, and only occur in existing stars by virtue of being drawn together, as a part of previously "used" matter, for the formation of a new star. We also expect substantial variations in the composition of stars with respect to these heavier-than-iron elements, according to the number of supernovae which have exploded in the vicinity.

With these considerations in mind, we go to the compilation of chemical abundances from several sources (Figure 1).

Four groups of elements appear to be most abundant in the universe:

$$\mathrm{H, He} \,|\, \mathrm{C, N, O} \,|\, \mathrm{Mg, Al, Si} \,|\, \mathrm{Fe}\,.$$

Knowing the temperature and pressure ranges usually prevalent in the planetary atmospheres, one can grossly estimate the elements and molecules which will be present in gaseous form in a planetary atmosphere.

$$\mathrm{H_2, He, CH_4, NH_3, H_2O, N_2, O_2, CO_2, etc.}$$

Given this primitive composition the atmospheres of the planets will have developed with time through four well-recognized processes:

(1) The gravitational escape of gases into the interplanetary space;

(2) The exhalation of gases from the interior of the planets;

(3) Chemical reactions of gases in the atmosphere with the solid material of the planet;

(4) The photo-dissociation and ionization of gases by the solar radiation.

Also: (a) The capture of gaseous constituents from the interplanetary medium; and (b) The radioactive processes could also be considered as a possible mechanism contributing to the structure of a planetary atmosphere.

We shall now discuss the first and the most important of the processes determining the structure of an atmosphere, namely the rate of escape of gases from the planet.

At any level in an atmosphere a number of atoms will be moving upward with speeds in excess of that required for escape from the gravitational attraction of this planet. The critical escape velocity, v_c, from a planet is given by

$$v_c = \sqrt{\frac{2MG}{R}}$$

where M is the mass of the planet, R its radius and G the universal gravitational constant. For the earth

$$v_c = 11.3 \text{ km/sec}$$

and for other planets

$$v_c = 11.3 \sqrt{\frac{M}{R}} \text{ km/sec}$$

where M and R now are in units of the earth's mass and radius. For the moon, Mars, and Jupiter the values of v_c are 2.3 km/sec, 5.0 km/sec, and 61.0 km/sec, respectively.

The escape of gases can only take place effectively at high altitudes where the density and the probability of collisions are small. These requirements define the region of the atmosphere known as the *exosphere*, a region in which the mean free path for collisions is comparable to the radius of the planet. Atmospheric particles moving in the exosphere execute ballistic trajectories in the gravitational field of the planet.

Knowing the temperature and density at the base of the exosphere, one can calculate the number of particles which will be escaping from the planet per unit time. The classical formula (e.g., Spitzer, 1952) for the time of escape, t_e, in which the density of an atmospheric constituent of molecular or atomic weight 'm' will fall to e^{-1} of its original value is given by

$$t_e = \left[B(6\pi)^{1/2}\mathbf{v}/3g\right]\frac{e^y}{y} \tag{1}$$

where

$$B = \frac{n_0 T_0}{n_c T_c} \quad \text{and} \quad y = \frac{mv_c^2}{2kT}.$$

n_0 and T_0 are the particle number density and temperature respectively at the ground level, n_c is the hypothetical number density at ground if the whole atmosphere were isothermal at the escape level temperature T_c, R is the radius of the planet, $\mathbf{v}$ is the root mean square velocity of the escaping particles, and v_c is the velocity of escape.

In the case of the earth, the temperature of the exosphere has recently been determined to be ~ 1600 °K. Hydrogen would therefore have escaped from earth in $\sim 10^3$ years. As the age of the planets is $\sim 3 \times 10^9$ years, the absence of hydrogen in the atmosphere of earth is comprehensible. Hydrogenated gases, like CH_4 and NH_3 which are susceptible to dissociation by the solar ultraviolet radiation, would also lose their hydrogen, and the carbon and nitrogen thus liberated will probably remain in the atmosphere as CO_2 and N_2.

The case of water is different. Due to the low temperature of the tropopause, almost all of the terrestrial water vapour remains confined to the troposphere. An extremely minute amount, corresponding to the saturated water vapour pressure of 200 °K, will traverse the tropopause and will be susceptible to dissociation in the upper atmosphere. This special circumstance preserves the water on our planet.

In the case of helium, the time of escape from earth is of the order of 10^8 years which, due to the possible uncertainty in the rate of outgassing of He from the crust, cannot be interpreted as a definite indication of complete absence of helium in the earth's atmosphere. In fact, a layer of helium at an altitude of ~ 1600 km has recently been detected by satellite investigations.

From the above equation it can be concluded that exospheric constituents of molecular or atomic weight >6 will probably be retained by the earth up to the present time.

These considerations indicate that H_2 has escaped and hence the present atmosphere of earth should be in an oxidized state.

Nitrogen, oxygen and argon make up more than 99.9% of the earth's atmosphere. Though the almost complete absence of hydrogen and helium can perhaps be explained by the gravitational escape of gases, the presence of free oxygen and the relative absence of CO_2 in the terrestrial atmosphere are problems related to the presence of life on earth and the possible reactions of atmospheric gases with the crust.

The mean molecular weight of the earth's atmosphere ($N_2 \sim 79\%$, $O_2 \sim 20\%$) is ~ 29. The atmosphere is known to be "mixed" up to an altitude of ~ 80 km, and the composition remains nearly constant except for H_2O and O_3 which are present in small amounts and have their maximum concentration at the ground level and ~ 30 km respectively.

The total pressure at the surface due to this atmosphere is $\sim 10^6$ dyn/cm^2 (1000 mb) and the number density at the ground $\sim 2 \times 10^{19}$ mol/cm^3. In a nearly isothermal atmosphere the variation of density with height is given by the formula expressing hydrostatic balance:

$$\varrho = \varrho_0 \exp\left(\frac{-(h - h_0)\mathbf{m}g}{k\mathbf{T}}\right) \tag{2}$$

in which ϱ and ϱ_0 are the densities at height h and h_0, $\mathbf{m}$ is the average molecular weight per particle, $\mathbf{T}$ the average temperature between h and h_0, g the acceleration of gravity and k the Boltzmann constant.

The quantity $k\mathbf{T}/\mathbf{m}g$ is known as the scale height, H, of the atmosphere; at an altitude of one scale height the density being reduced by a factor of e. For a mean temperature of 250 °K in the first 80 km of the atmosphere, the atmospheric scale height is ~ 8 km.

Above 120 km "diffusive separation" sets in and the pressure of each constituent varies in accordance with a scale height calculated for its own molecular weight. At the boundary of this domain, at a height of 120 km, nitrogen and oxygen are the major constituents, with N_2 dominating in the ratio 4:1, as on the surface. At higher altitudes the relative concentration of oxygen rapidly increases, and by 300 km the surface proportions are reversed, with oxygen now dominating in the ratio of 3:1.

The reason for this circumstance is that O_2 can be dissociated by absorption of solar ultraviolet, with relatively high probability, yielding atomic oxygen with half the weight per particle and therefore with twice the scale height and a correspondingly smaller rate of decrease of concentration with increasing altitude. Molecular nitrogen also undergoes photo-dissociation, but with a much smaller probability than for oxygen; hence it remains in primarily molecular form up to greater altitudes. Thus, again the twice-as-heavy nitrogen settles out of the atmosphere with relative rapidity. At 300 km, for example, less than 1% of the N_2 is in atomic form, while 99.5% of the oxygen is atomic.

For this reason atomic oxygen becomes the principal atmospheric constituent above 200 km, and continues to dominate the composition of the upper atmosphere up to about 1000 km. Above that level, the lighter gases, hydrogen and helium, emerge as the principal constituents because of their very large scale heights, although they are present only in trace amounts at lower altitudes. At the greatest altitudes all gases except hydrogen have settled out of the atmosphere, and this lightest gas dominates, until finally the hydrogen atmosphere merges into the interplanetary medium at a distance of some 2 to 10 earth's radii.

B. TEMPERATURE

The temperature of a planet essentially depends on its distance from the sun. In the case of Earth, the solar radiation flux, with an effective black body temperature ~ 6000 °K, reaching the top of the earth's atmosphere, has a value of $\sim 1.4 \times 10^6$ ergs/cm^2/sec. Part of this radiation is immediately 'reflected' back to space by clouds and the atmosphere and does not play any role in the energy balance of the planet. The fraction of the solar radiation thus reflected back to space is known as the Albedo of the planet, and in the case of earth, it has been estimated to be 0.39. Most of the remaining 61% of the solar radiation, mainly composed of the visible part of the spectrum, penetrates down to the ground and heats the surface to a certain temperature denoted as the effective temperature of the planet. In the case of fast rotating planets (e.g. Earth, Mars, Jupiter)

$$T_e^4 = \frac{S}{4\sigma}(1 - A) \tag{3}$$

where S is the solar constant or the radiation flux received at the top of the atmosphere and A is the Albedo of the planet over the whole solar spectrum. For the values of S and Albedo given above, the T_e for earth is 245 °K.

The surface radiating at this relatively low temperature emits primarily in the far infra-red. A large fraction of the radiation emitted by the surface is immediately absorbed by molecules of CO_2 and H_2O present in the lower layers of the earth's atmosphere. A part of this absorbed energy in the infra-red is returned to the surface and provides extensive heating of the ground, raising the surface temperature to the observed value of ~290 °K.

This additional heating of the surface by the return of infra-red from the atmosphere is referred to as the "greenhouse effect" and is an essential element in understanding the radiation budget of the planet in general and of the lower atmosphere in particular.

The lower atmosphere of the earth is in radiative and convective equilibrium and the temperature decreases with height at a rate of ~6.5 °K/km.

The adiabatic temperature gradient is given by $\beta_a = -g/C_p$, where C_p is the specific heat of air at constant pressure. For the earth's atmospheric composition $\beta = -9.6$ °K/km, but because of the eventual condensation of atmospheric water vapour in the form of clouds, the latent heat released in the atmosphere reduces the actual temperature gradient to the observed value of −6.5 °K/km.

This region of negative temperature gradient extends up to an altitude of ~12 km and is referred to as the *troposphere*. The top of the troposphere is denoted as the *tropopause* and the temperature at this level is of the order of ~200 °K. Above the tropopause, however, the atmosphere is mainly in radiative equilibrium. Because the water vapour precipitates out at the low temperature of the tropopause, there is a very small amount of water vapour in the atmosphere above to provide any significant infra-red opacity. The temperature therefore remains in the neighbourhood of ~200 °K and then rises subsequently because of the direct absorption of the solar ultraviolet radiation by the traces of ozone present at ~50 km.

The atmospheric region between the tropopause and the level of the secondary temperature maximum at 50 km therefore has a positive temperature gradient and is stable against convection. It is referred to as the *stratosphere*.

Above the ozone layer the temperature decreases again with altitude to a minimum value of ~180 °K at ~80 km *(mesopause)*.

Above 100 km a region of strong heating sets in, which results from the photodissociation of O_2 and the photoionization of N_2 and O by the solar radiation in the far ultraviolet. This is the region of the *thermosphere*. At the top of the thermosphere the temperature approaches a constant value of ~1500 °K.

The *exosphere* is defined as the region where the atmospheric density and the probability of collisions is so small that the particles execute ballistic trajectories in the gravitational field of the planet. The base of the exosphere is a level above which there is only one collision for a particle moving vertically away from the planet. In the case of the earth the base of the exosphere is located at an altitude of ~700 km, where the density is of the order of ~10^6 particles/cm^3. The exospheric region extends up to the distance where the atmospheric density falls to the value of ~100 particles/cm^3, the average density of the interplanetary medium. On this criterion the outer boundary of the exosphere lies near 5000 km. This can be considered as the limit of the earth's neutral atmosphere.

Charged particles trapped in the earth's magnetic field, however, populate a region beyond this, called the *magnetosphere*, which has been found to extend to several earth radii and is more commonly known as the Van Allen radiation belt.

2. Other Planets

For a rough estimate of the possible composition of the atmospheres of other planets, we can calculate from Equation (1) the approximate times of escape of gases as we did before for the earth. It follows that

(1) The moon must have lost all but a trace of its atmosphere early in its history; what atmosphere it has must consist of argon, and perhaps the still heavier rare gas, xenon;

(2) Hydrogen and helium escape rapidly from all the terrestrial planets;

(3) On Mars oxygen may be retained or lost, depending on the temperature, the situation resembling that for helium on the earth; on Venus, oxygen is definitely retained;

(4) Jupiter probably retained all the gases of its primitive atmosphere, including hydrogen. There is, however, some observational evidence for a hydrogen deficiency in the Jovian atmosphere.

With these considerations in mind we shall now review the observed characteristics of the atmospheres of Mars and Venus.

3. Mars

The following table summarizes reliable physical data for Mars.

TABLE I

PHYSICAL DATA OF MARS

Mean distance from sun	1.52 AU
Mean equatorial diameter	6790 km
Length of day	1.0012 earth day
Length of year	1.8808 earth year
Mass	0.1078 (earth 1)
Mean density	3.90 g/cm^3
Gravity	377 cm/sec^2
Albedo	0.26
Effective black body temperature	209 °K

A. COMPOSITION AND SURFACE PRESSURE

The only gases that have been spectroscopically detected up to now in the atmosphere of Mars are carbon dioxide and water vapour. Their abundances have been estimated by several authors, but generally accepted values of the amounts of these gases in the atmosphere of Mars are not yet available. The best estimates, at the present time, vary within the following ranges:

$$CO_2: 70 \pm 30 \text{ m-atm.}$$

$$H_2O: 10^{-3} - 10^{-2} \text{ g/cm}^2$$

(m-atm is the thickness of a homogeneous atmosphere in meters at normal temperature and pressure, 0 °C and 760 mm).

Search for the presence of other gases in the atmosphere of Mars has failed to give positive results. The upper limits on their possible abundances are tabulated below.

TABLE II

UPPER LIMITS OF ABUNDANCES OF GASES IN THE MARTIAN ATMOSPHERE

Molecules	*m-atm*
O_2	< 2.4
N_2O, NO_2 or N_2O_4	< 2.0
NH_3	< 0.2
CH_4	< 0.1
O_3	$< 5 \times 10^{-4}$
SO_2	$< 3 \times 10^{-5}$

Earlier photometric studies by de Vaucouleurs and polarimetric measurements by Dollfus had given consistent estimates of the surface pressure on Mars as 85 ± 10 mb. Recently, however, from the analysis of pressure-dependent absorption bands of CO_2 observed in the Martian atmosphere, Kaplan suggests that the total pressure at the surface of Mars may be as low as 20 ± 10 mb. If this value is taken to be correct, then CO_2 will make up for $\sim 25\%$ of the total atmosphere. On the assumption that the rest of the atmosphere is made up mainly of nitrogen with traces of argon, both being undetectable by earth-bound spectroscopic observations, the probable composition of the Martian atmosphere is therefore as follows.

TABLE III

PROBABLE COMPOSITION OF THE MARTIAN ATMOSPHERE

Gas	*% Volume*
N_2	72
CO_2	25
A	2
O_2	< 0.5
H_2O	very small

This atmosphere has a mean molecular weight of 32. With a surface pressure of 20 mb and an approximate temperature of 210 °K, the surface density would be approximately 5×10^{17} particles/cm^3 which, for an isothermal atmosphere, should decrease exponentially with altitude with a scale height of $= 14.5$ km. The vertical distribution of pressure for this model is shown in Figure 2. Results of a similar computation for earth, assuming an isothermal atmosphere at 250 °K, are also shown in the figure and are in accord with recent rocket measurements. It is interesting to note that although the pressure at the surface of Mars is about 1/50 of the earth's, at an

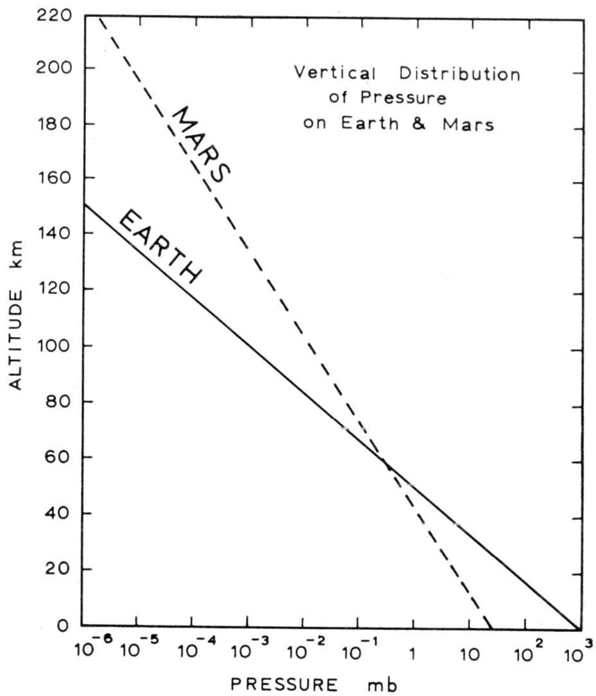

Fig. 2. Pressure versus altitude for Earth and Mars (Martian surface pressure 20 mb; atmospheric scale height = 15 km).

altitude of ~50 km the two atmospheres have the same pressure, and above this height the pressure in the Martian atmosphere is greater than the pressure at the corresponding height in the earth's atmosphere. Because the decrease of density with height of Mars is almost two times slower than in the earth, the levels of the ionosphere and thermosphere on Mars would be much higher than on the earth.

B. TEMPERATURE

Planetary temperatures are usually estimated by measuring the infra-red radiation emitted by the planet. A large part of the infra-red spectrum is, however, absorbed by the water vapour and CO_2 present in the earth's atmosphere. Ground based observations of the planets in the infra-red are therefore confined to the 8 – 12 μ region where the earth's atmosphere is relatively transparent. This spectral region in the infra-red is known as the atmospheric "window".

Extensive temperature measurements of Mars by infra-red radiometry in this "window" of the terrestrial atmosphere have been made since 1926, and the results to date are quite consistent. Since the Martian atmosphere should also be largely transparent in the 8–12 μ "window", it is assumed that the radiation intensity measurements in this wavelength region refer to the surface of the planet.

Some of the earlier investigators were also able to scan the planetary disc latitudinally and longitudinally so as to obtain the diurnal and seasonal variation of temperature as a function of latitude. The up-to-date information can be summarized as follows:

Maximum temperature at equator	~ 300 °K
Mean amplitude of diurnal variation noon to sunset (on earth in desert)	~ 60
	~ 30
Night side temperature cannot be measured but can probably be estimated at the equator	~ 200
Day side temperature at poles	~ 220
Mean temperature of day side	~ 260
Mean temperature of whole planet	~ 230

Several theoretical estimates for the vertical temperature structure of the Martian atmosphere are available in the literature.

Figure 3 shows the vertical distribution of temperature in the atmosphere of Mars. The temperatures up to an altitude of 100 km are based on the results recently obtained by Arking, while for above this altitude, the upper atmospheric model developed by Chamberlain has been used.

The results shown in Figure 3 do not take into account possible heating of the lower atmosphere by direct absorption of solar radiation in the ultraviolet. In the case of the terrestrial atmosphere, the ozone heating produces a temperature maximum at 50 km, but on Mars, with much less oxygen compared to earth, ozone would be con-

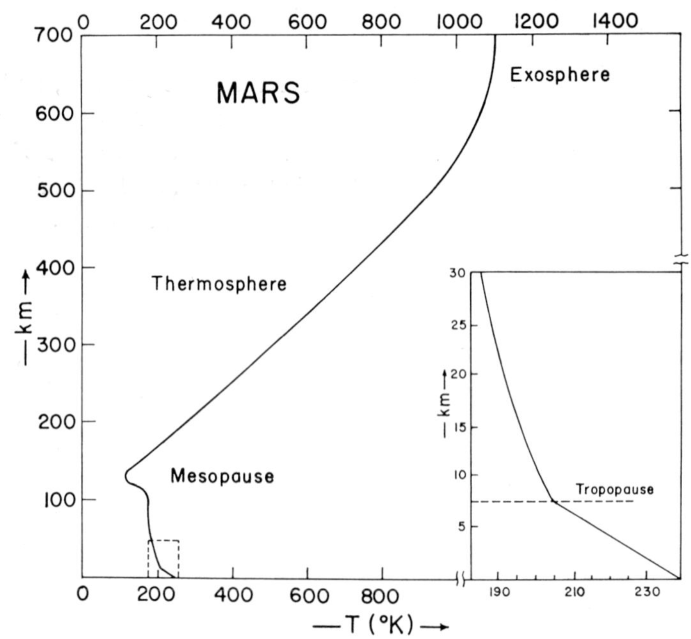

Fig. 3. Vertical temperature structure of the Martian atmosphere (after Arking and Chamberlain).

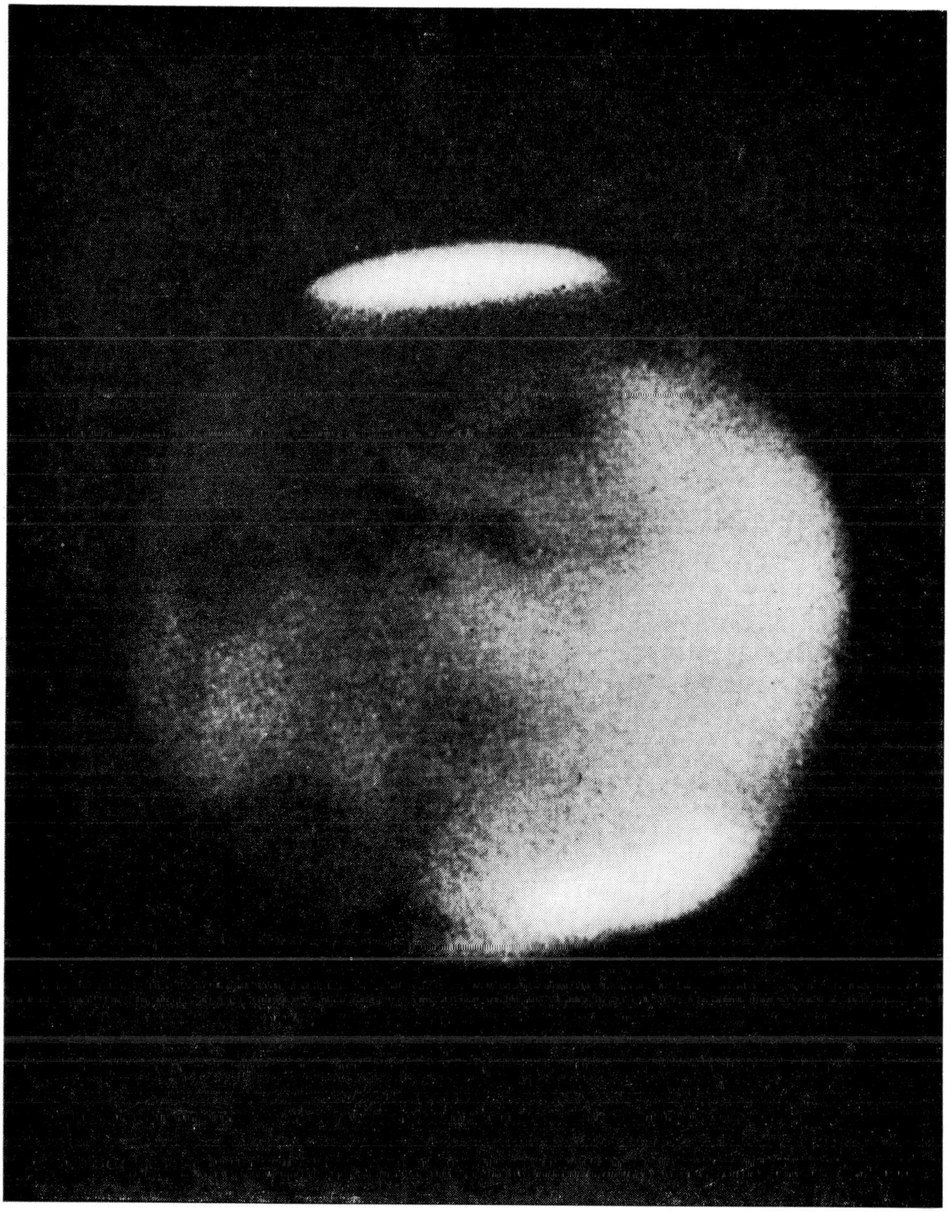

Plate 1. Photograph of Mars made on July 4, 1954 obtained by the National Geographic Society – Lowell Observatory expedition to the Lamont-Hussey Observatory, Bloemfontein, South Africa.

fined to lower layers of the atmosphere. The solar radiation in the region of 2500 Å, responsible for ozone dissociation and atmospheric heating, will therefore penetrate to a much lower depth in the Martian atmosphere. Due to a higher atmospheric density at this level (compared to the density in the earth's atmosphere at 50 km) and because of the lower intensity of the solar radiation at the distance of Mars, the heating rates will probably not be as important as in the ozonosphere on earth. Still they may affect the temperature gradients substantially and limit the extent of the convection zone.

C. SURFACE FEATURES OF MARS

As seen by telescope the outstanding features of Mars are (1) the dark areas (maria), (2) the general reddish-orange background ("desert"), and (3) the polar caps (Plate 1).

The nature of the dark maria is not very well understood because, apart from showing fairly regular seasonal changes, they are also subject to erratic variations which make the hypothesis of earth-type vegetation somewhat doubtful. Moreover, the infra-red spectrum does not have characteristic absorption bands of chlorophyll. The presence of organic material in these regions has, however, been reported by Sinton who detected C–H vibration absorptions in the 3.5 μ region which were absent in the reflection spectrum of the desert areas.

The probability of the existence of life on Mars is controversial and beyond the scope of this review; but it will be desirable to see the observations of C–H bands on Mars repeated and probably substantiated by infra-red spectroscopic measurements of the terrestrial "dark and bright" areas from an earth satellite.

The large orange areas of Mars are responsible for the reddish color of the planet seen by the naked eye. According to Kuiper they are composed of felsitic rhyolite.

The polar caps are rather reliably known to be a layer of H_2O frost a few centimeters thick deposited on the surface. These caps are observed to expand to lower latitudes during fall and winter and to recede to high latitudes during spring and summer. The observed evaporation of the polar caps in summer and the almost immediate transport of the water vapour towards the winter pole across the equator is a special meteorological phenomenon peculiar to Mars. In the earth's atmosphere the hemisphere to hemisphere mixing is extremely small.

The average speed of this "humidity wave" across the planet has been estimated to be about 45 km/day. This pole to pole circulation can perhaps only be understood by the greater length of the Martian year and the increased temperature differences between the cold and warm poles. In the summer the pole is the warmest region on the planet.

D. BLUE HAZE LAYER

Photographs of Mars taken through blue filters indicate no surface features, and it is therefore believed to be covered with a haze layer known as "blue haze", absorbing at these wavelengths. Sometimes during the opposition, however, it clears over certain regions, and surface features become observable in the blue. These "blue clearings"

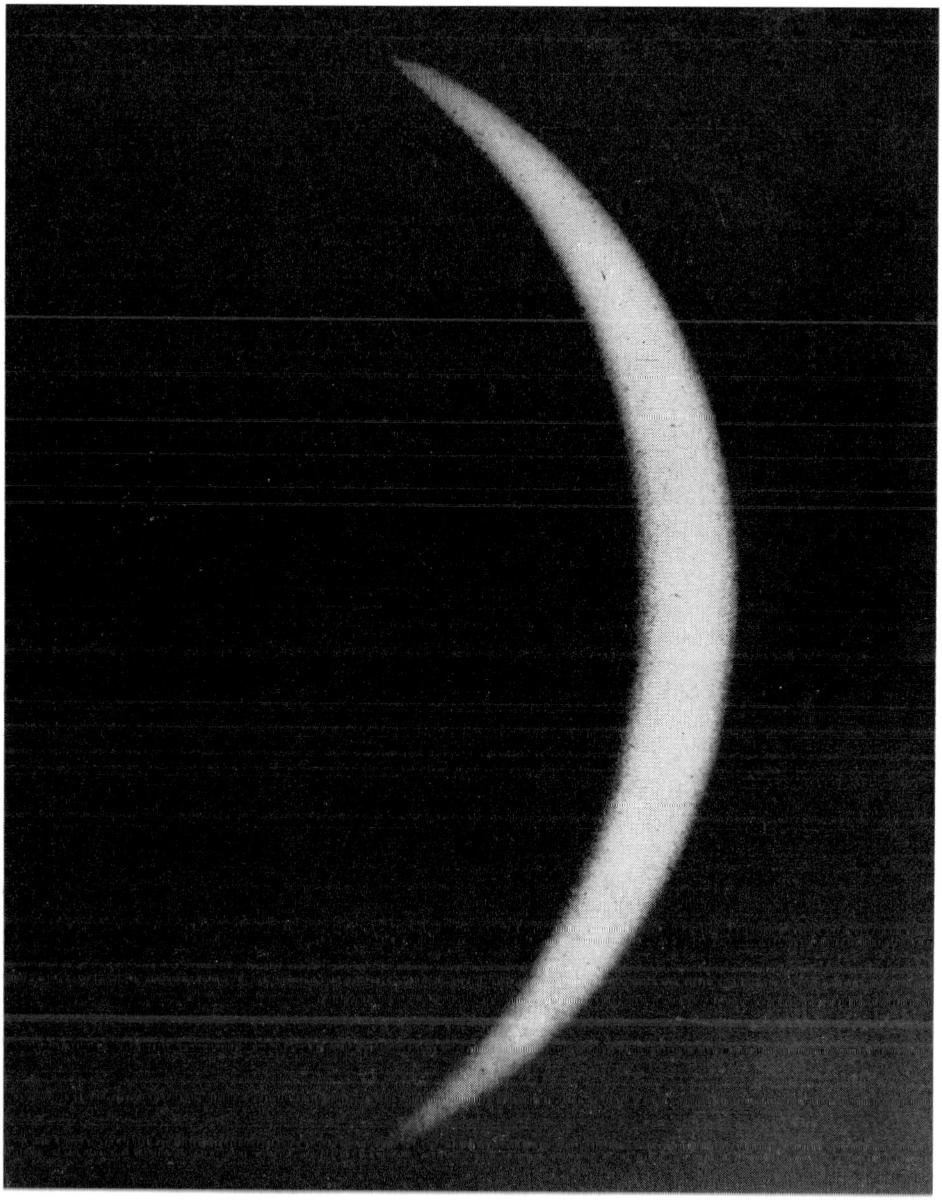

Plate 2. Venus in red light near the inferior conjunction (after M. L. HUMASON in *Planets and Satellites* (ed. by G. P. KUIPER); courtesy of the University of Chicago Press).

last usually a few days. Conflicting arguments have been extended to explain the nature of the blue haze, but an explanation which satisfies all the observed features is yet to be found.

4. Venus

Venus is our nearest planetary neighbour and, after the sun and moon, the brightest object in the sky. It has therefore attracted the attention of man since the beginning of civilization. Despite the great interest, very little is known about the atmosphere of this planet, especially when compared with the information we have about Mars. The main reason for this deficiency is that Venus is covered with a layer of white clouds and the surface remains permanently invisible (Plate 2).

Observations have, however, been made in the infra-red and radio-frequency regions, and new information regarding the composition and temperature distribution in the Venus atmosphere has been obtained in the last few years. These new results of temperature and pressure at the various levels in the Venus atmosphere have forced a complete revision of ideas regarding the atmospheric structure of this planet. We shall try to summarize our present-day knowledge regarding the Venus atmosphere by first giving the physical constants of the planet (Table IV) and then reviewing the various hypotheses relative to the structure of its atmosphere.

TABLE IV

COMPARISON OF PHYSICAL DATA OF EARTH AND VENUS

	Mass	*Rad.*	*Distance (AU)*	*Density*	*Albedo*	T_e	g *cm/sec*^{2}
Earth	1	1	1	5.5	0.39	245 °K	980
Venus	0.81	0.97	0.72	4.8	0.73	235 °K	842

A. COMPOSITION

From the analysis of reflected solar spectrum, the only atmospheric constituent so far detected in Venus is CO_2. Its abundance above the effective "reflecting level" of the sunlight has been estimated by several workers. Until recently, the amount of CO_2 above the cloud level was accepted to be $\sim 95\%$ of the total atmosphere above the clouds. Recent reinterpretations of the old spectra of Venus by Spinrad however give a CO_2 atmosphere ratio by mass of only 5%. Attempts have also been made to detect water vapour in the atmosphere of Venus. From high altitude balloon measurements some indications of the existence of a small amount of water vapour were actually obtained and recently Dollfus has also identified infra-red absorption by water vapour in the upper atmosphere of Venus. Quantitative interpretations of these measurements are still uncertain and it can be assumed that the mixing ratio of water vapour in the Venusian atmosphere above the clouds is $< 10^{-5}$ by mass. If, however, the clouds are composed of water drops or of ice crystals, a large abundance of water vapour can be expected below the clouds. The amount of water vapour above will be limited by the saturated vapour pressure at the temperature of the cloud tops.

No other constituent in the Venus atmosphere has been spectroscopically observed. Based on the new estimate of CO_2 mixing ratio, we shall therefore adopt the following composition for the Venus atmosphere.

TABLE V

ADOPTED COMPOSITION OF THE VENUS ATMOSPHERE

Gas	*Percentage by Mass*
N_2	95
CO_2	5
H_2O	$< 10^{-3}$

B. PRESSURE

There is conflicting evidence concerning the pressure at the cloud top level. A difference in the polarization of red and green light reflected from the cloud top has been interpreted by Dollfus as the result of molecular scattering in an atmosphere 800 m thick at STP, which corresponds to a cloud top pressure of 90 mb. Sagan has analyzed Spinrad's results on the temperature-pressure combinations indicated by CO_2 lines on different plates, and he concludes that the cloud top pressure on the illuminated side lies between 0.53 and 0.83 atm.

Estimates for the pressure at the surface of Venus range from 7 atm to 200 atm. These will be discussed in more detail in a later section.

C. TEMPERATURE

Cloud top. Sinton and Strong have repeatedly measured the planetary emission from the cloud top in the 8–12 μ region and find a temperature value of 235 ± 10 °K. This temperature probably refers to the cloud top level of the Venus atmosphere. Recent results from Mariner II seem to confirm this conclusion.

From this value of temperature it is possible to construct a model for the atmosphere of Venus above the clouds. Figure 4 shows a temperature profile computed by Rasool on the assumption that the atmosphere above the clouds is in radiative equilibrium.

The temperature should reach an asymptotic value of 197 °K above $z \sim 20$ km. The mesopause will probably occur above 60 km, and then the temperature will rise in the thermosphere owing to heating by photo-dissociation and photoionization reactions. Our lack of knowledge of the composition prevents us from predicting whether heating should occur below the mesopause by such mechanisms as ultraviolet absorption by ozone.

Surface. Determinations of the surface temperature of Venus by optical astronomy have been prevented by the cloud cover. However, the small amount of thermal radiation emitted by the planet in the centimeter wavelength region would penetrate through the clouds without significant attenuation, and therefore can be used to determine the temperature of the surface of the planet.

First attempts to measure radiation from Venus in the microwave region were made in 1956 by the radiotelescope of the Naval Research Laboratory. The temperature inferred from the measured radiation intensity was, however, unexpectedly high –

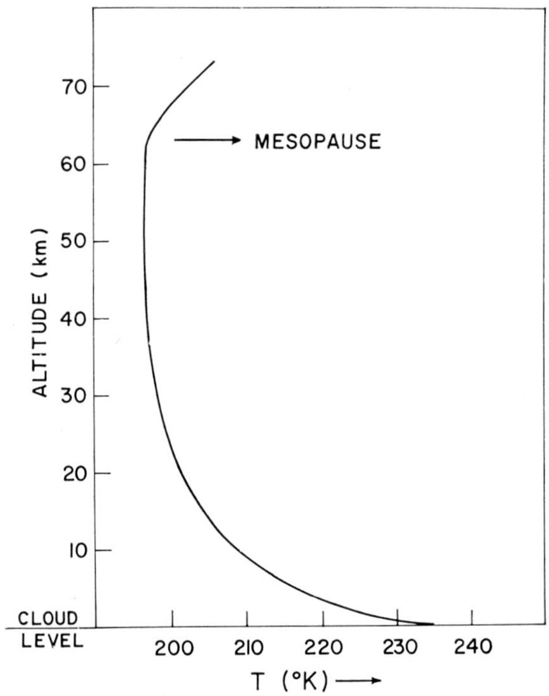

Fig. 4. Radiative temperature profile in the Venus atmosphere above the clouds (after RASOOL, 1963).

of the order of 600 °K, which is certainly too hot to support any imaginable form of life. Repeated measurements in the following years have confirmed this result.

The observed temperatures are plotted as a function of wavelength in Figure 5. At longer wavelengths the brightness temperature is observed to increase when the Venus-sun-earth angle becomes appreciable, indicating that the illuminated hemisphere is significantly hotter than the dark side. (The 10 cm brightness temperature may exceed 700 °K on the bright side).

Three models of Venus have been considerably discussed in connection with these measurements.

Greenhouse. First is the greenhouse model originally discussed by Sagan in which it is assumed that the 600 °K radio-brightness temperature actually exists at the surface of the planet and is maintained by a very effective "greenhouse effect". The solar radiation in the visible penetrates up to the surface; the planet, thereby being heated up to a temperature T_e, emits in the infra-red; but due to the presence of triatomic molecules like CO_2 and H_2O, which have strong absorption bands in the infra-red region, most of the radiation remains trapped in the atmosphere and heats up the surface.

Jastrow and Rasool have recently pointed out that in such a model an extremely high infra-red opacity of the atmosphere (optical thickness of 42, meaning a transmission of only 10^{-18}) will be required in order to obtain a Venus ground temperature of 600 °K.

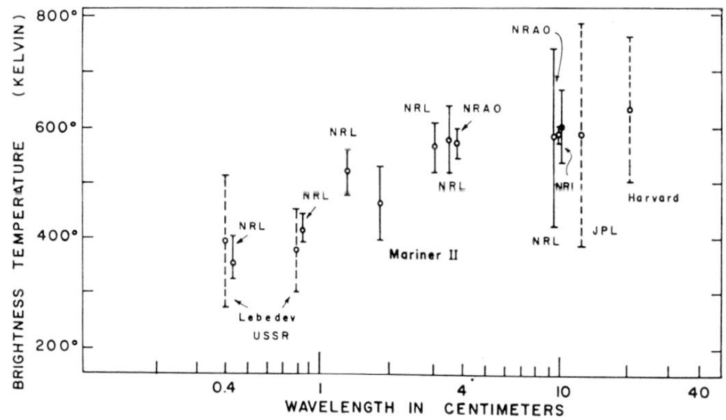

Fig. 5. Observed microwave brightness temperatures of Venus at different wavelengths.

It is very difficult to understand how the lower Venus atmosphere can be so opaque in the infra-red and still be transparent in the visible. Triatomic molecules are good infra-red absorbers, but they have "windows" at certain wavelengths which must be blocked by other molecules if the opacity is to be significant. Jastrow and Rasool have also shown that the combination of CO_2 and small amounts of H_2O, at total pressure of the order of 3 atm, still contains significant windows.

However, the situation alters if the atmospheric pressure at ground level is many tens or one hundred atmospheres and the amount of CO_2 is proportionally higher. Then pressure broadening may close all the windows. At the high temperatures on Venus there may also be many trace constituents in the atmosphere that significantly increase the opacity.

Aeolosphere. The second model, called the aeolosphere model, is due to Öpik who suggests that the blanketing must be caused by dust, wind friction of the surface being the main source of energy for the high temperature. The dust is probably made of calcium and magnesium carbonates and the atmosphere is mainly composed of CO_2 and N_2. No water vapor is necessarily present in the atmosphere. Due to the blanket of dust there is no sunlight penetrating to the surface. This model predicts that the microwave brightness temperatures of Venus should not change with the phase of Venus, which is apparently contradicted by the observations.

Ionospheric Emission. A third model invokes a dense ionosphere of Venus which is assumed to emit the observed microwave radiation. The electron density for the ionospheric thickness of about 300 km (comparable to Earth) for such a model, is about $10^9/\text{cm}^3$. This is about 1000 times greater than the maximum density in Earth's ionosphere. It is very difficult to envisage such high electron concentration, unless an ionization mechanism is found which is many orders of magnitude more effective than the solar ultraviolet and X-rays.

At the time of writing, the results of Mariner II fly-by to Venus have just been announced. The measurements made at the radio wavelengths ($\lambda = 1.9$ cm) indicate "limb darkening", thus ruling out the possibility that the high temperatures could be due to the free-free emission from a thick ionosphere. The brightness temperatures observed by Mariner are also of the order of about 500° (Figure 5), which confirms the earlier suggestions that a high temperature of the order of 600–700° may exist at the surface of Venus. The choice between the greenhouse and the aeolosphere models is still wide open and it is hoped that the continued interest and research of many workers in this field will soon bring us nearer to the solution.

General References

BARRETT, A. H. and LILLEY, E.: 1963, 'Mariner-2 Microwave Observations of Venus', *Sky and Telescope* **25,** No. 4.

CAMERON, A. G. W.: 1964, 'Physics of the Planets', *Space Physics,* University of California Engineering and Physical Sciences Extension Series (ed. by D. P. LEGALLEY), John Wiley and Sons, New York.

HESS, S. L.: 1961, 'Mars as an Astronautical Objective', *Adv. Space Sci. Techn.* (ed. by F. I. ORDWAY), Academic Press, New York, 3, 151.

JASTROW, R.: 1962, 'Results of Experiments in Space', *J. Aerospace Res.* **29,** 377.

JASTROW, R.: 1963, 'Physics of the Atmosphere', notes of lectures given at Columbia University, New York.

KELLOGG, W. W. and SAGAN, C.: 1961, 'The Atmospheres of Mars and Venus', Panel on Planetary Atmospheres of Space Science Board, Nat. Acad. of Sci. – Nat. Res. Council, Washington, D.C., Publ. 944.

KUIPER, G. P.: 1952, 'Planetary Atmospheres and Their Origin', *Atmospheres of the Earth and Planets* (ed. by G. KUIPER), University of Chicago Press, Ch. 12.

NEWBURN, R. L.: 1961, 'The Exploration of Mercury, the Asteroids, the Major Planets and Their Satellite Systems and Pluto', *Adv. Space Sci. Techn.* **3.**

RASOOL, S. I.: 1963, 'Structure of Planetary Atmospheres' *A.I.A.A. Journ.* **1,** 6.

RASOOL, S. I. and JASTROW, R.: 1964, 'The Atmospheres of Mars, Venus and Jupiter', Symposium on 'Extraterrestrial Biology and Organic Chemistry' at the Fourth COSPAR Meeting, Warsaw, Poland; North Holland Publishing Co., Amsterdam.

SAGAN, C. and KELLOGG, W. W.: 1963, 'The Terrestrial Planets', *Ann. Rev. Astron. Astrophys.* **1,** 235.

UREY, H. C.: 1959, 'The Atmospheres of the Planets', *Handbuch der Physik* **52,** 363.

VAUCOULEURS, G. de: 1954, *Physics of the Planet Mars,* Faber and Faber, London.

CHAPTER 38

AN INTRODUCTION TO LANGMUIR PROBES FOR SPACE RESEARCH

R. L. F. BOYD
Department of Physics, University College, London

1. Introduction

Early in the history of plasma physics Sir William Crookes employed a probing electrode which he termed an "idle pole" to explore the distribution of potential in a gas discharge. A quarter of a century later Langmuir showed (References E and G) that a study of the current-voltage characteristics of such a probe so transforms its usefulness that not only space potential but electron and ion concentrations and electron temperatures can be deduced. This paper discusses the theory and practice of Langmuir probes for instrumentation of spacecraft.

Because of inadequate theory and insufficient experimental study of probes in conditions where the mean free path is not large compared with the probe dimensions, such operation will not be considered here. This should be construed as a challenge to explore this use rather than as an indication of unsuitability. Some discussion of these problems will be found in Reference H.

Conventionally the electron current to a probe is taken as positive and so plotted. We shall follow that convention. This current arises from the random flux of electrons due to their thermal motions in the plasma. There is a similar random flux of positive ions, though generally it is about two orders of magnitude smaller because of the slower mean velocity of the ions.

2. The Retardation Characteristic

Consider a probe with a potential V somewhat negative to the space potential V_0. The fraction of the electron random flux able to reach it will be given by a Boltzmann factor so that the electron current collected will be:

$$i_e = A \cdot n_e \cdot e \cdot 1/4 \left(\frac{8kT_e}{\pi m_e}\right)^{1/2} \cdot \exp \frac{e(V - V_0)}{kT_e} \tag{1}$$

where A is the probe area, n_e the electron concentration, e the electronic charge and $(8kT_e/\pi m_e)^{1/2}$ is the mean velocity of the electrons.

Taking logarithms we may write:

$$\log i_e - \log A \cdot n_e \cdot e \left(\frac{kT_e}{2\pi m_e}\right)^{1/2} = \frac{1}{2.303} \cdot \frac{e(V - V_0)}{kT_e}. \tag{2}$$

so that the slope of the characteristic gives the temperature T_e. As the probe potential becomes positive with respect to space $[(V-V_0)>0]$ the Boltzmann factor is no longer valid and must be replaced by a weaker variation with V. The point of departure from a linear semi-logarithmic relationship thus indicates the value V_0 of V (the space potential). Further the value of i_e when $V=V_0$ leads to a value for the electron concentration n_e.

In principle the above remarks apply equally to the retardation characteristic for a positive probe in a plasma, where the positive ions approach the probe with a Maxwellian velocity distribution. For several reasons this condition may not hold in practice.

3. The Attraction Characteristic

When a probe is positive and is attracting electrons it becomes surrounded with a sheath of space charge which effectively localises its field within a distance determined by the properties of the plasma. The relevant parameter is the Debye length

$$h = \left(\frac{kT_e}{4\pi n_e e^2}\right)^{1/2}.$$

If $h=a$ (the probe radius), the probe collects the entire random flux of electrons which crosses the sheath and since under these circumstances the sheath area is virtually the probe area the current is given by Equation 2, with the exponential term replaced by unity for $V>V_0$. However, this situation is not common in the laboratory and is extremely improbable in space. More commonly the probe size is only a little larger than the Debye length or is smaller. The Debye length in various regions of the upper atmosphere has values of the following order.

Region	*D*	*E*	F_{max}	*Space 1 AU from Sun*
h (cm)	10	1	0.2	100

The actual sheath thickness depends on the potential difference $(V-V_0)$ and for a plane sheath is given by:

$$\text{sheath thickness} = 1.26\,h\left[\frac{e(V-V_0)}{kT_e}\right]^{3/4}. \tag{3}$$

When the sheath around a cylindrical or spherical probe is large $(h \gg a)$ the random current across the sheath may greatly exceed the current collected since the electrons perform orbits in the field of the probe. It is easy to show from the conservation of energy and angular momentum that the impact parameter, r, for an ion of energy E to be collected is:

$$r = a\left(1 + \frac{(V-V_0)}{E}\right)^{1/2}. \tag{4}$$

Thus r is the effective radius of the probe (for $h \gg a$) and the probe current for $V > V_0$ from Equation 2 becomes

$$i_e = A \cdot n_e \cdot e \cdot 1/4 \left(\frac{8kT_e}{\pi m_e}\right)^{1/2} \cdot \left(\frac{r}{a}\right)^l \tag{5}$$

where l is 1 for a cylinder (whose length is large compared with its radius) and 2 for a sphere. When the sheath is not large compared with the probe the value of r is less than that given by Equation 4. The characteristic for a sphere on both sides of space

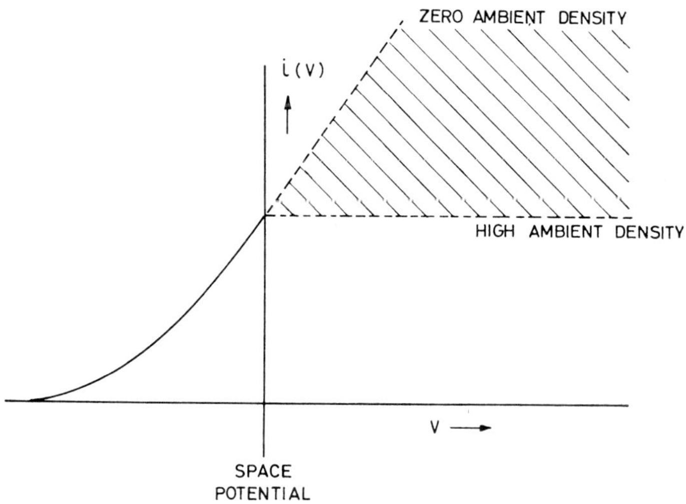

Fig. 1. Probe retardation and attraction characteristics.

potential is shown in Figure 1, where the shaded part represents the region between the extremes $h \ll a$ and $h \gg a$.

As for the retardation characteristic the above remarks apply equally to positive ion attraction with the appropriate changes in the symbols. For electrons the energy E in Equation 4 may in general be replaced with sufficient accuracy by the energy kT_e. For ions, however, the appropriate mean energy depends on the speed of the spacecraft.

4. The Langmuir Analysis

The net current to a probe for $V < V_0$ is obtained by adding Equation 1 to Equation 5 written for positive ions, and writing $n_+ = n_e$

$$i = i_e + i_+ = A \cdot n_e \cdot e \cdot \left(\frac{kT_e}{2\pi m_e}\right)^{1/2} \left[\exp\frac{e(V - V_0)}{kT_e} - \left(\frac{T_+ m_e}{T_e m_+}\right)^{1/2} \left(\frac{r}{a}\right)^l\right]. \tag{6}$$

The curve looks in general as shown in Figure 2. Analysis proceeds by removing the i_+ contribution by linearly extrapolating the positive ion collection at high negative potentials towards space potential, and then plotting the semi logarithmic characteristic referred to above. Alternatively a grid may be used to separate the i_e and i_+ components electrically (Reference B).

For $h \gg a$ the flux of particles of one sign may be isolated by making $(V - V_0)$ large enough then

$$i_e = A \cdot n_e \cdot e \cdot \left(\frac{kT_e}{2\pi m_e}\right)^{1/2} \cdot \left(1 + \frac{e(V - V_0)}{kT_e}\right)^{1/2}. \tag{7}$$

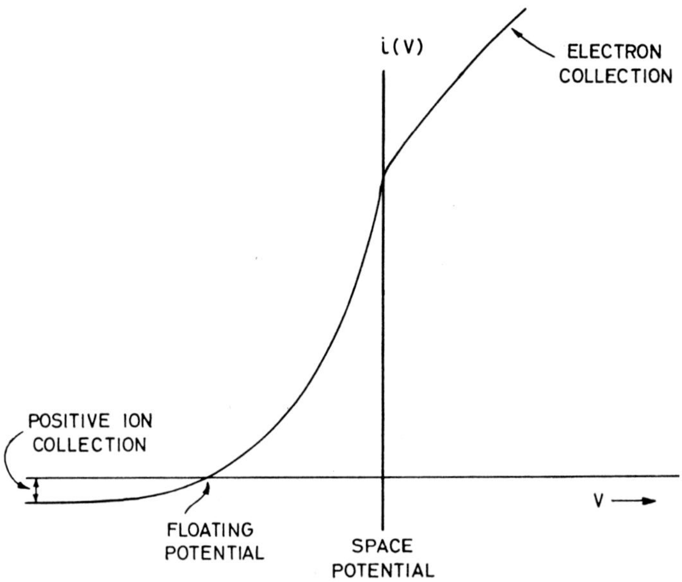

Fig. 2. Langmuir probe characteristics.

A similar equation may be written relating i_+, m_+ and n_+, though it will be necessary to replace the mean thermal energy by some other appropriate mean energy. This method of analysis provides a means of obtaining n_e and n_+ separately and hence in the presence of negative ions of obtaining their concentration from the difference $n_- = n_+ - n_e$.

If an attempt is made to analyse the positive ion part of the characteristic it is important to note that conditions at the sheath edge depend on the energy distribution of the positive ions. It is shown in (Reference A) for example that if $T_e \gg T_+$ the positive ion flux to the probe is controlled by T_e and not by T_+.

5. Spacecraft Potential

In no sense of the word can a satellite be said to be well earthed. Its potential must, therefore, adjust itself so that the net current to it is zero. In practice the potential of a spacecraft in sunlight will be influenced by the photo-electric emission. In the ionosphere this is normally less than the electron random flux, which itself exceeds the positive ion flux, so that the vehicle assumes a slightly negative potential. Deeper in space however the vehicle may be driven positive by the photo emission. The matter is adequately discussed in Reference 22.

Two modes of operation of simple (ungridded) Langmuir probes are commonly used in space as in the laboratory. Either the probe is made so small compared to the return electrode (the *clean* metallic surface of the vehicle) that the current taken by the probe can always be returned to the ionosphere as a result of a trivial change in potential of the satellite; or the return electrode (which may or may not be the body of the vehicle) is made comparable in size to the probe, and its variations in potential as the current varies are taken account of in the analysis. Although all probe systems are really double probe systems, the term is usually reserved for the latter.

In the absence of a photo current the equilibrium potential V of a sphere is given by setting $i=0$ in Equation 6

$$\exp\frac{e(V - V_0)}{kT_e} = \left(\frac{T_+}{T_e}\cdot\frac{m_e}{m_+}\right)^{1/2}\cdot\left(\frac{r}{a}\right)^2 \tag{8}$$

though it may be necessary to replace T_+ by some quantity other than the true ion temperature if the vehicle velocity approaches or becomes supersonic (see below) or if $T_e \gg T_+$ (Reference A). The values of the parameters commonly encountered are such that $(V - V_0) \sim 5\, kT_e/e$.

6. The Symmetrical Double Spherical Probe

Consider a pair of identical spheres separated by a distance large compared with the Debye length. The potential between them may be varied and the current may be measured.

The system will normally adopt a potential given by Equation 8. If now a small voltage difference ΔV be applied, we can think of $1/2\Delta V$ being applied to each sphere and the net increase of current to one being equal to the net decrease to the other.

Differentiating equation 6 and obtaining the slope at the potential given by Equation 8 leads to:

$$\frac{di}{dv} = 1/2\cdot A\cdot n_e\cdot e\left(\frac{kT_e}{2\pi m_e}\right)^{1/2}\cdot\frac{e}{kT_e}\cdot\left(\frac{T_+}{T_e}\cdot\frac{m_+}{m_e}\right)^{1/2}\cdot\left(\frac{r}{a}\right)^2 \tag{9}$$

where the factor 1/2 arises from half the voltage being applied to each sphere and where we have neglected the slope of the positive ion component compared with that of the electron component.

Equation 9 and Equation 5 written for the positive ion current when $\Delta V \gg kT_e/e$ together enable n_e and T_e to be found. However, it may be questioned whether the precision (especially with which T_e can be measured) is comparable to that obtainable by a highly unsymmetrical system in which the ratio of probe areas is greater than (say) 10^3.

7. The Magnitude of Currents to be Measured

It is useful to note that the random current crossing a sphere of one Debye radius is:

$$i_D = \left[\frac{1}{2\pi}\cdot\frac{e}{m}\cdot\left(\frac{kT}{e}\right)^3\right]^{1/2} = 1.49 \times 10^{-11}\ T^{3/2}\,\mathrm{A}\,. \tag{10}$$

For $T \sim 1000$ °K and $a \sim h$, $i_e \sim 10^{-6}$ A and $i_+ \sim 10^{-8}$ A. These therefore represent the magnitudes of currents to be measured if the probe is comparable to the Debye length. They must be compared with the photo emission current which usually lies between 10^{-8} and 10^{-9} A cm^{-2} (References 5 and 17). The photo-current may therefore frequently exceed the positive ion current.

Various means of removing the photo-current are employed, including the use of mesh electrodes to reduce the area exposed to the light, gridded systems, systems carried into shadow by the vehicle motion and electrical differentiation to remove the photo-electric component which varies little with V over much of the significant range.

8. Analysis of Energy Distribution

Druyvesteyn (Reference D) has shown that the energy distribution of particles in a plasma may be obtained by double differentiation of the retardation characteristics of a probe

$$f\,(E\cdot\mathrm{d}E) = \frac{2}{A\cdot e}\left(\frac{2m}{e}\right)^{1/2}\cdot |V - V_0|^{1/2}\,\frac{\mathrm{d}^2 i}{\mathrm{d}V^2}\cdot\mathrm{d}E\,. \tag{11}$$

This result was obtained for convex electrodes in an isotropic velocity distribution, but it applies also to spherical electrodes in anisotropic distributions. It is especially to be noted that the significant quantity is the second derivative of the current voltage curve. Clearly a method of obtaining this quantity directly is to be preferred to graphical or numerical differentiation.

Various electrical systems have been employed including the use of resistance-capacitance circuits and voltage scans aimed to convert the derivative to a time derivative. The most widely used methods, however, turn upon the fact, noted by Sloane and McGregor (Reference 1) that the i-V characteristic has a rectifying property dependent on $\mathrm{d}^2 i/\mathrm{d}V^2$.

Boyd and Twiddy (Reference C) used this property to demodulate a small 2000 c/s voltage modulated by 300 c/s superimposed on the probe potential. The system employed on the satellite Ariel used the same effect to mix two frequencies.

It may be noted that the effect of using an exponential characteristic such as the electron retardation curve to mix two small AC signals is to produce a percentage modulation m of one on the other such that

$$m \alpha \frac{d^2 i_e}{\mathrm{d}V^2}\bigg/\frac{di_e}{\mathrm{d}V} = -\frac{e}{kT_e} \tag{12}$$

which thus provides a measure of electron temperature insensitive to the exact point on the characteristic at which the mixing takes place (References 9, 10, 11 and 28). Further, space potential is given by the value of V at which $\mathrm{d}i/\mathrm{d}V$ is a maximum or $\mathrm{d}^2i/\mathrm{d}V^2=0$.

9. Energy Distribution of Ions in Frame of Spacecraft

The energy of an oxygen atomic ion at rest in the atmosphere as seen from a near Earth satellite is about 5 eV. This is to be compared to a mean thermal energy of about 0.1 eV. It is clear therefore that the energy distribution of ions arriving at a probe on a hypersonic vehicle is far from Maxwellian. This fact must be taken into

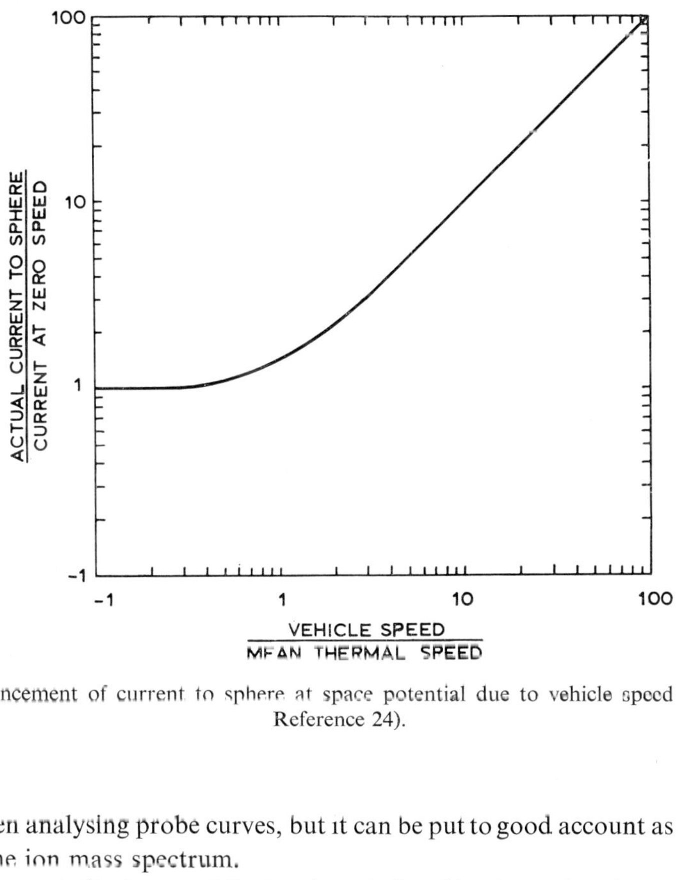

Fig. 3. Enhancement of current to sphere at space potential due to vehicle speed (according to Reference 24).

account when analysing probe curves, but it can be put to good account as a method of obtaining the ion mass spectrum.

SAGALYN *et al.* (Reference 24) give the relationship shown in Figure 3 for the increase in current to a sphere over the Maxwellian random current as the velocity of the sphere increases. It is to be noted that, for vehicle velocities less than about 0.25 the mean thermal speed, the current is very close to the random current; while for speeds in excess of 2.5 times the mean thermal speed, the current is very nearly independent of the thermal speed. The important case of plane probe systems moving at high speeds through plasma is dealt with by Whipple (Ref. 29) who gives the relationship:

$$i_+ = A \cdot n_+ \cdot e \cdot v \cos \Theta \left[1/2 + 1/2 \operatorname{erf}\left(-\frac{V \cos \Theta}{a} \right) + a\,(2\pi^{1/2} \cdot V \cos \Theta)^{-1} \exp - \left(\frac{V \cos \Theta}{a} \right)^2 \right] \quad (13)$$

where v is the vehicle velocity, Θ the angle between the probe normal and the vehicle velocity vector and a is the most probable thermal velocity of the ions.

The energy distribution function for ions whose thermal velocity is small compared with the vehicle velocity approximates to a gaussian curve centred on the energy $1/2m_+v^2$ and having a width at the $1/e$ value of $4(kT_+ \cdot 1/2m_+v^2)^{1/2}$. Both the position and width of the peak thus provide information on the ion mass while the temperature may be obtained from the width of the peak with an accuracy enhanced by the fact that the scale is multiplied by $(1/2m_+v^2)^{1/2}$.

The exact expression for the energy distribution is given by MEDICUS (Reference F) as:

$$G\left(\frac{E}{kT_e}\right) = \pi^{-1/2}\left(\frac{mv^2}{2kT}\right)^{-1/2} \cdot \exp - \left(\frac{E + 1/2mv^2}{kT}\right) \cdot \sinh \frac{2(E \cdot 1/2 \cdot mv^2)^{1/2}}{kT}. \quad (14)$$

10. Special Problems Introduced by Spacecraft

1. If the system is not specially designed as a double probe system, great care must be exercised to ensure that an adequate very clean metal surface is available on the spacecraft to provide a low impedance current return path.

2. In sunlight special precautions are generally necessary to allow for photo emission.

3. The equilibrium potential of the vehicle must be carefully considered. It is likely to vary with altitude, aspect, RF power on antennae and potentials on external electrodes.

4. At supersonic speeds probe current may be significantly dependent on aspect unless a spherical geometry is used. At hypersonic speed the plasma is greatly attenuated in the wake.

5. Gas emitted by the craft from leaking seams, control jets or motors after burn-out may seriously affect ionization close to the craft.

6. Contact potential differences may readily give rise to several volts difference in potential between a probe (of some noble metal) and the spacecraft body. A further difference may arise as a result of the emf induced by motion of a satellite through the Earth's magnetic field.

11. Summary of Some Probe Experiments in Space

On page 464/465 a summary of some probe experiments in space is to be found.

References for Langmuir Probes for Ionospheric Studies

A. USE OF PROBES ON SPACE VEHICLES

1. AL'PERT, Ya. L.: 1962, 'The Study of the Ionosphere and Interplanetary Gas Employing Artificial Earth Satellites and Space Rockets', *Plan. Space Sci* **9,** 391.
2. BOGGESS, R. L., BRACE, L. H., and SPENCER, N. W.: 1959, 'Langmuir Probe Measurements in the Ionosphere', *J. Geophys. Res.* **64,** 1627.
3. BOURDEAU, R. E.: 1961, 'Ionospheric Results with Sounding Rockets and the Explorer VIII Satellite', *Proc. 2nd International Space Sci. Symposium,* Florence.
4. BOURDEAU, R. E.: 1963, 'Ionospheric Research from Space Vehicles', *Space Sci. Rev.* **1,** 683.
5. BOURDEAU, R. E., DONLEY, J. L., SERBU, G. P., and WHIPPLE, E. C. Jr.: 1960, 'Measurements of Sheath Currents and Equilibrium Potential on the Explorer VIII Satellite' (1960), NASA Technical Note D-1064.
6. BOURDEAU, R. E., DONLEY, J. L., and WHIPPLE, E. C. Jr.: 1961, 'The Ionosphere Direct Measurements Satellite Instrumentation (Explorer VIII)', National IAS and ARS Meeting, Los Angeles, June, 1961.
7. BOURDEAU, R. E., WHIPPLE, E. C., Jr., and CLARK, J. F.: 1959, 'Analytic and Experimental Conductivity between the Stratosphere and the Ionosphere', *J. Geophys. Res.* **64,** 1363.
8. BOURDEAU, R. E., WHIPPLE, E. C., Jr., DONLEY, J. L., and BAUER, S. J.: 1961, 'Experimental Evidence for the Presence of Helium Ions Based on Explorer VIII Satellite Data', *Proc. URSI Meeting,* Austin, Texas, October, 1961.
9. BOWEN, P. J., BOYD, R. L. F., HENDERSON, C. L., RAITT, W. J., and WILLMORE, A. P.: 1963, 'Ionospheric Results using Langmuir Probes in the Ariel I Satellite', *Space Research* **4.**
10. BOYD, R. L. F.: 1961, 'The use of Probing Electrodes in the Study of the Ionosphere', *J. Brit. I.R.E.* **22,** 405.
11. BOYD, R. L. F.: 1963, 'Plasma Probes on Space Vehicles', *Proc. 5th International Conf. on Ionization Phenomena in Gases,* North-Holland Publishing Company, Amsterdam, p. 1387.
12. GRINGAUZ, K. I.: 1963, 'The Structure of the Ionized Gas Envelope of Earth from Direct Measurements in the USSR of Local Charged Particle Concentrations', *Plan. Space Sci.* **11,** 281.
13. GRINGAUZ, K. I., BEZRUKIH, V. V., OZEROV, V. D., and RYBČINSKIJ, R. E.: 1962, 'The Study of Interplanetary Ionized Gas, High-Energy Electrons and Corpuscular Radiation of the Sun, Employing Three-Electrode' Charged Particle Traps on the Second Soviet Space Rocket', *Plan. Space Sci.* **9,** 103.
14. GRINGAUZ, K. I., KURT, V. G., MOROZ, V. I., and ŠKLOVSKIJ, I. S.: 1962, 'Ionized Gas and Fast Electrons in the Vicinity of the Earth and in Interplanetary Space', *Plan. Space Sci.* **9,** 21.
15. GUREVIČ, A. V.: 1962, 'Perturbations in the Ionosphere Caused by a Travelling Body', *Plan. Space Sci.* **9,** 321.
16. HANSON, W. B. and MCKIBBIN, D. D.: 1961, 'An Ion-Trap Measurement of the Ion Concentration Profile above the F^2 Peak', *J. Geophys. Res.* **66,** 1667.
17. HINTEREGGER, H. E., DAMON, K. R., and HALL, L. H.: 1959, 'Analysis of Photo-electrons from Solar Extreme Ultraviolet', *J. Geophys. Res.* **64,** 961.
18. HOK, G., SPENCER, N. W., and DOW, W. G.: 1953, 'Dynamic Probe Measurement in Ionosphere', *J. Geophys. Res.* **58,** 235
19. ICHIMIYA, T., TAKAYAMA, K., and AONO, Y.: 1959, 'Measurement of Positive Ion Density in the Ionosphere', *Report of Ionosphere and Space Research in Japan* **13,** 155.
20. ICHIMIYA, T., TAKAYAMA, K., DOTE, T., AONO, Y., HIRAO, K., MYAZAKI, S., SUGIYAMA, T., and MURAOKO, T.: 1961, 'Measurement of Positive Ion Density in the Ionosphere by Sounding Rocket', *Nature* **190,** 156.
21. KRASOVSKIJ, V. I.: 1961, 'Soviet Investigations of the Ionosphere Employing Rockets and Artificial Earth Satellites', *Plan. Space Sci.* **5,** 223
22. KURT, P. G. and MOROZ, V. I.: 1962, 'The Potential of a Metal Sphere in Interplanetary Space', *Plan. Space Sci.* **9,** 259.
23. MAEDA, K. and HIRAO, K.: 1962, 'A Review of Upper Atmosphere Rocket Research in Japan', *Plan. Space Sci.* **9,** 355.
24. SAGALYN, R. C., SMIDDY, M., and WISNIA, J.: 1963, 'Measurement and Interpretation of Ion Density Distributions in the Daytime F-region', *J. Geophys. Res.* **68,** 199.

TABLE I

SUMMARY OF SOME PROBE EXPERIMENTS IN SPACE

Team	*Reference*	*Objective*	*Geometry and Number of Grids*
Michigan	18	T_e, n_e, n_+	Conic frustum
Michigan	2, 26	T_e, n_+	Two equal spheres, 15 cm diameter
Japanese	19, 20, 23	n_+	Solid or mesh sphere, 3 cm diameter
U.C. London	27	n_+	Flat probe 75 cm^2 area
U.C. London	—	n_+	As above but with grid allowed to float
G.C.A. and N.A.S.A.	25, 3	T_e, n_e	Cone
U.C. London	—	n_+, n_e, T_e	Sphere 0.8 mm diameter
U.C. London	9, 10, 11, 28	T_e, n_e	A disc with guard ring. Modulation technique
N.A.S.A.	3, 4	T_e, n_e	Positive disc with one variable voltage grid
Russian	20	n_e, n_+	Spherical coarse grid ion trap. Small collector sphere
A.F.C.G.R.D.	24	n_+	Spherical coarse grid ion trap. Small collector sphere
U.C. London	9, 10, 11, 28	m_+, n_+, T_+	Spherical fine grid biased to repel electrons with large collectc sphere swept to retard positive ions
N.A.S.A.	3, 4, 6, 8	n_+, m_+, T_+	Plane. Floating outer grid. Negative inner suppressor grid an swept voltage collector
Russian	12, 13, 14	n_+, v_+	Hemi-spherical outer grid at various potentials. Negative inne suppressor grid (planar). Negative planar collector
Lockheed	16	n_+, T_+	Three grid planar system

* i_{phot} = photo-electric emission current ; $i' = \dfrac{di}{dV}$; $i'' = \dfrac{d^2i}{dV^2}$; i_+(sat) = saturation positive ion curren

25. Smith, L. G.: 1962, 'Rocket Measurements of Electron Density and Temperature in the Night-time Ionosphere', *GCA Technical Report* 62-1-N.
26. Spencer, N. W., Brace, L. H., and Carignan, G. R.: 1962, 'Electron Temperature Evidence for Non-Thermal Equilibrium in the Ionosphere', *J. Geophys. Res.* **67,** 157.
27. Wrenn, G. L., Willmore, A. P., and Boyd, R. L. F.: 1962, 'A Profile of Sporadic-E Ionization', *Plan. Space Sci.* **9,** 765.
28. Willmore, A. P., Boyd, R. L. F., and Bowen, P. J.: 1962, 'Some Preliminary Results of the Plasma Probe Experiments on the Satellite Ariel', *Proc. of the International Conf. on the Ionosphere. London,* 1962. Institute of Physics and the Physical Society.
29. Whipple, E. C.: 1959, 'The Ion Trap Results in Exploration of the Upper Atmosphere with the Help of the 3rd Soviet Sputnik', *Proc. I.R.E.* **47,** 2023.

B. THEORY AND LABORATORY EXPERIMENTS

A. Allen, J. E., Boyd, R. L. F., and Reynolds, P.: 1957, 'The Collection of Positive Ions by a Probe Immersed in a Plasma', *Proc. Phys. Soc.* **B70,** 297.
B. Boyd, R. L. F.: 1950, 'The Collection of Positive Ions by a Probe in an Electrical Discharge', *Proc. Roy. Soc.* **A201,** 329.
C. Boyd, R. L. F. and Twiddy, N. D.: 1959, 'Electron Emergy Distributions in Plasmas. I', *Proc. Roy. Soc.* **A250,** 53.
D. Druyvesteyn, M. J.: 1930, *Z. Phys.* **64,** 790.

Analysis Method	*Position*	*Comments*
Unsymmetrical double probe	Nose	Of pioneering significance
Double probe	Ejected	Vehicle influence eliminated. i_{phot} a source of error. T_e accuracy uncertain
No V scan. i_+ (sat)	Rod	Good relative data referred to ionosonde at layer max ⊙ to make absolute. i_{phot} reduced by mesh
No V scan. i_+ (sat)	Side	As above but i_{phot} removed by shadow as vehicle rolls. 2 or 4 probes used to correct for aspect
No V scan. i_+ (sat)	Side	As above but effort to remove effect of uncertain sheath size by floating grid
Langmuir retardation	Nose tip	A successful orthodox system. T_e surprisingly high. Effect of telemetry RF studied
Orbital attraction	Rod	A successful system. i_{phot} made unimportant. Aim to find $n^- = n_+ - n_e$
$i'/i'' = -e/kT$	Boom or body	A reliable system giving T_e to 4% and n_e to 15%; used on satellite Ariel I
Retardation	Body	A reliable system used on satellite Explorer VIII
Grid voltage swept	Stub	Reliable for positive ions; used on Sputnik III
Grid voltage swept or fixed	Boom	Similar to above but full analysis enabled use to be made on slower vehicle
Modulation to give i' and i''	Stub	A reliable system used on Ariel I. Resolves H^+, He^+ and O^+ and gives T_+
Retardation	Body	A reliable system used on Explorer VIII. Rather sensitive to aspect
Fixed retardation ion trap	Body	System specially devised for interplanetary space where $i_{phot} \gg i_+$
Retardation	Nose	Earlier than the N.A.S.A. 2 grid system. Probably no advantage in 3 grids

Rod means a long axial projection; boom means a long lateral projection; stub means a short mount.

E. Langmuir, I. and Mott-Smith, H., Jr.: 1924, 'Studies of Electric Discharges at Low Pressures', *Gen. Electric Rev.* **27**, 449, 538, 616, 726, 810.

F. Medicus, G.: 1961, 'Spherical Langmuir Probe in 'Drifting' and 'Accelerated' Maxwellian Distribution', *J. Appl. Phys.* **32**, 2521.

G. Mott-Smith, H., Jr., and Langmuir, I.: 1926, 'The Theory of Collectors in Gaseous Discharges', *Phys. Rev.* **28**, 727.

H. Schulz, G. J. and Brown, S. C.: 1955, 'Microwave Study of Positive Ion Collection by Probes', *Phys. Rev.* **98**, 1642.

I. Sloane, R. H. and MacGregor, E. I. R.: 1934, 'An Alternating Current Method for Collector Analysis of Discharge Tubes', *Phil. Mag.* **18**, 193.

CHAPTER 39

THE IONOSPHERIC MEASUREMENTS FROM ARIEL I

A. P. WILLMORE
University College, London

The Ariel I satellite, which was launched on April 26th, 1962, carried three sets of experiments:

(I) Four ionospheric experiments, to measure electron and ion concentration and density, electron and ion temperatures and the constitution of the ionization;

(II) Two experiments to measure the intensity of the solar spectrum near the Lyman-α line at 1216 Å and in the waveband from 4–13 Å and, finally;

(III) A cosmic ray experiment. The inclination of the orbit was 54°, and the apogee and perigee heights were respectively 1200 km and 400 km. The ionospheric measurements have each indicated to an unexpected extent the importance of the geomagnetic field for the morphology of the ionosphere above the F2 maximum, and it is this topic which forms the basis of this chapter.

That the geomagnetic field would cause the diffusion coefficients and the thermal and electrical conductivities of the ionosphere to be anisotropic has, of course, long been recognized, but the fact that the energy input and the ionization rate in the vicinity of, or above, the F2 maximum were each very small suggested that the consequences of this isotropy would not be striking. Moreover, the effects of particle bombardment of the ionosphere (where the distribution of the source might be expected to show strong geomagnetic control) were thought to be substantially confined to the auroral zone. The results of the experiments which are now becoming available suggest very strongly that in fact the effects of particles, presumably trapped and presumably electrons, are far more widespread than previously anticipated, and that both the ionization and the ionospheric heating which they produce strongly modify the structure of the upper ionosphere.

The details of the experiments will not be discussed here, since accounts of them can be found in Chapter 23 (p. 324) and the seminar by Professor R. L. F. Boyd (Chapter 38, p. 455). Electron densities were measured both by means of a radio-frequency probe and by Langmuir probes, which also measured electron temperature; all the positive ion measurements were made by means of a positive ion probe. The radio frequency probe was provided by Professor J. Sayers at University of Birmingham, England, whilst the remaining probes were the responsibility of Professor R. L. F. Boyd and the author, at University College London. The satellite contained a tape recorder which recorded partially processed data from all but the ion probe thus enabling an excellent geographical coverage of electron density and temperature to be obtained. The measurements of these quantities which are de-

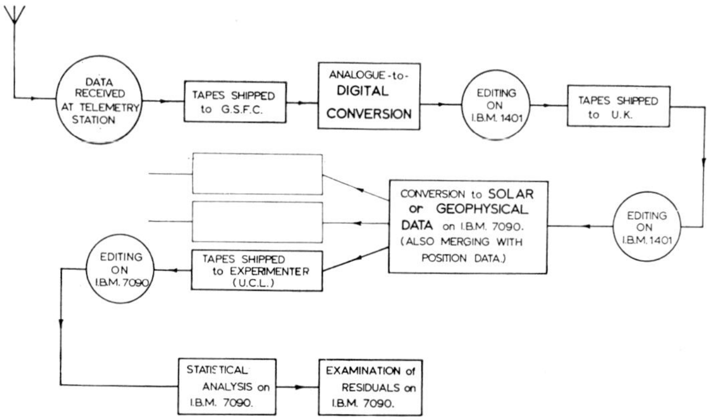

Fig. 1. Data processing system of the Ariel I.

scribed below were all obtained from tape recorder between 27th April and 18th July, 1962. Data from the positive ion probe were obtained only when the satellite was in the vicinity of a ground station so that good coverage was obtained only over the American continent where the majority of these stations is located. A very efficient data processing system has been set up for these measurements; the sequence of operations is shown in Figure 1.

1. Measurements of Electron Density

One of the striking results of the electron density measurements has been described by SAYERS (1963). It must be remembered that in the course of an orbit around the earth, the latitude, altitude and local solar time at the satellite vary simultaneously over a wide range. Successive orbits repeat this pattern with only slight alteration, the principal change occurring in the longitude of the observation corresponding to a given value of local time. Sayers found superimposed on the general orbital variation of electron density a number of peaks whose location (i.e. latitude) changed on successive orbits. Since only the orientation of the orbit with respect to the geomagnetic field changes on successive orbits, a correlation with the geomagnetic field is suggested and it was in fact found that the peaks could always be resolved into either or both of a single peak centred on the geomagnetic equator, and one or more pairs situated symmetrically about the equator. From this, and from the detailed variation of electron density at the peaks with longitude, Sayers concluded that the electron density was enhanced over a number of magnetic shells, those with $L = 1.09$, 1.27 and 1.75. Then as the satellite passes through the top of a shell, a single maximum of density over the equator results, whilst if it cuts the shell below the top, two maxima equispaced about the equator are observed. The lowest shell is quite strong even at the level of the F2 maximum, and has been well-known for many years from ionosonde measurements under the name of the "equatorial anomaly" of electron density. The existence of the shells has subsequently been confirmed by measurements with a top-side sounder in the Alouette satellite.

The origin of the shells is so far unknown, although it seems clear from the association with the geomagnetic field that particles, probably trapped, are involved. These may either produce additional ionization at low altitudes, or merely change the vertical distribution as a result of ionospheric heating and a consequent change of scale height.

2. Measurements of Electron Temperature

The main features of the electron temperature distribution have been established by means of a statistical analysis of a large sample totalling approximately 15000 observations. The method, which has been described in detail by WILLMORE, HENDERSON,

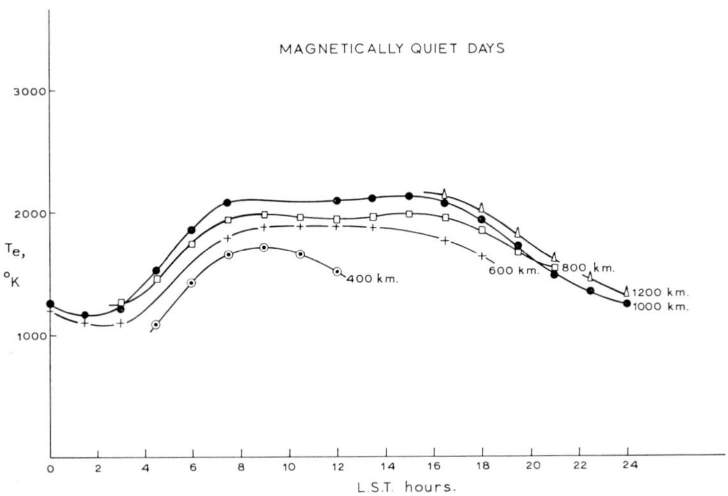

Fig. 2. Diurnal temperature at 40° geomagnetic latitude.

BOWEN and BOYD (1963), was to use a regression function (involving up to 63 unknown coefficients) to describe the variation of electron temperature with longitude, latitude, altitude and local solar time. Only data from magnetically quiet days ($K_p < 3$ throughout the day) was used. A comparison of three different co-ordinate systems showed that the best fit (i.e. the lowest residuals) were obtained with a given regression function when geomagnetic latitude was used, a change to geographic latitude increasing the residual by an amount which corresponded approximately to the observed change in temperature due to a latitude change of 10°. It was therefore concluded that the electron temperature distribution showed geomagnetic control. The error due to truncating the series expansions used in the regression functions was then investigated, and found to be less than 50 °K. Moreover, the probable error of the coefficients in the regression function was shown in general to be between 2% and 10% so that over-fitting had not occurred. The best function is plotted in Figures

2 and 3. The most striking result is that the electron temperature is found to increase sharply with both latitude and altitude.

In this altitude region, the neutral gas temperature is known to be independent of both altitude and latitude. Thus if the electron temperature is close to the gas temperature near the equator and the lowest altitude (400 km), it must be considerably greater at high altitudes and high latitudes. This requires an energy source for the electrons which is operative at least up to 1000 km. It has been shown theoretically (DALGARNO *et al.*, 1963) that the production of fast photo-electrons from atomic oxygen by sunlight will produce an increase in electron temperature, but this process must rapidly become unimportant at altitudes greater than 400 km. It appears once

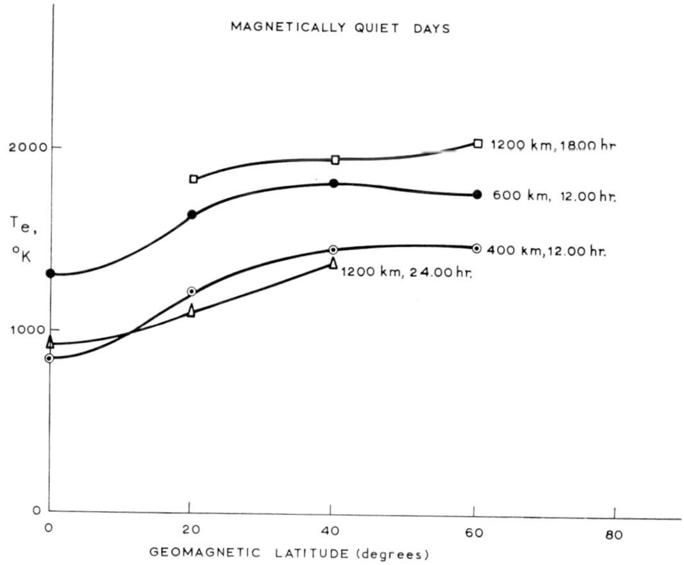

Fig. 3. Variation of electron temperature with geomagnetic latitude.

again that corpuscular radiation must be responsible, the heating occurring by collisions with the electrons, but the nature of the radiation is not clear. The strongest evidence for the corpuscular nature of the energy source is the fact that the latitude temperature variation, and so the heat source, are found to persist through the night (see Figure 3).

3. Measurements of Ion Composition

The presence of He^+ as a major constituent in the upper ionosphere was predicted theoretically e.g. by BAUER (1962), and first observed on Explorer VIII. The positive ion probe measurements on Ariel I have enabled the He^+–H^+ transition also to be observed and many details of the diurnal and geographical variations of the ionospheric composition to be studied. Figure 4 shows the diurnal variation of the transition altitudes of O^+–He^+, and He^+ to H^+. In general the satellite did not, of course, make observations precisely at the transition level, so the altitude of the latter was deduced from the composition at a slightly lower or greater altitude, on the assumption that the ionization was in simple diffusive equilibrium in a vertical direction. Whilst this assumption is certainly not exactly correct, when applied over small altitude ranges it will lead to negligible errors. It will be seen that the transition altitudes vary in the day in a way which reflects the diurnal variation of atmospheric temperature. The actual transition altitude for O^+ to He^+ is in approximate agreement, for daytime conditions, with that calculated by Bauer for a temperature of 1200 °K, a value probably slightly higher than the actual temperature at this period. This

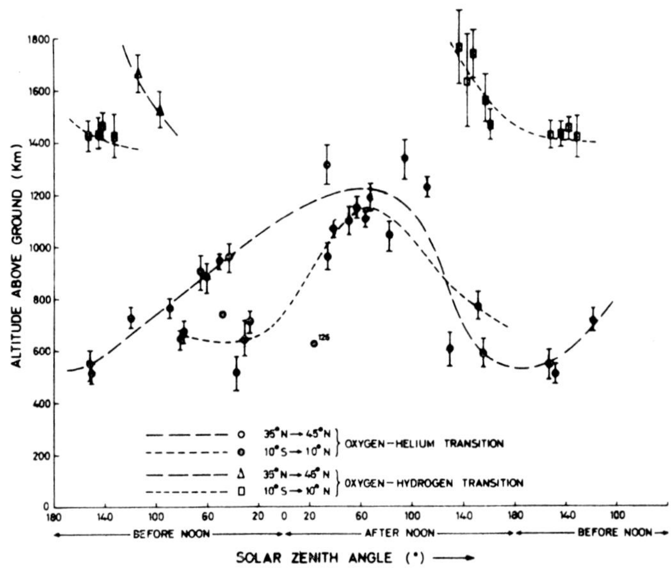

Fig. 4. Diurnal variation of ion transition altitudes.

discrepancy is probably the consequence of an error in the relative concentration of He^+ at 400 km as calculated by Bauer.

The observations of Figure 4 show a considerable scatter about the mean curve. The reason is evident from Figure 5 which shows the composition (calculated for an altitude of 1000 km from observations made near this altitude) as a function of position. It will be seen that the composition varies with geomagnetic latitude, although not in a symmetrical way about the magnetic equator. The explanation of this variation is not at the present time at all clear. It may be the consequence of the geographical variation of electron temperature coupled with the diffusive equilibrium of the ionization along the lines of force, or it may reflect similar composition variations at the level of the F2 maximum where the composition of the ionization is determined by photochemical processes. These points can only be clarified by further investigations.

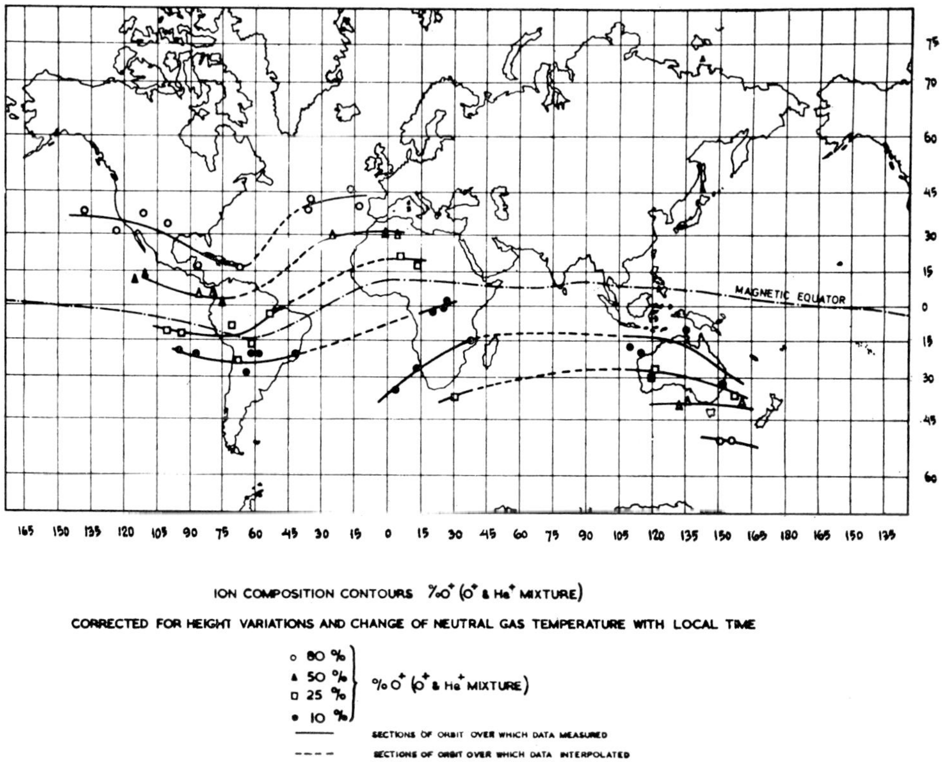

Fig. 5. The ion composition calculated for an altitude of 1000 km from observations made near this altitude, as a function of position.

4. Conclusions

Both the night-time variation of electron temperature with latitude and the magnetic-shell enhancements of density show that corpuscular radiation must produce either ionospheric heating or ionization or both, and that these effects are not confined, as previously supposed, to high latitudes and low altitudes, but occur over a wide altitude range and a wide latitude range. Evidently the radiation does not therefore consist of precipitated electrons from the outer Van Allen belt. It now seems possible that much of the anomalous behaviour of the F-region, may ultimately be traced to the incidence of corpuscular radiation. The origin of these particles, which lies presumably in the magnetosphere, will in itself be a topic of great interest.

References

Bauer, S. J.: 1962, *J. Atm. Terr. Sci.*, **19**, 276.

Dalgarno, A., McElroy, M. B. and, Moffat, R. J.: 1963, *Plan. Space Sci.* **11**, 463.

Sayers, J.: 1963, *Proc. Roy. Soc.* (in press).

Willmore, A. P., Henderson, C. L., Bowen, P. J., and Boyd, R. L. F.: 1963, *Proc. Roy. Soc.* (in press).

CHAPTER 40

SIMULTANEOUS MEASUREMENT OF PLASMA DENSITY BY THREE DIFFERENT METHODS

J. BÜCHAU, K. G. JACOBS, P. KAISER AND K. RAWER
Ionosphären Institut, Breisach, Germany

On May 1, 1963 a Véronique Rocket was fired at Hammaguir (Sahara) by the French "Centre National des Études Spatiales", which carried electronic equipment, developed for the measurement of physical parameters of the upper atmosphere by three groups: (1) the "Laboratoire de Physique de l'Atmosphère", Paris; (2) the French "Commissariat a l'Énergie Atomique"; and (3) the "Ionosphären Institut" Breisach, Germany.*

This seminar will describe the three experiments, developed or redesigned at Breisach in order to measure the electron density in the D- and E-layer by three different methods. The three experiments are: (1) A radiofrequency spectrometer; (2) Dispersive Doppler Measurements on 5 and 20 Mc/sec (Inversed Seddon-experiment); and (3) Impedance probe with variable frequency.**

Experiments (1) and (2) use propagation methods, where the influence of the ionosphere on a wave propagated through it is measured, while experiment (3) is a local measuring method.

On this first flight all three experiments behaved nearly as anticipated. For the following flights slight changes in the electronic design will have to be made to obtain data, which are better to evaluate or include more information. The basic theories, which are the object of this seminar, have proved to be correct.

1. The Radiofrequency Spectrometer

The ionosphere presents itself to an electromagnetic wave propagating through it, as a medium with a complex refractive index

$$n = \mu - i\chi \tag{1}$$

If we introduce this refractive index into the representation of a plane wave, travelling along z (which is the vertical axis), we get

$$\mathbf{E} = \mathbf{E}_0 \cdot \exp i\omega\left(t - \mu\frac{Z}{c}\right) \cdot \exp\left(-\omega\chi\frac{Z}{c}\right). \tag{2}$$

* The expedition was sponsored by "Deutsche Forschungsgemeinschaft".
** Sponsored by the Aero-Space Laboratories of the USAF through the European Office, Aerospace Research, US Air Force, Contract No. AF 61(052)–290.

This shows, that μ, the real part of the refractive index causes a phase shift, while χ, the imaginary part gives rise to an attenuation increasing with z. The absorption coefficient κ, defined by

$$\mathbf{E} = \mathbf{E}_0 \cdot \exp(-\kappa z) \tag{3}$$

can be derived from the Appleton-Hartree formula and is a function of the collision frequency γ and the electron density N_e.

By measuring the decrease of field strength at a given frequency one can determine κ and with a suitable $\nu = \nu(z)$-model or with ν taken from another experiment one can calculate the change of electron density with the height, the so-called "electron density

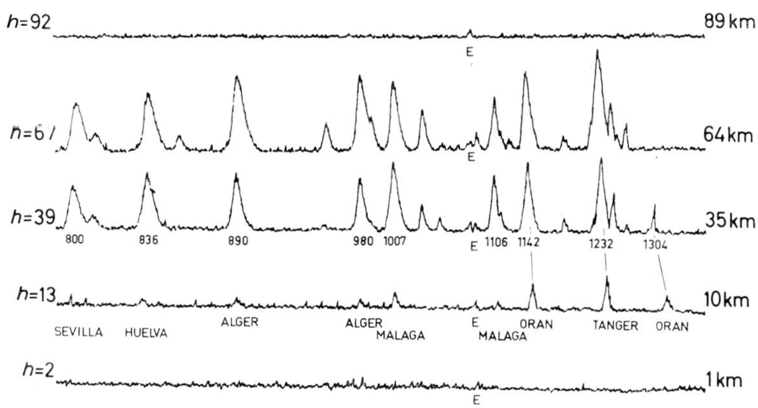

Fig. 1. Rf spectrometer recording (range 780–1380 kc/sec); Véronique V 27 (February 11, 1961).

profile". In this special experiment we do not work with our own transmitting station but use existing radio stations, which also allows us to use a great number of frequencies. With a three-channel sweep-frequency receiver we receive the following ranges. L=low frequency (160–240 kc/sec); M=medium frequency (880–1250 kc/sec); and H=high frequency (3.2–3.4 Mc/sec). Variable capacity diodes are used as tuning parts in the three variable input and oscillator stages. The frequency is swept through by an applied sawtooth voltage. Every 30 seconds the common IF-amplifier is switched to another input-circuit. The sensitivity-curve of the receiving system is logarithmic over 100 dB, and a low-pass filtered output voltage is used in the usual manner as a measure of the input-rf voltage. On board the rocket we use loop-antennae, their pattern-zero pointing in the direction of the axis of the rocket, thus receiving radio stations which are far away in relation to the height of the culmination point of the trajectory.

Figure 1 shows a part of the record of an earlier flight on February 11, 1961. It shows the mf-range from 780 to 1380 kc/sec. The rise in field strength with the height in the diffraction range (where there is no direct visibility between receiver and transmitter) and the complete absorption in the D region can easily be seen. The height resolution of this record is very poor because we used in this receiver a rotating capacitor, and the driving unit determined a limit to the sweep speed. The next picture shows a recording sample of the 1st May-flight, where all three experiments are recorded together (cf. Figure 2). The lines A, A' and A'' (recorded at three different times $H+21$ sec, $H+105$ sec, $H+186$ sec) show the recording of the rf-spectrometer. The different ranges are indicated by the dotted lines, the letters L, M, H and C refer to the three receiving ranges and a calibration-mark-range. The latter is

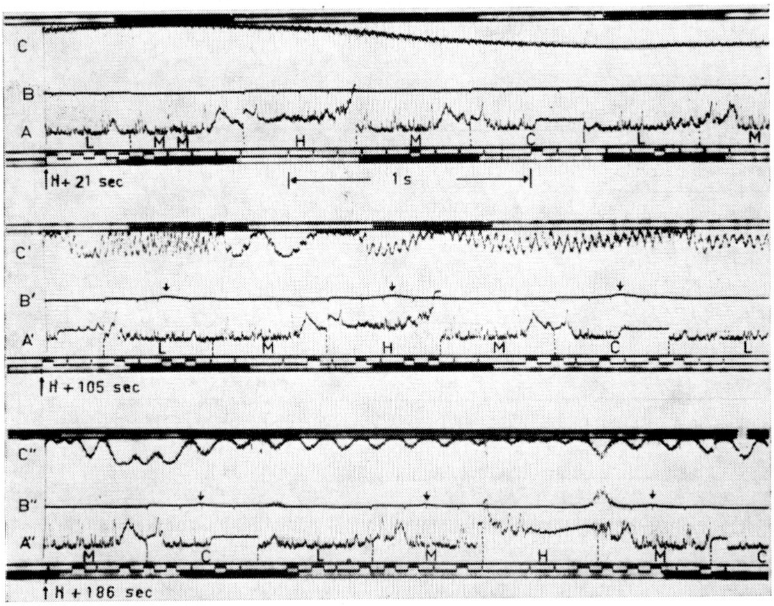

Fig. 2. Three samples of the registrations of the flight Véronique V 45 J1 (May 1, 1963).

used to check the stability of the telemetering system. Some of the peaks in the L, M and H ranges are radio stations. Identifications have not yet been done. Naturally the evaluation is not so easy as indicated in the basic theory, because there vertical propagation was assumed. As the received stations are at distances from 100 to almost 1000 km and the rocket is moving up to heights of 150–180 km; refraction influences the ray path between the rocket and the ground station, not only in the surrounding of the rocket but on its whole length. Thus a very complicated sheet by sheet evaluation is necessary to take account of the earth's magnetic field. This theory goes far beyond the range of this seminar and I want to refer to ARGENCE (1959) and ARGENCE and RAWER (1961), who have worked on this problem.

Anyway, we can obtain as direct results from these records

(a) Measurement of propagation by diffraction in the shadow-zone of the earth;
(b) Determination of true reflection level;
(c) Differential measurement of ionospheric absorption.

2. An Inversed Seddon-experiment

Since 1948 SEDDON (1953) has developed a method of measuring the electron density-profile by measuring the change of phase velocity of a radio wave in the ionosphere with rockets. In order to measure this change one transmits two harmonically related frequencies **f** and **a·f** from a rocket and measures the differential Doppler effect with a ground receiving station by multiplying the lower frequency by **a** and beating it against the high frequency. This eliminates the geometrical Doppler effect and shows a beat frequency which depends on the refractive indices on the two frequencies and the radial velocity **v** of the rocket.

$$f_{\text{beat}} = \frac{v}{c_0} \cdot af\,(\mu(af) - \mu(f)) \tag{4}$$

c_0 = velocity of light in vacuo.

Under the simplifying assumption, that $\mu(af) \approx 1$ (that means $af \gg f_{Ne}$, the plasma frequency) and neglecting the collisions and the earth magnetic field one derives

$$f_{\text{beat}} = \frac{v}{c_0} \cdot af \cdot \left(1 - \sqrt{1 - \frac{1}{\varepsilon_0} \cdot Ne \cdot \frac{e^2}{m} \cdot \frac{1}{4\pi^2} \cdot \frac{1}{f^2}}\right) \tag{5}$$

and this is a direct relation between the beat frequency and the electron density N_e. Standard values for N_e and v introduced into the equation (5) in our case, where we use 5 and 20 MHz, lead to a beat frequency of 20 c/sec for the E-layer maximum values used $N_e = 1.5 \times 10^{11}$ m^{-3}; $v = 1000$ m sec^{-1}). Seddon uses for his ground station a complicated antenna-system, which enables him to split the received signals into the o- and x-components which, with their different refractive indices, give different beat frequencies. This makes cross-checks possible.

In our case we work with the receiving system on board the rocket and the transmitters on the ground. In order to have clear results with a very simple antenna we must avoid the possibility of receiving both components. This is done by transmitting a "characteristic wave", that means a wave, which is circularly polarized at ground level and which does not split into two components.

But in any case the clarity of the picture obtained by mixing two frequencies is disturbed by different causes. Firstly in addition to the direct 5 and 20 Mc/sec propagation there may exist on 5 Mc/sec a reflection from the ionosphere. This leads to the fact, that there always exists a geometric Doppler effect and a second beat frequency. (There also exists the possibility of a great number of possible combinations between the direct and the reflected 5 Mc/sec wave in the mixing and multiplying processes of the receiver.) Secondly, appertaining to the simple theory, which leads to Equation (5) we used a vertical trajectory of the rocket and a constant electron density profile. Only under these assumptions do we obtain the derivative of $\int N_e \, dr$, so N_e. But since the trajectory of the rocket is inclined, the optical path length $\int \mu(f)\,dr$ changes and gives rise to an additive integrating behaviour of the beat frequency due to the effective change of $\int N_e \, dr$.

In Figure 2 the line C, C' and C'' show beat frequencies at the three indicated flight times ($H+21$ sec, $+105$ sec, $+186$ sec corresponding to altitudes of 4.7, 103 and 155 km respectively). In line C we see the very low beat frequency ($\sim 1/6$ c/sec) due to the rotation of the rocket. Line C' shows a high beat frequency when a sporadic E-layer (between 105.2 and 105.6 seconds) was traversed. The thickness of this layer corresponds to ~ 500 m. (The sporadic E was also observed by an ionosonde close to the launching place.) The line C'' shows a smaller beat frequency due to the very small radial velocity of the rocket in this height. The superimposed low frequency has not yet been identified, also definitive evaluation of beat frequency into electron density has not yet been made.

The receiving system is a fully transistorised two channel receiver, which gives over a 60 dB range an almost constant output amplitude, which is fed into the phase detector from where we derived the beat frequency. The gain control voltage is used as a measure of field strength. As antennae we use two tuned loops, the antenna-zeros pointing perpendicular to the rocket axis, thus receiving from the ground station underneath the rocket, where an 1-kW transmitter transmits on 5 Mc/sec and a 200-Watt transmitter on 20 Mc/sec circular polarized waves, thus avoiding double-refraction.

3. The Plasma Capacitor

The plasma capacitor is formed by a whip-antenna on top of the nose cone as one plate and the rocket body as the second plate.

To this antenna a sweep-frequency oscillator with a frequency range of 0.3–2.0 Mc/sec and with constant rf-amplitude is attached and its amplitude as a function of frequency is recorded.

According to KANE, JACKSON and WHALE (1962) the plasma capacity

$$C = C_0\left(1 - \frac{X}{U}\cdot\frac{U^2 - GY^2}{U^2 - Y^2}\right) \tag{6}$$

where

$$G = \int E^2 \cos^2\psi \,\mathrm{d}\tau / \int E^2 \,\mathrm{d}\tau \tag{7}$$

is a factor depending on the geometry of the capacitor and on the frequency. C_0=Vacuum capacity of the plasma capacitor; $\mathrm{d}\tau$=Volume element; $X=\omega_N{}^2/\omega^2$ (ω_N=plasma pulsation); $Y=\omega_H/\omega$ (ω_H=gyro pulsation, ω=oscillator pulsation); $U=1-j\nu/\omega$ (ν=collision frequency); ψ=angle between electric and terrestrial magnetic field.

In general, the variation with frequency is characterized by two serial resonance maxima (ω_s) and two parallel resonance minima (ω_p). If the variation of G with frequency is neglected we find:

$$\omega_s^2 = \frac{1}{2}\omega_H^2\left(1 + p\frac{\omega_N^2}{\omega_H^2} \pm \sqrt{\left(1 + p\frac{\omega_N^2}{\omega_H^2}\right)^2 - 4Gp\frac{\omega_N^2}{\omega_H^2}}\right) \tag{8}$$

$$\omega_p^2 = \frac{1}{2}\omega_H^2\left(1 + q\frac{\omega_N^2}{\omega_H^2} \pm \sqrt{\left(1 + q\frac{\omega_N^2}{\omega_H^2}\right)^2 - 4Gq\frac{\omega_N^2}{\omega_H^2}}\right) \tag{9}$$

where $p = C_0/(C_0 + C_1 + C_2)$ and $q = C_0/(C_0 + C_1)$; with C_2 = fixed capacitor in series with C_0; C_1 = capacitor in parallel with C_0 due to the connections inside the rocket.

From these four resonance frequencies we can determine ω_H, G, the capacity of the ionic sheath which forms at the electrodes and the plasma frequency.

In our tentative experiment (recording see Figure 2, lines B, B', B'') with a rather small frequency range only one serial resonance was measured (indicated by a little arrow). Substituting standard data for ω_H and G and neglecting the ionic sheath, an evaluation of the plasma pulsation ω_N was possible.

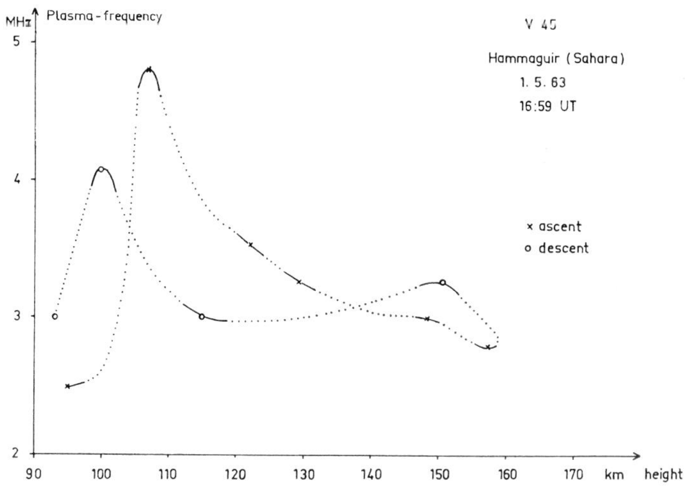

Fig. 3. Plasma frequency measured with the impedance probe Véronique V 45 Jl (May 1, 1963). Preliminary results: only the thick parts are obtained from our evaluation, the rest of the curves is interpolated.

With $G = 1/2$, $p = 1/9$, $f_H = 1.05$ Mc/sec;

$$\omega_N^2 = \frac{\omega_s^2}{p} \cdot \frac{\omega_s^2 - \omega_H^2}{\omega_s^2 - G\omega_H^2} = 9\omega_s^2 \cdot \frac{\omega_s^2 - 43.5 \cdot 10^{12}}{\omega_s^2 - 21.75 \cdot 10^{12}}.$$

Figure 3 shows the plasma frequency as determined at ascent and descent. The data are tentative as the precession of the rocket has been neglected and the orbit at descent was estimated from simple formulae. (It is thought that the corresponding corrections would tend to shift the descent curve towards greater heights). The tentative curve is only based on a few measuring points, it shows clearly a maximum ionization in the E-region and a rather important decrease in the "valley region" above.

Simultaneous ionosonde observations* from the ground at a distance of roughly

* Our thanks are due to Mr. Pron of the French "Centre National des Études Telecommunications" who has made these observations with his recently developed ionosonde.

20 km show the presence of sporadic E with a critical frequency of about 4 Mc/sec. The different maximum values found at ascent and descent are explained by the horizontal gradient of E_s-ionization. Knowing the rocket attitude a final evaluation should give details of the profile.

The accuracy of the present observations was limited by a jamming effect due to a telemetry transmitter. The uhf circuits introduced for protection decreased the sensitivity of the measuring device.

References

ARGENCE, E.: 1959, Thesis Paris, Faculte des Sciences.
ARGENCE, E. and RAWER, K.: 1961, *Space Research* **2**.
JACKSON, J. E.: 1954, *J. Geophys. Res.* **59,** 377–390 (and later publications in the same journal).
KANE, J. A., JACKSON, J. E., and WHALE, H. A.: 1962, *J. Nat. Bur. Stand.* **66D,** No. 6, 641–648.
SEDDON, J. C.: 1953, *J. Geophys. Res.* **58,** 323–335.
SEDDON, J. C., PICKAR, A. D., and JACKSON, J. E.: 1954, *J. Geophys. Res.* **59,** 513–514.

CHAPTER 41

ROCKET MEASUREMENTS OF ELECTRON DENSITY AND COLLISION FREQUENCY IN THE LOWER D-REGION

OVE E. PETERSEN

Ionosphere Research Laboratory, Technical University of Denmark

Abstract. Following a short review of the principle used to determine the electron density and collision frequency of the ionosphere by the inverse Seddon method and by differential absorption measurements, an account is given of the measuring technique, featuring high accuracy of the phase measurement and elimination of the effect of rocket spin. The rocketborne instrument package provides simultaneous data to deduce electron density, collision frequency and rocket trajectory.

1. Introduction

Ionosphere research with rocketborne equipment is being carried out from research stations the world over, through the application of a number of different methods, for example: radio propagation techniques, impedance probes, ionconductance probes and particle counters of various types.

The present seminar concerns itself with a group of propagation experiments, jointly undertaken by Norwegian and Danish research groups, to measure ionospheric parameters of the lower D-region in the auroral zone during disturbed conditions of the ionosphere.

Measurements of electron density and collision frequency in the auroral zone are best accomplished by radio propagation techniques, since high electron densities may be expected at low altitudes with comparatively high air density, i.e. down to 50 km. A prerequisite for the use of probes for measurements of these parameters is that the mean free path between the air molecules is great compared to the physical dimensions of the probe, and this is not easily achieved below about 80 km.

Furthermore, different types of propagation experiments may be performed with practically the same equipment in the rocket. For example, measurements of absorption, propagation velocity (phase shift) and Faraday rotation.

The experiment and instrumentation to be described is a combined differential absorption and phase experiment, with signal transmission from ground to rocket. The instrumentation provides means for eliminating the effect of rocket spin on the phase measurement and increased accuracy in the phase record.

2. Application of the Magneto-Ionic Theory

The general equation for the complex refraction coefficient for wave propagation in the ionosphere (the Appleton-Hartree formula) is:

$$n^2 = [\mu - j\chi]^2 = 1 - \frac{X}{1 - jZ - \frac{1}{2}Y_T^2/(1 - X - jZ) \pm \left[\frac{1}{4}Y_T^4/(1 - X - jZ)^2 + Y_L^2\right]^{1/2}}$$

where: $X = \dfrac{\omega_p^2}{\omega^2} = \dfrac{N \cdot e^2}{\varepsilon_0 m \omega^2}$

$\mathbf{Y} = \dfrac{\omega_c}{\omega} = \dfrac{e}{m\omega}\mathbf{B}$

$Z = \dfrac{\nu}{\omega}$

μ = refraction index

$\chi = \dfrac{c_0}{\omega} \cdot \kappa$

κ = absorption coefficient

N = electron density
ε_0 = free space dielectric constant
m = electron mass
e = electron charge
ν = collision frequency
$\mathbf{B}$ = magnetic field vector
ω = transmitted frequency
ω_p = plasma frequency
ω_c = gyro frequency
c_0 = propagation velocity in free space

In the present propagation experiment μ is measured according to the expression

$$\mu = \frac{c_0}{c},$$

where c is the phase velocity of the wave.
κ is derived from the attenuation of the transmitted signal, since

$$A = A_1 \cdot \exp\left(-\int_0^h \kappa \, \mathrm{d}h\right),$$

where A_1 = the initial amplitude and dh is an element along the propagation path.

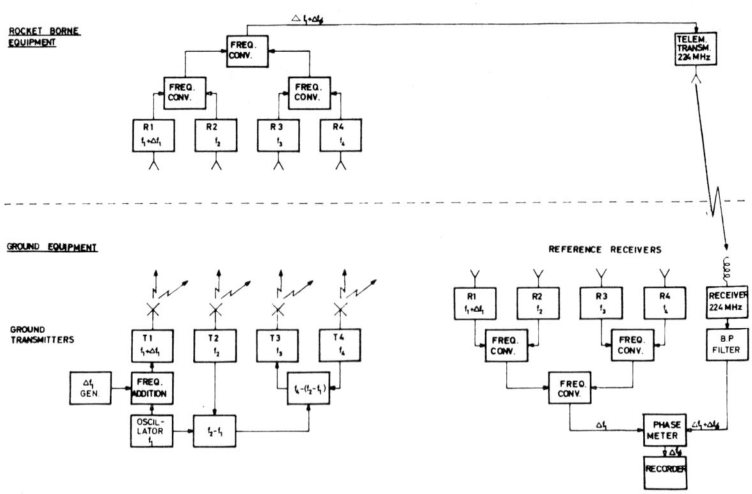

Fig. 1. Principle of phase experiment.

The earth's magnetic field causes the wave to be split into two components, the ordinary and extraordinary wave, with different refraction and absorption index. The plus and minus sign in the general equation correspond to the ordinary and extraordinary component respectively.

In the Appleton equation Y_T and Y_L correspond respectively to the magnetic field component perpendicular to and along the direction of propagation.

In the case of launching a rocket in the auroral zone, it may be assumed with good

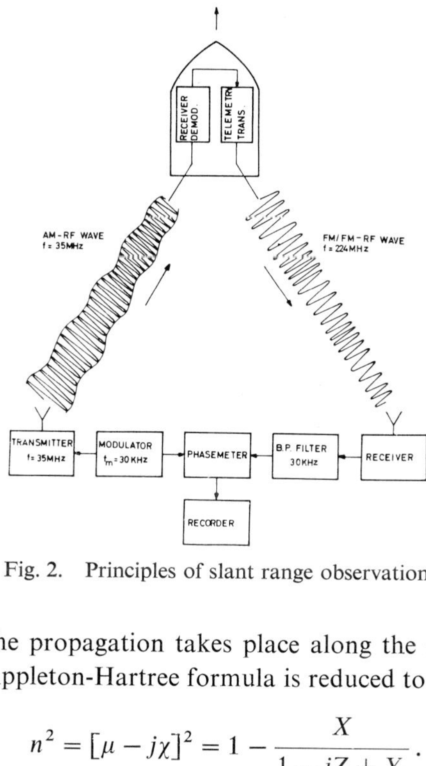

Fig. 2. Principles of slant range observations.

approximation, that the propagation takes place along the magnetic lines of force. Thus $Y_T = 0$, and the Appleton-Hartree formula is reduced to:

$$n^2 = [\mu - j\chi]^2 = 1 - \frac{X}{1 - jZ \pm Y_l}.$$

From this equation μ and χ may be expressed explicitly in terms of N and ν and, provided the magnetic field is known along the propagation path, N and ν can be computed from measurements of μ and $\kappa = \chi\omega/c_0$

Due to difficulties in determining the absolute signal amplitudes A and A_1, measurements are made of the differential absorption, that is, the ratio between the signal amplitudes of the ordinary and extraordinary components, A_0 and A_X:

$$A_0/A_x = \exp \int_0^h (\kappa_x - \kappa_0)\, dh .$$

This is achieved by transmitting alternately circularly polarized waves in the ordinary and extraordinary mode. Consequently, μ will be measured for both waves.

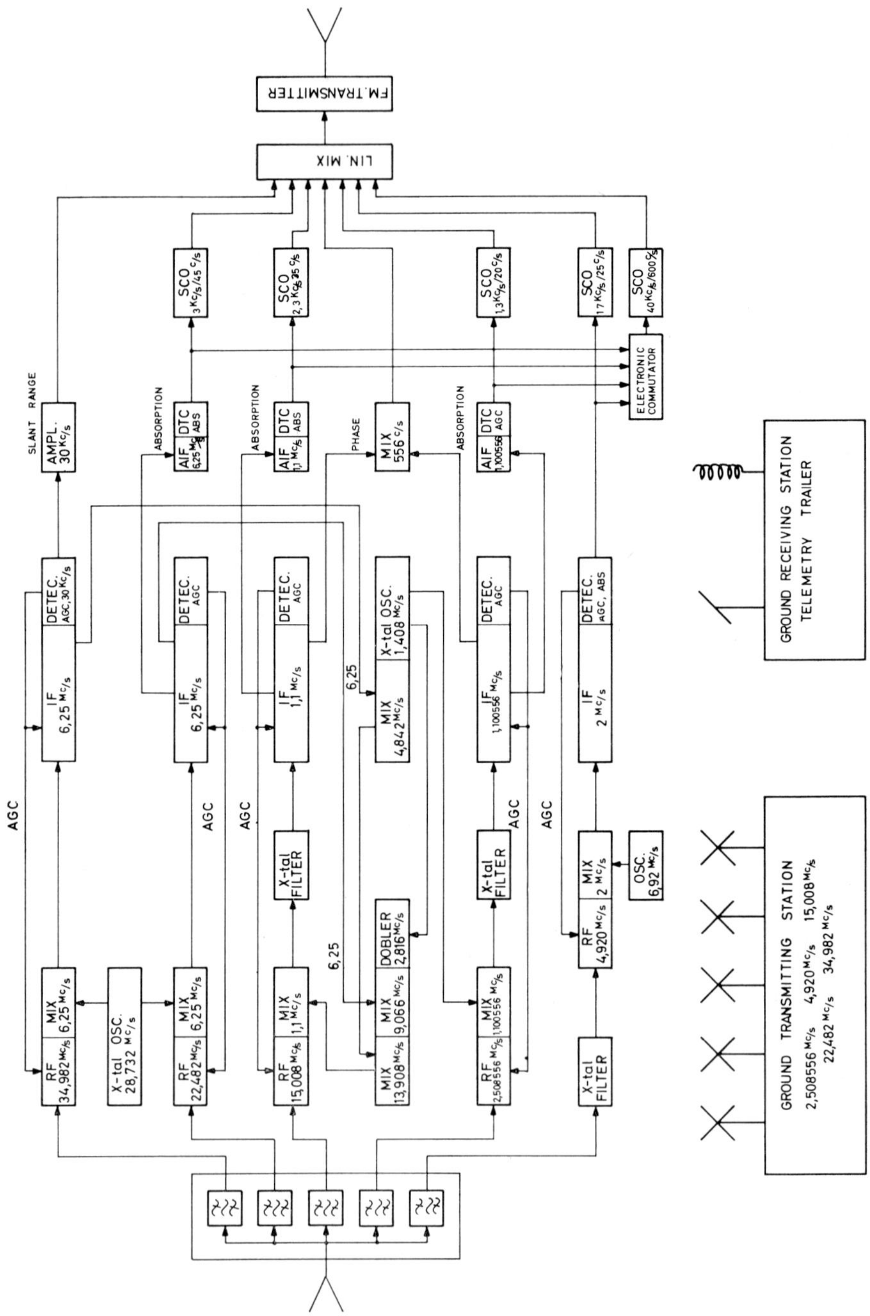

Fig. 3. Payload block diagram.

3. Experimental Method

The experiment consists of a modified inverse Seddon experiment combined with differential absorption measurements. The system is shown in the block diagram in Figure 1. The basic Seddon experiment requires the transmission (from the rocket) of wo harmonically related frequencies, the higher of which is assumed to have $\mu=1$.

Fig. 4. Payload for Nike-Cajun rocket.

The lower one is chosen sufficiently low to suffer a reasonable phase change ($\mu<1$) and yet penetrates the ionosphere to high altitudes without excess attenuation. After reception (on the ground) the low frequency is multiplied by the harmonic ratio between the high and low frequency and then phase compared with the high frequency. From the measured phase shift μ can be determined.

The system in Figure 1 differs from the Seddon principle in two ways:

1. Transmission takes place from ground to rocket to allow better transmitting antennas and greater transmitter power.

2. The phase measurement is accomplished by comparing the difference between two high frequencies f_4 and f_3 (both having $\mu=1$) with the difference between two low frequencies $f_2(\mu=1)$ and $f_1(u<1)$. The addition of the two auxiliary frequencies to the system provides: (a) elimination of the effect of rocket-spin on the phase

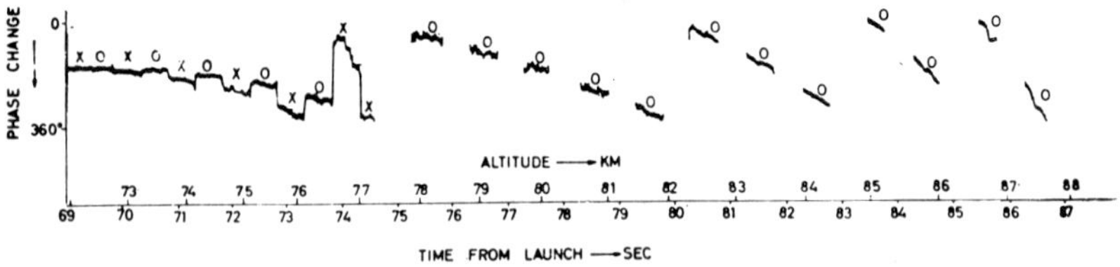

Fig. 5. Phase recording obtained from Ferdinand II. Alternate transmissions in the ordinary and extraordinary mode.

measurement; and (b) elimination of the effect of local oscillator drift on the phase measurement.

The signals f_3 and f_4 are circularly polarized in the same direction, while f_1 and f_2 are circularly polarized and switched between the ordinary and extraordinary mode twice a second.

For an ascending rocket velocity v and spin frequency f_s, we can now write the following expressions for the signals received in the rocket:

$$\begin{aligned} f_4' &= f_4 - f_4 \cdot \mu_4 (v/c_0) + f_s = f_4(1 - v/c_0) + f_s \\ f_3' &= f_3 - f_3 \cdot \mu_3 (v/c_0) + f_s = f_3(1 - v/c_0) + f_s \\ f_2' &= f_2 - f_2 \cdot \mu_2 (v/c_0) \pm f_s = f_2(1 - v/c_0) \pm f_s \\ f_1' &= f_1 - f_1 \cdot \mu_1 (v/c_0) \pm f_s = f_1(1 - \mu_1 v/c_0) \pm f_s . \end{aligned}$$

By subtracting the equations, we get:

$$\begin{aligned} f_h &= f_4' - f_3' = (f_4 - f_3)(1 - v/c_0) \\ f_e &= f_2' - f_1' = (f_2 - f_1)(1 - v/c_0) + f_1 (1 - \mu_1) v/c_0 . \end{aligned}$$

If we establish $f_4 - f_3 = f_2 - f_1$, we measure the phase shift per second of f_1:

$$\Delta f_\varphi = f_e - f_h = f_1 (1 - \mu_1) v/c_0 .$$

In order to improve the accuracy by performing the actual phase comparison on the

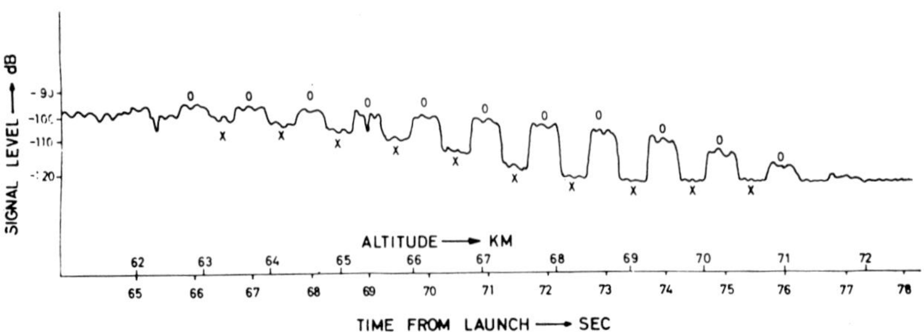

Fig. 6. Observed amplitude variations at 2.5 MHz. The signals are transmitted from the ground and received in the rocket. Alternating transmissions in the ordinary and extra-ordinary modes.

ground, a very low frequency $\Delta f_1 \ll f_1$ is added in the transmitter to the low frequency f_1. The output phase signal from the rocket receivers then becomes $\Delta f_1 + \Delta f_\varphi$. This signal is now telemetered back to ground and phase-compared with the original Δf_1. Δf_1 is chosen as the centre frequency of channel 2 (560 c/sec) in the IRIG standard telemetering system. This eliminates the need for a telemetry subcarrier oscillator for the phase experiment. By adding Δf_1 to one of the frequencies in the system a negligible error is introduced, since $f_4 - f_3$ is no more exactly equal to $f_2 - f_1$. Actually the frequency is 555.5...c/sec, obtained by dividing a crystal controlled 30 kc/sec oscillator by 6×9.

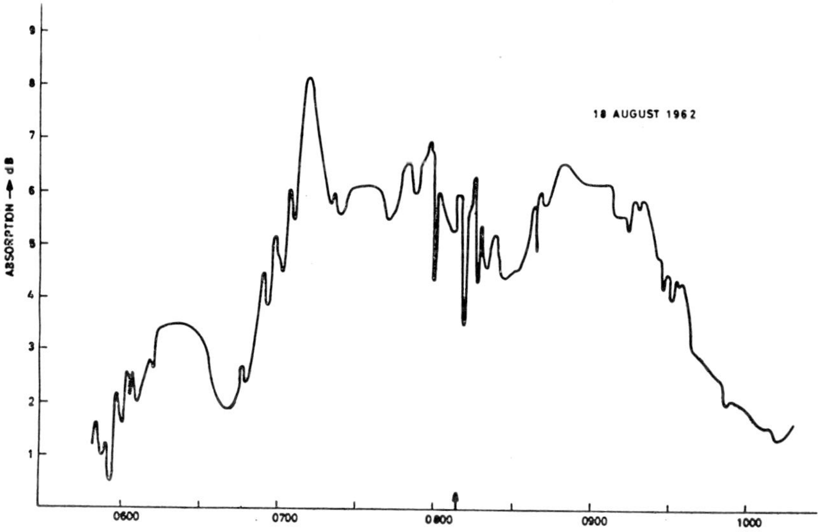

Fig. 7. Riometer record, Andøya, Norway, August 18, 1962.

4. Rocket Trajectory

The 30 kc/sec signal is AM-modulated on the highest frequency $f_4 = 35$ Mc/sec and serves the purpose of measuring the slant range to the rocket, i.e. the distance to the rocket as a function of time. Through knowledge of the ballistic behaviour of the rocket, it is possible from the slant range data to compute the rocket trajectory. As shown in Figure 2, the modulation frequency suffers a Doppler shift; this applies to both the AM-modulated signal on its way to the rocket and the FM-modulated telemetry signal back to ground. A phase comparison on the ground of the detected 30 kc/sec signal with the original 30 kc/sec signal will indicate a phase shift of the received signal. Since 30 kc/sec corresponds to a wavelength of 10 km, 360° phase shift will be observed for every 5 km the distance to the rocket is increased (or decreased). Similar to the technique in the phase experiment, the 30 kc/sec frequency corresponds to channel 15 in the IRIG system.

5. Instrumentation

The instrument package, which has been flown in three Nike-Cajun rockets from Andøya in Northern Norway in August and December 1962, is shown in the block diagram of Figure 3. It is seen that a fifth frequency (4.920 Mc/sec) is added to the system. This signal is independent of the other signals and is used to measure either Faraday rotation or differential absorption.

Differential absorption is measured on 2.5 Mc/sec and 15 Mc/sec and the relative

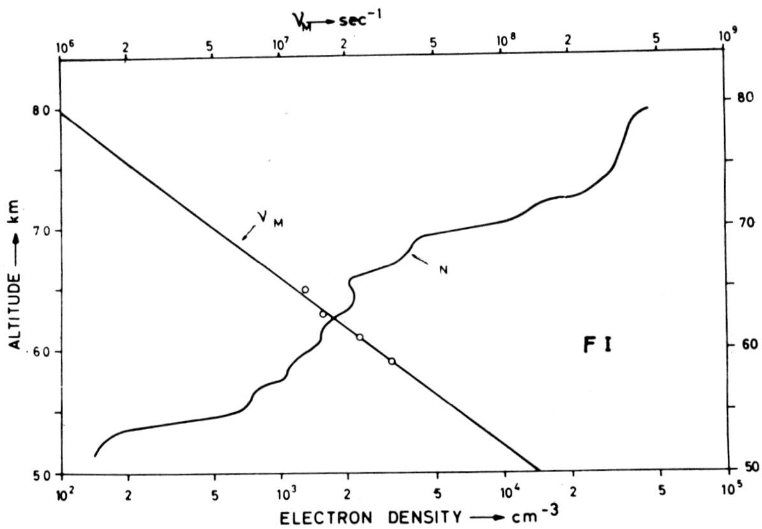

Fig. 8. Profiles of electron density N and collision frequency ν_M obtained from Ferdinand I.

absorption on 22.5 Mc/sec, thus providing some overlapping as the rocket ascends through the height ranges.

Through an efficient AGC system and auxiliary IF stages for the field strength measurements, the receiver system is capable of supplying both phase and differential absorption information.

The sensitivity of the system with respect to the phase measurement is a small fraction of a cycle. This is primarily achieved by special local oscillator arrangements for the superheterodyne type receivers. The local oscillators are common for the two high and for the two low frequencies, which means that any oscillator drift will be eliminated in the output.

Narrow band crystal filters are provided in the receivers for the three lower frequencies to avoid interference.

The receiver antenna system consists of one of the two crossed dipoles for the telemetry transmitter. The antenna is coupled to the five different receivers through five networks, matching the extremely low antenna impedance to the 50 Ohm receiver inputs.

The experiment payload is shown in Figure 4. It is built as a rigid mechanical construction, containing three decks, an upper passive one with the crystal filters and two active decks housing the five transistorized receivers. The compact receivers consist of a number of module units, corresponding in circuitry to the blocks in Figure 3.

The payload is designed to operate under severe environmental conditions with respect to acceleration, vibration and temperature. According to the test specifications, the equipment shall be capable of operating at linear acceleration up to 70 g and under vibration in three perpendicular directions, one being the axis of the rocket, at accelerations of 30 g in the frequency band 40–5000 c/sec.

6. Results from First Launching

On Figure 5 is shown a typical phase record obtained with the experiment. The phase recording is shifted back to zero, when it reaches 360° phase change, thus indicating a total measured phase shift of approximately $5 \times 360°$, before reaching the noise level. The illustration shows the greater phase change and also the rapid attenuation of the extraordinary wave as compared with the ordinary wave.

Figure 6 gives an amplitude recording from the 2.5 Mc/sec receiver. It is clearly seen how the extraordinary wave is absorbed before the ordinary wave.

Finally, in Figure 7 is given the riometer record at the first launching of the payload, and in Figure 8 the corresponding measured values of electron density N and collision frequency ν as a function of height.

For detailed analysis of the first three firings the reader is referred to: Ionosphere Research Laboratory, Report No. 16: "Electron Density and Collision Frequency Observations in the Lower D-Region during Auroral Absorption".

References

Jackson, I. E., Radicella, S. M.: 1962, NASA Report No. X-615-62-139, pp. 62–139.

Ratcliffe, J. A.: 1959, *The Magneto Ionic Theory and Its Applications to the Ionosphere*, Cambridge University Press.

Seddon, I. C.: 1953, *J. Geophys. Res.* **58**, 323.

CHAPTER 42

ELECTRON ENERGY SPECTRUM MEASUREMENTS DURING AURORAL ABSORPTION

J. ORTNER

*European Space Research Organisation, Paris**

1. Introduction

A few rocket measurements of the energy spectrum of auroral electrons have been made during recent years. McIlwain (1960) fired sounding rockets into visual aurora and a Canadian Group (McDiarmid *et al.*, 1961) studied electrons and protons associated with auroral radio absorption. These investigations, which were carried out from the Canadian range at Fort Churchill, were limited to the energy range for electrons above 40 keV.

The purpose of the experiments which were performed from the Swedish rocket range at Kronogard (situated about 200 km south of Kiruna) was to study the electron energy spectrum below 50 keV during aurora. These were cooperative experiments between the Air Force Cambridge Research Laboratories and the Kiruna Geophysical Observatory under the auspices of NASA.

Nike-Cajun rockets were used for these experiments, the payload of which included also sample cups to collect noctilucent cloud matter, and photographic emulsions for the detection of protons and heavier particles. Four rockets were fired, two of which were flown during auroral absorption events. However, only one of the auroral shots was successful, since the equipment failed during the second flight.

2. Instrumentation for the Auroral Experiment

An electrostatic analyser was developed by a division of American Science and Engineering, Inc., which was capable of measuring the electron energy spectrum in the range 3–50 keV. Its block diagram is reproduced in Figure 1.

The analyser consists of two concentric spherical plates to which a DC voltage is applied, modulated at 1200 cps. The voltage is exponentially swept between 7000 and 0 volts within 250 milliseconds. Electrons transversing the space between the plates are accelerated by a constant post-acceleration potential of 7000 volts applied to the surface of a scintillator. The output of the photomultiplier is then fed into the pre-amplifier circuit and passes through a band pass filter turned to 1200 cps. Thus, any noise which does not exist at 1200 cps is eliminated. The signal then enters the com-

* Formerly at the Kiruna Geophysical Observatory, Kiruna C, Sweden.

pression circuit, the output of which is telemetered to the ground. By comparing this signal with the exponential voltage sweep the electron fluxes are obtained at any time as function of energy.

The main specifications of the electrostatic analyser are the following: the radius of the inner analyser plate is 4.8 cm and the distance between the plates 0.3 cm. The solid angle is determined by the spherical segment 90° by 5°, the area of entrance aperture 0.25 cm^2.

The electron energy range is 3–50 keV and the energy resolution was estimated to be about 8%.

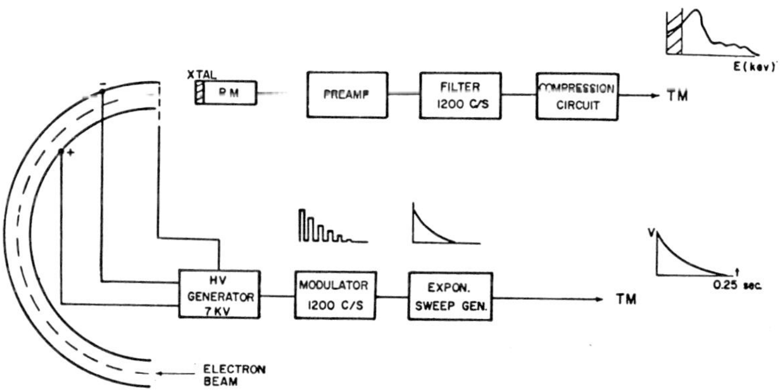

Fig. 1. Electrostatic analyser.

3. Rocket Experiments and Results

The first rocket was launched on 7 August 1962 at 0047 UT when the 40 Mc/sec riometer at Kronogard recorded about 1 db absorption of the auroral type. At the same time the 27 Mc/sec riometer at Kiruna measured an attenuation in signal strength of 3 db and the magnetic X-component showed a deflection of about 300 gammas. These observations indicated an auroral event of medium intensity.

The rocket reached a peak altitude of 102.5 km which, unfortunately, was too low for obtaining complete information about the electron fluxes at low energies within the aurora. However, the analysis of the data shows a number of electron bursts in the region of the peak altitude attained by the rocket. Because of arcing between the analyser plates at the maximum exponential sweep voltage, data on electron energies greater than 37 keV were lost. Electrons with energies up to 37 keV were recorded in the bursts, with a maximum intensity of about 10^9 electrons/cm^2 sec ster occurring at energies 20 keV. These bursts did not seem to originate from any particular direction.

The duration of the observed electron bursts was rather short, since each burst lasted not longer than a few seconds. Thus, these bursts seemed to be caused by some sudden disturbances precipitating large numbers of electrons into the atmosphere in relatively short periods of time.

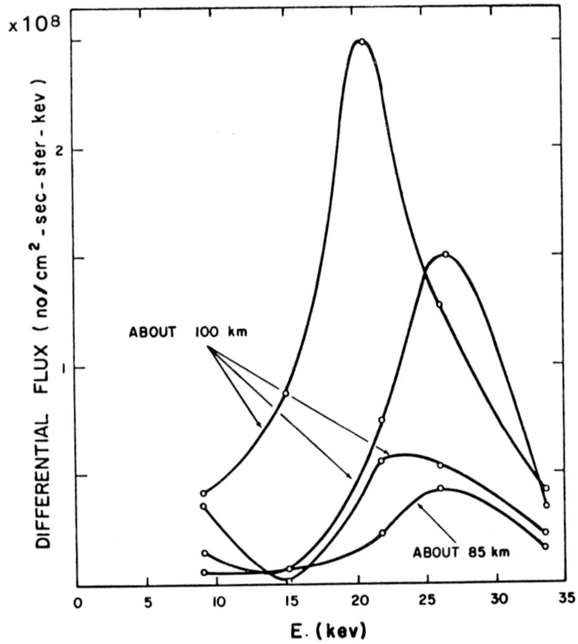

Fig. 2. Energy spectra at various times.

Figure 2 presents the differential electron fluxes observed at different times as function of energy. As can be seen, fluxes up to 2.5×10^8 electrons/cm^2 sec ster keV have been obtained at a height of 100 km. It is, however, rather doubtful if the peaks in electron fluxes between 18 and 28 keV were due to a smaller primary flux of electrons below these energies. The decreasing electron flux with energies less than 20 keV may possibly be explained by atmospheric absorption.

No electron bursts could be observed during the subsequent rocket flight of August 11 which was performed during quiet ionospheric conditions and was devoted to studies of the composition of noctilucent clouds. This fact proves the significance of the electron bursts obtained during disturbed ionospheric (auroral) conditions. Nevertheless, the results described above should still be considered as preliminary until further measurements studying other auroral events have been performed.

References

McDiarmid, I. B., Rose, D. C., and Budzinski, E.: 1961, *Can. J. Phys.* **39**, 1888–1900.
McIlwain, C. E.: 1960, *J. Geophys. Res.* **65**, 2727–2747.

INDEX OF SUBJECTS

INDEX OF NAMES